Trends in the Representation Theory of Finite Dimensional Algebras

CONTEMPORARY MATHEMATICS

229

Trends in the Representation Theory of Finite Dimensional Algebras

1997 Joint Summer Research Conference on
Trends in the Representation Theory of
Finite Dimensional Algebras
July 20–24, 1997
Seattle, Washington

Edward L. Green
Birge Huisgen-Zimmermann
Editors

American Mathematical Society
Providence, Rhode Island

The 1997 AMS-IMS-SIAM Joint Summer Research Conference in the Mathematical Sciences on "Trends in the Representation Theory of Finite Dimensional Algebras" was held at the University of Washington on July 20–24, 1997, with support from the National Science Foundation, Grant DMS-9618514, and the National Security Agency, Grant MDA904-97-1-0071.

1991 *Mathematics Subject Classification.* Primary 14D15, 14L30, 16D30, 16D60, 16E10, 16E30, 16E40, 16G10, 16G20, 16G30, 16G50, 16G60, 16G70, 16P10, 16P20, 16P40, 16S80, 16W30, 17B37, 81R50.

Any opinions, findings, and conclusions or recommendations expressed in this material are those of the authors and do not necessarily reflect the views of the National Science Foundation.

Library of Congress Cataloging-in-Publication Data

Trends in the representation theory of finite dimensional algebras : proceedings of the AMS-IMS-SIAM joint summer research conference, trends in the representation theory of finite dimensional algebras, July 20–24, 1997, University of Washington / Edward L. Green, Birge Huisgen-Zimmermann, editors.
p. cm. — (Contemporary mathematics, ISSN 0271-4132 ; 229)
Includes bibliographical references.
ISBN 0-8218-0928-8 (soft : alk. paper)
1. Associative algebras—Congresses. 2. Representations of algebras—Congresses. I. Green, Edward L., [DATE]. II. Huisgen-Zimmermann, Birge, [DATE]. III. Series: Contemporary mathematics (American Mathematical Society) : v. 229.
QA251.5.T74 1998
512′.24—dc21

98-44526
CIP

---

10 9 8 7 6 5 4 3 2 1 03 02 01 00 99 98

Contents

Preface

The American Mathematical Society, the Institute of Mathematical Statistics, and the Society for Industrial and Applied Mathematics sponsored a joint summer research conference at the University of Washington, from July 20–24, 1997, on Trends in the Representation Theory of Finite Dimensional Algebras. The conference was organized by E. L. Green (Virginia Polytechnic Institute and State University) and B. Huisgen-Zimmermann (University of California at Santa Barbara).

The lectures focused on the following three topics: (1) Interactions between representation theory and algebraic geometry; (2) Homological methods; (3) Applications of representation theory to the study of quantum groups. All three lines were strongly represented in the conference by lectures providing keynote reports and talks addressing current research, as well as by contributions to this volume.

(1) The analysis of representations via algebraic varieties is comparatively young. It goes back to the 1970's, in particular to Gabriel, Bernstein, I. M. Gel'fand, S. I. Gel'fand, and Ponomarev. Their representation-theoretic work, based heavily on geometric argumentation, proved the varieties of modules of fixed dimension to store an enormous amount of accessible algebraic information. Another triggering event was the appearance of two pivotal papers by Kac in the early 1980's, broadening insights of Gabriel and Nazarova among others and completing, via invariant theory, the description of the dimension vectors of the finite dimensional indecomposable representations of quivers. Since then, numerous far-reaching geometric paths have been cut through the subject, with a great deal of movement having occurred in the recent past.

(2) Homological methods constitute one of the most time-honored lines within representation theory. Among the pioneers in this direction were Cartan, Eilenberg, Nakayama, MacLane, Nagata, Auslander, Buchsbaum, Serre. One would have to list many others, in particular if one were to go back to the commutative or topological roots of the subject. As for the commutative origins, let us just point to Hilbert, whose Syzygy Theorem, proved last century, was one of the first ring-theoretic results of an unmistakably homological nature. In the

meantime, a stately list of fundamental conjectures has been accumulated in the noncommutative situation, and an even more extensive list of partial and full solutions which have furnished strong structural building blocks for the present understanding of important classes of algebras.

(3) Much of the interest in quantum groups revolves around algebras such as $U_q(\mathfrak{g})$, a deformation of the universal enveloping algebra of a semisimple Lie algebra $\mathfrak{g}$, and $\mathcal{O}_q(G)$, a deformation of the coordinate ring of a semisimple Lie group G. In the case when the parameter q is a root of unity, the irreducible representations of these algebras are finite dimensional, and so the finite dimensional representation theory of the algebras is of key importance. As a consequence, classification techniques specifically developed for the finite dimensional situation provide powerful tools. For example, Ringel has brought Auslander-Reiten theory to bear in studying the positive part of $U_q(\mathfrak{g})$. Similarly, many open problems concerning Sklyanin algebras could be solved if their finite dimensional Koszul duals were better understood.

Many of the talks, especially those in connection with topics (2) and (3), were implicit homages to Maurice Auslander, one of the field's most prominent leaders up to his death in November of 1994.

We would like to express our thanks to the AMS for its generous support, as well as to all participants who contributed to the success of the conference. In particular, our thanks go to Dr. James Maxwell, the director of the Joint Summer Research Conference program, Romy Cascella, the on-site AMS conference manager, and Christine Thivierge, the coordinator of this volume, for their investment of energy, which went far beyond routine involvement.

Edward L. Green
Birge Huisgen-Zimmermann

List of Talks

Ibrahim Assem, Université de Sherbrooke, Endomorphism algebras of exceptional sequences over path algebras of type A_n

Michael Bardzell, Salisbury State University, Computational techniques for Hochschild cohomology

Michael Barot, Universidad Nacional Autónoma de México, Lifting derived equivalences to one-point extensions

Raymundo Bautista, Universidad Nacional Autónoma de México, Representations of coalgebras and algebras

Frauke Bleher, University of Pennsylvania, Group algebras, Hecke algebras, and special biserial algebras

Sheila Brenner, University of Liverpool, Wild sectional subquivers of Auslander-Reiten quivers of tame algebras

Kenneth Brown, University of Glasgow, Representation theory of noetherian P.I. Hopf algebras

Thomas Brüstle, Universität Bielefeld, Parabolic subgroups acting with a finite number of orbits on their unipotent radicals

Ragnar-Olaf Buchweitz, University of Toronto, Simple singularities, liaison of lines, and Hochschild cohomology

Michael Butler, University of Liverpool, Minimal resolutions of algebras

William Chin, DePaul University, Representations of $SL_q(2)$ at a root of 1

Riccardo Colpi, Università di Padova, Tilting in Grothendieck categories

William Crawley-Boevey, University of Leeds, Preprojective algebras, simple representations, and singularities

Andrew Dean, Bishops University, The concept of rank for modules over finite dimensional algebras

José Antonio de la Peña, Universidad Nacional Autónoma de México, Tame algebras: some geometrical and combinatorial aspects

Peter Dräxler, Universität Bielefeld, Tame algebras, matrix problems, and derived categories

Yuriy Drozd, Kiev Taras Shevchenko University, Vector bundles on projective curves and Cohen-Macaulay modules on surface singularities

Rolf Farnsteiner, University of Wisconsin – Milwaukee, AR-components for reduced enveloping algebras of reductive Lie algebras

Kent Fuller, University of Iowa, *-Modules over ring extensions

Christof Geiss, Universidad Nacional Autónoma de México, Deformations and degenerations of finite dimensional algebras

Murray Gerstenhaber, University of Pennsylvania, An overview of algebraic deformation theory

Iain Gordon, University of Glasgow, The cohomology of quantized function algebras at roots of unity

Robert Guralnick, University of Southern California, Prehomogeneous spaces for algebraic groups and representations of finite dimensional algebras

Dieter Happel, Technische Universität Chemnitz, On subcategories associated with tilting modules

Michael Kauer, Universität Stuttgart, Derived equivalence of graph algebras

Bernhard Keller, Université de Paris VII, Derived categories and homological invariants

Otto Kerner, Heinrich-Heine-Universität Düsseldorf, Basic results on wild hereditary algebras

Mark Kleiner, Syracuse University, Homologically finite subcategories over Artin algebras

Henning Krause, Universität Bielefeld, Finitistic dimension and Ziegler spectrum

Helmut Lenzing, Universität Paderborn, Weighted projective lines and canonical algebras

Edward Letzter, Texas A&M University, Prime ideals in coordinate rings of quantum matrices

Lawrence Levy, University of Wisconsin – Madison, Tameness versus wildness of commutative noetherian rings

Shiping Liu, Université de Sherbrooke, Tilted string algebras

Eduardo Marcos, Universidade de São Paolo, Hochschild cohomology of truncated quiver algebras

Roberto Martínez-Villa, Universidad Nacional Autónoma de México, Self-injective Koszul algebras and their Yoneda algebras

Hagen Meltzer, Technische Universität Chemnitz, Exceptional vector bundles and tilting complexes on weighted projective lines

Hector Merklen, Universidade de São Paolo, On the representation type of some schurian vector space categories

Susan Montgomery, University of Southern California, Classifying finite-dimensional semi-simple Hopf algebras

Daniel Nakano, Utah State University, Representation type of Schur algebras

Idun Reiten, Norges Teknisk-Naturvitenskapelige Universitet, Some aspects of derived categories

Jeremy Rickard, University of Bristol, Bousfield localization in homological algebra

Christine Riedtmann, Universität Bern, Geometry of modules

Claus Ringel, Universität Bielefeld, Quiver varieties, preprojective algebras and the canonical bases of quantum groups

Manuel Saorín, Universidad de Murcia, Grading techniques in the study of homological properties of algebras

Daniel Simson, Uniwersytet Mikołaja Kopernika w Toruniu, Tameness, Tits reduced quadratic forms, and orbit spaces for lattices over orders

Andrzej Skowroński, Uniwersytet Mikołaja Kopernika w Toruniu, Geometric and homological aspects of representation theory of tame algebras

Sverre Smalø, Norges Teknisk-Naturvitenskapelige Universitet, Finitistic dimensions and contravariantly finite subcategories

Øyvind Solberg, Norges Teknisk-Naturvitenskapelige Universitet, Minimal projective resolutions

Martha Takane-Imay, Universidad Nacional Autónoma de México, The directing modules of finite dimensional k-algebras

Luise Unger, Universität Paderborn, Tilting modules and homological conjectures

Takayoshi Wakamatsu, Saitama University, On graded Frobenius algebras

Dan Zacharia, Syracuse University, Minimal projective resolutions

James Zhang, University of Washington, Curves on noncommutative schemes

Contemporary Mathematics
Volume **229**, 1998

POSTPROJECTIVE PARTITIONS FOR TILTING TORSION PAIRS

IBRAHIM ASSEM AND FLÁVIO ULHOA COELHO

Postprojective partitions for the module category mod A of an artin algebra A (or for an additive subcategory of mod A) were introduced by Auslander and Smalø in [2], under the name of preprojective partitions. The situation we consider is the following: let T_A be a tilting module and $B = \text{End}\, T_A$, we wish to describe these partitions for each of the classes of the torsion pair induced by T in mod B, especially when A is hereditary (thus B is tilted). In doing so, we use essentially the relative postprojective partitions induced by the tilting functors, as defined by Betzler in [4].

We need some notation. Let (A, T, B) be a tilting triple and denote by $(\mathcal{T}, \mathcal{F})$ and $(\mathcal{X}, \mathcal{Y})$ the torsion pairs induced by T_A in mod A and mod B, respectively [1]. We use sometimes the notation $(X, -)$ and ${}^1(X, -)$ for the functors $\text{Hom}_A(X, -)$ and $\text{Ext}^1_A(X, -)$, respectively. We let $\mathcal{S}$ denote the additive subcategory of mod B generated by $\tau_B^{-1}\text{Hom}_A(T, \mathrm{D}A)$ (where D denotes the standard duality between mod A and mod A^{op}) and, for an indecomposable module $X \in \mathcal{X}$, we let $n_s(X)$ be the length of a shortest chain of irreducible morphisms from a module in $\mathcal{S}$ to X if such a path exists, and infinity otherwise. We prove the following theorem.

THEOREM. *Let (A, T, B) be a tilting triple, then:*

(a) The functors $Hom_A(T, \ )$ and $\ \bigotimes_B T$ induce quasi-inverse category equivalences between $\mathcal{P}_n^{(T,-)}(\mathcal{T})$ and $\mathcal{P}_n(\mathcal{Y})$ for each n, and the functors $Ext^1_A(T, -)$ and $Tor_1^B(-, T)$ induce quasi-inverse category equivalences between $\mathcal{P}_n^{{}^1(T,-)}(\mathcal{F})$ and $\mathcal{P}_n(\mathcal{X})$ for each n.

(b) If T_A is a splitting tilting module, then $\mathcal{P}_n(\mathcal{Y}) = \mathcal{P}_n(\text{ mod } B) \cap \mathcal{Y}$ for each n.

(c) If A is hereditary, then $\mathcal{P}_n(\mathcal{X}) = \{X \in \mathcal{X} | n_s(X) = n\}$ for each n.

This paper consists of two sections, the first being devoted to preparatory results, and the second to the proof of our theorem.

1. Relative postprojective partitions

Throughout this work, A denotes an artin algebra, and mod A the category of finitely generated right A-modules. We assume that $\mathcal{C}$ is an additive subcategory of mod A, with an epiclass $\mathcal{E}$ (in the sense of [4]) such that $\mathcal{C}$ admits an $\mathcal{E}$-postprojective partition. For an $\mathcal{E}$-postprojective module X, lying in the class $\mathcal{P}_n^{\mathcal{E}}$, we write $\pi^{\mathcal{E}}(X) = n$, and call n the *$\mathcal{E}$-level* of X. For an A-module M, we denote by add M the additive subcategory of mod A consisting of the direct sums

1991 *Mathematics Subject Classification.* 16G20, 16G70, 16G99.

(*) This work was started when the second author was visiting the first at the University of Sherbrooke, Québec. The authors gratefully acknowledge partial support from NSERC of Canada and CNPq of Brazil.

of indecomposable summands of M. For an indecomposable module X in $\mathcal{C}$, a chain of $\mathcal{E}$-epimorphisms $X_0 \xrightarrow{f_1} X_1 \xrightarrow{f_2} X_2 \longrightarrow \dots \xrightarrow{f_t} X_t = X$ ending with X is called a *non-repetitive chain* if add $X_{i-1} \cap$ add $X_i = \emptyset$ for each $i = 1, \dots, t$, and a *radical chain* if f_i lies in the radical of $\mathcal{C}$ for each $i = 1, \dots t$. The integer t is called the *length* of the chain.

PROPOSITION 1.1. *Let $0 < n < \infty$, and X be an indecomposable module in $\mathcal{C}$. The following conditions are equivalent:*

(a) X belongs to $\mathcal{P}_n^{\mathcal{E}}$.

(b) There exists a non-repetitive chain ending with X of length n, and the length of any non-repetitive chain ending with X is at most n.

(c) There exists a radical chain ending with X of length n, and the length of any radical chain ending with X is at most n.

Proof. (a) implies (b). We construct a non-repetitive chain ending with X of length n by descending induction, choosing, for each j, the morphism $f_j : X_{j-1} \longrightarrow X_j$ to be a $\mathcal{P}_{j-1}^{\mathcal{E}}$- cover of X_j. Let now

$$Z_0 \xrightarrow{h_1} Z_1 \xrightarrow{h_2} Z_2 \longrightarrow \dots \xrightarrow{h_t} Z_t = X$$

be a non-repetitive chain ending with X. Fix an i and write $Z_i = \bigoplus_{j=1}^{n_i} Z_{ij}$ with Z_{ij} indecomposable. Denoting by p_j the projection $Z_i \longrightarrow Z_{ij}$, the $\mathcal{E}$-epimorphism $p_j h_i : Z_{i-1} \longrightarrow Z_{ij}$ does not split because add $Z_{i-1} \cap$ add $Z_i = \emptyset$. It follows easily that there exists $Z'_{i-1} \in$ add Z_{i-1} indecomposable such that $\pi^{\mathcal{E}}(Z'_{i-1}) <$ min $\{\pi^{\mathcal{E}}(Z_{ij}) | j = 1, \dots n_i\}$. Proceeding inductively, we find, for each $i, Z'_i \in$ add Z_i indecomposable such that $\pi^{\mathcal{E}}(Z'_i) \leq n - t + i$, leading to a contradiction if $t > n$.

(b) implies (c). Clearly, a non-repetitive chain ending with X of length n is also a radical chain of the same length. Let

$$Z_0 \xrightarrow{h_1} Z_1 \xrightarrow{h_2} Z_2 \longrightarrow \dots \xrightarrow{h_t} Z_t = X$$

be a radical chain. If $t > n$ then the same argument as above yields that $\pi^{\mathcal{E}}(X) \geq t$. Therefore there exists a non-repetitive chain ending with X of length $t > n$, a contradiction.

(c) implies (a). The hypothesis yields that $\pi^{\mathcal{E}}(X) \geq n$. On the other hand, if $\pi^{\mathcal{E}}(X) > n$ then, by the implication (a) implies (b) implies (c), we get a longer chain, a contradiction. □

Let $\mathcal{A}$ be the category of abelian groups, and $F : \mathcal{C} \longrightarrow \mathcal{A}$ be a covariant functor. As observed in [4], the class of morphisms $\mathcal{E} = \{f | Ff$ is an epimorphism$\}$ is an epi-class in $\mathcal{C}$. Also, if F is finitely generated, then $\mathcal{C}$ has an $\mathcal{E}$-postprojective partition. In this situation, we write F-epimorphism instead of $\mathcal{E}$-epimorphism.

LEMMA 1.2. *Let $\mathcal{C}$ and $\mathcal{D}$ be additive categories and $G : \mathcal{C} \to \mathcal{D}$ be an equivalence with quasi-inverse $G' : \mathcal{D} \to \mathcal{C}$. Let $F : \mathcal{C} \to \mathcal{A}$ be a finitely generated covariant functor and $F' = FG'$. Then:*

(a) A morphism f in $\mathcal{E}$ is an F-epimorphism (or a split F-epimorphism) if and only if $G(f)$ is a F'-epimorphism (or a split F'-epimorphism, respectively).

(b) For each $n, G\mathcal{P}_n^F(\mathcal{C}) = \mathcal{P}_n^{F'}(\mathcal{D})$.

(c) $G\{M \in \mathcal{C} | FM = 0\} = \{X \in \mathcal{D} | F'X = 0\}$.

Proof. Only (b) needs a proof. The statement is trivial if $n = 0$ and follows from (1.1) for $n > 0$. (see also [3], (3.4.2)). □

It is useful to characterise the epiclasses given by tilting functors.

PROPOSITION 1.3. *Let (A, T, B) be a tilting triple, and $f : M \longrightarrow N$ be a morphism in mod A.*

(a) Suppose $M, N \in \mathcal{T}$, then f is a $(T, -)$-epimorphism if and only if f is an epimorphism and $Ker\ f \in \mathcal{T}$.

(b) Suppose $M, N \in \mathcal{F}$, then f is an ${}^1(T, -)$-epimorphism if and only if $Coker f \in \mathcal{T}$.

Proof. (a) Assume $\mathrm{Hom}_A(T, f)$ is an epimorphism. Applying the right exact functor $- \bigotimes_B T$ yields that f is an epimorphism, so that we have a short exact sequence $0 \longrightarrow K \longrightarrow M \xrightarrow{f} N \longrightarrow 0$. Applying $\mathrm{Hom}_A(T, -)$ to it, and using that $\mathrm{Hom}_A(T, f)$ is an epimorphism and that $\mathrm{Ext}^1_A(T, M) = 0$ (because $M \in \mathcal{T}$) yields that $\mathrm{Ext}^1_A(T, K) = 0$ so $K \in \mathcal{T}$. Conversely, consider the short exact sequence $0 \longrightarrow K \longrightarrow M \xrightarrow{f} N \longrightarrow 0$ which is entirely in $\mathcal{T}$. Applying $\mathrm{Hom}_A(T, -)$ yields the result.

(b) Let $C=$ Coker f and $K =$ Im f and consider the short exact sequence $0 \longrightarrow K \xrightarrow{f} N \longrightarrow C \longrightarrow 0$ and the epimorphism $h : M \longrightarrow K$ such that $f = gh$. Assume $\mathrm{Ext}^1_A(T, f)$ is an epimorphism, then so is $\mathrm{Ext}^1_A(T, g)$. Applying $\mathrm{Hom}_A(T, -)$ to the previous short exact sequence yields $\mathrm{Ext}^1_A(T, C) = 0$ so $C \in \mathcal{T}$. Conversely, if $C \in \mathcal{T}$, then $\mathrm{Ext}^1_A(T, g)$ is an epimorphism. Since $\mathrm{Ext}^1_A(T, h)$ is an epimorphism (because h is and pd $T \leq 1$), so is $\mathrm{Ext}^1_A(T, f)$. □

2. Proof of the theorem.

Let (A, T, B) be a tilting triple, with A hereditary. Our first lemma gives an algorithm for the calculation of the postprojective partition $\mathcal{P}_n(\mathcal{X})$ similar to the one obtained in the hereditary case by Todorov [6].

LEMMA 2.1. *Let (A, T, B) be a tilting triple, with A hereditary, X be an indecomposable in $\mathcal{X}$ and $n \geq 0$. Then $X \in \mathcal{P}_n(\mathcal{X})$ if and only if*

(a) for each irreducible morphism $Y \longrightarrow X$ with Y indecomposable, then $Y \in \mathcal{P}_{n-1}(\mathcal{X}) \cup \mathcal{P}_n(\mathcal{X})$ and

(b) there exists $Y \in \mathcal{P}_{n-1}(\mathcal{X})$ and an irreducible morphism $Y \longrightarrow X$ if $n \geq 1$.

Proof. (a) Assume $X \in \mathcal{P}_n(\mathcal{X})$. We prove (a) and (b) by induction on n. Clearly, $\mathcal{P}_0(\mathcal{X}) = \mathcal{S}$ so the result holds for $n = 0$. We only sketch the inductive step since it follows the ideas in [5, 6]. Assume $n > 0$, then the almost split sequence $0 \longrightarrow \tau_B X \xrightarrow{f} E \xrightarrow{g} X \longrightarrow 0$ lies in $\mathcal{X}$. Let $Y \in$ add E be indecomposable and assume $\pi_{\mathcal{X}}(Y) \leq \pi_{\mathcal{X}}(X) - 2$. Then there exists $Z \in \mathcal{P}_{n-1}(\mathcal{X})$ and an epimorphism $h : Z \longrightarrow X$. Since h does not split, there exists $h' : Z \longrightarrow E$ such that $gh' = h$. Then $[f\ h'] : \tau_B X \bigoplus Z \longrightarrow E$ is an epimorphism, implying that $\pi_{\mathcal{X}}(\tau_B X) < \pi_{\mathcal{X}}(Y)$ (I). Let E' be the direct sum of all the indecomposable summands of E which lie in $\bigcup_{j=0}^{n-2} \mathcal{P}_j(\mathcal{X})$ and write $E = E' \bigoplus E''$. Since the Auslander-Reiten component of mod B containing $\mathcal{S}$ is a connecting component, it is standard. From the induction hypothesis, we infer that $\mathrm{Hom}_B(Z, E') = 0$ and then Im $h' \subseteq E''$ and $g|_{E''}$ is an epimorphism. Hence there exists $M \in$ add E'' indecomposable lying in $\mathcal{P}_{n-1}(\mathcal{X})$ (II). From (I) and (II), we get an irreducible morphism $\tau_B X \longrightarrow M$ and $\pi_{\mathcal{X}}(\tau_B X) < \pi_{\mathcal{X}}(Y) \leq n - 2 = \pi_{\mathcal{X}}(M) - 1$, a contradiction to the induction hypothesis and so $\pi_{\mathcal{X}}(Y) \geq \pi_{\mathcal{X}}(X) - 1$. There clearly exists $Y' \in$ add E indecomposable with

$\pi_{\mathcal{X}}(Y') = n-1$. Suppose now $\pi_{\mathcal{X}}(Y) > n$. By the induction step, $\pi_{\mathcal{X}}(\tau_B X) \leq n-1$. Applying inductively the arguments above, we infer the existence of an irreducible morphism $M \longrightarrow N$ with $\pi_{\mathcal{X}}(N) - \pi_{\mathcal{X}}(M) \geq 2$, a contradiction to the induction hypothesis. $\square$

Proof of the theorem. (a) This follows easily from (1.2).

(b) We prove it by induction, the case $n = 0$ being trivial because the projective B-modules are in $\mathcal{Y}$. Suppose $n > 0$ and let $Y \in \mathcal{P}_n(\mathcal{Y})$ be indecomposable. Let

$$Y_0 \xrightarrow{f_1} Y_1 \xrightarrow{f_2} Y_2 \longrightarrow \dots \quad \xrightarrow{f_n} Y_n = Y$$

be a non-repetitive chain of length n (see (1.1)). Clearly, $Y \in \mathcal{P}_m(\text{mod } B) \cap \mathcal{Y}$ for some $m \geq n$. If $m > n$, then, by (1.1) again, there exists a non-repetitive chain (which, since T is splitting, lies entirely in $\mathcal{Y}$) of length m, a contradiction to the maximality of n. Thus, $\mathcal{P}_n(\mathcal{Y}) \subseteq \mathcal{P}_n$ (mod B) $\cap \mathcal{Y}$. Similarly, $\mathcal{P}_n$ (mod B) $\cap \mathcal{Y} \subseteq \mathcal{P}_n(\mathcal{Y})$.

(c) We must show that, for each $j \leq \infty$, we have $X \in \mathcal{P}_j(\mathcal{X})$ if and only if $n_s(X) = j$. We first note that $\mathcal{P}_0(\mathcal{X}) = \mathcal{S}$. Also, it follows from [2] that the modules of $\bigcup_{n<\infty} \mathcal{P}_n(\mathcal{X})$ belong to the Auslander-Reiten component of mod B containing $\mathcal{S}$, which is a connecting component. We claim that, if $X \in \mathcal{P}_j(\mathcal{X})$, then $n_s(X) \leq j$. Indeed, using (2.1), it is not hard to construct a path of irreducible morphisms from some module of $\mathcal{S}$ to X with length j, which implies the claim. We now prove the statement by induction on $n_{\mathcal{S}}(X) = n$. The case $n = 0$ follows trivially from the fact that $\mathcal{P}_0(\mathcal{X}) = \mathcal{S}$. Let $n > 0$ and $f : Z \to X$ be right minimal almost split. There exists $Z' \in \text{add } Z$ indecomposable with $n_{\mathcal{S}}(Z') = n - 1$. By induction, $Z' \in \mathcal{P}_{n-1}(\mathcal{X})$. By (2.1), we have $X \in \mathcal{P}_{n-1}(\mathcal{X}) \cup \mathcal{P}_n(\mathcal{X})$. Using the claim above, we conclude that $X \in \mathcal{P}_n(\mathcal{X})$, as required. $\square$

Examples. (a) Our first example shows that the assumption that T_A is splitting is necessary for the statement (b) of the theorem to hold. Let k be a commutative field, A be the finite dimensional k-algebra given by the commutative quiver

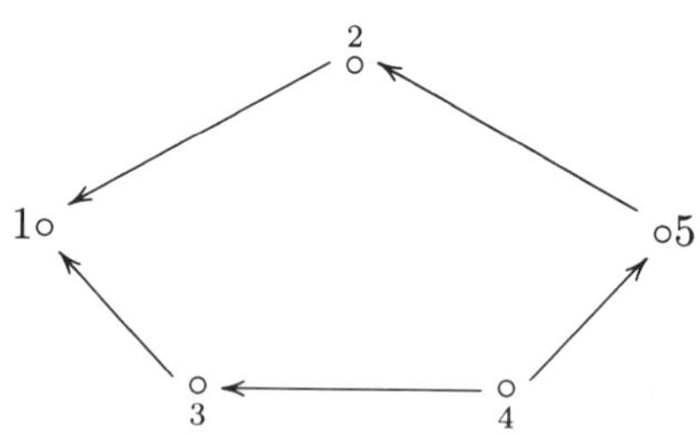

and B be the endomorphism algebra of the (non-splitting) tilting module $T_A = P_4 \bigoplus P_5 \bigoplus \tau_A^{-1} P_2 \bigoplus \tau_A^{-2} P_1 \bigoplus \tau_A^{-2} P_4$(where P_i represents the indecomposable projective A-module corresponding to the point i). The module $Y = \text{Hom}_A(T, \tau_A^{-3} P_3)$ lies in $\mathcal{Y}$, in $\mathcal{P}_1(\mathcal{Y})$, but also in $\mathcal{P}_2$ (mod B) $\bigcap \mathcal{Y}$.

(b) Our second example shows the partitions for the classes $\mathcal{X}$ and $\mathcal{Y}$ in a tilted algebra. Let k be a commutative field and A be the path algebra of the quiver

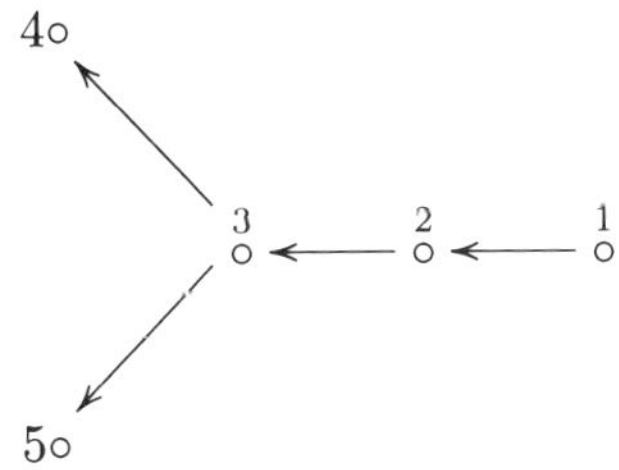

the Auslander Reiten quiver of the endomorphism algebra B of the tilting module $T_A = P_1 \bigoplus P_2 \bigoplus \tau_A^{-2} P_1 \bigoplus \tau_A^{-2} P_4 \bigoplus \tau_A^{-2} P_5$ has the following shape

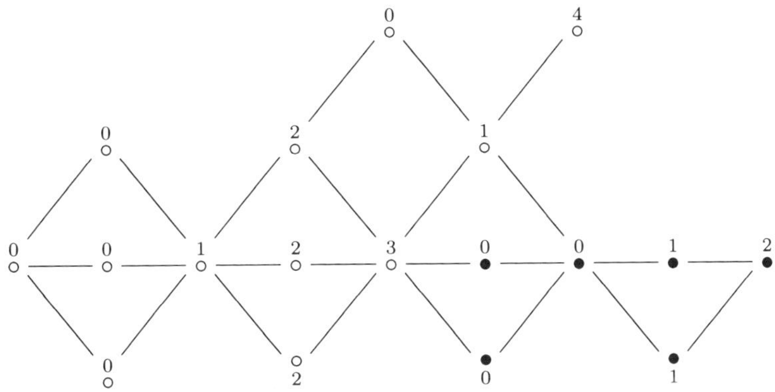

where the modules marked with $\circ$ are the modules in $\mathcal{Y}$, and those marked with $\bullet$ are the modules in $\mathcal{X}$. The number next to a module indicates the level of the module in the corresponding postprojective partition of $\mathcal{Y}$ or $\mathcal{X}$, respectively.

References

[1] I. Assem : Tilting theory-an introduction, in : Topics in Algebra, Banach Center Publications, vol. 26, PWN, Warsaw (1990) 127-180.

[2] M. Auslander, S.O. Smalø : Preprojective modules over artin algebras, J. Algebra 66 (1980) 61-122.

[3] R. Betzler : Relative präprojektive Zerlegungen, Ph.D. Thesis, Universität München, 1993.

[4] R. Betzler : Relative preprojective partitions, J. Algebra 180 (1996) 662-678.

[5] F.U. Coelho : A generalization of Todorov's theorem on preprojective partitions, Comm. Algebra 18 (1990) 1401-1423.

[6] G. Todorov : A note on preprojective partitions over hereditary artin algebras, Proc. Amer. Math. Soc. 85 (1982) 523-528.

Ibrahim Assem
Département de mathématiques et d'informatique
Université de Sherbrooke
Sherbrooke, Québec, J1K 2R1, Canada
ibrahim.assem@dmi.usherb.ca

Flávio Ulhoa Coelho
Departamento de Matemática-IME
Universidade de São Paulo
CP 66281, São Paulo, SP CEPO5315-970, Brazil
fucoelho@ime.usp.br

Contemporary Mathematics
Volume **229**, 1998

Derived Canonical Algebras as One-Point Extensions

Michael Barot and Helmut Lenzing

ABSTRACT. Canonical algebras have been intensively studied, see for example [**12**], [**3**] and [**11**] among many others. We are interested in the question when a one-point extension of a finite-dimensional algebra Σ by a Σ-module M is derived canonical, i.e. derived equivalent to a canonical algebra. We give necessary conditions on the algebra Σ and the module M. If the canonical algebra associated with Σ is tame the conditions are even sufficient.
As a further result we obtain that, if Σ is derived canonical then the one-point extension of Σ by M is derived canonical again if and only if M is derived simple, i.e. M is indecomposable and belongs to the mouth of a tube in the bounded derived category of Σ-modules.

1. Prerequisites

We work over an algebraically closed base field k. By algebra we mean always a basic, finite dimensional k-algebra and by module we mean finitely generated right module. For an algebra Σ we denote by $\operatorname{mod}\Sigma$ the category of Σ modules.

An algebra Σ which is derived equivalent to a canonical algebra will be called *derived canonical.* This terminology replaces the less suggestive term quasi-canonical used in [**10**]. For each derived canonical algebra there exists a weighted projective line $\mathbb{X} = \mathbb{X}(\underline{p}, \underline{\lambda})$ such that the (bounded) derived category $D^b(\Sigma)$ of $\operatorname{mod}\Sigma$ is equivalent to the (bounded) derived category of $\operatorname{coh}\mathbb{X}$, the category of coherent sheaves over $\mathbb{X}$, see [**3**]. Two derived canonical algebras are derived equivalent if and only if their associated weighted projective lines are isomorphic, see [**4**] and [**10**] for further details. Quite important information on $\mathbb{X}$, the weight type $\underline{p} = (p_1, \dots, p_t)$, can already be read from the Coxeter polynomial of Σ which has the shape

$$(T-1)^2 \cdot \prod_{i=1}^{t} \frac{T^{p_i} - 1}{T-1}.$$

Note that, if $t \leq 3$, then $\mathbb{X}$ is completely determined by its weight type. The *genus* of $\mathbb{X}$, hence of Σ, defined by

$$g_{\mathbb{X}} = g_{\Sigma} := 1 + \frac{1}{2}\left((t-2)p - \frac{p}{p_1} - \cdots - \frac{p}{p_t}\right), \text{ where } p = \text{l.c.m.}(p_1, \dots, p_t),$$

1991 *Mathematics Subject Classification.* 16G20,16E30.

The first author gratefully acknowledges support from University of Paderborn and PADEP, UNAM.

determines the representation type of $\operatorname{coh}\mathbb{X}$ and of the canonical algebra Λ associated with $\mathbb{X}$, see [**3**]: for $g_{\mathbb{X}} \leq 1$ both $\operatorname{coh}\mathbb{X}$ and Λ are tame, whereas $\operatorname{coh}\mathbb{X}$ and Λ are wild if $g_{\mathbb{X}} > 1$. If $g_{\mathbb{X}} = 1$ the algebra Λ is *tubular*, see [**12, 10**]. This happens if and only if the weight type is — up to permutation — one of $(2,2,2,2)$, $(3,3,3)$, $(2,4,4)$ or $(2,3,6)$. Note, in this context, that passing from the canonical algebra Λ to a derived equivalent algebra Σ may simplify the representation type.

For a hereditary category $\mathcal{C}$ like $\operatorname{coh}\mathbb{X}$ or $\operatorname{mod} H$, H hereditary, the derived category $\mathrm{D}^b(\mathcal{C}) = \bigvee_{n\in\mathbb{Z}} \mathcal{C}[n]$, the additive closure of the union of all $\mathcal{C}[n]$'s, is known as well as $\mathcal{C}$. Here, each $\mathcal{C}[n]$ is a copy of $\mathcal{C}$ with objects written $X[n]$, $X \in \mathcal{C}$, and morphisms are given by

$$\operatorname{Hom}_{\mathrm{D}^b(\mathcal{C})}(X[m], Y[n]) = \operatorname{Ext}^{n-m}_{\mathcal{C}}(X, Y).$$

Note that we have natural identifications for the Grothendieck-groups

$$\mathrm{K}_0(\mathbb{X}) = \mathrm{K}_0(\Sigma) = \mathrm{K}_0(\mathrm{D}^b(\Sigma)),$$

and that the identifications preserve the *Euler forms* given on classes of Σ-modules (coherent sheaves, objects from the derived category, respectively) by the formula

$$\langle [X], [Y] \rangle = \sum_{i=-\infty}^{\infty} (-1)^i \dim_k \operatorname{Hom}(X, Y[i]).$$

Let A be an algebra of finite global dimension. We recall from [**5**] that the bounded derived category $\mathrm{D}^b(A)$ of finite dimensional A-modules has Auslander-Reiten triangles. We say that an indecomposable A-module M is *derived peripheral* if the "middle term" E of the Auslander-Reiten triangle $\tau M \to E \to M \to \tau M[1]$ is indecomposable. An indecomposable A-module M is further called *derived simple* or also *derived simple regular*, if M is derived peripheral and τ-periodic for the Auslander-Reiten translation τ of $\mathrm{D}^b(A)$. Moreover, still assuming that M is indecomposable, we call M *derived quasi-simple* if M is derived peripheral and lies in a component of the form $\mathbb{Z}\mathbb{A}_\infty$ in $\mathrm{D}^b(A)$.

Assuming that Σ is derived canonical, it follows from [**3, 11**] and [**10**] that M is derived simple if and only if there exists a self-equivalence φ of $\mathrm{D}^b(\Sigma) = \mathrm{D}^b(\operatorname{coh}\mathbb{X})$ and a simple sheaf S on $\mathbb{X}$ such that $M = \varphi(S)$. For $g_\Sigma \neq 1$ the situation simplifies, and we may choose φ to be a translation functor $X \mapsto X[n]$.

Assuming that the algebra Σ is derived equivalent to a wild hereditary algebra H, a Σ-module M is derived quasi-simple if and only if there exists an integer n such that $M[n]$ is a regular H-module of quasi-length one.

For a representation-finite connected hereditary algebra H we fix an identification of the Auslander-Reiten quiver of $\mathrm{D}^b(H)$ with the translation quiver $\mathbb{Z}\Delta$ of the Dynkin diagram Δ attached to H. Relative to such an identification we define the *derived type* of an indecomposable Σ-module, with Σ derived equivalent to H, as the vertex v of Δ such that M belongs to the τ-orbit of v in $\mathbb{Z}\Delta$.

2. Derived canonical one-point extensions

Given an algebra Σ and a Σ-module M, we will consider the *one-point sink extension* or just *sink extension* $[M]\Sigma$ which is given as

$$[M]\Sigma = \begin{bmatrix} \Sigma & 0 \\ M & k \end{bmatrix}$$

with the corresponding matrix operations. Dually we define the *one-point source extension* or just *source extension* $\Sigma[M]$ (they are also called *one-point coextension* and *one-point extension* in the literature).

We are now going to investigate when a one-point extension is derived canonical. Since an algebra is derived canonical if and only if its opposite algebra is it suffices to investigate under which conditions a sink extension $[M]\Sigma$ is derived canonical.

We need to introduce some notation for the special case where Σ is derived equivalent to $k[\mathbb{A}_n]$, i.e. Σ is a branch with n points, see [**8, 1**]. In this case, the indecomposable objects in the derived category $\mathrm{D}^b(\Sigma)$ form a single Auslander-Reiten component of type $\mathbb{Z}\mathbb{A}_n$. A slice S in $\mathrm{D}^b(\Sigma)$ is called a *(p,q)-slice* if in the quiver of S, there are p arrows pointing upwards and the remaining $n-p-1=q$ arrows point downwards.

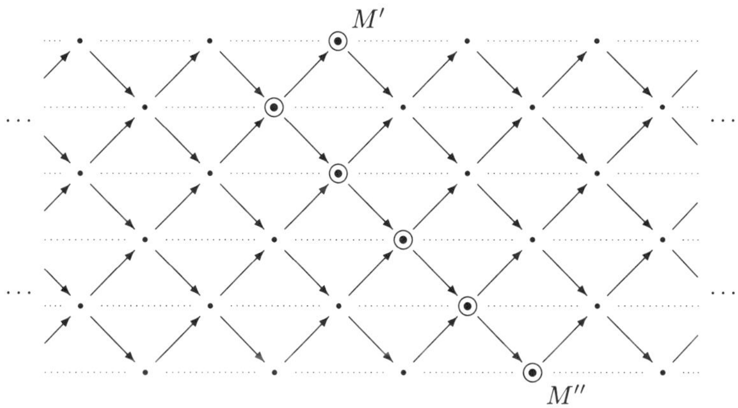

In the above figure, we have marked a $(1,4)$-slice, and the two peripheral objects belonging to that slice.

Back in the general situation, where we consider the sink extension $[M]\Sigma$, we denote by $\overline{M}$ the indecomposable projective module corresponding to the sink vertex of $\overline{\Sigma}=[M]\Sigma$, thus the radical of $\overline{M}$ is just M. We view $\operatorname{mod}\Sigma$ as the full exact subcategory of $\operatorname{mod}\overline{\Sigma}$ consisting of all $\overline{\Sigma}$-modules X with $\operatorname{Hom}_{\overline{\Sigma}}(\overline{M},X)=0$. The inclusion $\operatorname{mod}\Sigma\subset\operatorname{mod}\overline{\Sigma}$ induces an inclusion $\mathrm{D}^b(\Sigma)\subset\mathrm{D}^b(\overline{\Sigma})$. In particular, identifying modules with stalk complexes in the corresponding derived category, $\operatorname{mod}\Sigma$ and $\operatorname{mod}\overline{\Sigma}$ become full subcategories of $\mathrm{D}^b(\overline{\Sigma})$.

THEOREM 1. *Let Σ be an algebra and M a Σ-module. Assume that $\overline{\Sigma}=[M]\Sigma$ is derived canonical. Then M is the middle term of the Auslander Reiten triangle $\tau\overline{M}\longrightarrow M\longrightarrow\overline{M}\longrightarrow\tau\overline{M}[1]$ in $\mathrm{D}^b(\overline{\Sigma})$ associated with the projective module $\overline{M}$ attached to the sink vertex of $\overline{\Sigma}$.*

If further $\Sigma=\Sigma_1\times\cdots\times\Sigma_s$ is the decomposition of Σ into connected algebras, and $M=M_1\times\cdots\times M_s$ is the corresponding decomposition of M into Σ_i-modules M_i, then $s\leq 4$ and exactly one of the following cases happens:

1. a. *Σ is derived equivalent to $k[\mathbb{A}_\ell]$ and $M=M'\oplus M''$ is the direct sum of two indecomposable modules M' and M'' forming the periphery of a $(p-1,q-1)$-slice, $\ell=p+q-1$, of the component $\mathbb{Z}\mathbb{A}_\ell$ of $\mathrm{D}^b(\Sigma)$. Conversely, for each such choice for Σ and M the sink extension $[M]\Sigma$ is derived canonical of weight type (p,q).*

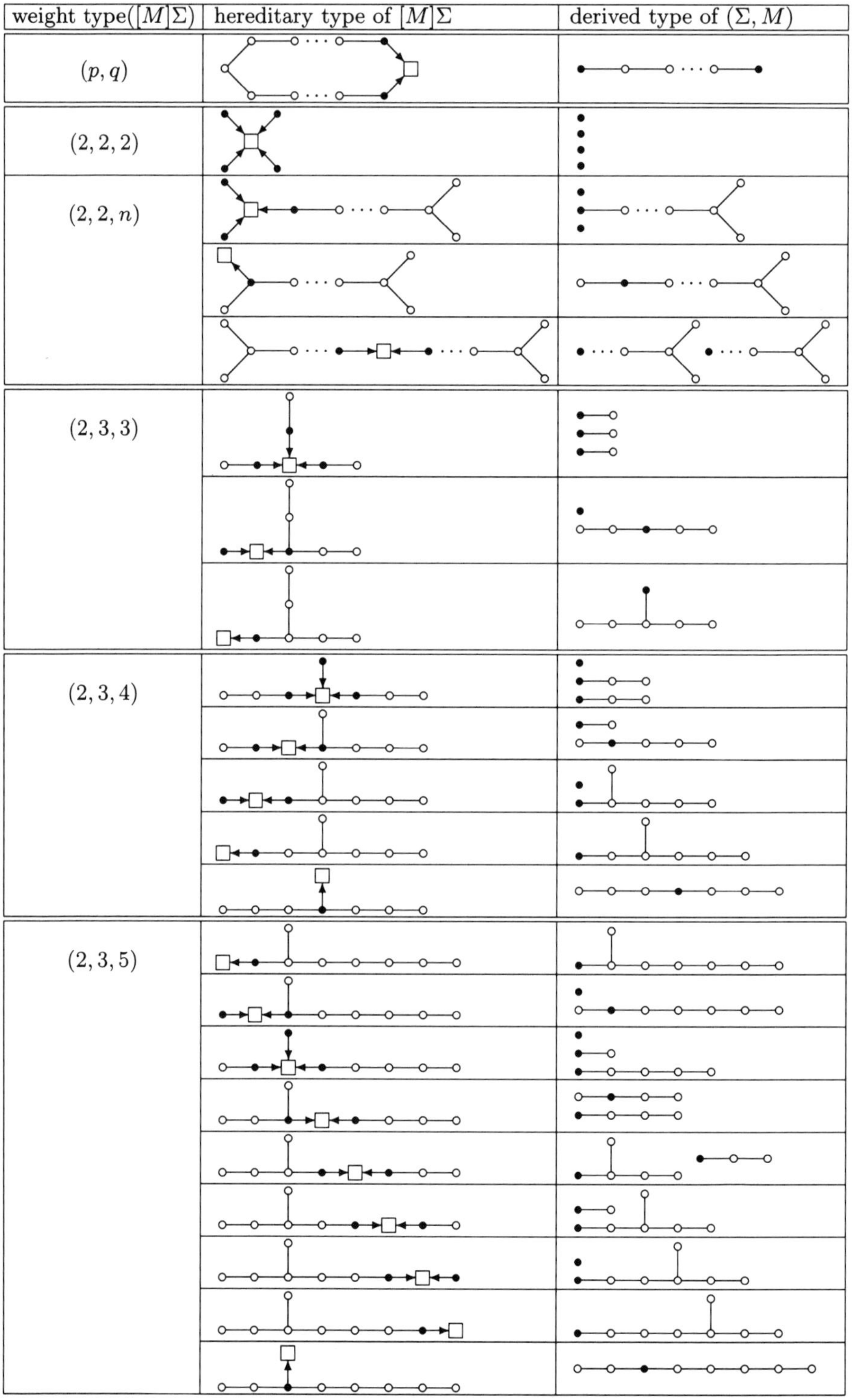

Table 1: Choices for (Σ_i, M_i)

b. *Each* Σ_i *is derived equivalent to a representation-finite connected hereditary algebra* H_i, *and each* M_i *is an indecomposable* Σ_i*-module, where the Dynkin types for the* Σ_i*'s and the derived types for the* M_i*'s are listed in the table below.*
Conversely, for each choice of (Σ, M) *conforming to the table, the sink extension* $[M]\Sigma$ *is derived canonical with the type given by the first column of the table.*

2. $\Sigma = \Sigma_1 \times \Sigma_2$, *where* $\Sigma_1 \neq 0$ *is derived canonical and* Σ_2 *is derived equivalent to a hereditary algebra of type* $\mathbb{A}_\ell$, $0 \leq \ell$. *Moreover,* M_1 *is derived simple over* Σ_1, *and — if* Σ_2 *is non-zero — the* Σ_2*-module* M_2 *is derived peripheral.*
Conversely, for each such choice of (Σ, M), *the sink extension* $[M]\Sigma$ *is derived canonical of weight type* $(p_1, \dots, p_{t-1}, p_t + \ell)$, *if* Σ_1 *has weight type* $(p_1, \dots, p_t)$ *and* M_1 *has* τ*-period* p_t *in the derived category.*
3. $\Sigma = \Sigma_1 \times \Sigma_2$, *where* $\Sigma_1 \neq 0$ *is derived equivalent to a connected wild hereditary algebra* H, *and* Σ_2 *is derived equivalent to a hereditary algebra of type* $\mathbb{A}_\ell$, $0 \leq \ell \leq 5$. *Moreover, the* Σ_1*-module* M_1 *is derived quasi-simple, and — if* Σ_2 *is non-zero — the* Σ_2*-module* M_2 *is derived peripheral.*

Concerning statement 3 it is an interesting open question when the sink extension $[M]\Sigma$ of a wild hereditary algebra Σ by a regular quasi-simple module M is derived canonical. The problem is related to the question when sink and source extension algebras are quasi-tilted [**6**].

PROOF. The first assertion is proved in [**2**]. Next we show that the conditions listed in 1, 2 and 3 are necessary, and exhaust all possible cases. Since $\overline{\Sigma}$ is derived canonical, there is a weighted projective line $\overline{\mathbb{X}}$ such that $\overline{\Sigma}$ can be realized as a *tilting complex* in the derived category $\mathrm{D}^b(\overline{\mathcal{C}})$, where $\overline{\mathcal{C}} = \operatorname{coh} \overline{\mathbb{X}}$, i.e. $\overline{\Sigma}$ is a full subcategory of $\mathrm{D}^b(\overline{\mathcal{C}})$ consisting of indecomposable objects $\overline{\Sigma}_j$, $j = 1, \dots, n+1$, which satisfy the condition

$$\mathrm{Hom}_{\mathrm{D}^b(\overline{\mathcal{C}})}(\overline{\Sigma}_i, \overline{\Sigma}_j[m]) = 0 \text{ for all } m \in \mathbb{Z} \setminus \{0\} \text{ and all } i, j = 1, \dots, n+1$$

and generate $\mathrm{D}^b(\overline{\mathcal{C}})$ as a triangulated category. In the present setting this latter condition is satisfied if and only if $n+1$ equals the rank of $\mathrm{K}_0(\overline{\mathcal{C}})$.

We denote by $\overline{M}$ the indecomposable object $\overline{\Sigma}_{n+1}$ corresponding to the coextension vertex, and by Σ the full subcategory consisting of $\overline{\Sigma}_1, \dots, \overline{\Sigma}_n$. By translation we may moreover assume that $\overline{M}$ lies in $\overline{\mathcal{C}}$. Notice that $\overline{M}$ is an *exceptional object* of $\overline{\mathcal{C}}$, i.e. has trivial endomorphism ring and no self-extensions. Let $\mathcal{H} = \overline{M}^{\perp}_{\overline{\mathcal{C}}}$ denote the perpendicular category

$$\mathcal{H} = \{X \in \overline{\mathcal{C}} \mid \mathrm{Hom}_{\overline{\mathcal{C}}}(\overline{M}, X) = 0 = \mathrm{Ext}^1_{\overline{\mathcal{C}}}(\overline{M}, X)\}$$

of $\overline{M}$ formed in $\overline{\mathcal{C}}$, and let $\mathcal{D} = \overline{M}^{\perp}_{\mathrm{D}^b(\overline{\mathcal{C}})}$ denote the perpendicular category

$$\mathcal{D} = \left\{X \in \mathrm{D}^b(\overline{\mathcal{C}}) \mid \mathrm{Hom}_{\mathrm{D}^b(\overline{\mathcal{C}})}(\overline{M}, X[m]) = 0 \text{ for all } m \in \mathbb{Z}\right\}$$

of $\overline{M}$ formed in the derived category $\mathrm{D}^b(\overline{\mathcal{C}})$. It is easily checked that $\mathcal{H}$ is an abelian hereditary category, and that $\mathcal{D} = \bigvee_{n \in \mathbb{Z}} \mathcal{H}[n]$. Moreover, Σ is a tilting complex in $\mathcal{D}$.

The structure of $\mathcal{H}$ is given as follows:

Case (i): If $\overline{M}$ has finite length n, then $\overline{M}$ lies in an exceptional tube and $\mathcal{H} \cong \mathcal{C} \times \operatorname{mod} H$, where $\mathcal{C} = \operatorname{coh} \mathbb{X}$ for a weighted projective line of a weight type dominated by the weight type of $\overline{\mathbb{X}}$ and where $H = k[1 \to 2 \to \cdots \to n-1]$. This follows from [**4**, Thm. 9.5] invoking an argument of Strauß [**13**]. The figure below shows the relevant part of the component of $\mathrm{D}^b(\overline{\mathcal{C}})$ containing $\overline{M}$:

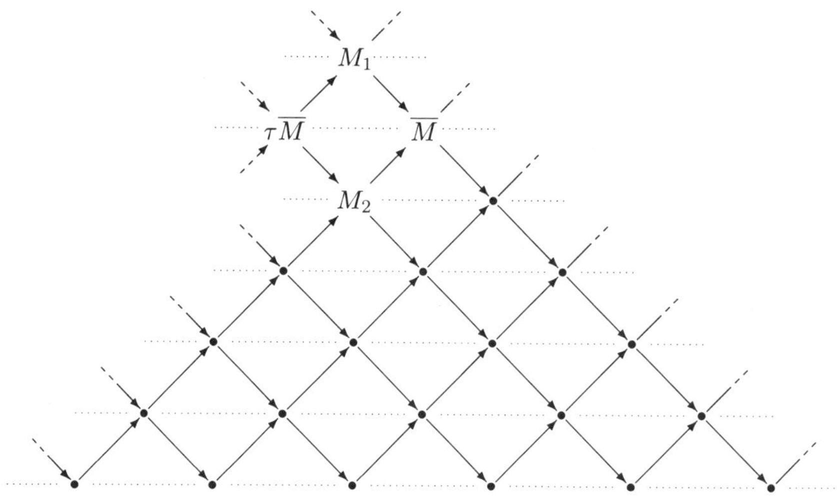

Here, the indecomposable H-modules form the subwing with "top" M_2. According to the decomposition $\mathrm{D}^b(\mathcal{H}) = \mathrm{D}^b(\mathcal{C}) \times \mathrm{D}^b(\mathrm{mod}\, H)$, the algebra Σ decomposes into two connected algebras Σ_1 and Σ_2, where Σ_1 is a tilting complex in $\mathrm{D}^b(\mathcal{C})$ and Σ_2 is a tilting complex in $\mathrm{D}^b(\mathrm{mod}\, H)$. Note that M_2 is derived peripheral over H, hence over Σ_2.

Further the object M_1 at the top of the figure is a simple object in $\mathcal{C}$, and therefore M_1 becomes a derived simple Σ_1-module. This proves the first part of the statement 2.

Case (ii): $\overline{M}$ is an exceptional vector bundle in $\mathrm{coh}\,\overline{\mathbb{X}}$. Here, it follows from [**7**] that $\mathcal{H}$ is equivalent to a module category over a (not necessarily connected) hereditary algebra. For a more complete analysis, we need to distinguish the various representation types for $\overline{\mathcal{C}} = \mathrm{coh}\,\overline{\mathbb{X}}$:

1. $\overline{\mathbb{X}}$ has genus < 1, i.e. the weight type $\Delta = (p, q, r)$ is of Dynkin type. Here the vector bundles form one component of type $\mathbb{Z}\overline{\Delta}$, where $\overline{\Delta}$ is the extended Dynkin diagram corresponding to Δ. To calculate the perpendicular category of $\overline{M}$ in $\mathrm{D}^b(\overline{\Sigma})$ we choose a slice $\overline{H}$ of $\mathbb{Z}\overline{\Delta}$ such that $\overline{M}$ becomes a sink in $\overline{H}$, which is a tilting object of $\overline{\mathcal{C}}$ whose endomorphism ring, here identified with $\overline{H}$, is a tame hereditary algebra. Since $\mathrm{D}^b(\overline{\mathcal{C}}) = \mathrm{D}^b(\mathrm{mod}\,\overline{H})$, the perpendicular category of $\overline{M}$ in $\mathrm{D}^b(\overline{\mathcal{C}})$ equals the derived category of $\overline{M}^{\perp}_{\mathrm{mod}\,\overline{H}}$. This category $\overline{M}^{\perp}_{\mathrm{mod}\,\overline{H}}$ is equivalent to the module category of a not necessarily connected hereditary algebra H, whose indecomposable objects consist of the objects of the slice $\overline{H}$ different from $\overline{M}$. The arising cases for H are listed in the table. Moreover, the almost-split sequence $0 \to \tau\overline{M} \to M \to \overline{M} \to 0$ in $\overline{\mathcal{C}}$, obtained from the first assertion, yields the

derived types of the M_i, $M = \bigoplus_{i=1}^{s} M_i$, as marked in the table. Since

$$M_{\mathcal{D}}^{\perp} = \mathrm{D}^b\,(\overline{M}_{\mathrm{mod}\,\overline{H}}^{\perp}) = \mathrm{D}^b\,(\mathrm{mod}\,H) = \prod_{i=1}^{s} \mathrm{D}^b\,(\mathrm{mod}\,H_i)$$

the tilting complex Σ decomposes into s connected pieces Σ_i, where each Σ_i as a tilting complex in $\mathrm{D}^b\,(\mathrm{mod}\,H_i)$ is derived-equivalent to H_i. This shows the first part of statement 1.

2. $\overline{\mathbb{X}}$ has genus one. By an automorphism of the derived category we can in this case achieve that $\overline{M}$ has finite length, see [**10**]. The assertion thus reduces to case (i).
3. $\overline{\mathbb{X}}$ has genus > 1. In this case, $\overline{M}$ belongs to a component of $\overline{\mathcal{C}}$ having type $\mathbb{Z}\mathbb{A}_\infty$ [**11**], and it is known that the quasi-length ℓ of $\overline{M}$ is at most 5 [loc. cit.]. Invoking arguments of [**13**], it further follows from [**7**] that $\overline{M}_{\overline{\mathcal{C}}}^{\perp}$ is equivalent to the product of $\mathrm{mod}\,A_{\ell-1}$, where $A_{\ell-1} = k[1 \to \cdots \to \ell - 1]$, with the module category $\mathrm{mod}\,H$ over a connected wild hereditary algebra H. Accordingly Σ decomposes into a product $\Sigma_1 \times \Sigma_2$, where Σ_1 is connected and derived wild hereditary, and where Σ_2 is derived equivalent to $A_{\ell-1}$, i.e. a branch in the sense of [**8, 1**]. Following arguments of [**13**] and [**9**] it follows moreover that M_1 is derived quasi-simple and M_2 is derived peripheral. This proves the first part of the statement 3.

Now we show the second part of the statements 1 and 2. So, first let Σ be derived equivalent to $k[\mathbb{A}_\ell]$ and $M = M' \oplus M''$ a Σ-module such that M' and M'' are indecomposable and form the periphery of a $(p-1, q-1)$-slice S of the component $\mathbb{Z}\mathbb{A}_\ell$ of $\mathrm{D}^b\,(\Sigma)$, where p and q are such that $\ell = p + q - 1$. By the first statement of the theorem, we have that $S \oplus \overline{M}$ is a tilting complex in $\mathrm{D}^b\,([M]\Sigma)$ with endomorphism algebra isomorphic to $k[\tilde{\mathbb{A}}_\ell]$. Thus $[M]\Sigma$ is derived canonical.

With the same argument we show that $[M]\Sigma$ is derived canonical, when $\Sigma = \Sigma_1 \times \cdots \times \Sigma_s$ and $M = M_1 \times \cdots \times M_s$ where Σ_j is derived hereditary and M_j is an indecomposable Σ_j-module $(j = 1, \ldots, s)$ such that the pair (Σ, M) is listed in Table 1.

Let now $\Sigma = \Sigma_1 \times \Sigma_2$, where Σ_1 is derived canonical and Σ_2 is derived equivalent to $k[\mathbb{A}_\ell]$ for some $\ell \geq 0$. Further let $M = M_1 \times M_2$, where M_1 is derived simple and, if $l > 0$, then let M_2 be deriphed peripheral. Let $\mathbb{X}(\underline{p}, \underline{\lambda})$ be the weighted projective line associated to Σ_1, where $\underline{p} = (p_1, \ldots, p_t)$ is its weight type. Let $\overline{\mathbb{X}}$ be the weighted projective line with weight type $(p_1, \ldots, p_{t-1}, p_t + \ell)$ and with the same parameter sequence $\underline{\lambda}$ as $\mathbb{X}$. We fix an indecomposable sheaf E of length $\ell + 1$ concentrated at λ_t, and form the perpendicular category $\mathcal{H} = E^\perp$ of E in $\overline{\mathcal{C}} = \mathrm{coh}\,\overline{\mathbb{X}}$. Then $\mathcal{H} = \mathcal{C} \times \mathrm{mod}\,H$, where $\mathcal{C} = \mathrm{coh}\,\mathbb{X}$ and $H = k[1 \to \cdots \to \ell]$. Moreover, the middle term of the almost-split sequence $0 \to \tau E \to S \oplus M_2 \to E \to 0$ decomposes into a simple sheaf S in $\mathcal{C}$, concentrated at λ_t and in the indecomposable projective-injective H-module M_2. Next, we realize Σ_1 as a tilting complex in $\mathrm{D}^b\,(\mathcal{C})$ so that, by means of the identification $\mathrm{D}^b\,(\Sigma_1) = \mathrm{D}^b\,(\mathcal{C})$ the module M_1 corresponds to the simple sheaf S. Further, we realize the branch Σ_2 as a tilting complex in $\mathrm{D}^b\,(H)$ such that, in the identification $\mathrm{D}^b\,(\Sigma_2) = \mathrm{D}^b\,(H)$, the H-module M_2 becomes a (derived peripheral) module over Σ_2. Following [**2**], it is easily checked that E together with Σ_1 and Σ_2 forms a tilting complex in $\mathrm{D}^b\,(\mathrm{coh}\,\overline{\mathbb{X}})$ with endomorphism algebra $\overline{\Sigma} = [M_1 \times M_2]\Sigma$. Hence $\overline{\Sigma}$ is derived canonical of type $\overline{\mathbb{X}}$.

This completes the proof of the Theorem. □

In Theorem 1 we have seen that the request for an algebra Σ to admit a derived canonical source or sink extension is very restrictive for Σ and for the "extension module" M. The information is even more specific if we start the extension procedure with a derived canonical algebra Σ.

COROLLARY 1. *Let Σ be derived canonical, and let M be a finite dimensional not necessarily indecomposable Σ-module.*

Then the sink extension $[M]\Sigma$ is derived canonical if and only if M is derived simple, in particular indecomposable.

PROOF. By Theorem 1, we only need to show that the condition is necessary. So let us assume that $[M]\Sigma$ is derived canonical. As a derived canonical algebra Σ is connected, hence Theorem 1 implies that $\overline{M}$ is derived simple. This uses that a wild hereditary or a representation-finite hereditary algebra is never derived equivalent to a canonical algebra since wild hereditary algebras always have spectral radius greater one and representation-finite hereditary algebras do not admit 1 as a root of their Coxeter polynomial, while canonical algebras have spectral radius one, and 1 is a root of their Coxeter polynomial. Thus cases 1 and 3 of Theorem 1 will not occur. □

We now consider the case where a branch $\mathcal{B}$ is attached to the extension vertex of $[M]\Sigma$, see [**1**] for definitions. We denote by $[\mathcal{B}, M]\Sigma$ the resulting algebra.

COROLLARY 2. *Let Σ be a derived canonical algebra, M a Σ-module and $\mathcal{B}$ a branch.*

Then $[\mathcal{B}, M]\Sigma$ is derived canonical if and only if M is derived simple.

PROOF. First let $\overline{\Sigma} = [\mathcal{B}, M]\Sigma$ be derived canonical. Clearly, $\overline{\Sigma}$ is derived equivalent to $\overline{\Sigma}' = [\mathcal{B}', M]\Sigma$, where $\mathcal{B}'$ denotes the linearly ordered branch with its sink α as root point. Thus we may identify $\overline{\Sigma}'$ with the algebra $[M \times P_\alpha](\Sigma \times \mathcal{B}')$, where P_α denotes the projective indecomposable associated to the point α. Therefore, we may apply Corollary 1. Once again, only the case 2 remains possible, and we infer that M is a derived simple Σ-module.

The converse is covered by Theorem 1, part 2. □

2.1. Criteria for derived canonical algebras. Theorem 1 provides us with a necessary condition for an algebra Σ to be derived canonical. So we might use this in order to prove that certain algebras are not derived canonical. We shall exhibit this in an example. Let A_n be the algebra given by the linear bound quiver with $n \geq 8$ vertices

$$\bullet \xrightarrow{x} \bullet \xrightarrow{x} \bullet \longrightarrow \cdots \longrightarrow \bullet \xrightarrow{x} \bullet$$

which satisfies the $n-6$ relations $x^7 = 0$. It is not difficult to check that A_9, A_10 and A_{11} are derived canonical of weight type $(2,3,5)$, $(2,3,6)$ and $(2,3,7)$, respectively. Also the Coxeter polynomial of A_{22} has canonical type $(2,7,14)$. However, the algebra A_{22} is not derived canonical.

Using Theorem 1, this can be seen as follows: First, we write A_{22} as a sink extension of the algebra A_{21} by a module M over A_{21}, thus $A_{22} = [M]A_{21}$. By Theorem 1, it suffices to show that A_{21} is neither derived equivalent to a representation-finite hereditary algebra, nor derived canonical, nor derived equivalent to a wild hereditary algebra:

Introducing the polynomials

$$V_n = \frac{T^n - 1}{T - 1},$$

the Coxeter polynomial $\mathbf{C}(A_{21})$ of A_{21} is seen to be $(T-1)^2 V_2 V_7 V_8 V_9 / V_4$. In particular, $\mathbf{C}(A_{21})$ is not of canonical type and hence A_{21} is not derived canonical. Since all roots of $\mathbf{C}(A_{21})$ lie on the unit circle in the complex plane and further 1 is a root of $\mathbf{C}(A_{21})$, the algebra A_{21} cannot be derived equivalent to a hereditary algebra which is wild or representation-finite.

References

[1] I. Assem and A. Skowroński: *Algebras with cycle-finite derived categories.* Math. Ann. **280** (1988), 441–463.

[2] M. Barot and H. Lenzing: *One-point extensions and derived equivalence.* To appear.

[3] W. Geigle and H. Lenzing: *A class of weighted projective lines arising in representation theory of finite-dimensional algebras.* In: Singularities, Representations of Algebras, and Vector Vector Bundles, Springer Lecture Notes in Math. **1273**, Springer (1987), 265–297.

[4] W. Geigle and H. Lenzing: *Perpendicular categories with applications to representations and sheaves.* J. Algebra **144** (1991), 273–343.

[5] D. Happel: *Triangulated categories in the representation theory of finite dimensional algebras.* London Math. Soc, Lecture Notes Series **119** (1988).

[6] D. Happel, I. Reiten, S. O. Smalø: *Tilting in abelian categories and quasitilted algebras.* Mem. Amer. Math. Soc. **575** (1996).

[7] T. Hübner and H. Lenzing: *Categories perpendicular to exceptional bundles.* Preprint Paderborn.

[8] B. Keller and D. Vossieck: *Aisles in derived categories.* Bull. Soc. Math. Belg. **40** (1988), 239–253.

[9] O. Kerner: *Stable components of wild algebras.* J. Algebra **142** (1991), 37–57.

[10] H. Lenzing and H. Meltzer: *Sheaves on a weighted projective line of genus one, and representations of a tubular algebra.* In: Representations of Algebras, Sixth International Conference, Ottawa 1992, CMS Conf. Proc. **14** (1993), 313–337.

[11] H. Lenzing and J. A. de la Peña: *Wild canonical algebras.* Math. Z. **224**, 403–425 (1997).

[12] C. M. Ringel: *Tame algebras and integral quadratic forms.* Springer Lecture Notes in Math. **1099**, Springer (1984).

[13] H. Strauß: *On the perpendicular category of a partial tilting module.* J. Algebra **144** (1991), 43–66.

Michael Barot, Instituto de Matemáticas, UNAM, Ciudad Universitaria, México, 04510, D.F., Mexico

E-mail address: barot@gauss.matem.unam.mx

Helmut Lenzing, Fachbereich Mathematik-Informatik, Universität Paderborn, D-33095 Paderborn, Germany

E-mail address: helmut@math.uni-paderborn.de

Contemporary Mathematics
Volume **229**, 1998

Special Biserial Algebras and Their Automorphisms

Frauke M. Bleher

ABSTRACT. In this paper, ring automorphisms and their operations on the Auslander–Reiten quiver are studied. First the general case of special biserial algebras is examined. Then a complete description of the simultaneous action of every ring automorphism of a special biserial algebra on the 1–tubes is given.

1. Introduction

Special biserial algebras form an important class of finite dimensional algebras of finite or tame representation type. For example, every socle quotient of a block of a group algebra whose defect groups are cyclic or dihedral 2–groups is special biserial. Also every socle quotient of a restricted enveloping algebra of sl_2 belongs to this class of algebras. Special biserial algebras are often used to test conjectures or to develop new methods.

The indecomposable modules of a special biserial algebra Λ can be determined using the representation theory of a quotient algebra of Λ which is a string algebra. For string algebras, which are certain special biserial algebras, there exists a complete description of the indecomposable modules and of the Auslander–Reiten quiver (cf. [**3, 7**]). According to this description, all indecomposable modules are given as so–called string and band modules. If the underlying field K is algebraically closed, the modules lying in 1–tubes correspond to 1–parameter families of bands B whose parameter sets K_B are either $K \setminus \{0\}$, K, or $K \cup \{\infty\}$. Additionally, the morphisms between the indecomposable modules are determined in [**4, 5**].

In this article, ring automorphisms of special biserial algebras and their operations on the corresponding Auslander–Reiten quiver are studied. In case that Λ is a string algebra and σ is a ring automorphism of Λ, the possible operations of σ on all components of the Auslander–Reiten quiver have already been described in [**1, 2**]: For every σ there exists a quiver automorphism ϕ of the quiver of Λ, which can be extended to a K–algebra automorphism of Λ, such that $\sigma(M)$ and $\phi(M)$ are isomorphic for all indecomposable Λ–modules M which do not lie in 1–tubes. Further, there exists a field automorphism κ of K, and for each band B there exists an automorphism ρ_B of the projective line $P^1(K)$ such that the band modules $\sigma(M(B,\lambda,m))$ and $\phi(M(B,\rho_B(\kappa(\lambda)),m))$ are isomorphic for all $\lambda \in K_B$ and

1991 *Mathematics Subject Classification.* Primary 16W20; Secondary 16G70, 16D90.

$m \geq 1$. The first result of this article is the extension of this result to arbitrary special biserial algebras Λ. Except for the determination of the quiver automorphism ϕ, we can use the same argumentation as in [**1, 2**].

Given these possible operations, the question arises which operations are really induced by a ring automorphism of Λ. Because all possible ϕ and κ induce ring automorphisms of Λ, we can restrict ourselves to the examination of the actually occurring operations on K_B, given by ρ_B. The results in [**2**] give an answer to this question in case that Λ is domestic, which means that Λ has only finitely many bands B and therefore only finitely many parameter sets K_B. In this article we examine the general case. We show that even if there are infinitely many bands, only finitely many parameters are needed to determine all actually occurring operations. If r is the number of arrows in the quiver of Λ, t is the number of bands B with $K_B = K$ and u is the number of bands B with $K_B = K \cup \{\infty\}$, then an upper bound of the number of parameters needed is given by $r + t + 2u$. Note that $t + 2u \leq r$. We can even make this more precise as follows: The different actually occurring operations on K_B for all bands B correspond bijectively to $(K \setminus \{0\})^s \times K^t \times (\mathrm{PGL}(2, K))^u$, where s is the rank of the $\mathbb{Z}$–linear span of the images of the bands in the maximal abelian quotient algebra of Λ.

This paper is organized as follows. After introducing special biserial algebras and string algebras, we give the precise definition of the indecomposable modules in Section 2. The main results about the possible operations of ring automorphisms of special biserial algebras on the Auslander–Reiten quiver and the actually occurring operations on the 1–tubes are proved in Section 3.

Throughout this article, K denotes an algebraically closed field and K^* denotes the set of non–zero elements of K. All algebras are finite dimensional K–algebras, all modules are finitely generated left modules. Λ-mod denotes the category of finitely generated Λ–modules. Maps are written on the left, but maps are composed as if they were written on the right.

2. Special biserial algebras, string algebras and their indecomposable modules

Let Λ be a finite dimensional K–algebra. Then the corresponding basic algebra is of the form $K\mathcal{Q}^{\mathrm{op}}/I$ for a unique finite quiver $\mathcal{Q}$ and an admissible ideal I. Here an admissible ideal is an ideal I of the path algebra $K\mathcal{Q}^{\mathrm{op}}$ such that the generators are linear combinations of paths of length at least 2 and such that for some $n \geq 2$ all paths of length n are contained in I.

DEFINITION 2.1. A finite dimensional K–algebra Λ is called *special biserial* if the corresponding basic algebra $K\mathcal{Q}^{\mathrm{op}}/J$ satisfies the following conditions:

(1) Each vertex of $\mathcal{Q}$ is starting point of at most two arrows and end point of at most two arrows.
(2) For each arrow β, there is at most one arrow γ such that $\beta\gamma \notin J$, and there is at most one arrow α such that $\alpha\beta \notin J$.

If additionally J is generated by zero relations, Λ is called a *string algebra.*

REMARK 2.2. (i) If $\Lambda = K\mathcal{Q}^{\mathrm{op}}/J$ is a basic special biserial algebra then the generating relations of J are either zero relations or of the form $a_1p_1 = a_2p_2$ where $a_1, a_2 \in K^*$ and p_1, p_2 are paths. By replacing arrows by scalar multiples, if necessary, we can assume that $a_1 = a_2 = 1$. If all the generating relations are either zero relations or of the form $p_1 = p_2$ then we say that Λ is in *normal presentation.*

The generating relations of J describe the projective indecomposable and the injective indecomposable Λ–modules. The relations of the form $p_1 = p_2$ correspond to projective indecomposable Λ–modules which are also injective and not uniserial. In this case $p_1 = p_2$ is a generator of the socle of this projective indecomposable Λ–module.

(ii) If Λ is a special biserial algebra and $\mathcal{P}$ is a full set of representatives of the projective indecomposable Λ–modules which are also injective and not uniserial, then $\bar{\Lambda} = \Lambda/(\oplus_{P\in\mathcal{P}}\mathrm{soc}(P))$ is a string algebra. Furthermore, the indecomposable $\bar{\Lambda}$–modules are exactly the indecomposable Λ–modules which are not isomorphic to any $P \in \mathcal{P}$.

For string algebras, the indecomposable modules are described with the aid of strings and bands. As a reference, use for example [**3**, §3]. Because of the structure of the indecomposable modules, it follows that special biserial algebras have either finite or tame representation type. Since we will need the precise definition of string and band modules, we give a brief introduction into these notions.

Let $\Lambda = K\mathcal{Q}^{\mathrm{op}}/I$ be the basic algebra of a string algebra. Given an arrow β of $\mathcal{Q}$ with starting point $s(\beta)$ and end point $e(\beta)$, denote by β^{-1} a formal inverse of β, with $s(\beta^{-1}) = e(\beta)$, $e(\beta^{-1}) = s(\beta)$, and write $(\beta^{-1})^{-1} = \beta$.

A word w is a sequence $w_1 \cdots w_n$, where w_i is either an arrow or a formal inverse such that $s(w_i) = e(w_{i+1})$ for $1 \leq i \leq n-1$. Define $s(w) = s(w_n)$, $e(w) = e(w_1)$ and $w^{-1} = w_n^{-1} \cdots w_1^{-1}$. For each vertex u of $\mathcal{Q}$ there exists an empty word 1_u of length 0 with $e(1_u) = s(1_u) = u$ and $(1_u)^{-1} = 1_u$. Denote the set of all words by $\mathcal{W}$, and the set of all non–empty words w with $e(w) = s(w)$ by $\mathcal{W}_r$. In the following, Greek letters inside words always denote arrows.

2.3. Strings and bands. STRINGS: Let $\sim_s$ be the equivalence relation on $\mathcal{W}$ such that $w \sim_s w'$ if and only if $w = w'$ or $w^{-1} = w'$. Then *strings* are representatives $w \in \mathcal{W}$ of the equivalence classes under $\sim_s$ with the following property: either $w = 1_u$, or $w = w_1 \cdots w_n$ where $w_i \neq w_{i+1}^{-1}$ for $1 \leq i \leq n-1$ and no subword of w or its formal inverse belongs to I.

A string $C = \beta_1 \cdots \beta_n$ is called *directed*, $C = \gamma_1^{-1} \cdots \gamma_n^{-1}$ is called *inverse.*

BANDS: Let $w = w_1 \cdots w_n \in \mathcal{W}_r$. Then, for $0 \leq i \leq n-1$, the i–th rotation of w is defined to be the word $\rho_i(w) = w_{i+1} \cdots w_n w_1 \cdots w_i$. Let $\sim_r$ be the equivalence relation on $\mathcal{W}_r$ such that $w \sim_r w'$ if and only if $w = \rho_i(w')$ for some i or $w^{-1} = \rho_j(w')$ for some j. Then *bands* are representatives $w \in \mathcal{W}_r$ of the equivalence classes under $\sim_r$ with the following property: $w = w_1 \cdots w_n$, $n \geq 1$, with $w_i \neq w_{i+1}^{-1}$ for $1 \leq i \leq n-1$ and $w_n \neq w_1^{-1}$, such that w is not a power of a smaller word and, for all positive integers m, no subword of w^m or its formal inverse belongs to I.

A band $B = w_1 \cdots w_n$ is called *balanced* if for each arrow ζ the number of w_i with $w_i = \zeta$ is equal to the number of w_j with $w_j = \zeta^{-1}$.

2.4. String modules and band modules. STRING MODULES: Let $C = w_1 \cdots w_n$ be a string of length n and let $\mathcal{Q}_C$ be the linear quiver

$$\mathcal{Q}_C = \cdot \overset{w_1}{—} \cdot \ \cdots \ \cdot \overset{w_n}{—} \cdot$$

where $\cdot \overset{w_i}{—} \cdot = \cdot \overset{\beta}{\longleftarrow} \cdot$ if $w_i = \beta$ is an arrow, and $\cdot \overset{w_i}{—} \cdot = \cdot \overset{\beta}{\longrightarrow} \cdot$ if $w_i = \beta^{-1}$ is a formal inverse. Then the representation of $\mathcal{Q}_C$ which assigns to each vertex as vector space the field K and to each arrow as map the identity map defines a Λ–module. This module is called the *string module* $M(C)$ corresponding to the string C. More precisely, since $M(C)$ has K–dimension $(n+1)$ as K–vector space, let

$$z_0, \dots, z_n$$

be base vectors generating $M(C)$. Then $\varphi_C : \Lambda \longrightarrow \mathrm{Mat}(n+1, K)$ which denotes the representation of Λ corresponding to $M(C)$ is given as follows:

Let $\varphi_C(u) = T_u$ for vertices u of $\mathcal{Q}$ and $\varphi_C(\alpha) = T_\alpha$ for arrows α of $\mathcal{Q}$. Set $v(i) = e(w_{i+1})$ for $0 \le i \le n-1$, $v(n) = s(w_n)$. Then

$$T_u \cdot z_i = \begin{cases} z_i & , \text{ if } v(i) = u \\ 0 & , \text{ else} \end{cases} \quad \text{and } T_\alpha \cdot z_i = \begin{cases} z_{i-1} & , \text{ if } w_i = \alpha \\ z_{i+1} & , \text{ if } w_{i+1} = \alpha^{-1} \\ 0 & , \text{ else} \end{cases}.$$

In the following, the representation φ_C is called the *canonical representation* corresponding to $M(C)$.

The string modules $M(1_u)$, u a vertex of $\mathcal{Q}$, correspond bijectively to the isomorphism classes of the simple Λ–modules.

BAND MODULES: Let $B = w_1 \cdots w_n$ be a band of length n. We may assume that w_1 is an arrow, by rotating and possibly inverting. Let $\mathcal{Q}_B^c$ be the circular quiver

$\mathcal{Q}_B^c =$ 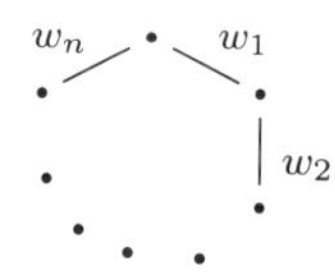

such that $\overset{w_i}{\cdot\!-\!\cdot} = \overset{\beta}{\cdot\!\longleftarrow\!\cdot}$ where the arrow points counterclockwise if $w_i = \beta$ is an arrow, and the arrow points clockwise if $w_i = \beta^{-1}$ is a formal inverse.

Let $V_m(\lambda)$ be the m–dimensional indecomposable $K[x, x^{-1}]$–module such that x acts as the indecomposable Jordan matrix $J_m(\lambda)$, $\lambda \in K^*$. Then the representation of $\mathcal{Q}_B^c$ which assigns to each vertex as vector space K^m, to the arrow w_1 as map $J_m(\lambda)$, and to all other arrows as map the identity map defines a Λ–module. This module is called the *band module* $M(B, \lambda, m)$ corresponding to the band B and $V_m(\lambda)$. More precisely, since $M(B, \lambda, m)$ has K–dimension $(n \cdot m)$ as K–vector space, let

$$z_{0,0}, \dots, z_{0,m-1}, z_{1,0}, \dots, z_{1,m-1}, \dots, z_{n-1,0}, \dots, z_{n-1,m-1}$$

be base vectors generating $M(B, \lambda, m)$. Then $\varphi_{B,\lambda,m} : \Lambda \longrightarrow \mathrm{Mat}(n \cdot m, K)$ which denotes the representation of Λ corresponding to $M(B, \lambda, m)$ is given as follows:

Let $\varphi_{B,\lambda,m}(u) = T_u^\lambda$ for vertices u of $\mathcal{Q}$ and $\varphi_{B,\lambda,m}(\alpha) = T_\alpha^\lambda$ for arrows α of $\mathcal{Q}$. Set $v(i) = e(w_{i+1})$ for $0 \le i \le n-1$. Then

$$\begin{aligned} T_u^\lambda \cdot (z_{i,j})_{0 \le j \le m-1} &= \begin{cases} (z_{i,j})_{0 \le j \le m-1} & , \text{ if } v(i) = u \\ 0 & , \text{ else} \end{cases}, \\ T_{w_1}^\lambda \cdot (z_{1,j})_{0 \le j \le m-1} &= J_m(\lambda) \cdot (z_{0,j})_{0 \le j \le m-1}, \end{aligned}$$

and, for $(i, \alpha) \ne (1, w_1)$:

$$T_\alpha^\lambda \cdot (z_{i,j})_{0 \le j \le m-1} = \begin{cases} (z_{i-1,j})_{0 \le j \le m-1} & , \text{ if } w_i = \alpha \\ (z_{i+1,j})_{0 \le j \le m-1} & , \text{ if } w_{i+1} = \alpha^{-1} \\ 0 & , \text{ else} \end{cases}.$$

In the following, the representation $\varphi_{B,\lambda,m}$ is called the *canonical representation* corresponding to $M(B,\lambda,m)$.

Note: Every indecomposable Λ–module is isomorphic to either a string or a band module.

2.5. Auslander–Reiten components consisting of band modules. Let B be a band, $\lambda \in K^*$ and let m be a positive integer. Then the almost split sequence starting in $M(B,\lambda,m)$, which is unique up to isomorphism, has the form

$$0 \to M(B,\lambda,m) \to M(B,\lambda,m-1) \oplus M(B,\lambda,m+1) \to M(B,\lambda,m) \to 0.$$

Especially, all band modules lie in 1–tubes. The Auslander–Reiten component which contains the module $M(B,\lambda,m)$ consists of all modules $M(B,\lambda,k)$, $k \geq 1$, where $M(B,\lambda,1)$ lies at the mouth of this component.

2.6. String modules which are band modules and lie in 1–tubes. The string modules which lie at the mouths of 1–tubes are exactly those corresponding to directed strings C, such that βC and $C\gamma$ lie in I for all arrows β and γ, and, additionally, there exists an arrow α with $C\alpha^{-1}C$ is again a string. This follows from the description of the almost split sequences containing string modules and especially of those having an indecomposable middle term as given in [**3**, §3]. Then all modules in the Auslander–Reiten component of $M(C)$ are string modules corresponding to strings of the form $(C\alpha^{-1})^{m-1}C$, m a positive integer.

These string modules can be interpreted as certain band modules with parameter $\lambda = 0$, namely $M((C\alpha^{-1})^{m-1}C) \cong M(\alpha C^{-1}, \lambda = 0, m)$. The isomorphism can be established explicitly by using standard matrix calculations and the canonical representations corresponding to these modules as described in 2.4.

In case $\mathcal{Q}$ contains a subquiver of the form $\cdot \overset{\alpha}{\underset{\beta}{\rightrightarrows}} \cdot$ and both string modules $M(\alpha)$ and $M(\beta)$ lie in 1–tubes, it follows that $M((\beta\alpha^{-1})^{m-1}\beta) \cong M(\alpha\beta^{-1}, \lambda = 0, m)$ and $M((\alpha\beta^{-1})^{m-1}\alpha) \cong M(\alpha\beta^{-1}, \lambda = \infty, m)$. Here $M(\alpha\beta^{-1}, \lambda = \infty, m)$ is isomorphic to $M(\beta\alpha^{-1}, \lambda = 0, m)$.

Note: The number of 1–tubes of the Auslander–Reiten quiver which consist of string modules is bounded by the number of arrows of $\mathcal{Q}$.

In the following, the string modules lying in 1–tubes are regarded as band modules, and the parameter sets corresponding to these bands are extended accordingly. Thus if B is a band the corresponding parameter set K_B is either K^*, K, or $K \cup \{\infty\}$.

3. Automorphisms of special biserial algebras

In [**1**, Theorem 3.1] and in [**2**, Theorem 3.1 and Proposition 3.4], the possible operations of ring automorphisms of a string algebra on the Auslander–Reiten quiver have been described. For arbitrary special biserial algebras we get a similar result.

THEOREM 3.1. *Let $\Lambda = K\mathcal{Q}^{\text{op}}/J$ be a basic special biserial algebra in normal presentation, and let σ be an autoequivalence of* Λ-mod. *Then there exists a quiver automorphism ϕ of $\mathcal{Q}$ which preserves J and can therefore be extended to a K–algebra automorphism of Λ, and there exists a field automorphism κ of K such that*

(i) *$\sigma(M) \cong \phi(M)$ for all indecomposable Λ–modules M which do not lie in 1–tubes of the Auslander–Reiten quiver corresponding to Λ.*

(ii) *For every band B there exists an automorphism ρ_B of the projective line $P^1(K)$, which maps the parameter set K_B of B onto itself, such that*

$$\sigma(M(B,\lambda,m)) \cong \phi(M(B,\rho_B(\kappa(\lambda)),m))$$

for all $\lambda \in K_B$ and for all positive integers m.

(iii) *If B is a balanced band then only $\rho_B = \mathrm{id}_{P^1(K)}$ occurs in (ii); otherwise every ρ_B with $\rho_B(K_B) = K_B$ can occur in (ii).*

Note: Since every autoequivalence is additive and since Λ is basic, σ corresponds to a ring automorphism of Λ.

PROOF. Let $\mathcal{P}$ be a full set of representatives of the projective indecomposable Λ–modules which are also injective and not uniserial, and let $\bar{\Lambda} = K\mathcal{Q}^{\mathrm{op}}/I = \Lambda/(\oplus_{P\in\mathcal{P}}\mathrm{soc}(P))$ be the string algebra corresponding to Λ. Since the indecomposable Λ–modules which do not lie in $\mathcal{P}$ are exactly the indecomposable $\bar{\Lambda}$–modules, we can use the same proofs as in [**1, 2**] if we have shown that σ induces a quiver automorphism ϕ which preserves the ideal J.

The autoequivalence σ of Λ-mod induces an autoequivalence $\bar{\sigma}$ of $\bar{\Lambda}$-mod. Further $\bar{\sigma}$ induces a quiver automorphism ϕ of $\mathcal{Q}$ which preserves I. We want to show that ϕ preserves J.

Let ϕ map the vertices u to $\tilde{u}$ and the arrows α to $\tilde{\alpha}$. The only relations in J which are not zero relations come from the projective–injective modules in $\mathcal{P}$. Let $P \in \mathcal{P}$ be given as

$$P = \begin{array}{cccc} \alpha_n & \swarrow & \searrow & \beta_m \\ \alpha_{n-1} & \downarrow & \downarrow & \beta_{m-1} \\ & \vdots & \vdots & \\ \alpha_1 & \downarrow & \downarrow & \beta_1 \\ \alpha_0 & \searrow & \swarrow & \beta_0 \end{array}.$$

Since Λ is in normal presentation the relation corresponding to P in J is given by

$$\alpha_0\alpha_1\cdots\alpha_n = \beta_0\beta_1\cdots\beta_m.$$

In the ideal I of $\bar{\Lambda}$ this relation is replaced by

$$\alpha_0\alpha_1\cdots\alpha_n = 0 = \beta_0\beta_1\cdots\beta_m.$$

Since σ is an autoequivalence, σ maps P to another module $P' \in \mathcal{P}$. By using similar arguments as in the proof of [**1**, Theorem 3.1], it follows that

$$\begin{aligned} \mathrm{rad}(P') &\cong M(\tilde{\alpha}_{n-1}^{-1}\cdots\tilde{\alpha}_0^{-1}\tilde{\beta}_0\cdots\tilde{\beta}_{m-1}) \text{ and} \\ P'/\mathrm{soc}(P') &\cong M(\tilde{\alpha}_1\cdots\tilde{\alpha}_n\tilde{\beta}_m^{-1}\cdots\tilde{\beta}_1^{-1}). \end{aligned}$$

Thus P' is the module

$$P' = \begin{array}{cccc} \tilde{\alpha}_n & \swarrow & \searrow & \tilde{\beta}_m \\ \tilde{\alpha}_{n-1} & \downarrow & \downarrow & \tilde{\beta}_{m-1} \\ & \vdots & \vdots & \\ \tilde{\alpha}_1 & \downarrow & \downarrow & \tilde{\beta}_1 \\ \tilde{\alpha}_0 & \searrow & \swarrow & \tilde{\beta}_0 \end{array}$$

and therefore

$$\tilde{\alpha}_0\tilde{\alpha}_1\cdots\tilde{\alpha}_n = \tilde{\beta}_0\tilde{\beta}_1\cdots\tilde{\beta}_m$$

is a relation in J. Thus ϕ preserves the ideal J, and the theorem is proved. □

We want to look more precisely at the actually occurring operations of autoequivalences of Λ-mod on the 1–tubes. From [**2**, Theorem 3.6] it follows in case $\Lambda = K\mathcal{Q}^{\mathrm{op}}/J$ is domestic that, given a quiver automorphism ϕ of $\mathcal{Q}$ which preserves J and a field automorphism κ of K, and given for each band B an automorphism ρ_B with $\rho_B(K_B) = K_B$, there exists a ring automorphism σ of Λ such that for all bands B simultaneously

$$\sigma(M(B,\lambda,m)) \cong \phi(M(B,\rho_B(\kappa(\lambda)),m)).$$

Thus all possible operations actually occur in this case. However, in the general case this is not true as examples show [**2**, Example 3.5].

We want now to study the general case and state precisely which combinations of ρ_B can occur simultaneously in the operation of a single autoequivalence of Λ-mod. Since every autoequivalence of Λ-mod corresponds to a ring automorphism and since for every ring automorphism σ of Λ there exists a field automorphism κ of K such that $\kappa\sigma$ is a K–algebra automorphism, we can restrict ourselves to K–algebra automorphisms of Λ. Further we can assume that the K–algebra automorphisms induce the trivial quiver automorphism ϕ, because every quiver automorphism which preserves the relations induces a K–algebra automorphism of Λ.

THEOREM 3.2. *Let $\Lambda = K\mathcal{Q}^{\mathrm{op}}/J$ be a basic special biserial algebra and let σ be a K–algebra automorphism of Λ such that σ induces the trivial quiver automorphism ϕ on $\mathcal{Q}$. Then for all arrows ζ,*

$$\sigma(\zeta) = a_\zeta\zeta + \sum_{\substack{T \text{ directed string}\\ l(T)\geq 1, T\neq\zeta}} a^\zeta_T T \qquad \text{with } a_\zeta \in K^* \text{ and } a^\zeta_T \in K.$$

Let B be a band with $B = w_1\cdots w_n$. Then

$$\sigma(M(B,\lambda,m)) \cong M(B,\mu,m)$$

such that

(i) *if $K_B = K^*$ then $\mu = a_{w_1}\cdots a_{w_n}\lambda$.*
(ii) *if $K_B = K$ and $B = \alpha C^{-1}$ then $\mu = a_{w_1}\cdots a_{w_n}\lambda + a_{w_2}\cdots a_{w_n}a^{w_1}_{(w_2\cdots w_n)^{-1}}$.*
(iii) *if $K_B = K\cup\{\infty\}$ and $B = \alpha\beta^{-1}$ then $\mu = (a_\alpha\lambda + a^\alpha_\beta)/(a^\beta_\alpha\lambda + a_\beta)$.*

For (i) and (ii), we set $a_{w_i} = a_\zeta$ if $w_i = \zeta$, and $a_{w_i} = a_\zeta^{-1}$ if $w_i = \zeta^{-1}$.

PROOF. The quiver automorphism ϕ is trivial, thus $a_\zeta \neq 0$. Since there exists s with $\zeta^s = 0$, it follows that $a^\zeta_{1_u} = 0$ for all vertices u of $\mathcal{Q}$. Let $B = w_1\cdots w_n$ be a band. It is enough to determine the isomorphism class of $\sigma(M(B,\lambda,1))$ for all $\lambda\in K_B$. To do this, we have to find an invertible matrix Θ with

$$\Theta^{-1}\sigma(\varphi_{B,\lambda,1})\Theta = \varphi_{B,\mu,1}$$

where μ has to be determined. Thus we have for all arrows ζ

$$\sigma(\varphi_{B,\lambda,1})(\zeta)\Theta = \Theta\varphi_{B,\mu,1}(\zeta).$$

Let now $\Theta = (\theta_{i,j})_{0\leq i,j\leq n-1}$.

(a) $B = \alpha\beta^{-1}$. Then

$$\begin{aligned} \sigma(\varphi_{B,\lambda,1})(\alpha) &= \begin{pmatrix} 0 & a_\alpha\lambda + a_\beta^\alpha \\ 0 & 0 \end{pmatrix} \quad \text{and} \\ \sigma(\varphi_{B,\lambda,1})(\beta) &= \begin{pmatrix} 0 & a_\alpha^\beta\lambda + a_\beta \\ 0 & 0 \end{pmatrix}. \end{aligned}$$

So it follows immediately that $\mu = (a_\alpha\lambda + a_\beta^\alpha)/(a_\alpha^\beta\lambda + a_\beta)$.

(b) $B = \alpha\beta_1^{-1}\cdots\beta_{n-1}^{-1}$, $n \geq 3$.
Then the second row of $\sigma(\varphi_{B,\lambda,1})(\beta_1)$ has the form

$$(0, a_{\beta_1}, 0, \cdots, 0).$$

Thus we get as condition for Θ:

$$a_{\beta_1}\theta_{1,1} = \theta_{2,2}.$$

For all $i \leq n-2$, we get in a similar way

$$a_{\beta_i}\theta_{i,i} = \theta_{i+1,i+1}.$$

Let for example $i = 2$. If $n \geq 4$ then the third row of $\sigma(\varphi_{B,\lambda,1})(\beta_2)$ has the form

$$(0, a_{\beta_2\beta_1}^{\beta_2}, a_{\beta_2}, 0, \cdots, 0).$$

So the condition for Θ is

$$a_{\beta_2\beta_1}^{\beta_2}\theta_{1,2} + a_{\beta_2}\theta_{2,2} = \theta_{3,3}.$$

Since the first row is zero for all $\sigma(\varphi_{B,\lambda,1})(\beta_i)$, it follows that $\theta_{1,j} = 0$ for all $j \neq 1$. Thus we get as condition

$$a_{\beta_2}\theta_{2,2} = \theta_{3,3}.$$

In a similar way we get for $i = n-1$:

$$a_{\beta_{n-1}}\theta_{n-1,n-1} = \theta_{0,0}.$$

The zero-th row of $\sigma(\varphi_{B,\lambda,1})(\alpha)$ has the form

$$(0, \lambda a_\alpha + a_{\beta_{n-1}\cdots\beta_1}^\alpha, a_{\beta_{n-1}\cdots\beta_2}^\alpha, \cdots, a_{\beta_{n-1}}^\alpha).$$

Thus it follows, using similar arguments as before, that

$$(\lambda a_\alpha + a_{\beta_{n-1}\cdots\beta_1}^\alpha)\theta_{1,1} = \mu\theta_{0,0}.$$

Using these conditions together we get

$$\begin{aligned} \mu\theta_{0,0} &= (\lambda a_\alpha + a_{\beta_{n-1}\cdots\beta_1}^\alpha)\theta_{1,1} = (\lambda a_\alpha + a_{\beta_{n-1}\cdots\beta_1}^\alpha)a_{\beta_1}^{-1}\theta_{2,2} \\ &= \cdots = (\lambda a_\alpha + a_{\beta_{n-1}\cdots\beta_1}^\alpha)a_{\beta_1}^{-1}\cdots a_{\beta_{n-1}}^{-1}\theta_{0,0} \end{aligned}$$

and then since $\theta_{j,j} \neq 0$ for all j,

$$\begin{aligned} \mu &= (\lambda a_\alpha + a_{\beta_{n-1}\cdots\beta_1}^\alpha)a_{\beta_1}^{-1}\cdots a_{\beta_{n-1}}^{-1} \\ &= a_\alpha a_{\beta_1}^{-1}\cdots a_{\beta_{n-1}}^{-1}\lambda + a_{\beta_1}^{-1}\cdots a_{\beta_{n-1}}^{-1}a_{\beta_{n-1}\cdots\beta_1}^\alpha. \end{aligned}$$

(c) $B = w_1 \cdots w_n = \alpha B_1 \beta^{-1}$, not as in (a) or (b).
To find μ, we consider subwords of B of the form $\gamma^{-1}\zeta_1 \cdots \zeta_s \delta^{-1}$ or of the form $\tau\xi_r^{-1} \cdots \xi_1^{-1} v$, where $\zeta_1 \cdots \zeta_s$ and $\xi_1 \cdots \xi_r$ are directed. First we consider the case: $w_i \cdots w_{i+s+1} = \gamma^{-1}\zeta_1 \cdots \zeta_s \delta^{-1}$, where we set $w_0 = w_n$.

Let first $s > 1$. Then the $(i+s-1)$–th row of $\sigma(\varphi_{B,\lambda,1})(\zeta_s)$ has the form

$$(0, \cdots, 0, a_{\zeta_s}, 0, \cdots, 0),$$

where a_{ζ_s} is at the $(i+s)$–th position. Thus we get as condition for Θ:

$$a_{\zeta_s}\theta_{i+s,i+s} = \theta_{i+s-1,i+s-1}.$$

If $s \geq 3$ then the $(i+s-2)$–th row of $\sigma(\varphi_{B,\lambda,1})(\zeta_{s-1})$ has the form

$$(0, \cdots, 0, a_{\zeta_{s-1}}, a^{\zeta_{s-1}}_{\zeta_{s-1}\zeta_s}, 0, \cdots, 0),$$

where $a_{\zeta_{s-1}}$ is at the $(i+s-1)$–th position. Therefore we get

$$a_{\zeta_{s-1}}\theta_{i+s-1,i+s-1} + a^{\zeta_{s-1}}_{\zeta_{s-1}\zeta_s}\theta_{i+s,i+s-1} = \theta_{i+s-2,i+s-2}.$$

Analogously to case (b), it follows that $\theta_{i+s,i+s-1} = 0$. Thus we have

$$a_{\zeta_{s-1}}\theta_{i+s-1,i+s-1} = \theta_{i+s-2,i+s-2}.$$

Similarly, we get

$$a_{\zeta_{s-l}}\theta_{i+s-l,i+s-l} = \theta_{i+s-l-1,i+s-l-1}$$

for all $2 \leq l \leq s-2$. If $i \neq 0$, we also have $a_{\zeta_1}\theta_{i+1,i+1} = \theta_{i,i}$. For $i = 0$ we get $\lambda a_\alpha \theta_{1,1} = \mu\theta_{0,0}$.

Let now $s = 1$. Suppose $w_{i-t} \cdots w_{i+2} = \rho\gamma_t^{-1} \cdots \gamma_1^{-1}\zeta_1\delta^{-1}$. Then the i–th row of $\sigma(\varphi_{B,\lambda,1})(\zeta_1)$ has the form

$$(0, \cdots, 0, a^{\zeta_1}_{\gamma_1\cdots\gamma_t}, \cdots, a^{\zeta_1}_{\gamma_1}, 0, a_{\zeta_1}, 0, \cdots, 0),$$

where a_{ζ_1} is at the $(i+1)$–th position. Thus we get as condition for Θ:

$$a_{\zeta_1}\theta_{i+1,i+1} + a^{\zeta_1}_{\gamma_1}\theta_{i-1,i+1} + \cdots + a^{\zeta_1}_{\gamma_1\cdots\gamma_t}\theta_{i-t,i+1} = \theta_{i,i}.$$

Using similar arguments as above, it follows that $\theta_{i-k,i+1} = 0$ for $1 \leq k \leq t$. And therefore, it follows again that $a_{\zeta_1}\theta_{i+1,i+1} = \theta_{i,i}$.

Next we consider the case: $w_j \cdots w_{j+r+1} = \tau\xi_r^{-1} \cdots \xi_1^{-1}v$, where we set $w_{n+1} = w_1$. Similarly as above, it follows that

$$a_{\xi_{r-l}}\theta_{j+l,j+l} = \theta_{j+l+1,j+l+1}$$

for all $0 \leq l \leq r-1$. If $j + r = n$, we set $\theta_{n,n} = \theta_{0,0}$.

If we now set $a_{w_i} = a_\zeta$ if $w_i = \zeta$, and $a_{w_i} = a_\zeta^{-1}$ if $w_i = \zeta^{-1}$, then we get the following conditions for $\theta_{i,i}$:

$$\begin{aligned} \lambda a_{w_1}\theta_{1,1} &= \mu\theta_{0,0} \\ a_{w_i}\theta_{i,i} &= \theta_{i-1,i-1} \qquad 2 \leq i \leq n-1 \\ a_{w_n}\theta_{0,0} &= \theta_{n-1,n-1} \end{aligned}$$

Therefore it follows that

$$\mu\theta_{0,0} = \lambda a_{w_1}\theta_{1,1} = \lambda a_{w_1}a_{w_2}\theta_{2,2} = \cdots = \lambda a_{w_1} \cdots a_{w_n}\theta_{0,0},$$

and thus, since $\theta_{j,j} \neq 0$ for all j, we get that $\mu = a_{w_1} \cdots a_{w_n}\lambda$.

□

Theorem 3.2 shows that the number of parameters needed to describe all possible operations on $\{K_B \mid B \text{ a band}\}$, induced by K–algebra automorphisms, is finite. More precisely, we get the following result:

Let r be the number of arrows in $\mathcal{Q}$ and let z be the number of maximal directed strings C such that $M(C)$ lies in a 1–tube. Note that $z \leq r$.

COROLLARY 3.3. *If Λ and σ are as in Theorem 3.2, then the operation of σ on the band modules is determined by*

$a_\zeta \in K^*$, *ζ an arrow in $\mathcal{Q}$,*

$a_C^\alpha \in K$, *C a maximal directed string such that αC^{-1} is a string and $M(C)$ lies in a 1–tube,*

and thus by $r+z$ parameters in $(K^*)^r \times K^z$.

In the next result we describe which operations on $\{K_B \mid B \text{ a band}\}$, induced by K–algebra automorphisms, actually occur and give the minimal number of parameters needed to determine all different operations.

THEOREM 3.4. *Let $\Lambda = K\mathcal{Q}^{\mathrm{op}}/J$ be a basic special biserial algebra such that $\mathcal{Q} = \{\zeta_1, \ldots, \zeta_r\}$ and such that the maximal directed strings corresponding to modules lying in 1–tubes are given by $\{C_1, \ldots, C_z\}$. Then for every tuple*

$$(a_{\zeta_1}, \cdots, a_{\zeta_r}, b_{C_1}, \cdots, b_{C_z}) \in (K^*)^r \times K^z$$

there exists a K–algebra automorphism σ of Λ with

$$\sigma(M(B, \lambda, m)) \cong M(B, \mu, m)$$

for all bands B such that if $B = w_1 \cdots w_n$ and

(i) *if $K_B = K^*$ then $\mu = a_{w_1} \cdots a_{w_n} \lambda$.*
(ii) *if $K_B = K$ and $B = \alpha C^{-1}$ then $\mu = a_{w_1} \cdots a_{w_n} \lambda + a_{w_2} \cdots a_{w_n} b_C$.*
(iii) *if $K_B = K \cup \{\infty\}$ and $B = \alpha\beta^{-1}$ then $\mu = (a_\alpha \lambda + b_\beta)/(b_\alpha \lambda + a_\beta)$.*

For (i) and (ii), we set $a_{w_i} = a_\zeta$ if $w_i = \zeta$, and $a_{w_i} = a_\zeta^{-1}$ if $w_i = \zeta^{-1}$.

The parameters corresponding to different operations can be found as follows: For B as in (i) or (ii), let $a_B = a_{w_1} \cdots a_{w_n}$. Then we can write a_B as

$$a_B = a_{\zeta_1}^{i_1} \cdots a_{\zeta_r}^{i_r}$$

with $i_j \in \mathbb{Z}$. Since $i_1, \ldots, i_r$ do not depend on the value of a_B but only on B, each B corresponds to a tuple $x_B = (i_1, \ldots, i_r)$ in $\mathbb{Z}^r$. Let X be the $\mathbb{Z}$–linear span of all x_B such that B is as in (i) or (ii). Then X is a free $\mathbb{Z}$–submodule of $\mathbb{Z}^r$. Denote the rank of X by s. Let t be the number of bands B in (ii) and let u be the number of bands B in (iii), which means that $z = t + 2u$. Then the different operations on $\{K_B \mid B \text{ a band}\}$, induced by K–algebra automorphisms, correspond bijectively to

$$(K^*)^s \times K^t \times (\mathrm{PGL}(2, K))^u,$$

where $(\mathrm{PGL}(2, K))^u$ is the direct product of u copies of the projective general linear group $\mathrm{PGL}(2, K)$.

PROOF. For the first part of the theorem we define the K–algebra automorphism σ as follows. If ζ is an arrow such that there exists a maximal directed string C with ζC^{-1} is again a string and $M(C)$ lies in a 1–tube, then we set

$$\sigma(\zeta) = a_\zeta \zeta + b_C C.$$

Otherwise we set

$$\sigma(\zeta) = a_\zeta \zeta.$$

This can be extended to a K–algebra automorphism which operates on the band modules as desired.

The different operations on K_B for bands B as in (iii) correspond to the projective general linear group $\mathrm{PGL}(2, K)$, since $\mathrm{PGL}(2, K)$ is the automorphism group of the projective line $P^1(K)$.

For the different operations on K_B for bands B as in (i) and (ii), we only have to show that the different possibilities for $\{a_B \mid B \text{ a band}\}$, correspond to $(K^*)^s$. For each B let x_B be the corresponding r–tuple in $\mathbb{Z}^r$, as described in the statement of the theorem, and let S be the set of these r–tuples x_B. Then we can define a map $f : S \to K^*$ by $f(x_B) = a_B$. The different possibilities for $\{a_B \mid B \text{ a band}\}$ are then given by the different possible maps $f : S \to K^*$. By Theorem 3.2, it follows that whenever $c_1 x_{B_1} + c_2 x_{B_2} \in S$ for $c_1, c_2 \in \mathbb{Z}$ and $x_{B_1}, x_{B_2} \in S$, then

$$f(c_1 x_{B_1} + c_2 x_{B_2}) = f(x_{B_1})^{c_1} \cdot f(x_{B_2})^{c_2}.$$

It also follows from Theorem 3.2 that each such f can be uniquely extended to the $\mathbb{Z}$–linear span X of S, and is thus a group homomorphism

$$f : (X, +) \to (K^*, \cdot).$$

Therefore we have to determine the different possible group homomorphisms from X to K^*. Since each such group homomorphism is uniquely determined by the images of a basis of X and since X has rank s, it follows that the different possibilities for f and thus for $\{a_B \mid B \text{ a band}\}$ correspond to $(K^*)^s$. This proves the theorem. □

References

[1] F. M. Bleher, *Automorphisms of string algebras*, J. Algebra **201** (1998), 528–546.

[2] ———, *Operations of automorphisms on 1–tubes*, accepted by J. Pure Appl. Algebra, 1998.

[3] M. C. R. Butler and C. M. Ringel, *Auslander–Reiten–Sequences with few middle terms and applications to string algebras*, Comm. Algebra **15** (1987), 145–179.

[4] W. W. Crawley–Boevey, *Maps between representations of zero–relation algebras*, J. Algebra **126** (1989), 259–263.

[5] H. Krause, *Maps between tree and band modules*, J. Algebra **137** (1991), 186–194.

[6] C. M. Ringel, *The indecomposable representations of the dihedral 2–groups*, Math. Ann. **214** (1975), 19–34.

[7] B. Wald and J. Waschbüsch, *Tame biserial algebras*, J. Algebra **95** (1985), 480–500.

Department of Mathematics, University of Pennsylvania, Philadelphia, Pennsylvania 19104-6395, U.S.A.

E-mail address: frauke@math.upenn.edu

Contemporary Mathematics
Volume **229**, 1998

Wild subquivers of the Auslander-Reiten quiver of a tame algebra

Sheila Brenner and Michael C. R. Butler

ABSTRACT. Let Λ be a finite dimensional algebra over an algebraically closed field and

$$0 \to X \to \oplus_{i=1}^{n} Y_i \to Z \to 0$$

be an almost spit sequence of Λ-modules. We prove that, if Λ is tame and $Y_1 \simeq Y_2$, then $n \leq 3$ and, if $n = 3$, then Y_3 is both projective and injective.

We also show, under some apparently fairly restrictive conditions (for example, if X and Z are Schurian or if, for all i, $\dim Y_i < \sum_{j \neq i} \dim Y_j$), that Λ is tame implies that $n \leq 5$ and, if $n = 5$, then one of the Y_i is both projective and injective.

In a final section we show that if no module M over a tame algebra satisfies either $\lim_{r \to \infty} \sqrt[r]{\dim \tau^r M} > 1$ or $\lim_{r \to \infty} \sqrt[r]{\dim \tau^{-r} M} > 1$, then for each finite wild quiver Δ there is a finite bound on the number of parallel sectional subquivers isomorphic to Δ which can occur in the Auslander-Reiten quiver of a tame algebra.

1. Introduction

Let Λ be a finite dimensional algebra over an algebraically closed field and

$$0 \to X \to Y = \oplus_1^n Y_i \to Z \to 0$$

be an almost split sequence of Λ-modules in which $Y_1, \ldots, Y_n$ are indecomposable. It has long been known that, if Λ has finite representation type, then no two of the Y_i are isomorphic, $n \leq 4$ and, if $n = 4$, then one of the Y_i is both projective and injective. These are the most easily stated results of the replication theory developed in [**4**] to describe precisely the ways in which an infinite representation type quiver can appear as a subquiver of the Auslander-Reiten quiver of an algebra of finite representation type. In this paper, we consider similar results aimed at distinguishing algebras of tame representation type from those of wild representation type. Our main objective is to discuss the following rather widely believed conjecture concerning the algebra and almost split sequence above.

CONJECTURE 1. *If either* (I) $n \geq 3$ *and two of the* Y_i *are isomorphic, or* (II) $n \geq 5$, *then* Λ *is wild unless one of the* Y_i *is both projective and injective and either* $n = 3$ *in case* (I) *or* $n = 5$ *in case* (II).

1991 *Mathematics Subject Classification.* 16E30, 16G60, 16G70.

We announced a proof of this conjecture at the Tsukuba International Conference on Representations of Algebras (ICRA), 1990, only to find later that there was an error in the combinatorics. In this paper we will prove (I) in Section 2. In Section 3 we prove (II), but only under some apparently fairly restrictive conditions (one sufficient condition is that, for each i, $2\dim Y_i < \dim Y$; another is that both X and Z be Schurian).

At the Gerainger ICRA in 1996, Marthe Takane announced a proof of the full result, in joint work in [**10**] with de la Peña, under the assumption that Y is directing. Their work stimulated us to write up our present results, which seem to go a little further than theirs and which exploit quite different methods. We prove that certain algebras have wild representation type by explicitly constructing infinite families of pairwise non-isomorphic indecomposable modules of the same dimension, none of which belong to homogeneous tubes. That this implies wildness follows from a well-known theorem of Crawley-Boevey [**6**].

Sections 2 and 3 are concerned with algebras whose Auslander-Reiten quivers have finite sectional subgraphs of one or other of two special wild types. In Section 4 we suggest possible generalisations of these results to other wild quivers. Subsection 4.1 shows how to generalise our constructions of infinite families of modules of the same dimension, though we are unable to prove that there are infinitely many isoclasses of indecomposables. In Subsection 4.2, we conjecture that for each wild quiver, there is an upper bound to the number of parallel sectional copies of it which can occur in the Auslander-Reiten quiver of a tame algebra. We also show that this conjecture is a consequence of the widely believed conjecture that, for a tame algebra, there is no module M which satisfies either $\lim_{r\to\infty} \sqrt[r]{\dim\tau^r M} > 1$ or $\lim_{r\to\infty} \sqrt[r]{\dim\tau^{-r} M} > 1$.

In Section 2 we prove Conjecture 1 Part (I) by means of the following remarkably precise theorem.

THEOREM 1.1. *Let Λ be a tame finite dimensional algebra over an algebraically closed field and*

$$0 \to X \to Y \oplus Y \oplus P \to Z \to 0$$

an almost split sequence of Λ-modules with Y indecomposable and $P \neq 0$. Then P is an indecomposable projective-injective module of composition length 4, Y is simple, and the connected component of the Auslander-Reiten quiver of Λ which contains these modules is as follows:

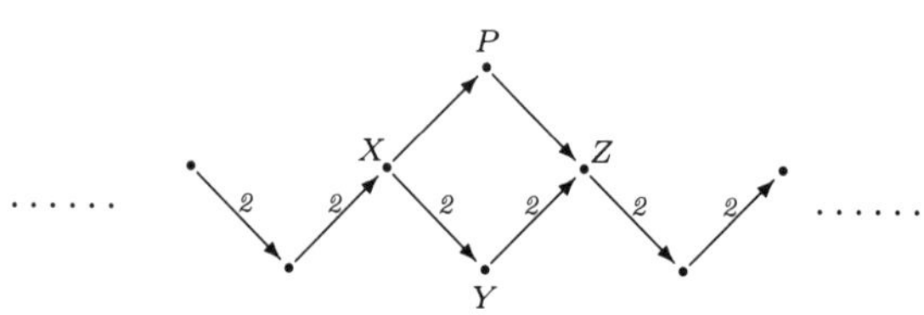

(where $\xrightarrow{2}$ denotes an arrow with valuation $(2, 2)$ which, from now on, we refer to as a double arrow).

This has other implications including the following corollary, pointed out to us by Shiping Liu.

COROLLARY 1.2. *A regular component in the Auslander-Reiten quiver of a tame algebra over an algebraically closed field cannot contain a double arrow.*

Further, Proposition 2.9 gives a fairly complete description of the tame basic algebras which can have double arrows in their Auslander-Reiten quivers, but no component of the type covered by Theorem 1.1. We will show that the only such local algebra is of dimension 3, whilst for any other, the ordinary quiver must contain a Kronecker subquiver whose arrows compose to zero with any other arrows in the quiver.

For the discussion of part (II) of Conjecture 1 in Section 3, we denote by $\mathcal{E}$ the set of all almost split sequences

$$E : 0 \to X \to Y = \oplus_{i=1}^{n} Y_i \to Z \to 0$$

in which the Y_i are indecomposable and either $n > 5$, or $n = 5$ and none of the Y_i is both projective and injective. Thus the conjecture, also called the five-terms-in-the-middle conjecture, is that the algebra is wild if $\mathcal{E}$ is non-empty. Of course this follows from Theorem 1.1 if $\mathcal{E}$ contains an almost split sequence with 2 or more isomorphic Y_i. Otherwise, our best results are given below in Proposition 1.3, which makes use of the observation that $\mathcal{E}$ is a union of subsets $\mathcal{L}$ and $\mathcal{R}$ defined as follows:

$\mathcal{L}$ contains all almost split sequences $E \in \mathcal{E}$ in which either

(i) $\dim X \geq \dim Z$ and at least 5 of the Y_i are not projective,
or
(ii) one of the Y_i is injective and not projective;

and, dually, $\mathcal{R}$ contains all almost split sequences $E \in \mathcal{E}$ in which either

(i) $\dim X \leq \dim Z$ and at least 5 of the Y_i are not injective,
or
(ii) one of the Y_i is projective and not injective.

PROPOSITION 1.3. *Suppose that Λ is a finite dimensional algebra over an algebraically closed field k. Let $E \in \mathcal{E}$. Then Λ is wild if one of the following conditions is satisfied*:
(A) *for all i*

$$\dim Y_i < \sum_{j \neq i} \dim Y_j\,;$$

(B) *$E \in \mathcal{L}$ and* $\mathrm{End}(X) = k$;
(B)′ *$E \in \mathcal{R}$ and* $\mathrm{End}(Z) = k$.

REMARK 1.4. The proposition shows that Λ is wild if $\mathcal{E}$ contains some E in which *both* X and Z are Schurian, for then one of (B) and (B)′ is satisfied. This implies the de la Peña - Takane result, because if Y is a directing module, then X and Z are Schurian.

REMARK 1.5. Condition (A) of Proposition 1.3 can be replaced by the seemingly weaker requirement

(A)′ *There is a subsequence* $i_1 < i_2 < \ldots < i_t$ *of* $1, 2, \ldots, n$ *with* $t \geq 5$ *such that*

$$\dim Y_{i_s} < \sum_{j \neq s} \dim Y_{i_j} \qquad \text{for} \qquad 1 \leq s \leq t,$$

and either

$$\text{(a)} \quad \sum_{r=1}^{t} \dim Y_{i_r} > 2 \dim Z \qquad \text{or} \qquad \text{(b)} \quad \sum_{r=1}^{t} \dim Y_{i_r} > 2 \dim X.$$

(We may even replace the condition $t \geq 5$ above by $t \geq 4$ provided we also require that either $t < n$ or there is an irreducible map from a module other than X to one of the Y_i.) Suppose that $E \in \mathcal{L}$. Then, as we shall show in Lemma 3.1, $\dim \tau X \geq \dim \tau Z$. It now follows easily that (b) is satisfied for $\tau^2 E$ with $t \geq 5$ (and for $t = 4$, provided one of the conditions in parenthesis is satisfied). Similarly, if $E \in \mathcal{R}$, then (a) is satisfied for $\tau^2 E$. Thus this apparently gives many more conditions for the five-terms-in-the-middle conjecture to be true.

Throughout this paper all algebras are finite dimensional over an algebraically closed base-field k.

2. Double arrows

This section contains the proof of Theorem 1.1 and the discussion of its consequences. We begin with some lemmas which will imply that the theorem need be proved only for modules satisfying some special assumptions. Throughout the next two subsections, the dimension of a module M will quite often be denoted by m, and that of $\dim \tau^r M$ by $m^{(r)}$.

2.1. Combinatorics.

All modules in the lemmas in this subsection are assumed to be indecomposable.

LEMMA 2.1. *Let* $Y \oplus Y \oplus U \to Z$ *be an irreducible map and suppose that* $2y > z$ *and* $y + u > z$. *Then, for each* $r \geq 0$, *none of* $\tau^r Y$, $\tau^r U$ *and* $\tau^r Z$ *is projective,* $2y^{(r)} > z^{(r)}$, $y^{(r)} + u^{(r)} > z^{(r)}$ *and, when* r *is sufficiently large* , $y^{(r+1)} - z^{(r+1)} > y^{(r)} - z^{(r)} > \dim \Lambda$.

PROOF. Clearly Z is not projective and $z^{(1)} = 2y + u - z + \epsilon$ for some $\epsilon \geq 0$. Since $2z^{(1)} - y > 0$ and $z^{(1)} - u > 0$, neither Y nor U is projective. Now $y^{(1)} = 3y + 2u - 2z + 2\epsilon + \eta$ for some $\eta \geq 0$ and $u^{(1)} = 2y - z + \epsilon + \zeta$ for some $\epsilon \geq 0$. Hence $2y^{(1)} - z^{(1)} = 4y + 3u - 3z + 3\epsilon + \eta > 0$ and

$$y^{(1)} + u^{(1)} - z^{(1)} = 3y + u - 2z + 2\epsilon + \eta + 2\zeta > 0.$$

Also $y^{(1)} - z^{(1)} = y + u - z + \epsilon + \eta > y - z$.

The result follows by induction on r. □

LEMMA 2.2. *Let* $X \to Y \oplus Y \oplus U$ *be an irreducible map with* X *injective. Then, for each* $r \geq 0$, *none of* $\tau^r X$, $\tau^r Y$ *and* $\tau^r U$ *is projective. Furthermore,* $2y^{(r+1)} > x^{(r)}$ *and* $y^{(r+1)} + u^{(r+1)} > x^{(r)}$.

PROOF. Since X is injective, none of X, Y, and U can be projective and $x > 2y+u$. Hence $y^{(1)} = 2x - y + \eta$, for some $\eta > 0$, and $u^{(1)} = x - u + \zeta$ for some $\zeta > 0$. Thus $2y^{(1)} > x$, and $y^{(1)} + u^{(1)} > x$. The result follows by induction on r, using Lemma 2.1. □

LEMMA 2.3. *Let $Y \oplus Y \oplus U \to Z$ be an irreducible map with Y injective. Then, for each $r \geq 0$, none of $\tau^r Y$, $\tau^r U$ and $\tau^r Z$ is projective, $2y^{(r)} > z^{(r)}$ and $y^{(r)} + u^{(r)} > z^{(r)}$.*

PROOF. Since Y is injective, neither Z nor Y can be projective and $y > z$, hence $2y > z$ and $y + u > z$. The result follows by induction on r, using Lemma 2.1. □

LEMMA 2.4. *Let $Y \oplus Y \oplus U \to Z$ be an irreducible map with U injective but not projective. Then, for each $r \geq 0$, none of $\tau^r Y$, $\tau^r U$ and $\tau^r Z$ is projective, and for each $r \geq 1$, $2y^{(r)} > z^{(r)}$ and $y^{(r)} + u^{(r)} > z^{(r)}$.*

PROOF. Since U is injective, $u > z$ and therefore Z is not projective and $z^{(1)} = 2y + u - z + \epsilon > y$ for some $\epsilon \geq 0$. Hence also Y is not projective and so $y^{(1)} = 2z^{(1)} - y + \eta$ for some $\eta \geq 0$. Now $y^{(1)} - z^{(1)} = y + u - z + \epsilon + \eta > 0$, and so $2y^{(1)} > z^{(1)}$ and, using the non-projectivity of U, $y^{(1)} + u^{(1)} > z^{(1)}$. The result now follows by applying Lemma 2.1 to the irreducible map $\tau Y \oplus \tau Y \oplus \tau U \to \tau Z$. □

There is an obvious dual to each of the above lemmas. The remaining possibility we need to consider is that there is no projective or injective module in the τ orbit of any of the modules under discussion.

LEMMA 2.5. *Suppose that there is an irreducible map $Y \oplus Y \oplus U \to Z$ and that there is no projective or injective module in the τ-orbits of Y, U or Z. Choose Z with z minimal in its τ orbit. Then $2y^{(r)} > z^{(r)}$ and $y^{(r)} + u^{(r)} > z^{(r)}$ for each $r \geq 1$.*

PROOF. Since Z is not projective, $z^{(1)} = 2y + u - z + \epsilon$ for some $\epsilon \geq 0$. Since neither Y nor U is projective, $y^{(1)} = 3y + 2u - 2z + 2\epsilon + \eta$ for some $\eta \geq 0$ and $u^{(1)} = 2y - z + \epsilon + \zeta$ for some $\zeta \geq 0$. Hence $2y^{(1)} - z^{(1)} = 2(z^{(1)} - z) + u + z + 2\eta + \epsilon > 0$ and $y^{(1)} + u^{(1)} - z^{(1)} = (z^{(1)} - z) + y + \eta + \zeta + \epsilon > 0$.

The result follows by induction on r, using Lemma 2.1. □

2.2. Proof of Theorem 1.1. We shall make use of the following result from [**5**].

PROPOSITION 2.6. *Suppose that*

$$0 \to A \xrightarrow{f} B \xrightarrow{g} C \to 0$$

is a short exact sequence in which A has no projective direct summand and which has the property that any map from A to the radical of a projective module or the socle factor of an injective module, factors through f. Then there is a short exact sequence

$$0 \to \tau A \xrightarrow{\tau f} \tau B \xrightarrow{\tau g} \tau C \to 0.$$

If g is irreducible, then A is indecomposable and, if $h : A \to X$ is not a split monomorphism, then either f factors through h or h factors through f [**1**, V, Proposition 5.6]. Hence the conditions of this proposition are satisfied if g is irreducible and $\dim A > \dim \Lambda$.

The following lemma shows that the module P appearing in Theorem 1.1 has to be both projective and injective.

LEMMA 2.7. *If there is an irreducible map $Y \oplus Y \oplus U \to Z$ in which Y, Z and U are indecomposable and U is not both projective and injective, then Λ is wild.*

PROOF. Given such an irreducible map, it follows from Lemmas 2.1, 2.2, 2.3, 2.4, 2.5 and their duals, that we may assume that either $\dim \tau Y - \dim \tau Z > \dim Y - \dim Z > \dim \Lambda$ or, dually, that $\dim \tau^{-1} Z - \dim \tau^{-2} Y > \dim Z - \dim \tau^{-1} Y > \dim \Lambda$. We shall assume the former, and use an argument we learnt from Bill Crawley-Boevey to prove that Λ has to be wild; of course, a similar argument works in the dual case. By assumption, there are two irreducible maps, $f, g : Y \to Z$ such that for all $(\mu, \nu) \in \mathbb{P}^1(k)$, $\mu f + \nu g$ is irreducible. Since $\dim Y > \dim Z$, this map is an epimorphism, so there is a short exact sequence

$$0 \to W \xrightarrow{h} Y \xrightarrow{\mu f + \nu g} Z \to 0$$

in which $W = W_{(\mu,\nu)}$ is indecomposable (as above). Suppose that

$$0 \to W \xrightarrow{h'} Y \xrightarrow{\mu' f + \nu' g} Z \to 0$$

is another presentation of W where (μ', ν') represents a point of $\mathbb{P}^1(k)$ different from that represented by (μ, ν). Then either h factors through h' or h' factors through h. Suppose that $h' = h\theta$. Since Y is indecomposable and (μ, ν) and (μ', ν') are different points of $\mathbb{P}^1(k)$, θ cannot be an isomorphism and so has to be nilpotent. An easy argument then shows that h cannot simultaneously factorise through h', so the set of sequences which present W as above is totally ordered by the existence of such nilpotent factors in $\mathrm{End}(Y)$; the evident transitivity of this ordering then implies that the number of such presentations is at most the nilpotence index of the radical of $\mathrm{End}(Y)$. Otherwise expressed, the number of points of $\mathbb{P}^1(k)$ which can represent the same kernel W is finite. Thus these kernels give an infinite family of pairwise non-isomorphic indecomposable modules W all of the same dimension. Since $\dim W = \dim Y - \dim Z > \dim \Lambda$, it follows from Proposition 2.6 that $\tau W_{(\mu,\nu)}$ is the kernel of $(\mu \tau f + \nu \tau g) : \tau Y \to \tau Z$ and since $\dim \tau W > \dim W$, certainly $\tau W \ncong W$. Hence, by Crawley-Boevey's Theorem [**6**], Λ is wild. This completes the proof of Lemma 2.7. □

PROOF OF THEOREM 1.1. We assume the notation and hypotheses of the theorem. Then the previous lemma shows that in the almost split sequence

$$0 \to X \to Y \oplus Y \oplus P \to Z \to 0,$$

the module P is projective-injective and indecomposable. Hence the elementary structure theorem [**1**, V Proposition 5.5] for almost split sequences containing a projective-injective module implies (using the simplicity of the top and of the socle of P) that $X = \mathrm{rad}\, P$, $Z = P/\mathrm{soc}\, P$, $Y \oplus Y = \mathrm{rad}\, P/\mathrm{soc}\, P$, and all the maps are given by natural inclusions or projections. We first show that the component of the Auslander-Reiten quiver which contains these modules is the quiver displayed. It suffices to exhibit almost split sequences

$$0 \to \tau^{n+1} Y \to \tau^n X \oplus \tau^n X \to \tau^n Y \to 0$$

for all $n \in \mathbb{Z}$, and

$$0 \to \tau^{n+1} X \to \tau^{n+1} Y \oplus \tau^{n+1} Y \to \tau^n X \to 0,$$

for all $n \neq -1$. Since P is projective and injective, Y is neither and there is an irreducible epimorphism $X \oplus X \to Y$, which has to be the sink map to Y; for if not, the tameness of Λ and Lemma 2.7 would imply that $Y = Q/\operatorname{soc}(Q)$ and $X \oplus X = \operatorname{rad}(Q)/\operatorname{soc}(Q)$ for some projective-injective module Q, contrary to the fact that $y < x$. This argument gives the required almost split sequence ending at Y, and furthermore shows that $y^{(1)} = 2x - y > x$, so that X is not projective. Again using the tameness of Λ and Lemma 2.7, we next see that the irreducible epimorphism $\tau Y \oplus \tau Y \to X$ is the sink map to X, so that the almost split sequence stopping at X is as asserted. The general assertions follow similarly by induction for $n \geq 1$ using dimension formulae $x^{(n)} - x = y^{(n)} - y = 2n(x - y)$, and by a dual argument for negative n. This completes the proof that P together with the τ-translates of X and Y are all the vertices of a single Auslander-Reiten component.

Finally we show by contradiction that Y is simple. Suppose not. Then the almost split sequence with Y as cokernel restricts to an exact sequence of socles, so that $\operatorname{soc}\tau Y \oplus \operatorname{soc} Y \cong \operatorname{soc} X \oplus \operatorname{soc} X$, which is of length 2 (since X is the radical of the indecomposable projective-injective module P). Hence τY has simple socle isomorphic to that of the injective module P, and so embeds in P, in contradiction to the fact that P has length $2y + 2 < 3y + 2 = y^{(1)}$. This shows that Y is, in fact, simple and hence that P has length 4. It completes the proof of Theorem 1.1. □

In the rest of this section we study conditions for the Auslander-Reiten quiver of an algebra to have a component $\mathcal{C}$ such that (1) there are no almost split sequences, with terms in $\mathcal{C}$, of the form

$$0 \to A \to B \oplus B \oplus B' \to C \to 0$$

with $B \neq 0 \neq B'$, and (2) there are double arrows in the quiver $\mathcal{C}$.

For brevity, we call such a component *special*, a term surely justified by the following easy result.

LEMMA 2.8. *A special component contains a double arrow* $X \xrightarrow{2} Y$ *in which either* X *is simple and* Y *projective of length 3, or* X *is injective of length 3 and* Y *is simple.*

PROOF. Choose the double arrow $X \xrightarrow{2} Y$ in the special component $\mathcal{C}$ so that the least of the lengths x and y of X and Y, respectively, is minimal amongst the lengths of indecomposables involved in double arrows in $\mathcal{C}$. Of course $x \neq y$, because there are irreducible maps from X to Y. Suppose $x < y$. If Y is not projective, $\mathcal{C}$ contains a double arrow $\tau Y \xrightarrow{2} X$, so that $y^{(1)} > x$; but by property (1) of a special component $y^{(1)} + y = 2x$, a contradiction. Hence Y is projective and has radical $X \oplus X$. Again by (1), the almost split sequence with kernel X has middle term $Y \oplus Y$, a projective module, so by a classical result, [**1**, V,Theorem 3.3], X is simple; hence Y is projective of length 3. The other possibility occurs when $x > y$. □

PROOF OF COROLLARY 1.2. By Theorem 1.1, a regular component of a tame algebra satisfies condition (1) for being special. On the other hand, Lemma 2.8 shows that special components are not regular. Hence regular components in tame algebras must fail condition (2) for being special.

□

We end this section with a rather detailed description of when and how special components occur. To exploit quiver technology, we assume our algebras to be basic.

A special component of the Auslander-Reiten quiver of an algebra has one of the following forms. It is formed by gluing the simple projective from a preprojective component of a Kronecker algebra to a simple injective from a (possibly empty) component of another algebra or, dually, it is formed by gluing the simple injective from a preinjective component of a Kronecker algebra to a simple injective from a (possibly empty) component of another algebra, or by doing both simultaneously with a component which contains both a simple projective and a simple injective. We include in this last case the component containing the simple module for the local algebra $k[x,y]/(x,y)^2$; here, of course the additional component is empty. In all cases, the 'additional component' must contain no multiple arrows. The corresponding algebras are formed by gluing suitable pieces to the quiver of the Kronecker algebra, source to sink or sink to source, so that all composites with arrows from the Kronecker algebra are zero.

The following Proposition makes this precise.

PROPOSITION 2.9. *Let $\mathcal{C}$ be a special component of the Auslander-Reiten quiver of a basic, finite-dimensional algebra, Λ, over an algebraically closed field. If Λ is local, then it is isomorphic to the algebra $k[x,y]/(x,y)^2$, and $\mathcal{C}$ is the component containing Λ. Otherwise the ordinary quiver of the algebra contains a Kronecker subquiver, $\mathcal{K}$, with vertices a and b and arrows $x, y : a \to b$, such that* (a) *x and y are the only arrows with a as source,* (b) *x and y are the only arrows with b as sink, and* (c) *all composites of each of these arrows with other arrows are zero in Λ. There is then a non-empty, connected sub-translation quiver, $\mathcal{D}$, of $\mathcal{C}$ such that one of the following three possibilities is the case:* (i) *there is a vertex u, which is a sink in $\mathcal{D}$, such that $\mathcal{C}$ is obtained from the disjoint union of $\mathcal{D}$ and $\mathbb{N}\mathcal{K}$ by identifying u with the simple projective vertex in $\mathbb{N}\mathcal{K}$;* (ii) *there is a vertex v, which is a source in $\mathcal{D}$, such that $\mathcal{C}$ is obtained from the disjoint union of $\mathcal{D}$ and $-\mathbb{N}\mathcal{K}$ by identifying v with the simple injective vertex in $-\mathbb{N}\mathcal{K}$;* (iii) *there is a sink u and a source v in $\mathcal{D}$ such that $\mathcal{C}$ is obtained from the disjoint union of $\mathcal{D}$, $\mathbb{N}\mathcal{K}$ and $-\mathbb{N}\mathcal{K}$ by identifying u and v with the vertices specified in* (i) *and* (ii)*, respectively.*

PROOF. Making use of Lemma 2.8, we consider only the case in which the component contains a double arrow $X \rightrightarrows Y$ with X simple and Y projective, leaving the other possibilities to the reader. So $Y = P(a)$ has both radical and socle equal to $X \oplus X$ with $X = S(b)$ for a subquiver $x, y : a \to b$ of the quiver of the algebra. An easy argument, like that used in the proof of Theorem 1.1, and using conditions (1) and (2) for being special, shows that the full subquiver of $\mathcal{C}$ with vertices $\tau^{-n}S(b)$ and $\tau^{-n}P(a)$ for all $n \geq 0$ is isomorphic to $\mathbb{N}\mathcal{K}$. If $S(b) = P(b)$ is a simple projective module, clearly $a \neq b$ and the component $\mathcal{C}$ is obviously just $\mathbb{N}\mathcal{K}$. If $S(b)$ is not projective, its sink map $Z \to S(b)$ is surjective and Z must be injective because the only irreducibles out of $S(b)$ have projective range. An indecomposable summand $I(c)$ of Z corresponds to an arrow $z : b \to c$ such that $xz = 0 = yz$ in Λ. If $a = b$, the formula $\operatorname{rad} P(a) = S(a) \oplus S(a)$ then requires that $c = a$ and $z = x$ or $z = y$, and we must then have $x^2 = xy = yx = y^2 = 0$. Hence $\mathcal{C}$ is the unique non-regular component of $k[x,y]/(x,y)^2$; this is, of course the situation if Λ is local. If $a \neq b$, we take $\mathcal{D}$ to be the set of vertices and arrows in $\mathcal{C}$ which are not in $\mathbb{N}\mathcal{K}$, together with a sink labelled u coinciding with $X = S(b)$. $\square$

3. Five Terms in the Middle

In this section Proposition 1.3 will be proved by constructing multiparameter families of indecomposable modules all of the same dimension. As in the case of double arrows, we start with a Lemma which enables us to restrict to cases where the starting almost split sequences satisfy some special conditions.

3.1. Combinatorics.

LEMMA 3.1. *If $\mathcal{L}$ is non-empty, then it contains an almost split sequence*

$$0 \to \tau X \to \oplus_{j=1}^{m} B_j \to X \to 0,$$

where each B_j is indecomposable, such that

$$\text{(a) for all } j,\ \sum_{k\neq j} \dim B_k > \dim X \qquad \text{and} \qquad \text{(b)} \quad \dim \tau X > \dim X.$$

PROOF. Suppose that $\mathcal{L}$ contains the almost split sequence

$$E : 0 \to X \to \oplus_{j=1}^{n} Y_j \to Z \to 0,$$

in which each Y_j is indecomposable. We will show that X is non-projective and that the almost split sequence with cokernel term X satisfies Lemma 3.1.

Let $\oplus_{j=1}^{m} B_j \to X$ be the sink map for X.

(i) Suppose that E satisfies the condition (i) for $\mathcal{L}$. Then $x \geq z$ and we may suppose, without loss of generality, that the first $l \geq 5$ of the Y_i are not projective and that $B_j = \tau Y_j$ for $1 \leq j \leq l$; if $l < m$, the remaining B_j are injective.

First we prove (a) which, of course, implies that X is non-projective. For each fixed j, $\sum_{k\neq j} b_k > x$ will hold trivially if $m > l + 1$, or if $m = l + 1$ and $j \leq l$, and only requires proof in the cases $m = l + 1 = j$, where it asserts that $\sum_{k\leq l} b_k > x$, and $m = l$ with any j, where it asserts that $\sum_{\substack{k\neq j\\ k\leq l}} b_k > x$. Now for $k \leq l$, $b_k = x - y_k + \eta_k$ for some $\eta_k \geq 0$. Hence

$$\begin{aligned} \sum_{\substack{k\neq j\\ k\leq l}} b_k - x &= (l-2)x - \sum_{\substack{k\neq j\\ k\leq l}} y_k + \sum_{\substack{k\neq j\\ k\leq l}} \eta_k \\ &\geq (l-4)x + x + z - \sum_{\substack{k\neq j\\ k\leq l}} y_k + \sum_{\substack{k\neq j\\ k\leq l}} \eta_k \\ &> (l-4)x > 0, \end{aligned}$$

as required.

Now we prove (b). Since $\sum_{k\neq j} b_k > x$, X is not projective. Also

$$\begin{aligned} x^{(1)} - x &= \sum_{1}^{m} b_k - 2x \\ &\geq (l-2)x - \sum_{k\leq l} y_k + \sum_{k\leq l} \eta_k \\ &\geq (l-4)x + x + z - \sum_{k\leq l} y_k + \sum_{k\leq l} \eta_k \\ &\geq (l-4)x > 0, \end{aligned}$$

and this is just (b).

(ii) Suppose now that condition (ii) for $\mathcal{L}$ is satisfied and that Y_1 is injective but not projective. Then $y_1 > z$ and no Y_k is projective. For $k \leq n$, we take $B_k = \tau Y_k$, so $b_k = x - y_k + \eta_k$, for some $\eta_k \geq 0$; for $n < k \leq m$, B_k is injective and so $b_k > x$. Hence, for any j,

$$\begin{aligned} \sum_{k\neq j} b_k - x &\geq (n-2)x - \sum_{k\neq j} y_k + \sum_{k\neq j} \eta_k \\ &\geq (n-3)x - z + y_j + \sum_{k\neq j} \eta_k. \end{aligned}$$

For $j = 1$, the last expression is positive as required because $y_1 > z$. For $j > 1$, we see that on replacing $x + \eta_1$ by $y_1 + b_1$ it becomes

$$(n-4)x + (y_1 - z) + b_1 + y_j + \sum_{k \neq 1} \eta_k,$$

which is also clearly positive. This establishes (a), and it follows that $\sum b_j > x$ and so X is not projective. Finally

$$\begin{aligned} x^{(1)} - x &\geq (n-2)x - \sum y_k + \sum \eta_k \\ &\geq (n-2)x - (x+z) + \eta_1 \\ &\geq (n-4)x + (y_1 - z) + b_1 > 0, \end{aligned}$$

which is (b).

The almost split sequence ending at X satisfies condition (i) for being in $\mathcal{L}$ because (a) implies that, for each j, there is an irreducible epimorphism from τX onto B_j and so B_j cannot be projective.

The dual lemma holds when $\mathcal{R}$ is non-empty. □

3.2. Reduction to a special case. Our proof of Proposition 1.3 requires two different, but dual, constructions of families of indecomposables, according to which of the hypotheses of the proposition are satisfied. We will give a construction which works when the set $\mathcal{E}$ contains an almost split sequence, E, such that *either* E satisfies Condition (A) and $\dim X \geq \dim Z$, *or* E satisfies Condition (B). One then sees that the following hypothesis holds.

Hypothesis (H). There is an irreducible epimorphism

$$f = (f_i) : Y = \oplus_{i=1}^n Y_i \to Z,$$

where $n \geq 5$, $\dim Y \geq 2 \dim Z$, and

either (H1) for all i, $\dim Y_i < \sum_{j \neq i} \dim Y_j$,

or (H2) for all i, $\sum_{j \neq i} \dim Y_j > \dim Z$, and $\mathrm{End}(Z) = k$.

(When Condition (A) holds and $\dim X \geq \dim Z$, the epimorphism in E satisfies (H1). When E satisfies Condition (B), the statement and proof of Lemma 3.1 shows that the epimorphism in τE satisfies (H2).)

In the rest of this section, we shall assume that (H) holds and show that Λ is wild. By Theorem 1.1, it will be sufficient to assume further that the Y_i are pairwise non-isomorphic. However, this assumption can be deferred until Subsection 3.6 where it is needed to show that we really have constructed an $(n-3)$–parameter family of pairwise non-isomorphic indecomposable modules of the same dimension.

The remaining cases of Proposition 1.3 are proved by a categorically dual construction.

3.3. Construction. The construction starts with a choice of distinct, non-zero elements $\lambda_1, \lambda_2, \ldots, \lambda_n$ of k. Such a choice determines irreducible maps $g_i = \lambda_i f_i$ and $g = (g_i) : Y \to Z$, and hence an exact sequence

$$0 \to W \xrightarrow{\alpha} Y \xrightarrow{(f,g)} Z \oplus Z.$$

Our strategy is to show that these kernels $W = W_{(f,g)}$ form a sufficiently large family of indecomposables to guarantee that Λ must be wild.

We shall meet several times maps of the form $g - f\zeta = (f_i(\lambda_i id_Z - \zeta))$, where ζ is an endomorphism of Z. Because the base field k is algebraically closed and Z is indecomposable, ζ has a unique eigenvalue, say λ, in k, and then $g - f\zeta$ remains irreducible unless $\lambda = \lambda_i$ for some i, in which case the ith component will no longer be irreducible; further, since the λ_j are all different, at most one component is so affected. We will use the conclusion that, *for any* ζ *and* (λ_i), *there is some index* i *such that the restriction of* $g - f\zeta$ *to* $\oplus_{j \neq i} Y_j$ *is still irreducible.*

LEMMA 3.2. *The map* $(f, g) : Y \to Z \oplus Z$ *is an epimorphism.*

PROOF. Let $U = \operatorname{im}(f, g)$. There are maps $u : Y \to U$ and $(p, q) : U \to Z \oplus Z$ such that $u(p, q) = (f, g)$. We need to show that (p, q) is an epimorphism. Suppose not. Since $\dim Y \geq 2 \dim Z$, it follows that u is a proper epimorphism and so not a split monomorphism. Hence the equation $up = f$ implies that p is a split epimorphism. We may write $U = Z \oplus V$, $p = \binom{1}{0}$, $q = \binom{\zeta}{r}$ and $u = (f, v)$. Then $g - f\zeta = vr$. Now r is not a split epimorphism, since (p, q) is not, and so using the observation preceding the lemma we conclude that there is an index i such that the restriction of v to $\oplus_{j \neq i} Y_j$ is a split monomorphism. Hence

$$\dim U \geq \dim Z + \sum_{j \neq i} \dim Y_j.$$

Now $2 \dim Z > \dim U$, so that $\dim Z > \sum_{j \neq i} \dim Y_j$ and the hypothesis (H2) fails. But also $\dim Y > \dim U$, so $\dim Y_i > \dim U - \sum_{j \neq i} \dim Y_j \geq \dim Z > \sum_{j \neq i} \dim Y_j$, and the hypothesis (H1) also fails. This contradiction completes the proof of the lemma. $\square$

REMARK 3.3. The above proof shows that $(f, g) : Y \to Z \oplus Z$ is an epimorphism if, for all i, $\sum_{j \neq i} \dim Y_j > \dim Z$. If $\mathcal{L}$ is non-empty, then it follows from Lemma 3.1 that it contains an almost split sequence E satisfying this condition, and so the corresponding map (f, g) will be an epimorphism, whether or not (H) holds. A dual construction of a monomorphism is, of course, possible whenever $\mathcal{R}$ is non-empty.

For the remainder of this subsection and the next, we fix the parameters (λ_i) and make use of the notation $g = (g_i)$, W, α introduced above. Here is a rather technical lemma, also independent of whether or not (H) holds.

LEMMA 3.4. *Let* $W \xrightarrow{\sigma} U$ *be a morphism and*

$$\begin{array}{ccc} W & \xrightarrow{\alpha} & Y \\ \downarrow{\scriptstyle\sigma} & & \downarrow{\scriptstyle m} \\ U & \xrightarrow{n} & M \end{array}$$

be the pushout of α *and* σ. *Assume that neither* m *nor* n *is a split monomorphism. Then there is an* $i \in 1, 2, \ldots, n$ *such that, if* $Y^i = \oplus_{j \neq i} Y_j$ *and* f^i *and* g^i *are the restrictions of* f *and* g *to* Y^i, *the pushout diagram takes the form*

$$\begin{array}{ccccccccc}
0 & \to & W & \xrightarrow{(\alpha_i,\alpha^i)} & Y_i \oplus Y^i & \xrightarrow{\begin{pmatrix} f_i & g_i \\ f^i & g^i \end{pmatrix}} & Z \oplus Z & \to & 0 \\
 & & \downarrow{\scriptstyle \sigma} & & \downarrow{\scriptstyle m = \begin{pmatrix} f_i & u & v \\ f^i & 1 & 0 \end{pmatrix}} & & \| & & \\
0 & \to & U & \xrightarrow[n=(0,b,c)]{} & M = Z \oplus Y^i \oplus N & \xrightarrow[r = \begin{pmatrix} 1 & \zeta \\ 0 & g^i - f^i\zeta \\ 0 & q \end{pmatrix}]{} & Z \oplus Z & \to & 0
\end{array}$$

with $g^i - f^i\zeta$ irreducible and $g_i - f_i\zeta$ reducible, so that $\zeta \in \operatorname{Aut} Z$, and q is not a split epimorphism.

PROOF. The pushout of α and σ is part of the exact and commutative diagram

$$\begin{array}{ccccccccc}
0 & \to & W & \xrightarrow{\alpha} & Y = Y_i \oplus Y^i & \xrightarrow{(f,g)} & Z \oplus Z & \to & 0 \\
 & & \downarrow{\scriptstyle \sigma} & & \downarrow{\scriptstyle m} & & \| & & \\
0 & \to & U & \xrightarrow[n]{} & M & \xrightarrow[(p,q)]{} & Z \oplus Z & \to & 0.
\end{array}$$

Since $mp = f$, which is irreducible, and m is not a split monomorphism, it follows that p is a split epimorphism. We may write $M = Z \oplus M'$, $p = \binom{1}{0}$, $q = \binom{\zeta}{q'}$, $m = (f, m')$, and $n = (0, a, b)$. Then $f\zeta + m'q' = g$. Now, for some i, the restriction of $g - f\zeta$ to $Y^i = \oplus_{j \neq i} Y_j$ is irreducible and so, since q' is not a split epimorphism (since (p, q) is not), it follows that the restriction of m' to Y^i is a split monomorphism. Thus we may write $M' = Y^i \oplus N$, $m' = \left(\begin{smallmatrix} u & v \\ 1 & 0 \end{smallmatrix}\right)$, and $q' = g^i - f^i\zeta$. This completes the proof. □

COROLLARY 3.5. *Let σ be a map from W to a module U. Suppose that*

$$\dim U < \min(\dim W, \dim Y^i - \dim Z) \quad \text{for} \quad 1 \leq i \leq n.$$

Then σ factors through f.

PROOF. The map σ is not a monomorphism and so neither is m. Since $\dim U < \dim Y^i - \dim Z$, the pushout M cannot have the form stated in Lemma 3.4 and so n must be a split monomorphism. Thus σ factors through f. □

3.4. Indecomposability of W. In this subsection we use the full force of (H) to prove that W is indecomposable. Suppose not. Let $W = U \oplus U'$ be a proper decomposition with projections $W \xrightarrow{\sigma} U$ and $W \xrightarrow{\sigma'} U'$. Form pushouts

$$\begin{array}{ccc}
W & \xrightarrow{\alpha} & Y \\
\downarrow{\scriptstyle \sigma} & & \downarrow{\scriptstyle m} \\
U & \xrightarrow{n} & M
\end{array}
\qquad \text{and} \qquad
\begin{array}{ccc}
W & \xrightarrow{\alpha} & Y \\
\downarrow{\scriptstyle \sigma'} & & \downarrow{\scriptstyle m'} \\
U' & \xrightarrow{n'} & M'.
\end{array}$$

Neither m nor m' is a monomorphism, so certainly neither is a split monomorphism. Suppose that n is a split monomorphism, say $np = 1_U$, for some p, and let $\iota : U \to W$ be a splitting of σ. Setting $e = mp\iota\alpha$, we find that $e = e^2$ and $\alpha e = (\sigma\iota)\alpha$; hence $U\iota\alpha$ is a summand of Y, which is impossible since the restrictions of f and g to any summand of Y are non-zero. Hence none of n, n', m and m' is a split monomorphism. For the first pushout we use the notation of Lemma 3.4 and for the second, we use a dashed version.

We claim first that $i' = i$ and $\mathrm{End}(Z) = k$. By Lemma 3.4, $\dim U + \dim Z = \dim Y^i + \dim N$ and $\dim U' + \dim Z = \dim Y^{i'} + \dim N'$. Also $\dim U + \dim U' = \dim W = \dim Y - 2\dim Z$. Hence $\dim Y = \dim Y^i + \dim Y^{i'} + \dim N + \dim N'$. But, if $i \neq i'$, then $\dim Y^i + \dim Y^{i'} = \sum_{j \neq i} \dim Y_j + \sum_{j \neq i'} \dim Y_j > \dim Y$. Thus $i' = i$ and, from the previous argument, $\dim Y_i = \dim Y^i + \dim N + \dim N' \geq \sum_{j \neq i} \dim Y_j$. Thus the hypothesis (H1) fails, so we must be in the case (H2) which includes the requirement that $\mathrm{End}(Z) = k$.

From now on, we suppose that $i = 1$ and write $Y^1 = \bar{Y}$, $f^1 = \bar{f}$, and $g^1 = \bar{g}$. Since $\mathrm{End}(Z) = k$, it follows that $\zeta = \lambda_1 = \zeta'$. We can now assemble the following exact and commutative diagram, in which (σ, σ') is an isomorphism, the middle row is the direct sum of the middle rows of the two pushout diagrams and $I_2 = \left(\begin{smallmatrix} 1_Z & 0 \\ 0 & 1_Z \end{smallmatrix}\right)$.

$$\begin{array}{ccccccccc}
 & & & & 0 & & 0 & & \\
 & & & & \downarrow & & \downarrow & & \\
0 & \to & W & \xrightarrow{(\alpha_1, \bar{\alpha})} & Y_1 \oplus \bar{Y} & \xrightarrow{\left(\begin{smallmatrix} f_1 & g_1 \\ \bar{f} & \bar{g} \end{smallmatrix}\right)} & Z \oplus Z & \to & 0 \\
 & & \downarrow{\scriptstyle (\sigma,\sigma')} & & \downarrow{\scriptstyle (m,m')} & & \downarrow{\scriptstyle (I_2,I_2)} & & \\
0 & \to & U \oplus U' & \xrightarrow[\left(\begin{smallmatrix} n & 0 \\ 0 & n' \end{smallmatrix}\right)]{} & Z \oplus \bar{Y} \oplus N \oplus Z \oplus \bar{Y} \oplus N' & \xrightarrow[\left(\begin{smallmatrix} r & 0 \\ 0 & r' \end{smallmatrix}\right)]{} & Z \oplus Z \oplus Z \oplus Z & \to & 0, \\
 & & & & \downarrow{\scriptstyle \left(\begin{smallmatrix} r \\ -r' \end{smallmatrix}\right)} & & \downarrow{\scriptstyle \left(\begin{smallmatrix} I_2 \\ -I_2 \end{smallmatrix}\right)} & & \\
 & & & & Z \oplus Z & = & Z \oplus Z & & \\
 & & & & \downarrow & & \downarrow & & \\
 & & & & 0 & & 0 & &
\end{array}$$

We now focus on the middle column of this diagram, which is

$$0 \to Y_1 \oplus \bar{Y} \xrightarrow{\left(\begin{smallmatrix} f_1 & u & v & f_1 & u' & v' \\ \bar{f} & 1_{\bar{Y}} & 0 & \bar{f} & 1_{\bar{Y}} & 0 \end{smallmatrix}\right)} Z \oplus \bar{Y} \oplus N \oplus Z \oplus \bar{Y} \oplus N' \xrightarrow{\begin{pmatrix} 1_Z & \zeta \\ 0 & \bar{g} - \bar{f}\zeta \\ 0 & q \\ -1_Z & -\zeta \\ 0 & -\bar{g} + \bar{f}\zeta \\ 0 & -q' \end{pmatrix}} Z \oplus Z \to 0.$$

Observe that $(1_Z, 0, 0, 1_Z, 0, 0)$ annihilates the cokernel map of this short exact sequence and so factors through the kernel. But this implies that 1_Z factors through the irreducible map f, which is impossible.

This contradiction shows that W must be indecomposable.

3.5. Isomorphic W's. Again using Hypothesis (H), we next obtain a necessary condition for two modules W and W', defined as in subsection 3.3 using different choices (λ_i) and (λ_i') of parameters, to be isomorphic. As described in Subsection 3.3 there are short exact sequences

$$0 \to W \xrightarrow{\alpha} Y \xrightarrow{(f,g)} Z \oplus Z \to 0 \tag{3.1}$$

and

$$0 \to W' \xrightarrow{\alpha'} Y \xrightarrow{(f',g')} Z \oplus Z \to 0, \tag{3.2}$$

in which $f' = f$.

LEMMA 3.6. *If W and W' are isomorphic, then either α factors through α' or α' factors through α.*

PROOF. We argue by contradiction. Assume that the lemma is false for the modules $W = W'$ with presentations given by (3.1) and (3.2). Since neither α nor α' factors through the other, it follows that the monomorphisms m and n in the pushout diagram

$$\begin{array}{ccc} W & \xrightarrow{\alpha} & Y \\ \downarrow{\scriptstyle \alpha'} & & \downarrow{\scriptstyle m} \\ Y & \xrightarrow{n} & M \end{array}$$

are not split monomorphisms. We start with the following reduction of this diagram, and will see that it already leads to the required contradiction if Z is Schurian.

ASSERTION 1. *The module M has a direct decomposition $M = Z \oplus M' \oplus Z$ with respect to which*

$$m = (f, m', 0) \qquad \textit{and} \qquad n = (0, n', g')$$

$$\operatorname{coker} n = \begin{pmatrix} 1 & \zeta \\ 0 & q' \\ 0 & \xi \end{pmatrix} \qquad \textit{and} \qquad \operatorname{coker} m = \begin{pmatrix} \sigma & 0 \\ r' & 0 \\ \tau & 1 \end{pmatrix}$$

where $\sigma, \xi \in \operatorname{rad}\operatorname{End}(Z)$, $\zeta, \tau \in \operatorname{Aut}(Z)$, and neither $g - f\zeta = m'q'$ nor $f' - g'\tau = n'r'$ is irreducible.

PROOF OF ASSERTION 1. Since n is a pushout of α, it follows from the short exact sequence (1) that $(f, g) = m(p, q)$ with $(p, q) : M \to Z \oplus Z$ a cokernel of n. Hence $f = mp$, and since f is irreducible and m is not a split monomorphism, p has to be a split epimorphism, so that there is a decomposition $M = Z \oplus M_1$ such that $(p, q) = \begin{pmatrix} 1 & \zeta_1 \\ 0 & q_1 \end{pmatrix}$, $m = (f, m_1)$ and so $n = (0, n_1)$. Now let $\operatorname{coker} m : Z \oplus M_1 \to Z \oplus Z$ be given by the matrix $\operatorname{coker} m = \begin{pmatrix} \sigma_2 & \sigma_1 \\ r_1 & s_1 \end{pmatrix}$. Then $n_1 s_1 = g'$ is irreducible, and so s_1 has to be a split epimorphism, because n, and hence n_1, is not a split monomorphism. On choosing a suitable decomposition $M_1 = M' \oplus Z$ with $M' = \ker(s_1)$, $\operatorname{coker} m$ takes the form

$$\begin{pmatrix} \sigma_2 & \sigma_1 \\ r' & 0 \\ \tau & 1 \end{pmatrix}$$

which then reduces to the form given in Assertion 1 on applying the automorphism $\begin{pmatrix} 1 & 0 & \sigma_1 \\ 0 & 1 & 0 \\ 0 & 0 & 1 \end{pmatrix}$ of M. The formulae for the other maps follow from commutativity and exactness, and their properties from the assumption that neither m nor n is a split monomorphism, so that neither of their cokernels is a split epimorphism. □

We can now obtain the contradiction required to prove the Lemma in the case when Z is Schurian and so, in particular, when the hypothesis (H2) holds. For Schurian Z we have $\sigma = 0 = \xi$, and so the injection $(1, 0, 0)$ of Z into M has to factorise through $m = (f, m', 0)$, an impossibility since f is irreducible.

To cover the case in which Z is not Schurian we will prove

ASSERTION 2. *The module M' has a summand isomorphic to $Y^i \oplus Y^j$ where $1 \leq i, j \leq n$.*

This gives the required contradiction to (H) when Z is not Schurian: for by construction $\dim M' = \dim Y$, so that Assertion 2 can only be valid if $i = j$ and $\dim Y_i \geq \dim Y^i$, which is precisely what the hypothesis (H1) excludes.

PROOF OF ASSERTION 2. We use Assertion 1. Since $m'q' = g - f\zeta$ is not irreducible, there is an index i such that $m^i q' = g^i - f^i\zeta : Y^i \to Z$ is irreducible, where m^i denotes the restriction to Y^i of $m' : Y \to M'$. Hence M' has a decomposition $M' = Y^i \oplus M''$ such that $m' : Y = Y_i \oplus Y^i \to M'$ has the form $\left(\begin{smallmatrix} * & * \\ 1 & 0 \end{smallmatrix}\right)$. With $M = Z \oplus Y^i \oplus M'' \oplus Z$, we obtain

$$m = \begin{pmatrix} f_i & * & * & 0 \\ f^i & 1 & 0 & 0 \end{pmatrix} \qquad \text{and} \qquad n = (0, n^i, n'', g')$$

$$\operatorname{coker} n = \begin{pmatrix} 1 & \zeta \\ 0 & g^i - f^i\zeta \\ 0 & q'' \\ 0 & \xi \end{pmatrix} \qquad \text{and} \qquad \operatorname{coker} m = \begin{pmatrix} \sigma & 0 \\ -f^i\sigma & 0 \\ r'' & 0 \\ \tau & 1 \end{pmatrix}.$$

From $(f', g') = n \operatorname{coker} m$ we see that

$$f' - g'\tau = n''r'' - n^i f^i \sigma.$$

Again using Assertion 1, we know that there is an index j such that the restriction to Y^j of $f' - g'\tau$ is irreducible. Since both f^i and σ are in $\operatorname{rad} \operatorname{mod} \Lambda$, we conclude that the restriction to Y^j of $n''r''$ is also irreducible. However r'' is certainly not a split epimorphism, so the restriction to Y^j of $n'' : Y \to M''$ has to be a split monomorphism. This completes the proof of the second assertion. □

The lemma follows. □

3.6. Proof of Proposition 1.3. To complete the proof of Proposition 1.3, we continue the analysis of the preceding subsection of when the two modules W and W' defined by (λ_i) and (λ'_i) are isomorphic, but will from now on assume, in addition to (H), that the Y_i are pairwise non-isomorphic. Let $W = W'$ be given by (3.1) and (3.2). By Lemma 3.6, we may suppose that $\alpha' = \alpha\theta$. Then θ induces the exact commutative diagram

$$\begin{array}{ccccccccc} 0 & \to & W & \xrightarrow{\alpha} & Y & \xrightarrow{(f,g)} & Z \oplus Z & \to & 0 \\ & & \| & & \downarrow \theta & & \downarrow \phi & & \\ 0 & \to & W' & \xrightarrow[\alpha']{} & Y & \xrightarrow[(f',g')]{} & Z \oplus Z & \to & 0 \end{array}$$

Let $\theta_{ij} : Y_i \to Y_j$ denote the components of θ with respect to the direct decomposition $Y = \oplus_{i=1}^n Y_i$, and let $\phi = \left(\begin{smallmatrix} \phi_{11} & \phi_{12} \\ \phi_{21} & \phi_{22} \end{smallmatrix}\right)$, where $\phi_{ij} \in \operatorname{End} Z$. Then it follows from the assumptions about the Y_i and Z that there are scalars $\bar\theta_{ii}, \bar\phi_{rs}$ in k such that $\theta_{ii} \equiv \bar\theta_{ii} id_{Y_i}$, $\theta_{ij} \equiv 0$ for $i \neq j$, and $\phi_{rs} \equiv \bar\phi_{rs} id_Z$, all modulo $\operatorname{rad} \operatorname{mod} \Lambda$. Now $f' = f$, $g_i = \lambda_i f_i$ and $g'_i = \lambda'_i f_i$ are all irreducible maps, so reducing the equation $\theta(f', g') = (f, g)\phi$ modulo $\operatorname{rad}^2 \operatorname{mod} \Lambda$ and equating coefficients of the f_i, we obtain scalar equations

$$\bar\theta_{ii} = \bar\phi_{11} + \lambda_i \bar\phi_{21} \tag{3.3}$$

and

$$\bar\theta_{ii}\lambda'_i = \bar\phi_{12} + \lambda_i \bar\phi_{22} \tag{3.4}$$

for $1 \leq i \leq n$.

We now use these equations to show that θ is either nilpotent, or is an automorphism, that is either all, or none, of the $\bar{\theta}_{ii}$ vanish. A first possibility is that $\bar{\theta}_{ii} = 0$ for two different i; since the λ_i are all different, it then follows at once that $\bar{\phi}_{rs} = 0$ for all r, s and so $\bar{\theta}_{ii} = 0$ for all i. This, then, is just the case in which θ is nilpotent.

Another apparent possibility is that, for some unique j we have

$$0 = \bar{\phi}_{11} + \lambda_j \bar{\phi}_{21} \tag{3.5}$$

and

$$0 = \bar{\phi}_{12} + \lambda_j \bar{\phi}_{22}, \tag{3.6}$$

and that $\bar{\theta}_{ii} \neq 0$ for all $i \neq j$. Then equation (3.3) implies that at least one of $\bar{\phi}_{11}$ and $\bar{\phi}_{21}$ is non-zero and then it follows from equation (3.5) that $\bar{\phi}_{11} : \bar{\phi}_{21} = -\lambda_j : 1$. Similarly $\bar{\phi}_{12} : \bar{\phi}_{22} = -\lambda_j : 1$. Hence, for $i \neq j$, elimination of $\bar{\theta}_{ii}$ gives $\lambda'_i = \bar{\phi}_{22}/\bar{\phi}_{21}$, contrary to the assumption that the λ'_i are all different.

Hence this possibility cannot arise, and we have shown that θ is either nilpotent or is an automorphism. Also, in the latter case, equations (3) and (4) above show that the collineation of the projective line $\mathbb{P}^1(k)$ induced by ϕ maps the ordered sequence of n different points $(1, \lambda_i)$ to the sequence of points $(1, \lambda'_i)$. Thus the isomorphism classes of short exact sequences (3.1) defining W are in $1-1$ correspondence with those sets of n different points on the projective line $\mathbb{P}^1(k)$ which, for example, have $(1,0), (1,1), (1,a)$ for some $a \neq 0, 1$ as their first three points.

Within any one isomorphism class of modules W defined by such ordered sequences of n points on the projective line, the short exact sequences of forms (3.1) and (3.2) can only be related by nilpotent θ, and the existence of such a θ provides a total ordering of the sequences. The sequences thus form a chain, and it follows from the transitivity of the ordering that its length, that is, the number of different sequences, is at most the nilpotence index, N say, of the radical of $\operatorname{End}(Y)$, and in particular is finite. Thus the set of all isomorphism classes of modules W is in $1-1$ correspondence with the members of a certain partitioning into subsets of cardinality at most N of the set of all sequences (λ_i) in which $\lambda_1 = 0$, $\lambda_2 = 1$, $\lambda_3 = a$ and $\lambda_1, \lambda_2, \ldots, \lambda_n$ are all different. We therefore obtain, as required, an $(n-3)$-parameter family of indecomposable modules W, all of the same dimension. Since $n \geq 5$, it follows that Λ is wild.

REMARK 3.7. Here is an alternative proof of wildness analogous to that used for Lemma 2.7. The argument above certainly shows that there are infinitely many pairwise non-isomorphic indecomposable modules W all of the same dimension $\dim Y - 2\dim Z$. Easy calculations, along the lines of the proof of Lemma 3.1, show that $(\dim \tau Y - 2\dim \tau Z) - (\dim Y - 2\dim Z) \geq (n-4)(\dim Y - \dim Z) > 0$ and $(\sum_{j\neq i} \dim \tau Y_j - \dim \tau Z) - (\sum_{j\neq i} \dim Y_j - \dim Z) > 0$ for $1 \leq i \leq n$. Hence we may assume, (possibly after repeated applications of τ) that $\dim W > \dim \Lambda$ and $\sum_{j\neq i} \dim Y_j - \dim Z > \dim \Lambda$ for $1 \leq i \leq n$. It follows from Corollary 3.5 and Proposition 2.6, that $\tau W = \ker(\tau f, \tau g)$ and that $\dim \tau W > \dim W$. Thus none of these modules belongs to a homogeneous tube and it follows from [**6**] that Λ is wild.

4. Generalisations

In this section we suggest possible generalisations of the preceding results for an algebra Λ whose Auslander-Reiten quiver, Γ_Λ, contains a finite sectional subquiver

Δ of wild type. By a sectional subquiver we mean a connected valued subquiver Δ such that the value of each arrow in Δ is at most its value in Γ_Λ and, whenever there are arrows $A \to B$ and $B \to C$ in Δ, then $A \not\cong \tau C$. We shall use the same symbol Δ for an abstract sectional subquiver and for its realisation by modules and morphisms in Γ_Λ.

4.1. Homological formulae for families of modules. The choice of a finite sectional subquiver Δ of an Auslander-Reiten quiver Γ_Λ canonically determines a left Λ- right $k\Delta$-bimodule B, namely, the representation of the quiver Δ which associates with its vertices and arrows the corresponding indecomposable Λ-modules and irreducible morphisms. Our purpose in this subsection is to draw attention to the fact that the functor $\mathrm{Tor}_1^{k\Delta}(B, -)$ evaluated on suitable $k\Delta$-modules gives the Λ-modules used in Sections 2 and 3 to prove wildness of Λ.

For Section 2 we take Δ to be the wild quiver with vertices 0, 1, 2 and arrows $a, b : 1 \to 0$ and $c : 2 \to 0$. The irreducible Λ-module map $(f, g, h) : Y \oplus Y \oplus U \longrightarrow Z$ postulated in Lemma 2.7 determines a left Λ-, right $k\Delta$-bimodule, B, which has restrictions Z, Y, U to 0, 1, 2 and f, g, h to a, b, c, respectively. For $\lambda = (\mu, \nu) \in \mathbb{P}^1(k)$, there is an indecomposable left $k\Delta$-module $L = L(\mu, \nu)$ with restrictions k to each vertex and $-\nu$, μ, 1 to the arrows a, b, c, respectively. A short calculation using a projective $k\Delta$ resolution of L shows that

$$B \otimes_{k\Delta} L = 0 \qquad \text{and} \qquad \mathrm{Tor}_1^{k\Delta}(B, L(\mu, \nu)) \cong \ker(\mu f + \nu g).$$

When $\dim Y > \dim Z$, the second of these is the module $W_{(\mu,\nu)}$ considered in the proof of Lemma 2.7.

For Section 3 we take Δ to be the quiver with $n + 1 \geq 6$ vertices, $0, 1, \ldots, n$, and n arrows $i \to 0$ for $1 \leq i \leq n$, and B to be the representation determined by an irreducible map $f : Y \to Z$, with components $f_i : Y_i \to Z$ for $1 \leq i \leq n$, satisfying Hypothesis (H) of Subsection 3.2. The algebra $k\Delta$ has dimension $2n + 1$, and a basis consisting of $n + 1$ pairwise orthogonal primitive idempotents, $e_0, e_1, \ldots, e_n$ such that $\sum e_i = 1$, and n elements, $e_{i0} = e_i e_{io} e_0$ for $1 \leq i \leq n$. It is of course wild, and for each $(\lambda_i) \in k^n$, has a left module, $L = L(\lambda_i)$ of dimension $n + 2$, with projective resolution

$$0 \to P_1 = \coprod_{i=1}^{n} k\Delta e_i \xrightarrow{\pi} P_0 = k\Delta e_0 \coprod k\Delta e_0 \to L = L(\lambda_i) \to 0,$$

where, for each i, $\pi(e_i) = (e_{i0}, \lambda_i e_{i0})$. If $\lambda_1, \ldots, \lambda_n$ are all different, one sees easily that this resolution is minimal and that L is indecomposable. It is easy to use this projective resolution to verify that

$$B \otimes_{k\Delta} L = 0 \qquad \text{and} \qquad W = W_{f,g} \cong \mathrm{Tor}_1^{k\Delta}(B, L(\lambda_i)).$$

4.2. Replication theory at the tame-wild border. We recall that two sectional subquivers Δ and $\bar{\Delta}$ of an Auslander-Reiten quiver Γ_Λ are said to be parallel if $\bar{\Delta} = \tau^r \Delta$ for some $r \in \mathbb{Z}$. Replication theory is concerned with the number of sectional subquivers parallel to a given one and, at the finite-infinite border, it is epitomised by the following theorem which summarises the more precise results of [**4**].

THEOREM 4.1. *For each Euclidean graph Δ there is an integer m such that, if Γ_Λ contains m parallel sectional subquivers of type Δ, then Λ is of infinite type.*

A plausible conjecture for the tame-wild border is as follows.

CONJECTURE 2. *For each wild graph Δ there is an integer n such that, if Γ_Λ contains n parallel sectional subquivers of type Δ, then Λ is wild.*

Our main objective in this subsection is the following Proposition which reduces Conjecture 2 to the widely believed Conjecture 3 below.

PROPOSITION 4.2. *For each wild graph Δ, there is an integer N such that, if Γ_Λ contains N parallel sectional subquivers of type Δ, then there exist Λ-modules M for which either*

$$\lim_{r\to\infty} \sqrt[r]{l(\tau^r M)} > 1 \tag{4.1}$$

or

$$\lim_{r\to\infty} \sqrt[r]{l(\tau^{-r} M)} > 1. \tag{4.2}$$

CONJECTURE 3. *The existence of a Λ-module satisfying* (4.1) *or* (4.2) *implies that Λ is wild.*

Obviously Theorem 1.1 shows that Conjecture 2 is true if Γ_Λ contains a wild sectional subquiver with a double arrow, and Proposition 1.3 shows that it is true, under certain conditions, for the five-subspace configuration.

Conjecture 3 is proved for tilted algebras by de la Peña and Takane [**10**] and is used by them as the basis for their proof of Conjecture 1 in the case where the middle term of the almost split sequence is directing.

The remainder of this subsection is devoted to the proof of Proposition 4.2. We start by recalling some technicalities needed to discuss replication.

For any sectional subquiver Δ, $\mathbf{l}(\Delta)$ is the vector with components indexed by the vertices of Δ, the ith component being the composition length $l(M_i)$ of the indecomposable module M_i at the vertex i. Since there are no oriented cycles in Δ [**1**, Chapter VII, Corollary 2.6], the Coxeter matrix $c = c_\Delta$ is defined, is invertible and has the following important properties.

If no M_i is projective, then $\mathbf{l}(\tau\Delta) = c\mathbf{l}(\Delta) + \mathbf{q}$, where c is the Coxeter matrix for Δ and $\mathbf{q}$ is the dimension vector of an injective $k\Delta$-module; further $\mathbf{q}$ is zero if and only if there is neither an arrow in $\Gamma_\Lambda \backslash \Delta$ with its tail in Δ nor an arrow in Δ which has a strictly larger valuation in Γ_Λ [**2**]. If M_i is projective, then $(c\mathbf{l})_i < 0$. Thus, if $c(\mathbf{l}(\Delta)) > 0$, then Δ contains no projective module. Dually, if no M_i is injective, then $\mathbf{l}(\tau^{-1}\Delta) = c^{-1}\mathbf{l} + \mathbf{p}$, where $\mathbf{p}$ is the dimension vector of a projective $k\Delta$-module (which is zero in analogous circumstances to those of the dual case above) and, if M_i is injective, then $(c^{-1}\mathbf{l})_i < 0$.

Additionally, if Δ is Euclidean, there exist a positive integer $m = m_\Delta$, a linear functional $\delta : \mathbb{Z}^{\Delta_0} \to \mathbb{Z}$ (called the defect of Δ) and a positive vector $\mathbf{n}$ (the minimal positive null vector of the quadratic form for Δ) such that $c\mathbf{n} = \mathbf{n}$ and, for any vector $\mathbf{x} \in \mathbb{Z}^{\Delta_0}$, $c^m\mathbf{x} = \mathbf{x} + \delta(\mathbf{x})\mathbf{n}$ and $\delta(c\mathbf{x}) = \delta(\mathbf{x})$ (see [**7**]). Clearly $\delta(\mathbf{n}) = 0$. Furthermore, if $\mathbf{p}$ is the dimension vector of a preprojective $k\Delta$-module and $\mathbf{q}$ is the dimension vector of a preinjective $k\Delta$-module, then $\delta(\mathbf{p}) < 0$ and $\delta(\mathbf{q}) > 0$. Notice that $\delta(\mathbf{l}(\tau\Delta)) \geq \delta(\mathbf{l}(\Delta))$.

We need the following lemma on the replication of Euclidean subquivers which, we note, already implies Theorem 4.1.

LEMMA 4.3. *Let Δ be a Euclidean sectional subquiver of Γ_Λ. Suppose that the defect of $\mathbf{l}(\Delta)$ is non-negative and that Γ_Λ contains $m = m_\Delta$ parallel sectional* *subquivers,* $\Delta, \tau\Delta, \dots, \tau^{m-1}\Delta$. *Then* Γ_Λ *contains sectional subquivers* $\tau^r\Delta$ *for all* $r \geq 0$. *The dual statement is true if the defect of* $\mathbf{l}(\Delta)$ *is non-positive.*

PROOF. We prove only the first statement. We shall show that the hypotheses are also satisfied for $\tau\Delta$ and so the result follows by induction. In order to show that $\tau^{m-1}\Delta$ contains no projective module, we shall show that $c\mathbf{l}(\tau^{m-1}\Delta) > 0$. Clearly $\mathbf{l}(\tau^{m-1}\Delta) = c^{m-1}\mathbf{l}(\Delta) + \mathbf{q}$, where $\mathbf{q}$ is the dimension vector of a preinjective $k\Delta$-module (which could be zero). Hence

$$c\mathbf{l}(\tau^{m-1}\Delta) = c^m\mathbf{l}(\Delta) + c\mathbf{q} = \mathbf{l}(\Delta) + \delta(\mathbf{l}(\Delta))\mathbf{n} + c\mathbf{q} \geq \mathbf{l}(\Delta) > 0,$$

as required.

Further $\mathbf{l}(\tau\Delta) = c\mathbf{l}(\Delta) + \mathbf{q}'$ where $\mathbf{q}'$ is the dimension vector of an injective $k\Delta$-module and so $\delta(\mathbf{l}(\tau\Delta)) \geq \delta(\mathbf{l}(\Delta)) \geq 0$. □

Next we extend this result to minimal wild subquivers with no double arrows. We note that the case of a minimal wild quiver containing a Kronecker subquiver is covered by Theorem 1.1.

LEMMA 4.4. *Let* Δ' *be a minimal wild sectional subquiver of* Γ_Λ *with no double arrow, so that*

$$\text{(a)}\ \ \Delta' = \Delta \bigcup \{i_0 \to i_1\} \quad \text{or} \quad \text{(b)}\ \ \Delta' = \Delta \bigcup \{i_1 \to i_0\},$$

where Δ *is a Euclidean subquiver with* $m = m_\Delta \geq 2$ *and* i_1 *is a vertex of* Δ. *Suppose that the defect of* $\mathbf{l}(\Delta)$ *is non-negative and that* Γ_Λ *contains, in case* (a), m *parallel sectional subquivers* $\tau^r\Delta'$ $(0 \leq r \leq m-1)$, *and in case* (b), $m+1$ *parallel sectional subquivers* $\tau^r\Delta'$ $(0 \leq r \leq m)$. *Then* Γ_Λ *contains parallel sectional subquivers* $\tau^r\Delta'$ *for all* $r \geq 0$.

The dual statement is true if the defect of $\mathbf{l}(\Delta)$ *is non-positive.*

PROOF. We consider first case (a). By Lemma 4.3, it is only necessary to prove that no τ-translate of M_{i_0} is projective. The proof in the first non-trivial case, $\tau^{m-1}M_{i_0}$, is clearly typical. By considering the almost split sequence ending at $\tau^{m-1}M_{i_1}$ and the construction of c_Δ as the composite of a sequence of reflections, one sees that $l(\tau^m M_{i_1}) \geq (c_\Delta \mathbf{l}(\tau^{m-1}\Delta))_{i_1} + l(\tau^{m-1}M_{i_0})$. The right hand side of this is, as in the proof of Lemma 4.3, at least $l(M_{i_1}) + l(\tau^{m-1}M_{i_0})$, and this gives the required result.

Case (b) may be reduced to case (a) by observing that $\Delta'' = \Delta \bigcup \{\tau i_0 \to i_1\}$ has the required form and satisfies the appropriate conditions.

Clearly the dual argument works for the dual situation. □

PROOF OF PROPOSITION 4.2. Suppose that Δ' and Δ are as in Lemma 4.4 and that Γ_Λ contains at least $2m_\Delta$ sectional subquivers parallel to Δ'. Then, in case (a), either there are at least m_Δ such sections satisfying $\delta(\mathbf{l}(\Delta)) \geq 0$ or there are at least $m_\Delta + 1$ such sections satisfying $\delta(\mathbf{l}(\Delta)) \leq 0$. Then it follows from Lemma 4.4 that either, for all $r \geq 0$, $\tau^r\Delta'$ contains no projective module or, for all $r \geq 0$, $\tau^{-r}\Delta'$ contains no injective module. In the first case, for each module M in Δ' the limit on the left-hand side of the formula (4.1) is the spectral radius of $c_{\Delta'}$ which is greater than 1 ([8], [9]). In the second case it is the limit on the left-hand side of the formula (4.2) which is equal to the spectral radius of ${c_{\Delta'}}^{-1}$, and is also greater than 1. The same conclusion holds in case (b). □

References

[1] M. Auslander, I. Reiten and S. Smalø, 'Representation theory of Artin algebras', Cambridge Studies in Advanced Mathematics 36 (1994).

[2] R. Bautista, 'Sections in Auslander-Reiten quivers', *Representation Theory II*, Lecture Notes in Mathematics 832 (Springer, Berlin, 1981), pp.26-96.

[3] R. Bautista and S. Brenner, 'On the number of terms in the middle of an almost split sequence', *Representations of algebras*, Lecture Notes in Mathematics 903 (Springer, Berlin, 1981), pp. 1-8.

[4] R. Bautista and S. Brenner, 'Replication numbers for non-Dynkin sectional subgraphs in finite Auslander-Reiten quivers and some properties of Weyl roots', Proc. London Math. Soc. (3), 47 (1983), 429-462.

[5] S. Brenner, 'The Auslander translate of a short exact sequence', Colloquium Mathematicum (to appear).

[6] W. W. Crawley-Boevey, 'On tame algebras and bocses', Proc.London Math.Soc. 56 (1988) 451-483.

[7] V. Dlab and C.M. Ringel, 'Indecomposable representations of graphs and algebras', Memoirs Amer. Math. Soc., 6 (1976), 1-57.

[8] V. Dlab and C.M. Ringel, 'Eigenvalues of Coxeter transformations and the Gelfand-Kirilov dimension of the preprojective algebras', Proc. Amer. Math. Soc. 83, 1981, 228-232.

[9] J. A. de la Peña and M. Takane, 'Spectral properties of Coxeter transformations and applications', Arch. Math., 55, 1990, 120-134.

[10] J. A. de la Peña and M. Takane, 'On the number of terms in the middle of an almost split sequence', Trans. Amer. Math. Soc. (to appear).

Department of Mathematical Sciences, The University of Liverpool, Liverpool, L69 3BX, UK

E-mail address: sbrenner@liv.ac.uk

Department of Mathematical Sciences, The University of Liverpool, Liverpool, L69 3BX, UK

E-mail address: mcrb@liv.ac.uk

Contemporary Mathematics
Volume **229**, 1998

REPRESENTATION THEORY OF NOETHERIAN HOPF ALGEBRAS SATISFYING A POLYNOMIAL IDENTITY

KENNETH A. BROWN

Abstract. A class of Noetherian Hopf algebras satisfying a polynomial identity is axiomatised and studied. This class includes group algebras of abelian-by-finite groups, finite dimensional restricted Lie algebras, and quantised enveloping algebras and quantised function algebras at roots of unity. Some common homological and representation-theoretic features of these algebras are described, with some indications of recent and current developments in research on each of the exemplar classes. It is shown that the finite dimensional representation theory of each of these algebras H reduces to the study of a collection $Alg(H)$ of (finite dimensional) Frobenius algebras. The properties of this family of finite dimensional algebras are shown to be intimately connected with geometrical features of central sub-Hopf algebras of H. A number of open questions are listed throughout.

1. Introduction

My aim in this paper is to review some common properties exhibited by four large and important classes of Noetherian Hopf algebras which are finite modules over their centres. In so doing I will point out ways

---

1991 *Mathematics Subject Classification.* 16E10,16G10,16P40,16R99,16W30, 17B35,17B37,20C07.

Key words and phrases. Hopf algebra, quantum group, representation theory, enveloping algebra, polynomial identity.

This paper is based on a talk given at the A.M.S. Joint Summer Research Conference on Trends in Representation Theory of Finite Dimensional Algebras, held at Seattle in July 1997. I am grateful to the organisers of the conference for the invitation to talk to the conference and to submit a paper to the Proceedings. This work has benefited from conversations and correspondence with K.R.Goodearl, T.H.Lenagan, A.Premet and (especially) my Glasgow colleagues I.Gordon and R.J.Marsh. Partial financial support was provided by Nato Grant CRG960250, held jointly with Goodearl and Lenagan.

in which each class behaves in ways distinctive from the others, but I also hope to highlight some traits which occur in all four classes, and suggest that there may be single theorems underlying at least some of these shared features. For the most part this paper is a survey of known results. A few new results are scattered through the text, but what novelty there is here rests primarily in the new perspective obtained by bringing together results from several previously distinct areas.

Throughout, K will denote an algebraically closed field. The four classes of examples I have in mind are the following:

(A) Let Γ be a finitely generated group with a normal Abelian subgroup A of finite index, and let H be the group algebra $K\Gamma$.

(B) Let K have positive characteristic p, let $\mathfrak{g}$ be a finite dimensional restricted Lie algebra over the field K, and let H be the enveloping algebra $U(\mathfrak{g})$.

(C) Let K have characteristic zero, and let $\mathfrak{g}$ denote a finite dimensional semisimple Lie algebra over K. Let ℓ be an odd integer which is greater than 2, and is prime to 3 if $\mathfrak{g}$ contains a factor of type G_2. Let ϵ be a primitive ℓth root of 1, and let H be the quantised enveloping algebra $U_\epsilon(\mathfrak{g})$ associated to $\mathfrak{g}$ and ϵ.

(D) Let K have characteristic zero and let G be a connected, simply connected, semisimple Lie group defined over K. Let ℓ and ϵ be as in (C). Let H be the quantised function algebra $O_\epsilon[G]$.

The above list is intended to be illustrative rather than exhaustive, and indeed there are other classes of examples which fall within the boundaries laid down by the axioms listed below - for example, certain subHopf algebras of the quantised enveloping algebras (C), and certain factor algebras of the function algebras (D). The basic references I shall use for these classes are: [45] for class (A); [29],[32] for (B); [10],[12] for (C); and [11] for (D). For the reader's convenience I recall some details of the relevant definitions in (2.1).

We begin by isolating the following properties exhibited by each of these classes of algebras. (Both the labelling of the classes above and the numbered hypotheses below will remain fixed throughout the paper.) Thus we shall consider a K-algebra H such that

(H1): *H is a K-affine K-Hopf algebra;*

(H2): *H satisfies a polynomial identity;*

(H3): *H has finite injective dimension as a (right and left) H-module;*

(H4): *H is (right and left) Noetherian.*

Our standard references for Hopf algebraic results will be [41], [54]. We'll denote the comultiplication, counit and antipode of a Hopf algebra by Δ, ϵ and S respectively, and follow the convention of writing $\Delta(h) = \sum h_1 \otimes h_2$. The antipode will be assumed to be a bijective map. (Note in passing that the existence of a bijective antihomomorphism from H to H ensures that hypotheses (H3) and (H4) follow from their one-sided analogues.)

The paper is organised as follows. In §2 I recall enough key properties of the algebras in classes (A),...,(D) to show that they satisfy the hypotheses (H1),...,(H4). The injective dimension hypothesis (H3) is considered further in §3 where I show that, when coupled with the other hypotheses, it yields strong homological restrictions on the algebras in question, namely the Auslander-Gorenstein and Cohen-Macaulay properties. §3 includes also the definitions of these properties and a discussion of their usefulness. The remaining two sections, §§4 and 5, are concerned with the representation theory of algebras satisfying (H1),...,(H4). In fact in order to make progress we're obliged at this stage to introduce various further (possibly superfluous) additional hypotheses, all satisfied by our four classes of examples (except for the need to exclude group algebras at certain points). We begin §4 by explaining how to reduce the finite dimensional representation theory of an algebra H satisfying our hypotheses to the study of a collection $Alg(H)$ of finite dimensional algebras attached to H. In practice, a geometric perspective is crucial for a proper understanding of $Alg(H)$ - we introduce this point of view in §4 and go on in §5 to review several topics - the Azumaya locus, the Kac-Weisfeiler conjecture, support varieties and representation type - from this angle. Scattered throughout the paper I've included some questions which - I hope - may help to motivate future research in the area described here.

2. Details of the classes

2.1. Definitions of the examples. (A) The *group algebra* $K\Gamma$ is a Hopf algebra [52] with

$$\Delta(\gamma) = \gamma \otimes \gamma, \quad S(\gamma) = \gamma^{-1}, \quad \epsilon(\gamma) = 1$$

for $\gamma \in \Gamma$.

(B) The *enveloping algebra* $U(\mathfrak{g})$ is a Hopf algebra [52] with

$$\Delta(x) = x \otimes 1 + 1 \otimes x, \quad S(x) = -x, \quad \epsilon(x) = 0$$

for $x \in \mathfrak{g}$.

(C) (See e.g. [12].) The *(simply connected form of the) quantised enveloping algebra* $U_\epsilon(\mathfrak{g})$ is the K-algebra on generators $\{E_i, F_i, K_i^{\pm 1} : 1 \leq i \leq n\}$, where $n = rank(\mathfrak{g})$, with the relations, for all i and j,

$$K_iK_j = K_jK_i, \quad K_iK_i^{-1} = 1 = K_i^{-1}K_i,$$

$$K_iE_jK_i^{-1} = \epsilon^{d_i\delta_{ij}}E_j,$$

$$K_iF_jK_i^{-1} = \epsilon^{-d_i\delta_{ij}}F_j,$$

$$E_iF_j - F_jE_i = \delta_{ij}(K_{\alpha_i} - K_{\alpha_i}^{-1})/(\epsilon^{d_i} - \epsilon^{-d_i}),$$

$$\sum_{s=0}^{1-a_{ij}} (-1)^s \begin{bmatrix} 1 & - & a_{ij} \\ & s & \end{bmatrix}_{d_i} E_i^{1-a_{ij}-s}E_jE_i^s = 0,$$

$$\sum_{s=0}^{1-a_{ij}} (-1)^s \begin{bmatrix} 1 & - & a_{ij} \\ & s & \end{bmatrix}_{d_i} F_i^{1-a_{ij}-s}F_jF_i^s = 0.$$

Here, $A = [a_{ij}]$ is the Cartan matrix of $\mathfrak{g}$ and $(d_1, \ldots, d_n)$ is the symmetrising vector of relatively prime positive integers (so that $[d_ia_{ij}]$ is a symmetric matrix). The symbols in square brackets denote Gaussian binomial coefficients for the parameter ϵ . The Hopf structure is given by

$$\Delta(E_i) = E_i \otimes 1 + K_{\alpha_i} \otimes E_i, \quad \Delta(F_i) = F_i \otimes K_{\alpha_i}^{-1} + 1 \otimes F_i,$$

$$\Delta(K_i) = K_i \otimes K_i;$$

$$S(E_i) = -K_{\alpha_i}^{-1}E_i, \quad S(F_i) = -F_iK_{\alpha_i}, \quad S(K_i) = K_i^{-1};$$

$$\epsilon(E_i) = \epsilon(F_i) = 0, \quad \epsilon(K_i) = 1,$$

for $i = 1, \dots, n$.

(D) (See e.g. [11].) For ϵ and G as stated in Section 1, $O_\epsilon[G]$ is obtained by first constructing a suitable integral form U_A (over $\mathbb{Q}[q^{\pm 1}]$) of $U_q(\mathfrak{g})$, then taking an appropriate sub-Hopf algebra of the Hopf dual of U_A, and finally specialising q to ϵ. For the case $G = SL_n(K)$, generators and relations for $O_\epsilon[G]$ are given in [28], for example.

2.2. **Legitimacy of the examples.** We indicate here why

(1) *each algebra in classes (A), (B), (C), (D) satisfies (H1), (H2), (H3), (H4).*

For (H1), this is a consequence of the definitions in (2.1). Hypotheses (H2) and (H4) are immediate consequences of the fact that

(2) *each algebra H in (A), (B), (C), (D) is a finite module over an affine commutative subalgebra S.*

In case (A) one takes $S = KA$; a transversal to A in Γ provides a finite set of generators for H as an S-module. When (B) H is an enveloping algebra $U(\mathfrak{g})$ with $\mathfrak{g} = \sum_{i=1}^{m\oplus} Kx_i$ and restriction map $x \mapsto x^{[p]}$ for x in $\mathfrak{g}$, one takes

(3) $$S = K < x_i^p - x_i^{[p]} : 1 \leq i \leq m >,$$

which is central in H by [32, Proposition 2.3]. The PBW theorem implies that H is free of rank p^m as an S-module.

For (C) $H = U_\epsilon(\mathfrak{g})$ the picture is similar to that for (B): the subalgebra S of $U_\epsilon(\mathfrak{g})$ generated by the ℓth powers of a suitably defined set of 'roots' $\{E_{\beta_1}^\ell, \dots, E_{\beta_N}^\ell; F_{\beta_1}^\ell, \dots, F_{\beta_N}^\ell\}$, (which include as a subset the ℓth powers of the simple roots $\{E_1^\ell, \dots, E_n^\ell; F_1^\ell, \dots, F_n^\ell\}$), together with $\{K_i^{\pm\ell} : 1 \leq i \leq n\}$, [12, Theorem 19.1(i)]. Again, the PBW theorem for $U_\epsilon(\mathfrak{g})$ ensures that $U_\epsilon(\mathfrak{g})$ is free of rank ℓ^{2N+n} over S.

Finally, in case (D) there is a central subalgebra S of $O_\epsilon[G]$ which is isomorphic to $O[G]$, such that $O_\epsilon[G]$ is a projective S-module of rank $\ell^{dim(G)}$ [11, Theorem 7.2].

Thus (2) is proved in all cases.

Note that in each of the above cases

$$\text{(4)} \qquad \begin{array}{c} \textit{the commutative subalgebra } S \textit{ is} \\ \textit{a normal sub-Hopf algebra of } H. \end{array}$$

Recall from [41, 3.4.1] that a non-empty subset X of a Hopf algebra H is *normal* if it is closed under the left and right adjoint actions of H; that is, for all $x \in X$ and $h \in H$,

$$(ad_\ell h)(x) := \sum h_1 x S(h_2) \in X$$

and

$$(ad_r h)(x) := \sum S(h_1) x h_2 \in X.$$

In cases (A) and (B) (4) is clear from the definitions. That S is a sub-Hopf algebra of H in cases (C) and (D) is proved in [12, Theorem 19.1(iii)] and [11, Proposition 6.4] respectively. Since a central subalgebra is certainly normal, (4) is proved in all four cases.

To complete the proof of (1) it remains to prove that (H3) holds for each class. This can be achieved on a class-by-class basis - for (A) and (B) the result is well-known, and for (C) and (D) a substantially stronger result is proved in [7, Propositions 2.2 and 2.7]. But I will instead deduce (H3) here in a more unified way, as an immediate consequence of (2) and (4) and Proposition 2.3 below (which is a generalisation of the well-known result that a finite dimensional Hopf algebra is self-injective).

2.3.

Proposition. *Let H be a K-Hopf algebra which is a finite module over a commutative normal affine sub-Hopf algebra S. Then H has finite injective dimension m, where m is the Krull dimension of S.*

The key ideas in the proof of this proposition are as follows. (A detailed proof will be given in [4].)

Step 1: *S, being a commutative affine K-Hopf algebra, is Gorenstein.* This is clear in characteristic zero, since then affine group schemes are smooth [54, Theorems 11.4, 11.6]. In positive characteristic p choose a positive integer n large enough so that the image C of S under the nth power of the Frobenius map has no non-zero nilpotent elements. Then C is smooth [54, Theorem 11.6] and S is a finite C-module, so that S is Gorenstein by a special case of Steps 2 and 3 below.

Recall that an inclusion $A \subseteq B$ of algebras is *Frobenius* if B is a finitely generated projective A-module, and there is an isomorphism $B \longrightarrow \mathrm{Hom}_A(B, A)$ of (B, A)-bimodules [15, p.242].

Step 2: *H is a Frobenius extension of S.*

This follows by standard Hopf-algebraic results [35, Theorem 1.7(1),(5)].

Step 3: *If a ring B is a Frobenius extension of a ring A then the (say, left) injective dimension of B is bounded above by the left injective dimension of A.*

This result dates back to the first paper on general Frobenius extensions [43, Corollary 9]; it's an immediate consequence of the fact that, for all $i \geq 0, \mathrm{Ext}^i_B(-, B)$ and $\mathrm{Ext}^i_A(-, A)$ are equivalent functors on left B-modules.

Remark: It's easy to remove the hypothesis that K is algebraically closed from the above proposition - just prove the result for $H \otimes_K \overline{K}$, and then descend to H.

3. Homological issues

3.1. In the light of Proposition 2.3 it's natural to ask:

Question A: Is (H3) a consequence of (H1), (H2) and (H4)? That is, does every affine Noetherian K-Hopf algebra satisfying a polynomial identity have finite injective dimension?

In fact I don't even have an example where the answer to Question A is 'No' if the words 'satisfying a polynomial identity' are omitted. Of course, it's conceivable that a positive answer to Question A could follow because all the algebras involved are already encompassed in Proposition 2.3. Thus, in the absence of any counterexamples, we should at least record

Question B: Let H be an affine Noetherian K-Hopf algebra satisfying a polynomial identity. Is H a finite module over a commutative normal sub-Hopf algebra S?

Note that if 'commutative normal' is strengthened to 'central' in Question B then essentially any choice of infinite non-Abelian group Γ in (A) yields a counterexample. For example, let $\Gamma = \quad < a, b : bab = a^{-1}, b^2 = 1 >$ and set $H = \mathbb{C}\Gamma$, so that $Z(H) = \mathbb{C}[a + a^{-1}]$, and $\mathbb{C}$ is the only sub-Hopf algebra of H contained in $Z(H)$. Question B is discussed further in [4].

3.2. An even more obvious question than Question A concerning (H3) is:

(5) *What is (H3) good for?*

To answer this we need to recall some important homological concepts.

Definitions: Let R be a ring and M a right R-module.

(i) The *grade* of M is

$$j(M) := \inf\{j : \mathrm{Ext}^j_R(M, R) \neq 0\}.$$

(ii) The ring R *satisfies the Auslander condition* if, for every Noetherian R-module M and for all $i \geq 0, j(N) \geq i$ for all submodules N of $\mathrm{Ext}^i_R(M, R)$.

(iii) The ring R is *Auslander-Gorenstein* if it is Noetherian, satisfies the Auslander condition, and has finite right and left injective dimension.

(iv) If R is Auslander-Gorenstein and has finite global dimension then R is said to be *Auslander-regular.*

(v) The ring R is said to be *Cohen-Macaulay* (with respect to the Krull dimension K-dim$(-)$) if, for all non-zero Noetherian R-modules M

$$j(M) + \text{K-dim}(M) = \text{K-dim}(R).$$

Useful sources for the basic ideas and results involving these concepts are [1], [2], [3], [37]. One should think of 'Auslander-Gorenstein Cohen-Macaulay' as the appropriate generalisation to a non-commutative setting of the Gorenstein property (that is, R_R has finite injective dimension) for a commutative Noetherian ring. Having finite injective dimension is a relatively weak attribute of a non-commutative ring, and it's only by imposing these stronger hypotheses that one can obtain analogues of the commutative results. Note that, in contrast to the commutative case, the Auslander-Gorenstein condition doesn't imply the Cohen-Macaulay condition - for example,

$$\begin{bmatrix} K & K \\ 0 & K \end{bmatrix}$$

is Auslander-regular but not Cohen-Macaulay.

The Cohen-Macaulay property provides a bridge between the homological properties of a module and its 'size'. Other measures of size can be - and have been - used here in place of Krull dimension to obtain variants of the above definition. The chief such variant uses Gelfand-Kirillov dimension, G-Kdim(−), as in [37, 5.8] for example. But since the two dimensions coincide for finitely generated modules over an affine Noetherian PI ring [40, 13.10.6], I won't discuss this issue further here.

The Auslander-Gorenstein and Cohen-Macaulay properties are powerful weapons to use in proving structural results about modules and rings; for example, in addition to the references already cited one should note [49] and [22]. A further application will be given in (5.4) below. The link between the categories of left and right R-modules which is implicit in the definition of the Auslander property is made explicit in [1, Proposition 5.4]: *If R is an Auslander-Gorenstein ring of injective dimension d then the categories of Noetherian left and right R-modules are in $(d+1)$-step duality.* (Here, categories $\mathfrak{C}$ and $\mathfrak{D}$ are in *n-step duality* if there are dense subcategories

$$0 = \mathfrak{C}_0 \subseteq \mathfrak{C}_1 \subseteq \ldots \subseteq \mathfrak{C}_n = \mathfrak{C}$$

and

$$0 = \mathfrak{D}_0 \subseteq \mathfrak{D}_1 \subseteq \ldots \subseteq \mathfrak{D}_n = \mathfrak{D}$$

such that the quotient categories $\mathfrak{C}_i/\mathfrak{C}_{i-1}$ and $\mathfrak{D}_i/\mathfrak{D}_{i-1}$ are in duality for all $i = 1, \ldots, n$.)

We can now provide an answer to (5). We'll use inj.dim.(M) to denote the injective dimension of a module M.

Proposition. [Brown-Goodearl, [7, Theorem 1.14]] *Let H satisfy (H1), (H2), (H3) and (H4). Then H is Auslander-Gorenstein and Cohen-Macaulay, and*

$$\text{inj.dim.}(H) = \text{K-dim}(H) = \text{G-Kdim}(H).$$

Here are the main points of the proof. Let $d = \text{inj.dim.}(H)$.

Step 1: *There exists an irreducible H-module V such that* $\text{Ext}_H^d(V, H) \neq 0$.

This follows from [5, Theorems A and D].

Step 2: *Let W be an H-module with $dim_K(W) = r < \infty$. Then*

$$\text{Ext}_H^d(W, H) \cong \text{Ext}_H^d(K, H)^{\oplus(r)}.$$

This is [7, Proposition 1.3]; it is easily deduced from elementary properties of Hopf algebras. The conclusions of Steps 1 and 2 together say that H is *injectively smooth*, and the conclusions of the proposition therefore follow from work of Stafford and Zhang in [49]. (If one is willing to assume here that H is a finite module over its centre, one can instead use the more elementary results of [8] for this last step.)

3.3. Given the definitions and proposition in (3.2) it is natural to ask

Question C: Is there an easy method to recognise when a Noetherian Hopf algebra has finite global dimension?

A Hopf algebra has finite global dimension provided the projective dimension of its trivial module is finite [38, §2.4]. The available evidence, scant though it is, tempts me to make the extravagant suggestion that a Noetherian Hopf algebra will have finite global dimension if and only if all its finite dimensional Hopf subalgebras satisfy this property. In the case of (all known) Noetherian group algebras this is true, and follows from a result of Serre [45, 10.3.12]. For the other test cases we in fact have the

Theorem. *The algebras in classes (B), (C) and (D) have finite global dimension.*

For the enveloping algebras (B) this is an old result which is easily deduced by filtered-graded techniques [40, Corollary 7.6.18]. Similar though slightly more involved methods will take care of $U_\epsilon(\mathfrak{g})$; but $O_\epsilon[G]$ requires separate treatment - see [7, Proposition 2.7].

4. Representation theory and finite dimensional algebras

4.1. For the remainder of this paper we shall assume that H is a K-algebra satisfying

(H1): *H is a K-affine K-Hopf algebra, and*

(H5): *H is a finite module over an affine commutative normal sub-Hopf algebra S.*

Thus H will also satisfy (H2), (H4) and (H3), the last of these by Proposition (2.3). Moreover, as we've already remarked in §2.3

(6) **(H1)** *and* **(H5)** *imply that H is a Frobenius extension of S*

and hence, in particular,

(7) **(H1)** *and* **(H5)** *imply that H is a faithful projective S-module.*

For technical reasons we are obliged to introduce a third hypothesis at this point, to accompany (H1) and (H5). Namely, we'll require that

(H6): *if P is a maximal ideal of H then $P \cap S$ is a semimaximal ideal of S.*

Note that if P is a maximal ideal of an algebra H satisfying (H1) and (H5) then $S/P \cap S$ is a finite dimensional K-algebra. If S is central rather than simply normal then of course $P \cap S$ is maximal and (H6) is superfluous. Similarly, (H6) is easily seen to hold in case (A) [44, Lemma 14.1], or indeed in any case where H is generated as an S-module by group-like elements. But it seems to be an open question whether (H6) is a consequence of (H1) and (H5) in general. We note for future use that, if (H1), (H5) and (H6) hold and P is a maximal ideal of H, one can show easily that

(8) *if* $P \cap S \subseteq I \lhd S$ *and* $IH = HI \neq H$, *then* $I = P \cap S$.

4.2. Let's fix the following

(Aim): *To understand the H-modules of finite composition length.*

Every H-module is a finite dimensional K-vector space by [40, 13.10.3], so we could equally well have stated our aim as being to understand the category of finite dimensional H-modules. And for the same reason (recalling also that we are assuming that K is algebraically closed), an ideal P of H is primitive if and only if $H/P \cong M_t(K)$, the ring of $t \times t$ matrices over K, for some t. Denote the space of primitive ideals of H by $\mathrm{Prim}(H)$, and let H-prim(S) be the space of ideals of S which are H-normal, H-prime, and have finite K-codimension in S. (Following [42] we say that an H-normal ideal P of S is H-prime if the product IJ of ideals I and J of S for which $IH = HI$ and $JH = HJ$ can lie in P only if $I \subseteq P$ or $J \subseteq P$.) Thus, when S is central in H, H-prim(S) is just the spectrum of maximal ideals of S; but this is not so in case (A), where H-prim(S) consists of the intersections of the Γ/A-orbits of maximal ideals of $S = KA$. Consider the map

$$\Psi : Prim(H) \longrightarrow H\text{-prim}(S) : P \longmapsto P \cap S.$$

Lemma. *Ψ is continuous (with respect to the Zariski topology induced in both spaces), surjective and has finite fibres. Indeed, for $\mathfrak{p} \in H$-prim(S),*

$$\Psi^{-1}(\mathfrak{p}) = \{P \in Prim(H) : \mathfrak{p}H \subseteq P\}.$$

Proof. A closed subset of H-prim(S) has the form $\mathcal{V}_S(I) := \{P \in H\text{-prim}(S) : \ I \subseteq P\}$, for some H-normal ideal I of S. But then IH is an ideal of H with $IH \cap S = I$ by [42, Lemma 1.3(2)]. Hence

$$\Psi^{-1}(\mathcal{V}_S(I)) = \mathcal{V}_H(IH).$$

Surjectivity follows from [42, Lemma 2.2(3)], while the finiteness of $\Psi^{-1}(\mathfrak{p})$ for $\mathfrak{p} \in H$-prim(S) follows from the fact that $H/\mathfrak{p}H$ is a finite dimensional K-algebra. Clearly $\Psi^{-1}(\mathfrak{p})$ is contained in the set $\{P' \in \text{Prim}(H) : \mathfrak{p}H \subseteq P'\}$. That equality in fact holds here is a consequence of (8) applied to $I := P' \cap S$, for $P' \in \mathcal{V}_H(\mathfrak{p}H)$. □

We now define

$$Alg(H) := \{H/\mathfrak{p}H : \mathfrak{p} \in H\text{-prim}(S)\},$$

a collection of finite dimensional algebras over H-Prim(S). In large measure, to achieve our (Aim) of understanding the finite dimensional representations of H it will be enough to understand $Alg(H)$. For *if V is an irreducible H-module then V is a module for one and only one of the algebras in $Alg(H)$*. This follows from the surjectivity of Ψ and the description of $\Psi^{-1}(\mathfrak{p})$ given in the lemma. Moreover, let $\mathfrak{p}$ be the H-primitive ideal of S with $\mathfrak{p}V = 0$. *Then every finitely generated H-module with socle V is annihilated by a power of $\mathfrak{p}H$*. This follows from [40, 4.2.2] once one observes (using an easy adaptation of the proof of [45, Theorem 11.2.3]) that $\mathfrak{p}H$ satisfies the Artin-Rees property.

4.3. **The restricted Hopf algebra.** Amongst the algebras in $Alg(H)$ there is one distinguished member, namely the *restricted algebra*, got by taking for $\mathfrak{p}$ the augmentation ideal S^+ of S. We denote this factor by $\overline{H}$, so

$$\overline{H} = H/S^+H = H/HS^+ = H/(H^+ \cap S)H = H//S,$$

where the notation on the extreme right was introduced in [52]. Thus $\overline{H}$ is a finite dimensional Hopf algebra. In case (A), recall from (2.2) that $S = KA$, so that $\overline{H}$ is $K(\Gamma/A)$, the group algebra of Γ/A. In case (B), $\overline{H}$ is the restricted enveloping algebra of $\mathfrak{g}$, [51, p.192]. In the case (C) of quantised enveloping algebras, $\overline{H} = \overline{U_\epsilon(\mathfrak{g})}$, (sometimes called *a restricted quantised enveloping algebra of* $\mathfrak{g}$), is given by generators and relations as in (2.1)(C), augmented by equating ℓth powers of K_i to 1, $(1 \leq i \leq n)$, and ℓth powers of all positive and negative roots to 0. The resulting Hopf algebra has dimension $\ell^{dim(\mathfrak{g})}$, thanks to the PBW theorem for $U_\epsilon(\mathfrak{g})$, [12,

Theorem 19.1]. Moreover, $\overline{U_\epsilon(\mathfrak{g})}$ is isomorphic to (a central quotient of) Lusztig's finite dimensional Hopf algebra attached to ϵ and $\mathfrak{g}$, [39, 6.5]. (A proof of this last statement is given in [26, Proposition 2 and Remark 3].) There is a similar story in case (D), the resulting finite dimensional Hopf algebra $\overline{H} = \overline{O_\epsilon[G]}$ having dimension $\ell^{dim(G)}$ [11, Theorem 7.2].

4.4. **The members of $Alg(H)$.** Experience with the four classes of examples (A), (B), (C) and (D) suggests that it is fruitful to view the members of $Alg(H)$ as deformations of the restricted algebra $\overline{H}$, with $\overline{H}$ itself having the most complex structure amongst the members of $Alg(H)$. There are however limits to the extent to which this viewpoint can be made precise - for example, when S is central and H is a free S-module of rank t then clearly every member of $Alg(H)$ has dimension t; but when (as in case (A)) S is not central the K-dimension of the members of $Alg(H)$ will not usually be constant, although certainly bounded above. Despite this caveat the members of $Alg(H)$ do share some common properties:

Theorem. *Assume (H1), (H5) and (H6). Then the algebras in $Alg(H)$ are Frobenius algebras.*

This can be deduced easily from (6) and the persistence of the property of being a Frobenius extension under passage to suitable factor rings. For example, one can use [15, Proposition 2.2(1) and Theorem 2.3].

4.5. **Stratification and the cardinality of $Alg(H)$.** Recall that *a Poisson bracket* on a commutative K-algebra A is a Lie bracket $\{-,-\}: A \otimes A \longrightarrow A$ which satisfies the Liebniz identity $\{fg, h\} = f\{g,h\} + \{f,h\}g$, for all $f, g \in A$. To such a bracket on a smooth affine $\mathbb{C}$-algebra A there is associated a decomposition of Maxspec(A) as a disjoint union of symplectic leaves [12, §11.1]. In particular, such a Poisson bracket is defined on the sub-Hopf algebra S in cases (B), (C) and (D). (Naturally, in such a situation one wants the Hopf and Poisson structures to be compatible - the conditions needed, which are fulfilled in the above cases, are set out in [12, §11.2], for example.)

An important mechanism whereby a Poisson bracket may be defined on the centre of an algebra H occurs when H is a specialisation, say $H = U/hU$, with h a central non-zero divisor of the K-algebra U. For $\bar{z} \in Z(H), \bar{z} = z + hU$, one defines a derivation $D_{\bar{z}}$ of H by setting

$$D_{\bar{z}}(\bar{u}) = [z, u]h^{-1},$$

where $\bar{u} \in H, \bar{u} = u + hU$. Changing the choice of lift z of $\bar{z}$ alters $D_{\bar{z}}$ by an inner derivation only. Defining, for $\bar{z}, \bar{w} \in Z(H)$,

$$\{\bar{z}, \bar{w}\} := D_{\bar{z}}(\bar{w})$$

yields a Poisson structure on $Z(H)$. Indeed this is precisely what happens in cases (C) and (D).

The relevance of Poisson structures in the present context is made clear by an observation of [12, §11.2] that if $K = \mathbb{C}$, H and S satisfy (H1) and (H5), (with S central), and there is a Poisson structure of the above type, then

$$H/m_1H \cong H/m_2H \tag{9}$$

for maximal ideals m_1 and m_2 of S in the same symplectic leaf. If in addition one can find a group T of algebra automorphisms of H which preserve S and respect its Poisson structure, then (9) will remain true for ideals m_1 and m_2 lying in the same T-orbit of symplectic leaves. Under favourable circumstances there will be only finitely many such T-orbits of leaves, and thus only finitely many isomorphism classes in $Alg(H)$. In particular one has the following table:

Class and ref.	**Permitted class of algebras**	**Symplectic leaves (or analogue)**	**Poisson autos T**	**T-orbits of leaves parametrised by**
(B) [32] (Rem. (1))	$U(\mathfrak{g}), \mathfrak{g} = \mathrm{Lie}(G)$ classical semisimple, char$K = p > 0$, p very good	Maxspec(S) parametrised by $\mathfrak{g}^*$; symplectic leaves are the coadjoint orbits	$\{1\}$	Conjugacy classes of $\mathfrak{g}$ under action of G
(C) [12, §16]	All members $U_\epsilon(\mathfrak{g})$ of (C)	Maxspec(S) is a solvable Poisson algebraic group H which is an unramified cover $\sigma : H \longrightarrow G_0$ of the big cell G_0 in G	$\{1\}$	$\sigma^{-1}(\mathcal{C} \cap G_0)$, $\mathcal{C}$ a non-central conjugacy class in G
(D) [28]	All $\mathcal{O}_\epsilon(G)$ in (D), $K = \mathbb{C}$	$S \cong \mathcal{O}_\epsilon(G)$ Poisson Lie groups; leaves given by Bruhat decomposition of $G \times G$	Maximal torus in G	$W \times W$, where W is the Weyl group of G

Remarks. (1) In case (B), by $\mathfrak{g}$ being *classical semisimple* we mean that $\mathfrak{g}$ is the Lie algebra of the connected, simply connected semisimple algebraic group G. *A very good prime* is one for which $\mathfrak{g}$ is a direct sum of simple Lie algebras, see [32, §6.4(b)]. In fact similar results can be obtained when p is merely *good* [32, §6.4(a)], although we can no longer identify

$\mathfrak{g}$ with $\mathfrak{g}^*$. Note that all primes greater than 5 are good.

(2) It follows from the table that

the number of isomorphism classes in $Alg(H)$ is finite

in class(D) and in class (B) with the stated restrictions. For (B), we argue as follows. First, identify $\mathfrak{g}$ with $\mathfrak{g}^*$ using the non-degenerate trace form for $\mathfrak{g}$ [32, §§6.3,6.4]. Recall that there are only finitely many conjugacy classes of nilpotent elements of $\mathfrak{g}$ [30]. By a theorem of Friedlander and Parshall [19, Theorem 3.2] every algebra H/mH in $Alg(H)$ is Morita equivalent to one parametrised by a nilpotent character. Since it's easy to check that there are only finitely many Morita equivalent K-algebras of a fixed finite dimension, the claim follows.

(3) It seems very reasonable to conjecture that an analogous result to that in the preceding remark applies also in case (C); but no such result has yet appeared in the literature. Nor do I know what happens in case (B) away from the classical semisimple Lie algebras.

(4) Case (A), $H = K\Gamma$ isn't included in the above table because it's not clear to me how to incorporate those group algebras (even for $K = \mathbb{C}$) in the same symplectic leaf formalism as the other cases. It should nevertheless be true that for a given group algebra H, $Alg(H)$ will also in this case contain only finitely many isomorphism classes, and in fact I've checked this when Γ is the semidirect product of A by a finite group. Another noteworthy special case: when the characteristic of K is coprime to the order of Γ/A all the algebras in $Alg(H)$ are semisimple, so it's trivial that there are only finitely many possible isomorphism classes.

It's natural to look for a common proof of the above finiteness results. We therefore ask

Question D: Given H satisfying (H1), (H5) and (H6), under what conditions are there only finitely many isomorphism classes of algebras in $Alg(H)$?

4.6. **The geometric triad.** We continue to assume that H is a K-algebra satisfying (H1), (H5) and (H6), as in (4.1). The discussion in the previous paragraphs suggests that our approach to the representation theory of H should be constructed using the interplay between the finite dimensional representation theory of H and each of:

the geometry of $Z(H)$

and

the geometry of S.

Note that the geometry of S should be taken to include its Hopf structure as well as whatever compatible symplectic structure it may possess. Of course, connections with $Z(H)$ will only be significant when H is a finitely generated $Z(H)$-module, as is the case in each of our classes of examples (A), ..., (D). We'll thus want to impose in the rest of this paper the hypothesis

(H7): *H is a finitely generated $Z(H)$-module.*

It's possible that (H7) is already a consequence of (H1), (H2) and (H4). That is, we ask

Question E: Is every affine Noetherian Hopf K-algebra which satisfies a polynomial identity a finitely generated module over its centre?

It may perhaps be more reasonable to ask if (H7) is always true for a Hopf algebra containing a large commutative sub-Hopf algebra - that is, does (H7) follow from (H1) and (H5)? A possible route from these hypotheses to (H7) would follow the group algebra case (A), where, when $K\Gamma$ is prime, the centre of $K\Gamma$ is precisely the K-algebra of Γ/A-invariants of A. Thus we're led to ask whether, in the notation of (4.1) and (4.3), S is a finite module over the algebra $S^{\overline{H}}$ of $\overline{H}$-invariants of S? Special cases of this question are known to have a positive answer. For instance it's true that S is a finite $S^{\overline{H}}$-module if $\overline{H}$ is cocommutative [41, Theorem 4.2.5] or if $\overline{H}$ is semisimple [41, Theorem 4.3.7]; but the general case is open.

5. Connecting representations with geometry

5.1. In §5 I will examine some of the strands connecting the three themes in (4.6). I will continue throughout to assume hypotheses (H1), (H5) and (H7). That is, H is a Hopf K-algebra which is a finite module over a commutative affine normal sub-Hopf algebra S, and H is a finitely generated module over its centre $Z(H)$. In addition, for technical simplicity in (5.2) and (5.3) we'll assume that

(H8): *H is a prime ring.*

This hypothesis could if we so wished be avoided with a little extra care. It is satisfied in cases (B), (C) and (D) (where in all cases H is a domain); it holds in case (A) if and only if Γ has no non-trivial finite normal subgroups [45, Theorem 4.2.10].

5.2. **The Azumaya locus.** Let m be the PI-degree of H, so that m is the maximal K-dimension of irreducible H-modules [40, 13.7.2], and by [40, 13.6.7, 13.3.5]

$$\dim_{Q(Z(H))} Q(H) = m^2, \tag{10}$$

where $Q(R)$ denotes the Goldie quotient ring of the (prime Noetherian) ring R. The *Azumaya locus* of H is

$$\mathcal{A}(H) := \{\mathfrak{p} \in \operatorname{Maxspec}(Z(H)) : H/\mathfrak{p}H \cong M_m(K)\},$$

(where $M_t(R)$ denotes the ring of $t \times t$ matrices with entries from the ring R). Since H is a Jacobson ring [40, 13.10.3] its maximal ideals intersect in zero and so it's easy to see that $\mathcal{A}$(H) is a non-empty open (and so dense) subset of Maxspec($Z(H)$). A first step towards achieving our (Aim) of (4.2) is thus to identify this subset, so that we can thereafter concentrate efforts on its complement, the (proper, closed) subset of Maxspec($Z(H)$) over which the representation theory is non-trivial.

5.3. It's worth noting that, for $\mathfrak{p} \in \operatorname{Maxspec}(Z(H))$,

$$\mathfrak{p} \in \mathcal{A}(H) \Longleftrightarrow H/\mathfrak{p}H \text{ is semisimple Artinian.} \tag{11}$$

For the left to right implication is trivial here; and the converse is also true because if $\mathfrak{p}H$ is semimaximal then the maximal ideals of H which lie over $\mathfrak{p}$ are in separate blocks by [33, Proposition 5.3.12], so that, by Mueller's theorem [23, Theorem 11.20], there can only be one such maximal ideal. That is, $H/\mathfrak{p}H$ is simple Artinian, so that $H/\mathfrak{p}H \cong M_t(K)$ for some $t \geq 1$. Since the PI-degree of H bounds above that of $H/\mathfrak{p}H$,

$$t \leq m;$$

on the other hand

$$\dim_K(H/\mathfrak{p}H) \geq m^2,$$

as follows from (10) by an application of Nakayama's lemma to the $Z(H)_{\mathfrak{p}}$-module $H_{\mathfrak{p}}$. This proves (11).

Consider for a moment the special case where S is a central subalgebra of H, as in cases (B), (C) and (D). Then it follows at once from (11) that

if $\mathfrak{p} \in$ *Maxspec*(S) *and* $H/\mathfrak{p}H$ *is semisimple Artinian then* $\hat{\mathfrak{p}} \in \mathcal{A}(H)$ *for all maximal ideals* $\hat{\mathfrak{p}}$ *of* $Z(H)$ *which lie over* $\mathfrak{p}$.

Here, since $S \subseteq Z(H) \subseteq H$, the map Ψ of (4.2) factorises as

$$\Psi : \begin{array}{ccccc} \mathrm{Prim(H)} & \xrightarrow{\pi} & \mathrm{Prim}(Z(H)) & \xrightarrow{\eta} & \mathrm{Prim}(S) \\ P & \longrightarrow & P \cap Z(H) = \mathfrak{p} & \longrightarrow & \mathfrak{p} \cap S; \end{array}$$

and if we set

$$\mathfrak{M}(S) := \{\mathfrak{p} \in \mathrm{Maxspec}(S) : \mathfrak{p}H \text{ semimaximal}\},$$

then the above statement can be reformulated as

$$\eta^{-1}(\mathfrak{M}(S)) \subseteq \mathcal{A}(H). \tag{12}$$

The inclusion (12) can be strict. For instance in case (B) let $\mathfrak{p}$ be a maximal ideal of S corresponding to a regular nilpotent character. Then $\eta^{-1}(\mathfrak{p}) \in \mathcal{A}(H)$, but $\mathfrak{p} \notin \mathfrak{M}(S)$ [19, Theorem 4.3].

5.4. In many cases the Azumaya locus can be linked with the geometry of $Z(H)$ using the following result. We write, for a commutative Noetherian ring A,

$$\mathfrak{S}(A) := \{\mathfrak{m} \in \mathrm{Maxspec}(A) : A_{\mathfrak{m}} \text{ not regular}\}$$

Proposition. (Lebruyn, [36, Proposition 5], [7, Theorem 3.8]) *Let Λ be a prime Noetherian ring which is a finite module over its centre Z. Suppose that Λ is Auslander-regular and Cohen-Macaulay. Suppose also that (†) $\Lambda_{\mathfrak{p}}$ is Azumaya over $Z_{\mathfrak{p}}$ for all height one primes $\mathfrak{p}$ of Z. Then*

$$\mathfrak{S}(Z) = \mathrm{Maxspec}(Z) \quad \setminus \quad \mathcal{A}(\Lambda).$$

The definitions of the Auslander-regular and Cohen-Macaulay conditions can be found in (3.2). In view of Proposition 3.2 and Theorem 3.3 the Hopf algebras in classes (B), (C) and (D) satisfy the homological conditions of the above proposition. It therefore becomes important to know whether (†) holds for the algebras in these classes. This question is answered in [7] for classes (C), (D) and part of class (B). The upshot of these case by case analyses is

Theorem. [7, Theorem C] *Let H be one of the algebras in classes (B), (C) or (D), and write Z for $Z(H)$. Then*

$$\mathfrak{S}(Z) = Maxspec(Z) \quad \setminus \quad \mathcal{A}(\Lambda),$$

provided H satisfies one of the following three hypotheses:

1. *$H \in$ (C);*
2. *$H \in$ (D);*
3. *H is the enveloping algebra of the Lie algebra $\mathfrak{g}$ of a connected, simply connected algebraic group G over K, such that*
   (a) p is a good prime for G, (so $p > 5$ will do),
   and (b) G has no component of type A_r with $r \equiv -1 \pmod{p}$.

5.5. **Remarks about Theorem 5.4.** (1) Theorem 5.4(i) verifies a conjecture of De Concini and Kac [10, Conjecture 5.2(c)], and Theorem 4.3(iii) partially answers a question of Premet [46, §4.4, Question 2]. Since this latter question remains open I restate it here as

Question F:(Premet) Determine those algebras in class (B) for which the conclusion of Theorem 5.4 is true.

(2) In connection with Theorem 5.4(3) and Question F note that not

all algebras in class (B) satisfy the conclusion of the theorem. For example, let K have characteristic 2 and $\mathfrak{g} = \mathbf{sl}(2, K)$, so that $H = U(\mathfrak{g}) = K[x, y, z : [x, y] = z, z \text{ central}]$, $\mathfrak{g}$ being isomorphic to the Heisenberg algebra. Then $Z(H) = K[x^2, y^2, z]$, a polynomial algebra, so that $\mathfrak{S}(Z(H)) = \emptyset$, but H is not Azumaya, so that

$$\mathfrak{S}(Z(H)) \subsetneqq \text{Maxspec}(Z(H)) \setminus \mathcal{A}(H).$$

(3) As mentioned in 5.4 part of the deduction of Theorem 5.4 from Proposition 5.4 consists in a case by case check that hypothesis (†) of the proposition is satisfied by the relevant algebras. This prompts

Question G: Is there a convenient "ring-theoretic" means of recognising when an algebra (prime, Noetherian, finite over its centre) satisfies hypothesis (†)?

5.6. **The (generalised) Kac-Weisfeiler conjecture.** Theorem (5.4) shows that, in certain circumstances, a geometric hypothesis on a point of Maxspec$(Z(H))$ yields a consequence for the vector space dimension of an irreducible H-module annihilated by the maximal ideal of that point. The Kac-Weisfeiler conjecture proposed a connection of a similar nature for *all* irreducible H-modules. In its original form of 1971 it is now a beautiful theorem:

Theorem. (Premet, [46, Corollary 3.11],) *Let G be a connected reductive group over the algebraically closed field K of characteristic $p > 0$. Assume G has a simply connected derived subgroup, and that p is a good prime for G, (e.g. $p > 5$). Let $\mathfrak{g} = Lie(G)$, and set $H = U(\mathfrak{g})$ and S the central subHopf algebra of H as in (3). Let V be an irreducible H-module which is faithful for $\mathfrak{g}$, let $\mathfrak{p}$ be the annihilator ideal of V in S and let $\Omega(\mathfrak{p})$ denote the orbit of $\mathfrak{p}$ under the coadjoint action of G on $\mathfrak{g}^*$. Then*

$$p^{\frac{1}{2}dim(\Omega(\mathfrak{p}))} | dim_K(V).$$

Analogues of the Kac-Weisfeiler conjecture are

true in Case (D), (De Concini, Procesi, [13]);

conjectured in case (C), (De Concini, Kac, Procesi [14, §6.8]).

Moreover, a similar result can be formulated and proved in case (A), at least when Γ splits over A [6]. We may thus ask

Question H: Is there a (true!) formulation of the Kac-Weisfeiler conjecture for Hopf algebras H satisfying (say) (H1) and (H5)?

5.7. **Irreducible modules.** Of course, an important problem subsumed in our (Aim) of (4.2) is that of obtaining a description of all the irreducible H-modules for an algebra H in one of our classes of interest. Such a description should - at least - say how many distinct isomorphism classes of irreducibles there are for each algebra in Alg(H), and give the dimensions of these modules. Naturally, at this level of detail an axiomatic approach of the sort discussed in this paper will quickly run into the sand, and one is theefore obliged to treat each class separately. Nevertheless it may well be that some common features will be found; so although it's not central to the theme of this paper I will briefly indicate where matters stand in this endeavour for classes (B), (C) and (D).

The irreducible modules of a restricted, completely solvable Lie algebra $\mathfrak{g}$ are all induced from 1-dimensional modules over restricted subalgebras of $\mathfrak{g}$ [34, Theorem 1b], [50, Satz 3b], and all the irreducible modules in a block have the same dimension [53]. For a survey of these and related results, see [16].

When $\mathfrak{g}$ is classical semisimple there are not many general results to put beside Premet's theorem and its immediate consequences, although there are many interesting calculations and partial results - see, for example, the excellent surveys [29] and [32]. At the present time the dimensions of the irreducible $\overline{U(\mathfrak{g})}$−modules are not known in general; nor is it known whether, for arbitrary $\chi \in \mathfrak{g}^*$, there exists an irreducible $U(\mathfrak{g})/m_\chi U(\mathfrak{g})$-module of "minimal" dimension - that is, equal to the lower bound given by Premet's theorem, (5.4). For simple Lie algebras of non-classical type almost nothing is known about the irreducible modules apart from isolated calculations - for example, even the PI degree of $U(\mathfrak{g})$ has not in general been determined.

For the quantised enveloping algebras (C) very little seems to be known when one moves away from the restricted quantised enveloping algebra $\overline{U_\epsilon(\mathfrak{g})}$ and from the Azumaya locus; but the expectation is that the description of the irreducibles eventually obtained will be similar to that for enveloping algebras of classical simple Lie algebras in positive characteristic.

In case (D), quantised function algebras, the task of describing the irreducible modules has essentially been completed, thanks to the work in [13]. Not only are the dimensions and numbers of the irreducible modules for a given member of $Alg(H)$ known (in terms of Weyl group data), but a sort of inductive procedure can be described for constructing the modules.

5.8. **Support varieties and representation type.** Premet's proof of Theorem 5.6 makes crucial use of the theory of support varieties. This technique has had considerable success in the modular representation theory of finite groups over the past twenty years at the hands of Benson, Carlson and others, building on ideas of Quillen from 1971. [9] is a recent survey of the theory in this setting. In a series of papers beginning in the mid-eighties with [18, 17, 19], Friedlander and Parshall adapted these ideas to the study of finite dimensional restricted Lie algebras. In fact, the basic ideas are quite straightforward and can be formulated in a fairly general setting. Thus we assume

(H1): *H is a K-affine K-Hopf algebra, and*

(H9): *H is a finite module over an affine central sub-Hopf algebra S.*

Note that such an H satisfies (H2), (H3), (H4) (using Proposition (2.3)), (H5), (H6) and (H7); but (H9) rules out case (A). We can construct, as in (4.3), the restricted Hopf algebra $\overline{H} = H//S = H/S^+H$, which is finite dimensional, and consider

$$E \quad := \quad \bigoplus_{i\geq 0} \operatorname{Ext}^i_{\overline{H}}(K, K),$$

which is a ring under Yoneda product. Since $\overline{H}$ is a Hopf algebra, E is a graded-commutative K-algebra [21, Corollary 5.6]. In particular, the subalgebra

$$E^{ev} \quad := \quad \bigoplus_{i \geq 0} \operatorname{Ext}^{2i}_{\overline{H}}(K, K)$$

is central in E. Suppose now that $\mathfrak{m}$ is a maximal ideal of S and M is a finite dimensional $H/\mathfrak{m}H$-module. Then

$$E_M \quad := \quad \operatorname{Ext}^*_{H/\mathfrak{m}H}(M, M) \quad \cong \quad \operatorname{Ext}^*_{\overline{H}}(K, M \otimes M^*)$$

(where the isomorphism follows easily from the fact that, if $X \longrightarrow I^{\cdot}$ is an $H/\mathfrak{m}H$-injective resolution of X then $X \otimes X^* \longrightarrow I^{\cdot} \otimes X^*$ is an $\overline{H}$-injective resolution of $X \otimes X^*$ [19, Proposition 5.1]). Via this isomorphism and the cup product E_M admits a structure as E^{ev}-module, through which we can hope to obtain information about M. The theory can however only take flight in cases where we have a positive answer to the following question.

Question I: Let A be a finite dimensional K-Hopf algebra and let M and N be finite dimensional A-modules. Is $E^{ev} = \bigoplus_{i \geq 0} \operatorname{Ext}^{2i}_A(K, K)$ an affine K-algebra, and is $\bigoplus_{i \geq 0} \operatorname{Ext}^i_A(M, N)$ a finitely generated E^{ev}-module?

We postpone to (5.8) a review of the current state of play for Question I, first indicating how to proceed in cases where we know that

(H10): *$\overline{H}$ affords positive answers to Question I.*

Given (H10) we can consider the subvariety of $\operatorname{Maxspec}(E^{ev})$ supported by E_M, namely

$$\mathcal{V}(M) := \operatorname{Maxspec}\big(E^{ev}/Ann(E_M)\big),$$

where $Ann(E_M)$ denotes the ideal of elements of E^{ev} which annihilate E_M. The *complexity* of the $H/\mathfrak{m}H$-module M is defined to be

$$c(M) := \dim\big(\mathcal{V}(M)\big).$$

Finally, we define the *complexity* of $H/\mathfrak{m}H$ to be

$$\mathcal{C}(H/\mathfrak{m}H) := \max\big\{c(M) : M \text{ a fin. dim. } H/\mathfrak{m}H - \text{module}\big\}.$$

It's not hard to relate the complexity $c(M)$ of an $H/\mathfrak{m}H$-module as just defined to the rate of growth of the sequence $\{\dim_K(P_j) : j \geq 0\}$, where

$$\ldots \longrightarrow P_j \longrightarrow \ldots \longrightarrow P_0 \longrightarrow M \longrightarrow 0$$

is a minimal $H/\mathfrak{m}H$-projective resolution of M. That is, one shows that there is a non-negative constant A such that

$$\dim_K(P_j) \leq Aj^{(c(M)-1)}$$

for all $j \geq 0$, and that $c(M)$ is the least integer with this property; see [18, Proposition 3.2] for the case where M is a restricted $U(\mathfrak{g})$-module, from which a general proof can easily be constructed.

The connection with representation type is afforded by a result of Rickard [48, Theorem 2] on the representation type of self-injective algebras. Adapted to the present setting, noting that the algebras $H/\mathfrak{m}H$ are self-injective by Theorem 4.4, Rickard's theorem translates to (i) of the following result. When coupled with the other - more elementary - parts of this theorem, which are culled from [47, Lemma 5.1] and [24, Proposition 10], this shows that the only uncertainties in connecting complexity with representation type occur at complexities 1 and 2.

Theorem. *Assume (H1), (H9) and (H10). Let* $\mathfrak{m}$ *be a maximal ideal of* S.

(i) Suppose that $\mathcal{C}(H/\mathfrak{m}H) \geq 3$. *Then* $H/\mathfrak{m}H$ *has wild representation type.*

(ii) $H/\mathfrak{m}H$ *is semisimple Artinian if and only if* $\mathcal{C}(H/\mathfrak{m}H) = 0$.

(iii)If $H/\mathfrak{m}H$ *has finite representation type then* $\mathcal{C}(H/\mathfrak{m}H) = 1$.

(iv) If $\mathcal{C}(H/\mathfrak{m}H) = 2$ *then* $H/\mathfrak{m}H$ *has tame or wild representation type.*

5.9. **Finite generation of cohomology.** I summarise the main cases where positive answers are known to Question I in the following table. Not all the classes of Hopf algebra listed in the left hand column are mutually exclusive - I've tried rather to acknowledge chronological developments to some extent, at the cost of some duplication.

Class of finite dim. Hopf algebra A	E^{ev} **affine and** E_M **a f.g.** E^{ev}**–module**	**Structure of** E^{ev} **(if known)**	**References**
$A = KG, G$ a finite group	✓		[9]
A commutative	✓		See [20, Introduction]
$A = \overline{U(\mathfrak{g})}$ $U(\mathfrak{g}) \in$ (B)	✓		[17, (1.4)]
$A = \overline{U(\mathfrak{g})}$, $\mathfrak{g}$ classical semisimple, $\mathrm{char}K = p \gg 0$	✓	$E^{ev} \cong K\mathcal{N}$, the functions on the nilpotent cone $\mathcal{N}$ of $\mathfrak{g}$	[31] [18]
$A = \overline{U_\epsilon(\mathfrak{g})}, U_\epsilon(\mathfrak{g})$ in (C), $\ell \gg 0$	✓	$E^{ev} \cong K\mathcal{N}$	[21]
$A = O_\epsilon[G], O_\epsilon[G]$ in (C), $\ell \gg 0$	✓	$E^{ev} \cong \mathrm{Sym}(V)$, vector space V of dim. $2N$	[25] [24]
A cocommutative	✓		[20]

When $H = U(\mathfrak{g})$ or $H = O_\epsilon[G]$ the relevant parts of the above table have been used in [47] and [27] respectively to work towards the determination of the representation type of the members of $Alg(H)$. In both cases the complexity of all members of $Alg(H)$ is known, and can be described in terms of (respectively) conjugacy classes in $\mathfrak{g}$ and elements of the Weyl group, thanks to the table in (4.5). All the occurrences of algebras of finite type in $Alg(H)$ are known and described for both these classes of algebra H, and the only remaining ambiguity at the time of writing lies in the possibility that some tame algebras might appear at complexity 2.

REFERENCES

[1] Smith S.P. Zhang J.J. Ajitabh, K. Auslander-Gorenstein rings. preprint, U. of Washington, 1997.

[2] J.-E. Bjork. The Auslander condition on Noetherian rings. In *Seminaire Dubreil-Malliavin 1987-88*, volume 1404 of *Lecture Notes in Math.*, pages 137–173, Heidelberg-Berlin-New York, 1989. Springer-Verlag.

[3] J.-E. Bjork and E.K. Ekstrom. Filtered Auslander-Gorenstein rings. In *Colloque en l'Honneur de J. Dixmier*, pages 424–448. Birkhauser, Basel, 1990.

[4] K.A. Brown. Hopf algebras satisfying a polynomial identity. in preparation.

[5] K.A. Brown. Fully bounded Noetherian rings of finite injective dimension. *Quart. J. Math. (Oxford)*, 41:1–13, 1990.

[6] K.A. Brown. Unpublished notes on skew group ring representations. 1996.

[7] K.A. Brown and K.R. Goodearl. Homological aspects of Noetherian PI hopf algebras, and irreducible modules of maximal dimension. *J. Algebra*, 1998. to appear.

[8] K.A. Brown and C.R. Hajarnavis. Injectively homogeneous rings. *J. Pure Appl. Algebra*, 51:65–77, 1988.

[9] J.F. Carlson. *Modules and Group Algebras.* Lectures in Mathematics. Birkhäuser, Basel, 1996.

[10] C. De Concini and V.G. Kac. Representations of quantum groups at roots of 1. In A. Connes et al, editor, *Operator Algebras, Unitary Representations, Enveloping Algebras and Invariant Theory*, number 92 in Progress in Mathematics, pages 471–506. Boston, Birkhäuser, 1990.

[11] C. De Concini and V. Lyubashenko. Quantum function algebras at roots of 1. *Adv. Math.*, 108:205–262, 1994.

[12] C. De Concini and C. Procesi. Quantum Groups. Springer Lecture Notes in Mathematics 1565. 31-140.

[13] C. De Concini and C. Procesi. Quantum Schubert cells and representations at roots of 1. In G I Lehrer, editor, *Algebraic groups and Lie groups*, number 9 in Australian Math. Soc. Lecture Series. Cambridge University Press, Cambridge, 1997.

[14] Kac V.G. De Concini, C. and C. Procesi. Quantum coadjoint action. *J. Amer. Math. Soc.*, 5:151–190, 1992.

[15] R. Farnsteiner. Note on Frobenius extensions and restricted Lie superalgebras. *J. Pure Appl. Algebra*, 108:241–256, 1996.

[16] J. Feldvoss. Homological topics in the representation theory of restricted Lie algebras. *Contemp. Math.*, 194:69–119, 1996.

[17] E. Friedlander and B. Parshall. Cohomology of infinitesimal and discrete groups. *Math. Ann.*, 273:353–374, 1986.

[18] E.M. Friedlander and B.J. Parshall. Geometry of p-unipotent Lie algebras. *J. Alg.*, 109:25–45, 1987.

[19] E.M. Friedlander and B.J. Parshall. Deformations of Lie algebra representations. *Amer. J. Math.*, 112(3):375–395, 1990.

[20] E.M. Friedlander and A. Suslin. Cohomology of finite group schemes over a field. *Invent. Math.*, 127(2):209–270, 1997.

[21] V. Ginzburg and S. Kumar. Cohomology of quantum groups at roots of unity. *Duke Mathematical Journal*, 69(1):179–198, 1993.

[22] K.R. Goodearl and T.H. Lenagan. Catenarity in quantum algebras. *J. Pure Appl. Algebra*, 111:123–142, 1996.

[23] K.R. Goodearl and R.B. Warfield. *An Introduction to Noncommutative Noetherian Rings*, volume 16 of *London Math. Soc. Student Texts*. Cambridge University Press, Cambridge, London, 1989.

[24] I. Gordon. Addendum to cohomology of quantized function algebras at roots of unity. University of Glasgow preprint series, no 57, 1997.

[25] I. Gordon. Cohomology of quantized function algebras at roots of unity. University of Glasgow preprint series, no.14, 1997.

[26] I. Gordon. Quantized function algebras at roots of unity, path algebras and representation type. University of Glasgow preprint series, no 48, 1997.

[27] I. Gordon. The representation theory of quantised function algebras. University of Glasgow preprint series, 1997.

[28] T.J. Hodges and T. Levasseur. Primitive ideals of $\mathbb{C}_q[sl(3)]$. *Comm. Math. Physics*, 156:581–605, 1993.

[29] J.E. Humphreys. Modular representation of simple lie algebras. Bull. Amer. Math. Soc., to appear.

[30] J.E. Humphreys. *Conjugacy Classes in Semisimple Algebraic groups*, volume 43 of *Math. Surveys Monographs*. Amer. Math. Soc., Providence, RI, 1995.

[31] J.C. Jantzen. Restricted Lie algebra cohomology. Springer Lecture Notes in Mathematics 1271, 1987. 91-108.

[32] J.C. Jantzen. Representations of Lie algebras in prime characteristic. In *Proceedings of Montreal Conf. in Lie Theory*, 1997. to appear.

[33] A.V. Jategaonkar. *Localization in Noetherian Rings*, volume 98 of *London Math. Soc. Lecture Note Series*. Cambridge University Press, Cambridge, London, 1986.

[34] V.G. Kac and B.J. Weisfeiler. Irreducible representations of Lie p-algebras. *Funct. Anal. Appl.*, 5:28–36, 1971.

[35] H.F. Kreimer and M. Takeuchi. Hopf algebras and Galois extensions of an algebra. *Indiana U. Math. J.*, 30:675–692, 1981.

[36] L. Le Bruyn. Central singularities in quantum spaces. *J. Algebra*, 177:142–153, 1995.

[37] T. Levasseur. Some properties of non-commutative regular graded rings. *Glasgow Math. J.*, 34:277–300, 1992.

[38] M.E. Lorenz and M. Lorenz. On crossed products of Hopf algebras. *Proc. Amer. Math. Soc.*, 123:33–38, 1995.

[39] G. Lusztig. Finite dimensional Hopf algebras arising from quantized enveloping algebras. *J. Amer. Soc.*, 3:257–296, 1990.

[40] J.C. McConnell and J.C. Robson. *Noncommutative Noetherian Rings.* Wiley-Interscience, New York, 1987.

[41] S. Montgomery. *Hopf Algebras and their Actions on Rings*, volume 82 of *CBMS Regional Conf. Series.* Amer. Math. Soc., Providence, R.I., 1993.

[42] S. Montgomery and H.-J. Schneider. Prime ideals in Hopf Galois extensions. preprint, 1997.

[43] T. Nakayama and T. Tsuzuku. On Frobenius extensions I. *Nagoya Math. J.*, 17:89–110, 1960.

[44] D. Passman. *Infinite Crossed Products.* Academic Press Inc., 1989.

[45] D.S. Passman. *The Algebraic Sructure of Group Rings.* Wiley-Interscience, New York, 1977.

[46] A. Premet. Irreducible representations of Lie algebras of reductive groups and the Kac-Weisfeiler conjecture. *Invent. Math.*, 121(1):79–117, 1995.

[47] A. Premet. Complexity of Lie algebra representations and nilpotent elements of the stabilizers of linear forms. Preprint, 1996.

[48] J. Rickard. The representation type of self-injective algebras. *Bull. London. Math. Soc.*, 22:540–546, 1990.

[49] J.T. Stafford and J.J. Zhang. Homological properties of (graded) Noetherian PI rings. *J. Algebra*, 168:988–1026, 1994.

[50] H. Strade. Darstellungen auflosbarer Lie -p-algebren. *Math. Ann.*, 232:15–32, 1978.

[51] H. Strade and R. Farnsteiner. *Modular Lie Algebras and their Representations*, volume 116 of *Monographs and Textbooks in Pure and Appl. Mathematics.* Marcel Dekker Inc., New York-Basel, 1988.

[52] M. Sweedler. *Hopf Algebras.* Benjamin, New York, 1969.

[53] D. Voigt. *Induzierte Darstellungen in der Theorie der Endlichen Algebraischen Gruppen*, volume 592 of *Lecture Notes in Math.* Springer-Verlag, Heidelberg-Berlin-New York, 1977.

[54] W.C. Waterhouse. *Introduction to Affine Group Schemes.* Number 66 in Graduate texts in Mathematics. Springer-Verlag, Heidelberg-Berlin-New York, 1979.

DEPARTMENT OF MATHEMATICS, UNIVERSITY OF GLASGOW, GLASGOW G12 8QW, UNITED KINGDOM

E-mail address: k.brown@maths.gla.ac.uk

Contemporary Mathematics
Volume **229**, 1998

Finite Representation Type and Periodic Hochschild (Co-)Homology

Ragnar-Olaf Buchweitz

Dedicated to the memory of Maurice Auslander

CONTENTS

Introduction

We want to understand the relationship between the work of Auslander-Reiten in [**7**] and that of Ringel-Schofield in [**34**] where either authors show that the Auslander-Reiten transformation on the stable category of the preprojective algebra of a Coxeter-Dynkin diagram is periodic of period dividing 6. The treatments given could hardly be more different at first glance: Auslander-Reiten pull the result out of the thin air called categorical algebra, whereas Ringel-Schofield give a very concrete, but unpublished, proof deducing the result from an explicit description of the minimal resolution of the preprojective algebra over its enveloping algebra, producing along the way examples of algebras with periodic Hochschild (co-)homology.

Thus one may ask whether there is a general connection between finite representation type and periodic Hochschild (co-)homology.

1991 *Mathematics Subject Classification.* 18A25, 16G60, 16E40.

The author was partially supported by NSERC grant 3-642-114-80, and part of the work was done under the RiP-program of the Volkswagen Foundation at the Math. Forschungsinstitut Oberwolfach.

One link is essentially provided by the following two facts: For a representation finite category $\mathcal{B} = \mathrm{add}(M)$, the projection from the Auslander algebra $\Lambda = \mathrm{End}_{\mathcal{B}}(M)$ onto the stable Auslander algebra $\Gamma = \underline{\mathrm{End}}_{\mathcal{B}}(M)$ satisfies $\mathrm{Tor}_1^{\Lambda}(\Gamma, \Gamma) = 0$, that is, the ring homomorphism is *pseudoflat* in the terminology of [**12**, p.323]. Secondly, for almost all examples of rings Γ, for which the periodicity of the Auslander-Reiten transformation is shown in [**7**], one can realize Γ in such a way that the corresponding Auslander algebra Λ is (the completion of) an algebra whose Hochschild dimension is known to be 2. Taken together, these facts easily imply that Γ admits a quasi-periodic projective resolution of period 3 over its enveloping algebra with periodicity factor $\mathrm{Tor}_2^{\Lambda}(\Gamma, \Gamma)$. If this latter bimodule is invertible and of finite order in $\mathrm{Pic}\,\Gamma$ one obtains indeed periodic Hochschild (co-)homology.

A notable exception are stable Auslander algebras of maximal Cohen-Macaulay modules over odd dimensional hypersurfaces. Their Auslander-Reiten translation is periodic of period (dividing) 3, but we have not found an ambient algebra of Hochschild dimension 2.

In section 1 we give a detailed account of the results and explore the examples mentioned. Section 2 treats the simple and basic categorical background needed for our version of the Auslander-Reiten structure theorem and section 3 reviews functor categories. Here our point of view differs slightly from that of Auslander-Reiten: The restriction functor from modules over the stable category $\underline{\mathcal{B}}$ to modules over $\mathcal{B}$ admits always left and right adjoints, the question is when these adjoints preserve coherent modules. This also sheds some new light on the role of contravariantly finite subcategories as we explain there. Section 4 contains the formulation and proof of the structure theorem and establishes pseudoflatness on the categorical level. In section 5 the special features of modules over exact Frobenius categories are reviewed, we exhibit the complete resolutions of coherent functors and describe a relative Nakayama equivalence. Finite representation type is dealt with in section 6 and section 7 proves the Ringel-Schofield theorem from our point of view here, shows that the deformed preprojective algebras of extended Coxeter-Dynkin diagrams, introduced by B.Crawley-Boevey and M.Holland [**15**], are of Hochschild dimension at most 2 and that the corresponding algebras for ordinary Coxeter-Dynkin diagrams have quasi-periodic Hochschild (co-)homology of period 3.

Acknowledgements: I want to thank I.Reiten for numerous valuable discussions and suggestions and her tireless efforts to explain the finer points of representation theory to an (essentially) commutative algebraist. I also owe thanks to W.Crawley-Boevey not only for sending me a copy of [**34**] but also for his recent preprint [**14**] that arrived just a few days before the ultimate deadline for this submission and that forcefully explains the importance of pseudoflatness.

Finally I wish to thank B.Huisgen-Zimmermann who did such a marvellous job in organizing this conference and who insisted with almost infinite patience that this contribution should be written up, by no means an easy task as many people can confirm.

1. Results and Examples

The Auslander-Reiten structure theorem, formal extension of results accumulated in the sequence of papers [**1**], [**2**], [**5**], [**6**] and [**7**], takes the following form in case of finite representation type. Its hypotheses are explained and explored in detail in sections 2 and 3, the proof follows in (6.3), once the corresponding "metatheorems" (4.1), (4.3), (5.5) in general functor categories have been established.

THEOREM 1.1. *Let $\mathcal{B}$ be a representation finite additive category that is closed under direct summands and embedded as a full additive subcategory with plenty of projectives into an abelian category $\mathcal{A}$.*

If M is a representation generator of $\mathcal{B}$, denote $\Lambda = \mathrm{End}_{\mathcal{B}}(M)$ its Auslander algebra and $\Gamma = \underline{\mathrm{End}}_{\mathcal{B}}(M)$ the corresponding stable Auslander algebra.

1. *The ring Γ is right coherent* [1], *each finitely presented right Γ-module is of projective dimension at most 2 over Λ and*
$$\mathrm{Ext}^i_\Lambda(\ ,\Lambda) = 0 \qquad \text{for } i \neq 2$$
as functors on right Γ-modules.
2. *The ring epimorphism $\Lambda \to \Gamma$ is pseudoflat* [2] *and so*
$$\mathrm{Tor}^\Lambda_i(\ ,\Gamma) = 0 \qquad \text{for } i \neq 0, 2$$
as functors on right Γ-modules.
3. *The Γ-bimodule $L = \mathrm{Tor}^\Lambda_2(\Gamma,\Gamma)$ is flat as a left Γ-module, finite projective as a right Γ-module, and the exact functor $(\ )\otimes_\Gamma L \cong \mathrm{Tor}^\Lambda_2(\ ,\Gamma)$ represents the third syzygy functor Ω^3_Γ on the category of coherent right Γ-modules.*

If the category $\mathcal{B}$ is furthermore exact Frobenius with enough projective-injectives, then (1) – (3) hold mutatis mutandis for left Γ-modules and L is an invertible Γ-bimodule with inverse $L^{-1} = \mathrm{Ext}^2_\Lambda(\Gamma,\Lambda)$. Moreover, there is a functorial isomorphism on finitely presented Γ-modules
$$\mathrm{Hom}_\Gamma(\ ,L^{-1}) \cong \mathrm{Ext}^2_\Lambda(\ ,\Lambda)$$
so that $\omega_{\Gamma/\Lambda} = L^{-1}$ constitutes a relative dualizing module for $\Lambda \to \Gamma$.

To explain the allusion to the Grothendieck duality theory in the last sentence, note that the last functorial isomorphism in the theorem can be written in a more fanciful way as a (quasi-)isomorphism of exact functors on the derived category $\mathrm{D}^-(\Gamma)$ of the abelian category of (left or right) coherent Γ-modules

$$\mathbb{R}\mathrm{Hom}_\Gamma(\ ,\omega_{\Gamma/\Lambda}) \cong \mathbb{R}\mathrm{Hom}_\Lambda(\ ,\Lambda)[2] \tag{1}$$

and in this form it represents the duality theorem for the finite morphism $\Lambda \to \Gamma$ of (cohomological) codimension 2, if compared to the general theory in the commutative case [**23**]. But for coherent Γ-modules the theorem above gives

$$\mathrm{Ext}^i_\Gamma(\ ,\omega_{\Gamma/\Lambda}) = \mathrm{Ext}^{i+2}_\Lambda(\ ,\Lambda) = 0 \quad \text{for } i \neq 0,$$

and so the quasi-isomorphism (1) reduces to the stated isomorphism of functors. Nevertheless, if one looks at the categorical incarnation, it suggests that there is a deeper connection between the Auslander-Reiten theory of almost split sequences and Grothendieck duality theory. Just compare the categorical version of the duality theorem, (5.6), with [**7**, Prop. 2.3] for the obvious similarity.

If one strengthens the hypotheses by requiring that $\mathcal{B}$ is a *resolving* subcategory of an abelian category $\mathcal{A}$ with enough projectives, (that is, $\mathcal{B}$ is closed under extensions and kernels of epimorphisms and contains the projectives in $\mathcal{A}$, [**7**]), then the theorem indeed essentially evaluates [**7**] on a representation generator. The

[1] The ring Γ is (right) *coherent* if any finitely generated submodule of a finitely presented (right) module is already finitely presented.

[2] The ring epimorphism $\Lambda \to \Gamma$ is *pseudoflat* if $\mathrm{Tor}^\Lambda_1(\Gamma,\Gamma) = 0$, see [**12**, p.323].

decision to weaken the hypotheses is mainly motivated by aesthetics: If $\mathcal{B}$ is for example the category of maximal Cohen-Macaulay modules over a local Gorenstein ring, it is exact Frobenius with enough projective-injectives as well as closed under kernels of epimorphisms. But the last property is no longer true for the opposite category as cokernels of monomorphisms of such modules are rarely ever maximal Cohen-Macaulay. Thus the stronger assumptions are not necessarily shared by the opposite category whereas the conclusion is true for $\mathcal{B}$ iff it is true for $\mathcal{B}^{op}$.

Of course we pay a price for the generalization: As a German proverb says, the devil hides in the detail, and so we need to reexamine carefully the proof of the Auslander-Reiten structure theorem under these weaker hypotheses. This will be done in sections 2 through 5, and we will phrase and prove the functorial version of the theorem that needs no assumption on representation type.

The result restricts considerably the possible structure of a stable Auslander algebra of a representation finite exact Frobenius category and the following corollary gives a sample. Note that an endomorphism γ is a *quasi-unipotent automorphism* if $\mathrm{id} - \gamma^n$ is nilpotent for some n. For an endomorphism of a finite dimensional vectorspace, this is equivalent to all eigenvalues being (n'th) roots of unity.

COROLLARY 1.2. *Let Γ be a stable Auslander algebra of a representation finite exact Frobenius category $\mathcal{B}$ that has enough projective-injectives.*

1. *If the ring Γ is noetherian, then it is selfinjective artinian. In particular, if Λ is finite over a commutative noetherian ring, then Γ is a selfinjective Artin algebra.*
2. *If $\mathcal{B}$ is a Krull-Schmidt category and if Γ is an Artin algebra, then the invertible Γ-bimodule $L = \mathrm{Tor}_2^{\Lambda}(\Gamma, \Gamma)$ satisfies $L \cong {}_1\Gamma_{\gamma}$ for some quasi-unipotent algebra automorphism γ of Γ.*
3. *If Γ is Morita equivalent to a commutative Artin algebra R, then R is a finite product of rings of the form $V/\mathfrak{m}^{\nu}$ where V is a discrete valuation ring with maximal ideal $\mathfrak{m}$ and the exponent ν is 1 or 2.*

REMARK 1.3. In case the representation finite category $\mathcal{B}$ has almost split sequences the structure of a stable Auslander algebra is of course even further restricted in general, as the stable Auslander-Reiten quiver of $\mathcal{B}$ becomes then the ordinary quiver of Γ so that only variants of preprojective algebras of Coxeter-Dynkin diagrams can occur, see e.g. [**33**], [**21**], [**22**], [**16**], [**8**] for more definite results along these lines.

To describe succinctly the relationship to Hochschild (co-)homology and the result of Ringel-Schofield we introduce the following notion.

DEFINITION 1.4. A complex $(\cdots C_i \xrightarrow{d_i} C_{i-1} \cdots \to C_0 \to 0)$ of (right) modules over a ring A is *quasi-periodic of period n with periodicity factor L* if L is an A-bimodule and there is an isomorphism of complexes of (right) A-modules

$$\begin{array}{ccccccccc}
\cdots \longrightarrow & C_i \otimes_A L & \xrightarrow{d_i \otimes_A L} & C_{i-1} \otimes_A L & \longrightarrow \cdots \longrightarrow & C_0 \otimes_A L & \longrightarrow & 0 \\
& \downarrow \cong & & \downarrow \cong & \cdots & \downarrow \cong & & \\
\cdots \longrightarrow & C_{n+i} & \xrightarrow{d_{n+i}} & C_{n+i-1} & \longrightarrow \cdots \longrightarrow & C_n & \longrightarrow & 0
\end{array}$$

If A is a K-algebra over a commutative ring K, its *Hochschild dimension* is

$$\mathrm{HH}\dim_K A = \mathrm{projdim}_{A \otimes_K A^{op}} A\,.$$

The following result provides the link between the Auslander-Reiten structure theorem and Hochschild (co-)homology.

THEOREM 1.5. *Let k be a field and Λ a k-algebra of Hochschild dimension 2. If $\Lambda \to \Gamma$ is a pseudoflat ring epimorphism, then, with respect to the induced k-algebra structure, Γ admits a quasi-periodic projective resolution over $\Gamma \otimes_k \Gamma^{op}$ of period 3 with periodicity factor $L = \mathrm{Tor}_2^\Lambda(\Gamma, \Gamma)$.*

In particular, if the periodicity factor L is an invertible Γ-bimodule of finite order n in the Picard group Pic Γ, *then the Hochschild (co-)homology of Γ over k is periodic of period dividing $3n$. The period must be* even, *unless the characteristic of k equals 2 or Γ is a separable k-algebra.*

That in case of characteristic different from 2 the order of L cannot be odd unless Γ is separable over k uses M.Gerstenhaber's celebrated theorem [**19**] that the Hochschild cohomology ring $\mathrm{HH}(\Lambda/k, \Lambda)$ for an associative k-algebra Λ is graded commutative.

1.6. In the examples arising from finite representation type, the morphism $\Lambda \to \Gamma$ is always pseudoflat and sometimes Λ is a k-algebra of global dimension 2, but usually we don't know its Hochschild dimension. Nevertheless, an Auslander algebra occurs sometimes as a flat extension of an algebra Λ' of Hochschild dimension 2 such that the composed ring map $\Lambda' \to \Lambda \to \Gamma$ is still an epimorphism of rings. This is enough, as then Γ is pseudoflat over Λ': The change-of-rings spectral sequence shows that the natural map $\mathrm{Tor}^{\Lambda'}(\Gamma, \Gamma) \to \mathrm{Tor}^{\Lambda}(\Gamma, \Gamma)$ is an isomorphism.

We now look at examples.

1.7. *Reflexive Modules over Rational Double Points.* Let k be a field and $G \subseteq \mathrm{GL}(2, k)$ a finite subgroup such that the characteristic of k does not divide the order of G, and let $R = k[[u, v]]^G$ be the ring of invariants with respect to the given action of G by linear homogeneous transformations. It was first shown by J.Herzog [**25**] that the category $\mathcal{B}$ of finite reflexive modules over R is a representation finite category with the R–module $S = k[[u, v]]$ as a representation generator. Clearly $\mathcal{B}$ is closed under direct summands and a full additive subcategory with plenty of projectives of the abelian category $\mathcal{A}$ of all finite R-modules. It is an exact Frobenius category with enough projective-injectives if $G \subseteq \mathrm{SL}(2, k)$.

If G contains no pseudo-reflections, for example if $G \subseteq \mathrm{SL}(2, k)$, then M.Auslander [**4**] showed that the Auslander algebra $\Lambda = \mathrm{End}_R(S)$ is of global dimension 2; indeed he proved that Λ is isomorphic to the skew group algebra $k[[u, v]] * G$. The stable algebra $\Gamma = \underline{\mathrm{End}}_R(S)$ is then isomorphic to the Artin algebra $k[[u, v]] * G/(e)$ where $e = \frac{1}{G}\sum_{g \in G} g$ is the canonical idempotent in the group algebra $k[G] \subseteq \Lambda$.

The k-algebra $\Lambda' = k[u, v] * G$ is of projective dimension 2 over its enveloping k-algebra, as follows for example from [**30**], and the inclusion $\Lambda' \to \Lambda$ fits (1.6), whence Γ is an Artin algebra with quasi-periodic projective resolution over its enveloping k-algebra of period 3.

If k is algebraically closed of characteristic zero and $G \subseteq \mathrm{SL}(2, k)$, I. Reiten and M. van den Bergh [**32**, Prop.2.4, 2.13] prove that the algebra Λ' is Morita equivalent to the preprojective algebra of the extended Coxeter-Dynkin diagram underlying the McKay quiver of G. Under the same Morita equivalence, the homomorphic image Γ corresponds to the preprojective algebra of the (nonextended) Coxeter-Dynkin diagram; see also [**15**, Lemma 2.2., Cor.3.5] for a more explicit proof and a generalization to deformed preprojective algebras.

Using [**7**, Prop.3.2, Prop.3.3], it follows furthermore that Γ is a Frobenius k-algebra, that there are Γ-bimodule isomorphisms $\mathrm{Hom}_k(\Gamma,k)\cong L\cong {}_1\Gamma_\gamma$ and that L is of order 2 in $\mathrm{Pic}\,\Gamma$ except for type A_1, in which case $\Gamma=k$. See also (6.5) below.

In particular, as also observed by Ringel-Schofield, tensor product with L defines the Nakayama equivalence $\mathcal{N}$ on the category of finite right Γ-modules and γ represents a Nakayama automorphism of Γ. So (the preprojective algebra Morita equivalent to) Γ is a symmetric k-algebra iff it belongs to the Coxeter-Dynkin diagram A_1.

Finally, the Auslander-Reiten translate on the stable category of finite right Γ-modules satisfies

$$\mathrm{DTr}\cong\Omega^2_\Gamma\mathcal{N}\cong\Omega^5_\Gamma\cong\Omega^{-1}_\Gamma\,,$$

whence the DTr-period is the same as the period of the Hochschild (co-)homology, that is either $1,2$ or 6, with 1 only occurring for type A_1.

To summarize it, whenever a preprojective algebra of an extended Coxeter-Dynkin diagram is Morita equivalent to a skew group algebra as above, (1.6) establishes the result by Ringel-Schofield [**34**] that Hochschild (co-)homology of the preprojective algebra of a Coxeter-Dynkin diagram is periodic of period dividing 6.

1.8. Of course this state of affairs is not really satisfactory, after all Ringel-Schofield make no assumption on the base field and it remains to extend the representation theoretic argument to arbitrary fields. In any characteristic there is a rational double point over an algebraically closed field whose stable Auslander algebras are Morita equivalent to the preprojective algebra of the corresponding Coxeter-Dynkin diagram, see [**7**, Prop.3.3], and it seems very likely that the completion of the preprojective algebra of the extended Coxeter-Dynkin diagram is Morita equivalent to the Auslander algebras of reflexive modules for some such singularity. This would then fully "explain" the validity of the Ringel-Schofield result.

But here we go a simpler route and deduce the general form directly from what we have so far. The finishing touch comes from the fact that the preprojective algebra of a quiver over an arbitrary commutative base ring K is a graded flat K-algebra as the Hilbert function of such an algebra stays constant under arbitrary base change $K\to K'$. In fact, we can extend the treatment as well to the deformations of preprojective algebras introduced and studied by W.Crawley-Boevey and M.Holland [**15**], [**14**]. Recall the setup:

To a finite graph Q one associates the *preprojective quiver* $\overline{Q}$ whose vertices are those of Q but that contains a pair of arrows $a\colon i\to j$ and $a^*\colon j\to i$ for each edge between i and j in Q. If I is the vertex set of Q or $\overline{Q}$, and if K is any commutative ring, then $e_i\in K\overline{Q}$ denotes the idempotent corresponding to the trivial path at $i\in I$ in the path algebra of $\overline{Q}$.

An element $\lambda\in K^I$ defines then the *deformed preprojective algebra*

$$\Pi^\lambda(KQ)=K\overline{Q}/(\sum_{a\in Q}[a,a^*]-\sum_{i\in I}\lambda(i)e_i)\,,$$

and the "classical" preprojective algebra of Q is $\Pi^0(KQ)$. According to [**15**, Lemma 2.2], up to isomorphism the (deformed) preprojective algebra for a given λ does only depend on the graph underlying Q whence the notation is justifiable.

The result is now the following.

THEOREM 1.9. *Let K be a commutative ring. If $\tilde{\Sigma}$ is an extended Coxeter-Dynkin diagram, then $\Pi^\lambda(K\tilde{\Sigma})$ is of projective dimension at most 2 over its enveloping K-algebra, and equality holds for $\lambda = 0$.*

If Σ is a Coxeter-Dynkin diagram, then $\Pi^\lambda(K\Sigma)$ has a quasi-periodic projective resolution over its enveloping K-algebra of period 3. If $\lambda = 0$ and 2 is a nonzero divisor in K, then the minimal graded resolution is periodic of period 2 or 6, unless Σ is of type A_1.

1.10. *Two-dimensional Tame Orders.* The case of rational double points reviewed above has been extended in [**32**, 5.3] to two-dimensional tame orders: Let k be an algebraically closed field of charcateristic zero, R a normal two-dimensional complete local noetherian domain with residue class field k, and finally A a tame R-order in some simple algebra. The category $\mathcal{B}$ of reflexive A-modules is exact with enough projectives, and if it is representation finite, then any Auslander algebra Λ of $\mathcal{B}$ is Morita equivalent to a crossed product $k[[u,v]] *_c G$ where G is a finite group acting linearly on $k[[u,v]]$ and $c\colon G \times G \to k^*$ is a cocycle. It follows then, e.g. again from [**30**], that $\Lambda' = k[u,v] *_c G$ is of Hochschild dimension 2 and so (1.6) applies. The category of reflexive A-modules is exact Frobenius with enough projective-injectives if A is a Frobenius R-algebra and one can certainly classify the corresponding crossed products. Can one get in this way k-algebras Γ with periodic Hochschild (co-)homology that are not already covered in (1.7)?

1.11. *Artin algebras of finite representation type.* Let k be a field and A an Artin k-algebra of finite representation type. With $\mathcal{B} = \mathcal{A}$ the abelian category of all finite (right) A-modules, any Auslander algebra $\Lambda = \mathrm{End}_A(M)$ is then itself an Artin algebra that is of global dimension 2.

As shown by D.Happel [**20**], the Hochschild dimension of an Artin algebra over k equals its global dimension, hence there is no need for an auxiliary algebra Λ' this time. It follows right away that the stable Auslander algebra Γ admits a quasi-periodic projective resolution of period 3 with periodicity factor $L = \underline{\mathrm{Hom}}_A(M, \Omega^1_A M)$, see (6.5). The Hochschild (co-)homology will thus be periodic as soon as the syzygy functor on $\underline{\mathrm{mod}}A$ is of finite order and examples of such, necessarily selfinjective, Artin algebras can be found in [**7**, 3.5-3.7].

1.12. *Counterexamples.* Lest the reader becomes too optimistic, let us point out that there are examples of finite representation type where there is no algebra of Hochschild dimension 2 in sight. For example, if one takes a simple hypersurface singularity of dimension n, then the category $\mathcal{B}$ of its maximal Cohen-Macaulay modules is of finite representation type [**28**], but the global dimension of the Auslander algebra is at least n. Not too much is lost though, the stable Auslander algebra Γ only depends on the *parity* of the dimension, cf. (loc.cit). Thus the even dimensional case can be dealt with by realizing Γ as the stable Auslander algebra of a rational double point.

More intriguing is the odd dimensional case. According to [**7**], the Auslander-Reiten translate has period dividing 3 and if it is equal to 3, this can not be the period of the Hochschild dimension of the stable Auslander algebra by (1.5), at least in characteristic different from 2. On the other hand, the syzygy functor on $\mathcal{B}$ is of order at most two, and the invertible Γ-bimodule $L = \mathrm{Tor}_2^\Lambda(\Gamma, \Gamma)$ is of the same order, see (6.5) below. So, is there periodicity for the Hochschild (co-)homology of Γ in this case?

2. Full Additive Subcategories with Plenty of Projectives

In this section we run through the basic background material needed to prove theorem (1.1) in its functorial form (4.1), (5.5).

2.1. The full subcategory of projectives in an abelian category $\mathcal{A}$ is denoted $\mathcal{P}(\mathcal{A})$, that of injectives $\mathcal{I}(\mathcal{A})$.

Embedding an additive category $\mathcal{B}$ into an abelian category $\mathcal{A}$ as a full subcategory endows $\mathcal{B}$ with an exact structure that one can use to define the notion of (*enough* or even *plenty of*) *projectives* or *injectives* in $\mathcal{B}$.

DEFINITION 2.2. Let $\mathcal{B}$ be an additive category, embedded as a full subcategory into an abelian category $\mathcal{A}$.

1. A morphism $\alpha\colon B \to A$ from $\mathcal{B}$ is an *admissible epimorphism* if it fits in $\mathcal{A}$ into an exact sequence $0 \to C \to B \xrightarrow{\alpha} A \to 0$ with C in $\mathcal{B}$. Equivalently, α is an epimorphism in $\mathcal{A}$ and its kernel is isomorphic to an object in $\mathcal{B}$.
2. An object P from $\mathcal{B}$ is *$\mathcal{B}$-projective* if it has the lifting property with respect to admissible epimorphisms. The full subcategory of $\mathcal{B}$-projectives is clearly additive and closed under direct summands; it is denoted $\mathcal{P} \subseteq \mathcal{B}$. (A more precise notation would be $\mathcal{P}_{\mathcal{B}\subseteq\mathcal{A}}$, and one should rather say "$\mathcal{B}$-projective with respect to the given embedding", but that is obviously too cumbersome.)
3. The category $\mathcal{B}$ has *enough $\mathcal{B}$-projectives* if for every A in $\mathcal{B}$ there exists an admissible epimorphism $P \to A$ from a $\mathcal{B}$-projective P.
4. The category $\mathcal{B}$ has *plenty of $\mathcal{B}$-projectives* if for every morphism $B \to A$ in $\mathcal{B}$ there is some admissible epimorphism $P \to A$ from a $\mathcal{B}$-projective such that the induced morphism $B \oplus P \to A$ is an admissible epimorphism. Taking $B = 0$, it follows that plenty is more than enough.

The dual notions of *admissible monomorphisms, (enough or plenty of) $\mathcal{B}$-injectives* are left to the reader to formulate.

We next collect simple but basic facts about these notions. For the most part, we only state the facts pertinent to $\mathcal{B}$-projectives, leaving it to the reader to dualize. The following is obvious.

LEMMA 2.3. *Let $\mathcal{B} \subseteq \mathcal{A}$ be a full additive subcategory of an abelian category. If $0 \to C \xrightarrow{\beta} B \xrightarrow{\alpha} A \to 0$ is an exact sequence in $\mathcal{A}$, then the following are equivalent.*

(i) *α is isomorphic to an admissible epimorphism in $\mathcal{B}$.*
(ii) *β is isomorphic to an admissible monomorphism in $\mathcal{B}$.*
(iii) *Each of A, B, C is isomorphic to an object in $\mathcal{B}$.* □

An exact sequence in $\mathcal{A}$ whose terms are objects from $\mathcal{B}$ will be called an *$\mathcal{A}$-exact sequence in $\mathcal{B}$*, or just an exact sequence in $\mathcal{B}$ if the ambient abelian category is understood. A *syzygy object* of an object A from $\mathcal{B}$ is the kernel of an admissible epimorphism $P \to A$ from a $\mathcal{B}$-projective P onto A. First we show that adding a $\mathcal{B}$-projective to a syzygy object along any morphism returns again a syzygy object. This implies the usual Schanuel Lemma.

LEMMA 2.4. *Let $\mathcal{B} \subseteq \mathcal{A}$ be a full additive subcategory of an abelian category.*

1. *If $\pi\colon P \to A$ is an admissible epimorphism from a $\mathcal{B}$-projective onto A and if $\eta\colon P' \to A$ is any morphism from a $\mathcal{B}$-projective, then $(\eta{+}\pi)\colon P' \oplus P \to A$*

is an admissible epimorphism from a $\mathcal{B}$-projective onto A and $\ker(\eta+\pi) \cong P' \oplus \ker \pi$.

2. (SCHANUEL'S LEMMA) *If $\pi\colon P \to A$ and $\pi'\colon P' \to A$ are admissible epimorphisms from $\mathcal{B}$-projectives onto A, then*

$$P \oplus \ker \pi' \cong \ker \pi \oplus P' .$$

PROOF. (1) Note that $Q = \ker(\eta + \pi\colon P' \oplus P \to A)$ is in $\mathcal{A}$ a fibre product of $-\eta$ with π over A, whence it appears in $\mathcal{A}$ in a commutative diagram with exact rows

$$\begin{array}{ccccccccc}
0 & \longrightarrow & \ker\pi & \longrightarrow & Q & \longrightarrow & P' & \longrightarrow & 0 \\
 & & \Vert & & \downarrow & & \downarrow{\scriptstyle -\eta} & & \\
0 & \longrightarrow & \ker\pi & \longrightarrow & P & \xrightarrow{\ \pi\ } & A & \longrightarrow & 0
\end{array} .$$

As π is an admissible epimorphism and P' is $\mathcal{B}$-projective, $-\eta$ lifts through π and so the exact sequence on top splits to give $Q \cong P' \oplus \ker \pi$. (2) follows by symmetry. □

2.5. If $\mathcal{B}$ has enough $\mathcal{B}$-projectives, we may (and do) choose for every object A in $\mathcal{B}$ an exact sequence $0 \to \Omega^1 A \to P_A \xrightarrow{\pi_A} A \to 0$ in $\mathcal{A}$ with π_A an admissible epimorphism, P_A in $\mathcal{P}$, and $\Omega^1 A$ in $\mathcal{B}$. The object $\Omega^1 A$ is "the" first syzygy of A.

Schanuel's lemma (2.4.2) shows that different choices of such admissible epimorphisms and their kernels return syzygy objects isomorphic up to $\mathcal{B}$-projective summands.

Now consider the following version of the "horseshoe lemma".

LEMMA 2.6 *Let $\mathcal{B} \subset \mathcal{A}$ be again a full additive subcategory of an abelian category. Assume $\mathcal{B}$ has enough $\mathcal{B}$-projectives and let $0 \to C \xrightarrow{\beta} B \xrightarrow{\alpha} A \to 0$ be an exact sequence in $\mathcal{B}$. There exists then a morphism $\widetilde{\pi}_A : P_A \to B$ that fits into the following commutative diagram that is exact in $\mathcal{A}$ and whose terms are in $\mathcal{B}$. In*

$$\begin{array}{ccccccccc}
 & & 0 & & 0 & & 0 & & \\
 & & \downarrow & & \downarrow & & \downarrow & & \\
0 & \longrightarrow & \Omega^1 C & \longrightarrow & P_C \oplus P_A \oplus \Omega^1 B & \longrightarrow & \Omega^1 A \oplus P_B & \longrightarrow & 0 \\
 & & \downarrow & & \downarrow & & \downarrow & & \\
0 & \longrightarrow & P_C & \xrightarrow{\ incl.\ } & P_C \oplus P_A \oplus P_B & \xrightarrow{\ proj.\ } & P_A \oplus P_B & \longrightarrow & 0 \\
 & & \downarrow{\scriptstyle \pi_C} & & \downarrow{\scriptstyle \beta\pi_C + \widetilde{\pi}_A + \pi_B} & & \downarrow{\scriptstyle \pi_A + \alpha\pi_B} & & \\
0 & \longrightarrow & C & \xrightarrow{\ \beta\ } & B & \xrightarrow{\ \alpha\ } & A & \longrightarrow & 0 \\
 & & \downarrow & & \downarrow & & \downarrow & & \\
 & & 0 & & 0 & & 0 & &
\end{array}$$

particular, each vertical epimorphism is admissible.

PROOF. By definition, P_A is $\mathcal{B}$-projective, and by assumption, α is an admissible epimorphism. So there is a morphism $\widetilde{\pi}_A : P_A \to B$ with $\alpha\widetilde{\pi}_A = \pi_A$ and the given exact sequence fits into the smaller commutative diagram whose rows are

$$\begin{array}{ccccccccc}
0 & \longrightarrow & P_C & \xrightarrow{incl.} & P_C \oplus P_A & \xrightarrow{proj.} & P_A & \longrightarrow & 0 \\
 & & \downarrow{\scriptstyle \pi_C} & & \downarrow{\scriptstyle \beta\pi_C + \widetilde{\pi}_A} & & \downarrow{\scriptstyle \pi_A} & & \\
0 & \longrightarrow & C & \xrightarrow{\beta} & B & \xrightarrow{\alpha} & A & \longrightarrow & 0
\end{array}$$

exact in $\mathcal{A}$. This is the usual horseshoe lemma. We need to go a step further, as there is no guarantee that the middle vertical morphism is admissible, i.e. has its kernel in $\mathcal{B}$. But at least $\beta\pi_C + \widetilde{\pi}_A$ is a morphism from a $\mathcal{B}$-projective, so we can use lemma (2.4.1.) twice to check first that both $\beta\pi_C + \widetilde{\pi}_A + \pi_B$ and $\pi_A + \alpha\pi_B$ are admissible epimorphisms, and secondly that the kernels are as indicated. $\square$

Next we study the condition for plenty of projectives in case that $\mathcal{B}$ is closed under direct summands.

LEMMA 2.7. *Let $\mathcal{B} \subseteq \mathcal{A}$ be a full additive subcategory of an abelian category and assume that $\mathcal{B}$ is closed under direct summands.*

1. *For a morphism $B \to A$ in $\mathcal{B}$, the following are equivalent:*
   (i) *For some admissible epimorphism $P \to A$ from a $\mathcal{B}$-projective the induced morphism $B \oplus P \to A$ is an admissible epimorphism.*
   (ii) *For every admissible epimorphism $P \to A$ from a $\mathcal{B}$-projective the induced morphism $B \oplus P \to A$ is an admissible epimorphism.*
2. *If $\mathcal{B}$ has plenty of $\mathcal{B}$-projectives, then $B \to A$ is an admissible epimorphism iff* $\mathrm{Hom}(P, B) \to \mathrm{Hom}(P, A)$ *is surjective for every $\mathcal{B}$-projective P.*

PROOF. (1) Let $\pi\colon P \to A$ and $\pi'\colon P' \to A$ be admissible epimorphisms with P, P' from $\mathcal{P}$. Assuming that $C = \ker(B \oplus P \to A)$ is in $\mathcal{B}$, we only need to show that $C' = \ker(B \oplus P' \to A)$ is in $\mathcal{B}$ as well. To this end, set $Q = \ker(\pi' - \pi\colon P' \oplus P \to A)$ and consider the following cube all of whose faces are cartesian squares in $\mathcal{A}$.

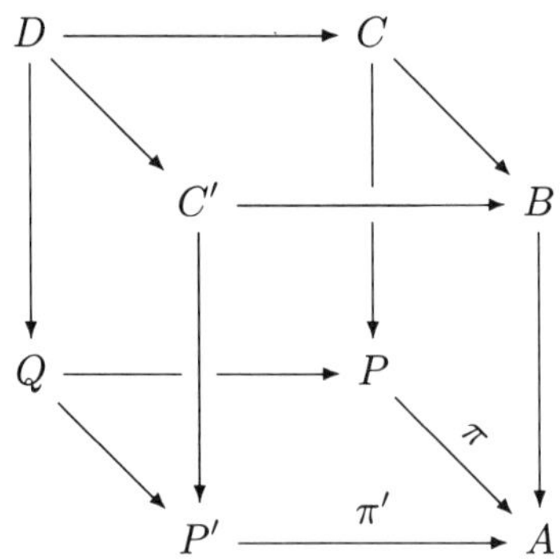

One has successively

$$\begin{array}{ll}
Q \cong \ker \pi' \oplus P \cong P' \oplus \ker \pi & \text{by } (2.4.2), \\
D \cong \ker \pi' \oplus C & \text{from the square } DCPQ, \\
D \cong C' \oplus \ker \pi & \text{from the square } DC'P'Q.
\end{array}$$

It follows that

$$C' \text{ in } \mathcal{B} \iff D \text{ in } \mathcal{B} \iff C \text{ in } \mathcal{B}$$

as by assumption $\ker\pi, \ker\pi'$ are in $\mathcal{B}$ and this category is closed under direct sums and summands.

(2) The "only if" part follows from the definition of a $\mathcal{B}$-projective. Conversely, assume given a morphism $\alpha\colon B \to A$ in $\mathcal{B}$ such that the induced map $\operatorname{Hom}(P, B) \to \operatorname{Hom}(P, A)$ is surjective for every $\mathcal{B}$-projective P. In particular, any admissible epimorphism $\pi\colon P \to A$ from a $\mathcal{B}$-projective onto A lifts through α, whence α is necessarily an epimorphism in $\mathcal{A}$ as soon as $\mathcal{B}$ has enough $\mathcal{B}$-projectives. It remains to see that $\ker\alpha$ is isomorphic to an object in $\mathcal{B}$ if there are plenty of $\mathcal{B}$-projectives. But in that case $C = \ker(\alpha - \pi\colon B \oplus P \to A)$ is (isomorphic to an object) in $\mathcal{B}$ by part (1). Writing C again as a fibre product in $\mathcal{A}$ of B with P over A gives rise in $\mathcal{A}$ to the exact diagram

$$\begin{array}{ccccccccc}
0 & \longrightarrow & \ker\alpha & \longrightarrow & C & \longrightarrow & P & \longrightarrow & 0 \\
 & & \Vert & & \downarrow & \swarrow & \downarrow{\scriptstyle \pi} & & \\
0 & \longrightarrow & \ker\alpha & \longrightarrow & B & \xrightarrow{\ \alpha\ } & A & \longrightarrow & 0
\end{array} \quad .$$

As π lifts through α, the extension C of P by $\ker\alpha$ splits, $C \cong \ker\alpha \oplus P$, and so $\ker\alpha$ is isomorphic to an object from $\mathcal{B}$ as that category is closed under direct summands. □

DEFINITION 2.8. A full additive subcategory $\mathcal{B}$ of an abelian category $\mathcal{A}$ is *extension closed in* $\mathcal{A}$, if in every exact sequence $0 \to C \to B \to A \to 0$ from $\mathcal{A}$ with A, C isomorphic to objects in $\mathcal{B}$, the middle term B is also isomorphic to an object in $\mathcal{B}$. The category $\mathcal{B}$ is *exact* (with respect to the given embedding) if it is closed under direct summands and extensions.

Note that these exact categories are the "abelian" ones of A. Heller [**24**], and that D. Quillen [**31**, §2.Examples] calls them "exact and Karoubian" as his notion of exactness does not include closure under direct summands.

In extension closed categories, life becomes slightly simpler.

LEMMA 2.9. *Let $\mathcal{B} \subseteq \mathcal{A}$ be an extension closed subcategory.*

1. *An object P in $\mathcal{B}$ is $\mathcal{B}$-projective iff $\operatorname{Ext}^1_{\mathcal{A}}(P, A) = 0$ for every A from $\mathcal{B}$.*
2. *There are enough $\mathcal{B}$-projectives iff there are plenty of them.*
3. *If $B \to A$ is any morphism in $\mathcal{B}$ and $\pi'\colon A' \to A$ is an admissible epimorphism, then the induced morphism $B \oplus A' \to A$ is an admissible epimorphism.*

PROOF. (1) Let P be $\mathcal{B}$-projective and let $0 \to A \to X \xrightarrow{\xi} P \to 0$ be an extension of P in $\mathcal{A}$ by an object A from $\mathcal{B}$. As $\mathcal{B}$ is extension closed in $\mathcal{A}$, the object X is (isomorphic to one) in $\mathcal{B}$ and ξ is thus (isomorphic to) an admissible epimorphism. As P is $\mathcal{B}$-projective, the identity on P lifts through ξ and so ξ is a split epimorphism. It follows that $\operatorname{Ext}^1_{\mathcal{A}}(P, A) = 0$, and the converse holds in any full additive subcategory of $\mathcal{A}$.

(2) Assume there are enough $\mathcal{B}$-projectives, let $\alpha\colon B \to A$ be a morphism in $\mathcal{B}$ and $\pi\colon P \to A$ an admissible epimorphism from a $\mathcal{B}$-projective onto A. The object $C = \ker(B \oplus P \to A)$ is in $\mathcal{A}$ the fibre product of B and P over A whence one

obtains in $\mathcal{A}$ the exact diagram

$$\begin{array}{ccccccccc}
0 & \longrightarrow & \ker\pi & \longrightarrow & C & \longrightarrow & B & \longrightarrow & 0 \\
 & & \| & & \downarrow & & \downarrow{\scriptstyle\alpha} & & \\
0 & \longrightarrow & \ker\pi & \longrightarrow & P & \xrightarrow{\pi} & A & \longrightarrow & 0
\end{array}$$

As B and $\ker\pi$ are (isomorphic to) objects in the extension closed category $\mathcal{B}$, the object C is isomorphic to an object from $\mathcal{B}$.

(3) Set $C = \ker(B \oplus A' \to A)$ and consider again the corresponding exact diagram in $\mathcal{A}$

$$\begin{array}{ccccccccc}
0 & \longrightarrow & \ker\pi' & \longrightarrow & C & \longrightarrow & B & \longrightarrow & 0 \\
 & & \| & & \downarrow & & \downarrow & & \\
0 & \longrightarrow & \ker\pi' & \longrightarrow & A' & \xrightarrow{\pi'} & A & \longrightarrow & 0
\end{array}$$

By assumption, $\ker\pi'$ is isomorphic to an object in $\mathcal{B}$ and thus C, being an extension of B by $\ker\pi'$, is also isomorphic to an object in $\mathcal{B}$. □

3. Functor Categories

Now we turn our attention to functor categories and recall first some well known facts. Our main aim here is to clarify the role and structure of adjoints to the fully faithful embedding $\operatorname{Mod}\underline{\mathcal{B}} \to \operatorname{Mod}\mathcal{B}$ where $\underline{\mathcal{B}} = \mathcal{B}/\mathcal{P}$ is the category of $\mathcal{B}$ modulo $\mathcal{B}$-projectives.

3.1. If $\mathcal{C}$ is any skeletally small additive category, $\operatorname{Mod}\mathcal{C} = (\mathcal{C}^{op}, \mathfrak{Ab})$, the *category of right $\mathcal{C}$-modules* in M.Auslander's terminology, denotes the abelian category of all contravariant additive functors from $\mathcal{C}$ to the category of abelian groups $\mathfrak{Ab}$.

This category is a full subcategory of the abelian category $\hat{\mathcal{C}}_{ab} = \operatorname{Fun}(\mathcal{C}^{op}, \mathfrak{Ab})$ of *all* contravariant functors from $\mathcal{C}$ to $\mathfrak{Ab}$, the category of *presheaves of abelian groups on $\mathcal{C}$* in A.Grothendieck's terminology [**35**, I.5.8.0]. Indeed, $\operatorname{Mod}\mathcal{C}$ can be viewed as the category of *sheaves in abelian groups on $\mathcal{C}$* with respect to the *"Pierce topology"*, the Grothendieck topology on $\mathcal{C}$ generated by direct sum decompositions, that is, by all those finite families of morphisms $(f_\iota\colon C_\iota \to C)_{\iota\in I}$ for which the induced morphism $f\colon \bigoplus_\iota C_\iota \to C$ is an isomorphism.

In $\operatorname{Mod}\mathcal{C}$ all (small) limits and colimits are representable and a sequence of functors $0 \to H \to G \to F \to 0$ is exact iff $0 \to H(C) \to G(C) \to F(C) \to 0$ is an exact sequence of abelian groups for every object C from $\mathcal{C}$.

The shorthand notation $(\ ,C)$ means the functor in $\operatorname{Mod}\mathcal{C}$ represented by C so that (C',C) stands for $\operatorname{Hom}_{\mathcal{C}}(C',C)$. Each such functor is a projective in $\operatorname{Mod}\mathcal{C}$, and $\operatorname{Hom}_{\operatorname{Mod}\mathcal{C}}((\ ,C),F) \cong F(C)$ by Yoneda's Lemma. Arbitrary but small direct sums of representable functors represent enough projectives.

3.2. A $\mathcal{C}$-module F is *finitely presented* if it admits a presentation $(\ ,D) \to (\ ,C) \to F \to 0$, and $\operatorname{mod}\mathcal{C} \subseteq \operatorname{Mod}\mathcal{C}$ stands for the full additive subcategory of such modules. Any functor in $\operatorname{Mod}\mathcal{C}$ is an inductive limit of finitely presented ones. The subcategory $\operatorname{mod}\mathcal{C} \subseteq \operatorname{Mod}\mathcal{C}$ is closed under direct summands, and, by a suitable version of the horseshoe lemma, it is also extension closed, thus exact. Moreover, this subcategory is closed under cokernels.

Each functor $(\ ,C)$ is a projective in $\operatorname{mod}\mathcal{C}$, and these functors constitute, up to isomorphism, precisely *all* projective objects in $\operatorname{mod}\mathcal{C}$ iff $\mathcal{C}$ is closed under direct summands. By definition, every object from $\mathrm{mod}\,\mathcal{C}$ admits an epimorphism from a projective in $\mathrm{mod}\,\mathcal{C}$ onto it, but such epimorphisms are not necessarily admissible with respect to the embedding into $\mathrm{Mod}\,\mathcal{C}$, as kernels of morphisms between finitely presented modules are not necessarily so. Thus there are *not necessarily enough* $\mathrm{mod}\,\mathcal{C}$*-projectives*.

Notions such as epi-/monomorphisms or exact sequences of finitely presented functors will always be meant with respect to the ambient abelian category $\mathrm{Mod}\,\mathcal{C}$.

3.3. Dually, $\mathcal{C}\,\mathrm{Mod} = (\mathcal{C}, \mathfrak{Ab}) \cong \mathrm{Mod}\,\mathcal{C}^{op}$ denotes the category of all (covariant, additive) functors from $\mathcal{C}$ to $\mathfrak{Ab}$ and $\mathcal{C}\,\mathrm{mod}$ is its full subcategory of finitely presented functors. One has as in [**6**, Sect.6] a canonical contravariant and left exact functor from $\mathrm{Mod}\,\mathcal{C}$ to $\mathcal{C}\,\mathrm{Mod}$ that associates to a right $\mathcal{C}$-module F the left $\mathcal{C}$-module $F^* : \mathcal{C} \to \mathfrak{Ab}$ given by $F^*(C) = \mathrm{Hom}_{\mathrm{Mod}\,\mathcal{C}}(F, (\ ,C))$. Since $(\ ,C)^* = (C,\ )$, a finitely presented functor $(\ ,D) \to (\ ,C) \to F \to 0$ is transformed into the finitely corepresented functor $0 \to F^* \to (C,\ ) \to (D,\ )$.

3.4. If $u\colon \mathcal{C} \to \mathcal{D}$ is an additive functor, the restriction functor $u_*\colon \mathrm{Mod}\,\mathcal{D} \to \mathrm{Mod}\,\mathcal{C}$ that maps F to Fu admits both a left adjoint u^* and a right adjoint $u^!$. This fact can be seen either directly from the corresponding construction of adjoints for the category of all presheaves as in [**35**, I.Prop.5.1.], or one may observe that an additive functor u is continuous and cocontinuous for the Pierce topologies on $\mathcal{C}$ and $\mathcal{D}$ respectively, to conclude that the adjoints between the categories of presheaves restrict to adjoints between the categories of sheaves, [**35**, Exp.III]. It is for the existence of this triple $(u^*, u_*, u^!)$ of adjoint functors that we mentioned the interpretation of $\mathrm{Mod}\,\mathcal{C}$ as sheaves in abelian groups with respect to the Pierce topology on $\mathcal{C}$.

3.5. In general, neither u_* nor $u^!$ will carry finitely presented modules to such. But u^* does restrict to a functor from $\mathrm{mod}\,\mathcal{C}$ to $\mathrm{mod}\,\mathcal{D}$; one has, functorially in G from $\mathrm{Mod}\,\mathcal{D}$ and in C from $\mathcal{C}$, that

$$\begin{aligned}
\mathrm{Hom}_{\mathrm{Mod}\,\mathcal{C}}((\ ,C), u_*G) &\cong u_*G(C) && \text{by Yoneda's Lemma}\\
&= G(uC) && \text{by definition of } u_*\\
&\cong \mathrm{Hom}_{\mathrm{Mod}\,\mathcal{D}}((\ ,uC), G) && \text{by Yoneda's Lemma again.}
\end{aligned}$$

It follows that $u^*(\ ,C) \cong (\ ,uC)$ and, since u^* is right exact as a left adjoint, a presentation $(\ ,C_2) \to (\ ,C_1) \to F \to 0$ is mapped under u^* to the presentation $(\ ,uC_2) \to (\ ,uC_1) \to u^*F \to 0$ whence u^*F is also finitely presented.

When does the restricted functor $u^*\colon \mathrm{mod}\,\mathcal{C} \to \mathrm{mod}\,\mathcal{D}$ admit a right adjoint? If such a right adjoint exists, say $v\colon \mathrm{mod}\,\mathcal{D} \to \mathrm{mod}\,\mathcal{C}$, then v is necessarily isomorphic to the restriction of u_* to $\mathrm{mod}\,\mathcal{D}$. Indeed, one has, functorially in F from $\mathrm{Mod}\,\mathcal{D}$ and in C from $\mathrm{mod}\,\mathcal{C}$, that

$$\begin{aligned}
vF(C) &\cong \mathrm{Hom}_{\mathrm{mod}\,\mathcal{C}}((\ ,C), vF) && \text{by Yoneda's Lemma}\\
&\cong \mathrm{Hom}_{\mathrm{mod}\,\mathcal{D}}(u^*(\ ,C), F) && \text{by adjunction}\\
&\cong \mathrm{Hom}_{\mathrm{Mod}\,\mathcal{D}}(u^*(\ ,C), F) && \text{as } \mathrm{mod}\,\mathcal{D} \subseteq \mathrm{Mod}\,\mathcal{D} \text{ is a full subcategory}\\
&\cong \mathrm{Hom}_{\mathrm{Mod}\,\mathcal{C}}((\ ,C), u_*F) && \text{by adjunction}\\
&\cong u_*F(C) && \text{by Yoneda's Lemma again.}
\end{aligned}$$

In other words, u^* as a functor between finitely presented modules admits a right adjoint iff the image of a finitely presented module under u_* is still finitely presented, and, if so, the restriction of u_* to the finitely presented modules is a right adjoint.

If u_* restricts to a functor between the full subcategories of finitely presented modules, then the same argument shows that it in turn admits a right adjoint iff the image of a finitely presented module under $u^!$ is still finitely presented, and that then the restriction of $u^!$ to the finitely presented modules is a right adjoint to the restriction of u_*.

3.6. If $\mathcal{Q} \subseteq \mathcal{C}$ is a full additive subcategory of an additive category $\mathcal{C}$, denote $\underline{\mathcal{C}} = \mathcal{C}/\mathcal{Q}$ the category $\mathcal{C}$ modulo (the additive relation generated by the full subcategory) $\mathcal{Q}$. By definition, $\underline{\mathcal{C}}$ has the "same" objects as $\mathcal{C}$. The natural functor $p\colon \mathcal{C} \to \underline{\mathcal{C}}$ will often also be indicated just by underlining. Thus, $\underline{C} = p(C)$ represents the object C from $\mathcal{C}$ but considered as an object in $\underline{\mathcal{C}}$. The functor p is full and additive, universal with respect to additive functors F on $\mathcal{C}$ that vanish on $\mathcal{Q}$, meaning that $F(Q)$ is a zero object for every Q from $\mathcal{Q}$.

Assuming from now on again that $\mathcal{C}$ is skeletally small, the projection functor p induces the restriction functor $p_*\colon \operatorname{Mod}\underline{\mathcal{C}} \to \operatorname{Mod}\mathcal{C}$ that maps $\operatorname{Mod}\underline{\mathcal{C}}$ by an isomorphism of categories onto the full subcategory of those functors that vanish on $\mathcal{Q}$, see [**6**, Prop.1.1.].

Furthermore, as the projection p is the identity on objects and exactness of sequences of functors is tested on objects, the restriction functor p_* is a *forgetful functor*, it is exact and detects exact sequences: A sequence $0 \to H \to G \to F \to 0$ of functors on $\underline{\mathcal{C}}$ is exact iff it, or rather its image under p_*, is exact as a sequence of functors on $\mathcal{C}$.

As a particular consequence, $\operatorname{Mod}\underline{\mathcal{C}}$ is exact as a subcategory of $\operatorname{Mod}\mathcal{C}$, and so is then $\operatorname{mod}\underline{\mathcal{C}}$ as the latter is exact in $\operatorname{Mod}\underline{\mathcal{C}}$.

In view of these properties, one usually identifies $\operatorname{Mod}\underline{\mathcal{C}}$ as a full subcategory of $\operatorname{Mod}\mathcal{C}$ and drops p_* from the notation. For example, $(\ ,\underline{C})$ will be considered ambiguously as a functor on $\underline{\mathcal{C}}$ as well as on $\mathcal{C}$ where its "correct" description would be either of $p_*(\ ,\underline{C}) = (p(\ ),\underline{C}) = (\_,\underline{C})$.

3.7. Thinking of functors into abelian groups as modules, the projection $p\colon \mathcal{C} \to \underline{\mathcal{C}} = \mathcal{C}/\mathcal{Q}$ becomes the analogue of a ring epimorphism with respect to which the restriction functor p_* then corresponds to "restriction of scalars". Accordingly, a left adjoint to this functor should be thought of as tensor product, whence we write also $p^* = (\ ) \otimes_{\mathcal{C}} \underline{\mathcal{C}}$. In this analogy, the right adjoint should be thought of as $p^! = \operatorname{Hom}_{\mathcal{C}}(\underline{\mathcal{C}},\ )$.

The adjunction morphism $(\ ,C) \to p_*p^*(\ ,C) = p_*(\ ,C) \otimes_{\mathcal{C}} \underline{\mathcal{C}} = (\ ,\underline{C})$ is an epimorphism for any object C from $\mathcal{C}$ precisely because p is a full functor. Right exactness of p_* and p^* shows then that the adjunction morphism $F \to p_*p^*F$ at any finitely presented functor F is also an epimorphism. As an arbitrary functor in $\operatorname{Mod}\mathcal{C}$ is a direct limit of finitely presented ones and as p_*, p^* commute with all direct limits, the adjunction morphism for such a functor is a direct limit of epimorphisms, thus an epimorphism itself.

Furthermore, as stated in (3.6) above, a functor F in $\operatorname{mod}\mathcal{C}$ vanishes on $\mathcal{Q}$ iff $F \cong p_*G$ for some functor G from $\operatorname{Mod}\underline{\mathcal{C}}$. Since the adjunction morphism $p_*G \to p_*p^*(p_*G)$ is always a split monomorphism, split by $p_*(p^*p_*G \to G)$, it follows that a functor F on $\mathcal{C}$ vanishes on $\mathcal{Q}$ iff the adjunction morphism $F \to p_*p^*F$ is an isomorphism. In terms of "modules", the right $\mathcal{C}$-module F is a $\underline{\mathcal{C}}$-module iff the natural $\mathcal{C}$-(homo-)morphism $F \to F \otimes_{\mathcal{C}} \underline{\mathcal{C}}$ is an isomorphism of right $\mathcal{C}$-modules — and the corresponding property characterizes precisely the epimorphisms in the category of rings, as these are those ring maps $R \to S$ for which the multiplication map $S \otimes_R S \to S$ is an isomorphism, see [**12**, p.323, Remark].

We now show that a finitely presented (contravariant) functor on $\underline{\mathcal{C}}$ restricts to a finitely presented functor on $\mathcal{C}$ iff the subcategory $\mathcal{Q}$ is contravariantly finite. In the analogy with rings, such projections $\mathcal{C} \to \mathcal{C}/\mathcal{Q}$ correspond exactly to the surjective ring homomorphisms. First recall the definition.

Definition 3.8 ([**9**]). The full subcategory $\mathcal{Q} \subseteq \mathcal{C}$ is *contravariantly finite* if for every object C in $\mathcal{C}$ there exists a morphism $\rho_C\colon Q_C \to C$ such that any morphism $\rho\colon Q \to C$ from an object of $\mathcal{Q}$ lifts through ρ_C. Such a morphism ρ_C is called a *right $\mathcal{Q}$-approximation* of C.

If ρ_C is a right $\mathcal{Q}$-approximation, then clearly

$$\mathrm{Hom}_{\mathcal{C}/\mathcal{Q}}(\ ,C) = \mathrm{Hom}_{\mathcal{C}}(\ ,C)/\rho_C\,\mathrm{Hom}_{\mathcal{C}}(\ ,Q_C)\,,$$

equivalently, the sequence of $\mathcal{C}$-modules

$$(\ ,Q_C) \xrightarrow{(\ ,\rho_C)} (\ ,C) \to (\ ,\underline{C}) \to 0$$

is exact. In particular, $p_*(\ ,C) = (\ ,\underline{C})$ is finitely presented in $\mathrm{Mod}\,\mathcal{C}$ for any C that admits a right $\mathcal{Q}$-approximation. The converse is true as well: If $(\ ,\underline{C})$ is finitely presented as a functor on $\mathcal{C}$, then there is a presentation of the form

$$(\ ,D) \xrightarrow{(\ ,\gamma)} (\ ,C) \to (\ ,\underline{C}) \to 0\,.$$

As $\gamma \in \mathrm{Hom}_{\mathcal{C}}(D,C)$ is the image of the identity from $\mathrm{Hom}_{\mathcal{C}}(D,D)$, it maps to 0 in $(D,\underline{C}) = \mathrm{Hom}_{\mathcal{C}/\mathcal{Q}}(\underline{D},\underline{C})$ and must thus factor through some Q_C from $\mathcal{Q}$, say $\gamma = \rho_C\gamma'$ with $\rho_C : Q_C \to C$. We claim that ρ_C is a right $\mathcal{Q}$-approximation of C. Indeed, if $(\rho\colon Q \to C) \in \mathrm{Hom}_{\mathcal{C}}(Q,C)$ is some morphism from an object in $\mathcal{Q}$ to C, it is mapped to zero in $(Q,\underline{C}) = \mathrm{Hom}_{\mathcal{C}/\mathcal{Q}}(Q,C) = 0$, and so it lifts through γ in view of the given presentation. In particular, ρ lifts through ρ_C.

The discussion so far can be summarized thus:

Proposition 3.9. *Let $\mathcal{Q} \subseteq \mathcal{C}$ be a full additive subcategory of an additive category $\mathcal{C}$, denote $p\colon \mathcal{C} \to \underline{\mathcal{C}} = \mathcal{C}/\mathcal{Q}$ the projection functor, and $p_*\colon \mathrm{Mod}\,\underline{\mathcal{C}} \to \mathrm{Mod}\,\mathcal{C}$ the corresponding forgetful restriction functor.*

1. *The left adjoint $p^* = (\ )\otimes_{\mathcal{C}} \underline{\mathcal{C}}$ maps $\mathrm{mod}\,\mathcal{C}$ to $\mathrm{mod}\,\underline{\mathcal{C}}$. One has $(\ ,C)\otimes_{\mathcal{C}} \underline{\mathcal{C}} \cong (\ ,\underline{C})$ and a finite presentation*

$$(\ ,D) \to (\ ,C) \to F \to 0$$

maps to

$$(\ ,\underline{D}) \to (\ ,\underline{C}) \to F \otimes_{\mathcal{C}} \underline{\mathcal{C}} \to 0\,.$$

The natural morphism $F \to F\otimes_{\mathcal{C}}\underline{\mathcal{C}}$ is an epimorphism for every F in $\mathrm{Mod}\,\mathcal{C}$, and it is an isomorphism iff $F(Q) = 0$ for every Q in $\mathcal{Q}$.

2. *The following are equivalent:*
   (i) *Every finitely presented right $\underline{\mathcal{C}}$-module restricts to a finitely presented right $\mathcal{C}$-module.*
   (ii) *The functor $(\ )\otimes_{\mathcal{C}} \underline{\mathcal{C}} : \mathrm{mod}\,\mathcal{C} \to \mathrm{mod}\,\underline{\mathcal{C}}$ admits a right adjoint.*

(ii') *The functor* $(\ )\otimes_{\mathcal{C}}\underline{\mathcal{C}}:\operatorname{mod}\mathcal{C}\to\operatorname{mod}\underline{\mathcal{C}}$ *admits* $p_*|_{\operatorname{mod}\underline{\mathcal{C}}}$ *as a right adjoint.*

(iii) *The subcategory* $\mathcal{Q}$ *is contravariantly finite in* $\mathcal{C}$. □

4. The Auslander-Reiten Structure Theorem

Now we return to the original setup where $\mathcal{B}\subseteq\mathcal{A}$ is a full additive subcategory of an abelian category and we denote accordingly $\underline{\mathcal{B}}=\mathcal{B}/\mathcal{P}$ the category $\mathcal{B}$ modulo (the additive relation generated by the) $\mathcal{B}$-projectives.

The choices made earlier define Ω^1 as a functor on $\underline{\mathcal{B}}$, the *syzygy functor*. Pedantically, one has $\underline{\Omega^1 A}=\Omega^1\underline{A}$ in $\underline{\mathcal{B}}$.

To study the impact of enough or plenty of $\mathcal{B}$-projectives on the structure of $\operatorname{mod}\underline{\mathcal{B}}$, note first that $\mathcal{P}$ is contravariantly finite in $\mathcal{B}$ if there are enough $\mathcal{B}$-projectives. In particular, proposition (3.9) applies.

Recall that a morphism $\alpha\colon B\to A$ in an additive category $\mathcal{D}$ admits a *pseudo-kernel* if there is a morphism $\beta\colon C\to B$ such that $(\ ,C)\xrightarrow{(\ ,\beta)}(\ ,B)\xrightarrow{(\ ,\alpha)}(\ ,A)$ is exact. Every morphism in a skeletally small additive category $\mathcal{D}$ admits a pseudo-kernel iff $\operatorname{mod}\mathcal{D}$ is abelian, see [**2**].

The following reviews and extends results obtained by Auslander-Reiten in the abelian case [**6**, Prop.4.1& 4.2].

THEOREM 4.1. *Assume that* $\mathcal{B}\subseteq\mathcal{A}$ *is a skeletally small and full additive subcategory of an abelian category* $\mathcal{A}$. *Consider the following statements for a finitely presented functor* F *on* $\mathcal{B}$.

(i) *If* $(\ ,B)\xrightarrow{(\ ,\alpha)}(\ ,A)\to F\to 0$ *is exact in* $\operatorname{Mod}\mathcal{B}$ *for some morphism* α *from* $\mathcal{B}$, *then* α *is an admissible epimorphism.*

(ii) *There exists an exact sequence* $0\to C\xrightarrow{\beta}B\xrightarrow{\alpha}A\to 0$ *in* $\mathcal{B}$ *such that*

$$0\to(\ ,C)\xrightarrow{(\ ,\beta)}(\ ,B)\xrightarrow{(\ ,\alpha)}(\ ,A)\to F\to 0$$

is a projective resolution of F *in* $\operatorname{Mod}\mathcal{B}$.

(iii) $F(P)=0$ *for every* $\mathcal{B}$*-projective* P.

(iv) $\operatorname{Ext}^i_{\operatorname{Mod}\mathcal{B}}(F,Q)=0$ *for* $i\neq 2$ *and every projective* Q *in* $\operatorname{mod}\mathcal{B}$.

(v) $\operatorname{Hom}_{\operatorname{Mod}\mathcal{B}}(F,Q)=0$ *for every projective* Q *in* $\operatorname{mod}\mathcal{B}$.

The implications among them are as follows.

1. *One has always* (i)$\Longrightarrow$(ii) $\Longrightarrow$(iii) *as well as* (ii)$\Longrightarrow$(iv)$\Longrightarrow$(v).
2. *If* $\mathcal{B}$ *is closed under direct summands and has plenty of* $\mathcal{B}$*-projectives, then* (i)$\Longleftrightarrow$(ii)$\Longleftrightarrow$(iii) *and* $\operatorname{mod}\underline{\mathcal{B}}$ *is abelian with enough projectives.*
3. *If* $\mathcal{B}=\mathcal{A}$ *is abelian with enough projectives, all five conditions are equivalent,* $\operatorname{mod}\mathcal{B}$ *is as well abelian with enough projectives, and the right adjoint* $p^!$ *to the restriction functor* p_* *maps* $\operatorname{mod}\mathcal{B}$ *to* $\operatorname{mod}\underline{\mathcal{B}}$.

PROOF. (1) The implications (i)$\Longrightarrow$(ii) $\Longrightarrow$(iii) follow from the definitions and clearly (iv)$\Longrightarrow$(v). The proof of (ii)$\Longrightarrow$(iv) carries over from [**6**, Prop.4.1,4.2]: As was said in (3.2), every projective in $\operatorname{mod}\mathcal{B}$ is isomorphic to a direct summand of a representable functor, whence it suffices to show that (ii) implies (iv) with $Q=(\ ,D)$ for objects D from $\mathcal{B}$. To calculate $\operatorname{Ext}^i_{\operatorname{Mod}\mathcal{B}}(F,(\ ,D))$, apply $\operatorname{Hom}_{\operatorname{Mod}\mathcal{B}}(\ ,(\ ,D))$ to the projective resolution of F given in (ii). As one has for any X in $\mathcal{B}$ that $\operatorname{Hom}_{\operatorname{Mod}\mathcal{B}}((\ ,X),(\ ,D))\cong\operatorname{Hom}_{\mathcal{B}}(X,D)=(X,D)$ by Yoneda's

Lemma, the resulting complex is (isomorphic to) the complex of abelian groups

$$0 \to (A, D) \xrightarrow{(\alpha,D)} (B, D) \xrightarrow{(\beta,D)} (C, D) \to 0 .$$

By assumption, the sequence $0 \to C \xrightarrow{\beta} B \xrightarrow{\alpha} A \to 0$ is exact in the abelian category $\mathcal{A}$, of which $\mathcal{B}$ is a full subcategory, and so the dualized sequence allows nontrivial cohomology at most at its right end where it equals

$$\mathrm{Ext}^2_{\mathrm{Mod}\,\mathcal{B}}(F, (\ \ , D)) = \mathrm{Hom}_{\mathcal{B}}(C, D)/\,\mathrm{Hom}_{\mathcal{B}}(B, D) .$$

(2) The proof of [**7**, Prop.1.1] shows that $\underline{\mathcal{B}}$ admits pseudokernels if there are plenty of projectives, in which case $\mathrm{mod}\,\underline{\mathcal{B}}$ is abelian [**2**]. In fact, we introduced the notion of plenty of projectives just for this purpose. The representable functors provide enough projectives in $\mathrm{mod}\,\underline{\mathcal{B}}$.

To see that (iii)$\Longrightarrow$(i), assume (iii) and consider a presentation as in (i). For any $\mathcal{B}$-projective P the map $(P, \alpha)\colon (P, B) \to (P, A)$ is then surjective as $F(P) = 0$, and so α is an admissible epimorphism by Lemma (2.7.2).

(3) This is the case treated in [**6**, Prop.4.1& 4.2] and we include the argument for completeness and to make evident where the abelian property is indeed needed. If $\mathcal{B} = \mathcal{A}$ is abelian, $\mathcal{B}$ itself admits (pseudo-)kernels whence $\mathrm{mod}\,\mathcal{B}$ is abelian, and again the representable functors provide enough projectives.

It remains to establish (v)$\Longrightarrow$(i). Applying $\mathrm{Hom}_{\mathrm{Mod}\,\mathcal{B}}(\ \ , (\ \ , D))$ to a presentation of F as in (i) results up to isomorphism in the exact sequence of abelian groups

$$0 \to \mathrm{Hom}_{\mathrm{Mod}\,\mathcal{B}}(F, (\ \ , D)) \to (A, D) \xrightarrow{(\alpha,D)} (B, D) .$$

Thus (v) holds true iff (α, D) is injective for every object D from $\mathcal{B}$ iff α is an epimorphism in $\mathcal{B}$. As $\mathcal{B}$ is abelian, every epimorphism is admissible, whence (v)$\Longrightarrow$(i); and as $\mathcal{B}$ has enough, thus plenty of projectives, all conditions (i) through (v) are equivalent.

The final assertion goes back to [**1**, Prop.3.4], see the comments following [**6**, Prop. 4.3.] where $p^!$ is denoted t and is explicitly described. □

If $\mathcal{B}$ is closed under direct summands and has plenty of $\mathcal{B}$-projectives as a full additive subcategory of some abelian category $\mathcal{A}$, the preceding theorem says that nonzero finitely presented right $\underline{\mathcal{B}}$-modules are of projective dimension 2 as $\mathcal{B}$-modules and that their projective resolutions of length 2 correspond precisely to the exact sequences in $\mathcal{B}$.

Viewing short exact sequences as short exact complexes, one may thus express the equivalence (ii)$\Longleftrightarrow$(iii) as follows.

COROLLARY 4.2. *With hypotheses as just stated,* $\mathrm{mod}\,\underline{\mathcal{B}}$ *is naturally equivalent to the homotopy category of short exact sequences in* $\mathcal{B}$. *The equivalence sends the exact sequence* $0 \to C \xrightarrow{\beta} B \xrightarrow{\alpha} A \to 0$ *in* $\mathcal{B}$ *to the functor* $F = \mathrm{Cok}(\ \ , \alpha)$ *and an inverse is obtained by choosing a projective* $\mathrm{mod}\,\mathcal{B}$*-resolution of length* 2. □

The correspondence between short exact sequences in $\mathcal{B}$ and projective resolutions of finitely presented $\underline{\mathcal{B}}$-modules in $\mathrm{Mod}\,\mathcal{B}$ allows to calculate $\mathrm{Tor}^{\mathcal{B}}(F, \underline{\mathcal{B}})$, the evaluation on F of the left derived functors of $(\ \ ) \otimes_{\mathcal{B}} \underline{\mathcal{B}}$. From this perspective, the general form of the horseshoe lemma (2.6) expresses that the projection $\mathcal{B} \to \underline{\mathcal{B}}$ is *pseudoflat*. The proof is the same as that of [**7**, Prop.1.2]. It is this result that builds the bridge to investigations of Hochschild (co-)homology for stable Auslander algebras.

THEOREM 4.3. *If $\mathcal{B}$ is closed under direct summands and has plenty of $\mathcal{B}$-projectives as a full additive subcategory of some abelian category $\mathcal{A}$, then*

$$\operatorname{Tor}_i^{\mathcal{B}}(\ ,\underline{\mathcal{B}}) = 0 \qquad \text{for } i \neq 0, 2\,.$$

If F is a finitely presented $\underline{\mathcal{B}}$-module with projective resolution

$$0 \to (\ ,C) \xrightarrow{(\ ,\beta)} (\ ,B) \xrightarrow{(\ ,\alpha)} (\ ,A) \to F \to 0$$

in $\operatorname{Mod}\mathcal{B}$, *then*

$$\begin{aligned}\operatorname{Tor}_2^{\mathcal{B}}(F,\underline{\mathcal{B}}) &\cong \operatorname{Cok}\left((\ ,\underline{\Omega^1\alpha})\colon (\ ,\underline{\Omega^1 B}) \to (\ ,\underline{\Omega^1 A})\right)\\ &\cong \operatorname{Ker}\left((\ ,\underline{\beta})\colon (\ ,\underline{C}) \to (\ ,\underline{B})\right)\end{aligned}$$

and a projective resolution of F in $\operatorname{mod}\underline{\mathcal{B}}$ *is given by*

$$\cdots(\ ,\underline{\Omega^1 B}) \xrightarrow{(\ ,\underline{\Omega^1\alpha})} (\ ,\underline{\Omega^1 A}) \to (\ ,\underline{C}) \xrightarrow{(\ ,\underline{\beta})} (\ ,\underline{B}) \xrightarrow{(\ ,\underline{\alpha})} (\ ,\underline{A}) \to F \to 0\,.$$

The functor $\operatorname{Tor}_2^{\mathcal{B}}(\ ,\underline{\mathcal{B}})$ *lifts the third syzygy functor Ω^3 from* $\underline{\operatorname{mod}\underline{\mathcal{B}}}$ *to* $\operatorname{mod}\underline{\mathcal{B}}$.

PROOF. Indeed, the functor $\mathcal{A} \to \operatorname{Mod}\mathcal{B}$ that sends A to $(\ ,A)$ is left exact, whence embedding the diagram from the horseshoe lemma (2.6) into $\operatorname{Mod}\mathcal{B}$ yields first of all the exact sequence of cokernels

$$(\ ,\underline{C}) \xrightarrow{(\ ,\underline{\beta})} (\ ,\underline{B}) \xrightarrow{(\ ,\underline{\alpha})} (\ ,\underline{A}) \to F \to 0\,.$$

This shows that $\operatorname{Tor}_1^{\mathcal{B}}(F,\underline{\mathcal{B}}) = 0$ and that the sequence

$$0 \to \operatorname{Tor}_2^{\mathcal{B}}(F,\underline{\mathcal{B}}) \to (\ ,\underline{C}) \xrightarrow{(\ ,\underline{\beta})} (\ ,\underline{B}) \xrightarrow{(\ ,\underline{\alpha})} (\ ,\underline{A}) \to F \to 0$$

is exact. Using the snake lemma on the embedded "horseshoe diagram" shows then that the sequence

$$(\ ,\underline{\Omega^1 C}) \xrightarrow{(\ ,\underline{\Omega^1\beta})} (\ ,\underline{\Omega^1 B}) \xrightarrow{(\ ,\underline{\Omega^1\alpha})} (\ ,\underline{\Omega^1 A}) \to \operatorname{Tor}_2^{\mathcal{B}}(F,\underline{\mathcal{B}}) \to 0$$

is exact as well, whence the result. □

COROLLARY 4.4. *One has* $\operatorname{Tor}_2^{\mathcal{B}}((\ ,\underline{A}),\underline{\mathcal{B}}) \cong (\ ,\underline{\Omega^1 A})$ *for any A in $\mathcal{B}$.* □

4.5. We finish this section with some remarks on duality. Denoting $\mathcal{I} \subseteq \mathcal{B}$ the full subcategory of $\mathcal{B}$-injectives, set $\overline{\mathcal{B}} = \mathcal{B}/\mathcal{I}$ and note that $(\overline{\mathcal{B}})^{op} \cong \underline{\mathcal{B}^{op}}$ which in general is different from $\underline{\mathcal{B}}^{op}$. Should $\mathcal{B}$ be *Frobenius*, in the sense that $\mathcal{I} = \mathcal{P}$, then clearly $\overline{\mathcal{B}} = \underline{\mathcal{B}}$.

Under the usual duality between $\mathcal{B}$ and $\mathcal{B}^{op}$, any exact sequence in $\mathcal{B}$ is as well an exact sequence in $\mathcal{B}^{op}$, considered as a full additive subcategory of the abelian category $\mathcal{A}^{op}$. Now $\mathcal{B}^{op}$ is clearly closed under direct summands iff $\mathcal{B}$ is, and $\mathcal{B}^{op}$ has plenty of $\mathcal{B}^{op}$-projectives iff $\mathcal{B}$ has plenty of $\mathcal{B}$-injectives. So the corollary (4.2) yields in this case a duality between $\operatorname{mod}\underline{\mathcal{B}}$ and $\overline{\mathcal{B}}\operatorname{mod}$. It can be described directly.

PROPOSITION 4.6. *If $\mathcal{B}$ is closed under direct summands and has plenty of $\mathcal{B}$-projectives as well as plenty of $\mathcal{B}$-injectives, associating to F in* $\operatorname{mod}\underline{\mathcal{B}}$ *the (covariant) additive functor*

$$F^{\vee}\colon \underline{\mathcal{B}} \to \mathfrak{Ab} \qquad \text{with} \qquad F^{\vee}(D) = \operatorname{Ext}^2_{\operatorname{Mod}\mathcal{B}}(F,(\ ,D))$$

defines a duality between $\operatorname{mod}\underline{\mathcal{B}}$ *and* $\overline{\mathcal{B}}\operatorname{mod}$.

Under this duality, a projective resolution

$$0 \to (\ ,C) \xrightarrow{(\ ,\beta)} (\ ,B) \xrightarrow{(\ ,\alpha)} (\ ,A) \to F \to 0$$

of F as right $\mathcal{B}$-module corresponds to the projective resolution

$$0 \to (A,\ ) \xrightarrow{(\alpha,\ )} (B,\ ) \xrightarrow{(\beta,\ )} (C,\ ) \to F^{\vee} \to 0$$

of $F^{\vee}$ as left $\mathcal{B}$-module.

PROOF. Applying $\mathrm{Hom}_{\mathrm{Mod}\,\mathcal{B}}(\ ,(\ ,D))$ for an object D from $\mathcal{B}$ to the given projective resolution of F returns as in the proof of (4.1.1.(ii)$\Longrightarrow$(iv)) the exact sequence of abelian groups

$$0 \to (A,D) \xrightarrow{(\alpha,D)} (B,D) \xrightarrow{(\beta,D)} (C,D) \to \mathrm{Ext}^2_{\mathrm{Mod}\,\mathcal{B}}(F,(\ ,D)) = F^{\vee}(D) \to 0\,.$$

As β is an admissible monomorphism in $\mathcal{B}$, thus an admissible epimorphism in $\mathcal{B}^{op}$, it follows that $F^{\vee}$ is indeed a right $\underline{\mathcal{B}^{op}}$-module and that $F^{\vee\vee} \cong F$. Using (4.1.2.(iii)$\Longrightarrow$(ii)) for $\mathcal{B}^{op}$ shows that every finitely presented left $\overline{\mathcal{B}}$-module is up to isomorphism of the form $F^{\vee}$ for some finitely presented right $\underline{\mathcal{B}}$-module F. □

COROLLARY 4.7. *If P is an object in $\mathcal{B}$ that is at the same time $\mathcal{B}$-projective and $\mathcal{B}$-injective, and if $P \to A$ is an admissible epimorphism onto an object A in $\mathcal{B}$, then*

$$(\ ,\underline{A})^{\vee} = (\overline{\Omega^1 A},\ ) = \overline{\mathcal{B}} \otimes_{\mathcal{B}} (\Omega^1 A,\ )\,.$$

□

REMARK 4.8. Any duality between abelian categories transforms noetherian objects into artinian objects and vice versa whence the categories $\mathrm{mod}\,\underline{\mathcal{B}}$ and $\overline{\mathcal{B}}\,\mathrm{mod}$ are both artinian iff they are both noetherian, as soon as they are abelian.

Note that this does not hold for exact Frobenius categories as the example of maximal Cohen-Macaulay modules over a local commutative Gorenstein ring shows.

5. Exact Frobenius Categories with Enough Projective-Injectives

Considered as a complex in $\mathrm{Mod}\,\mathcal{B}$, the exhibited projective resolution of F as right $\underline{\mathcal{B}}$-module can be extended to a "complete" complex through

$$\cdots(\ ,\underline{C}) \xrightarrow{(\ ,\underline{\beta})} (\ ,\underline{B}) \xrightarrow{(\ ,\underline{\alpha})} (\ ,\underline{A}) \to \mathrm{Ext}^1_{\mathcal{A}}(\ ,C) \to \mathrm{Ext}^1_{\mathcal{A}}(\ ,B) \to \mathrm{Ext}^1_{\mathcal{A}}(\ ,A)\cdots$$

but there is of course no guarantee that, for $i \geqslant 1$ and X in $\mathcal{B}$, a functor of the form $\mathrm{Ext}^i_{\mathcal{A}}(\ ,X)$

(a) will vanish on $\mathcal{B}$-projectives, and, if so,
(b) will be finitely presented in $\mathrm{Mod}\,\underline{\mathcal{B}}$, and, if that too,
(c) will be representable thus projective in $\mathrm{Mod}\,\underline{\mathcal{B}}$.

Fortunately, Auslander-Reiten [**7**, Prop.1.3], [**5**, Prop.3.2] show essentially that (a) through (c) will hold if $\mathcal{B}$ is an exact *Frobenius category with enough projective-injectives*, a result that we recall, along with the relevant terminology, after the following remark.

REMARK 5.1. Evaluating the complete complex on an object from $\mathcal{B}$ results in an exact sequence of abelian groups that extends the usual long cohomology sequence of Exts to the left and, as A. Martsinkovsky pointed out to us, in the (dual) context where $\mathcal{B}$ itself is abelian with enough injectives, these exact sequences seem to appear for the first time in P. Hilton's book [**26**, Thm.13.16] as "homotopy sequences". In turn, they are attributed there to unpublished work of C. J. Shaddock.

DEFINITION 5.2. A full additive subcategory $\mathcal{B}$ of an abelian category $\mathcal{A}$ is *Frobenius* if an object is $\mathcal{B}$-projective iff it is $\mathcal{B}$-injective. Not to discriminate against either, a $\mathcal{B}$-projective/injective object is then called a *projective-injective* in $\mathcal{B}$. There are *enough projective-injectives* if the projective-injectives constitute enough projectives as well as enough injectives. In the same way one obtains the notion of *plenty of projective-injectives*.

The conditions (a) through (c) above can now be dealt with as follows.

LEMMA 5.3. *Let $\mathcal{B} \subseteq \mathcal{A}$ be a full additive subcategory of an abelian category and assume that there are enough $\mathcal{B}$-injectives. Let X be an object in $\mathcal{B}$.*

1. *If there is an admissible monomorphism $X \to I_X$ into a $\mathcal{B}$-injective I_X with $\mathrm{Ext}^1_{\mathcal{A}}(A, I_X) = 0$ for all A from $\mathcal{B}$, then $\mathrm{Ext}^1_{\mathcal{A}}(\ ,X)$ is a finitely presented functor on $\mathcal{B}$ that vanishes on $\mathcal{B}$-projectives.*
2. *If I_X is furthermore $\mathcal{B}$-projective, then $\mathrm{Ext}^1_{\mathcal{A}}(\ ,X)$ is representable as a functor on $\underline{\mathcal{B}}$.*

PROOF. (1) As the monomorphism from X to I_X is admissible, there is an exact sequence $0 \to X \to I_X \to Y \to 0$ in $\mathcal{B}$. As $\mathrm{Ext}^1_{\mathcal{A}}(\ ,I_X) = 0$ on $\mathcal{B}$, the sequence of right $\mathcal{B}$-modules

$$0 \to (\ ,X) \to (\ ,I_X) \to (\ ,Y) \to \mathrm{Ext}^1_{\mathcal{A}}(\ ,X) \to 0$$

is exact and constitutes a projective resolution of $\mathrm{Ext}^1_{\mathcal{A}}(\ ,X)$ in $\mathrm{mod}\,\mathcal{B}$. As $I_X \to Y$ is an admissible epimorphism, $\mathrm{Ext}^1_{\mathcal{A}}(\ ,X)$ vanishes on $\mathcal{B}$-projectives.

(2) If the $\mathcal{B}$-injective I_X is as well $\mathcal{B}$-projective, applying the right exact functor $(\ )\otimes_{\mathcal{B}}\underline{\mathcal{B}}$ to the resolution of $\mathrm{Ext}^1_{\mathcal{A}}(\ ,X)$ returns an isomorphism $(\ ,\underline{Y}) \cong \mathrm{Ext}^1_{\mathcal{A}}(\ ,X)$ as $(\ ,\underline{I_X}) = 0$ and $\mathrm{Ext}^1_{\mathcal{A}}(\ ,X) \cong \mathrm{Ext}^1_{\mathcal{A}}(\ ,X) \otimes_{\mathcal{B}} \underline{\mathcal{B}}$ by (3.9.1.). □

LEMMA 5.4. (cf. [**7**, Prop.1.3]) *Let $\mathcal{B} \subseteq \mathcal{A}$ be an exact subcategory of an abelian category with enough $\mathcal{B}$-projectives and enough $\mathcal{B}$-injectives. The functors of the form $\mathrm{Ext}^1_{\mathcal{A}}(\ ,X)$ for X in $\mathcal{B}$ are then enough injectives in the abelian category* $\mathrm{mod}\,\underline{\mathcal{B}}$.

PROOF. With $\mathcal{B}$ in $\mathcal{A}$, also $\mathcal{B}^{op}$ in $\mathcal{A}^{op}$ is extension closed, whence $\mathrm{Ext}^1_{\mathcal{A}}(A, I) = 0$ for any $\mathcal{B}$-injective I and any A in $\mathcal{B}$ by (2.9.1). Thus (5.3.1) shows that each functor $\mathrm{Ext}^1_{\mathcal{A}}(\ ,X)$ for X in $\mathcal{B}$ is in $\mathrm{mod}\,\underline{\mathcal{B}}$. The remaining assertions follow then as in [**7**, Prop.1.3]. □

To formulate the next result, call a (generally not bounded) complex in an abelian category with enough projectives a *complete resolution* if it is an exact complex of projectives and if dualizing the complex into any projective object returns an exact complex of abelian groups.

THEOREM 5.5. (cf. [**7**, Prop.1.4]) *Let $\mathcal{B} \subseteq \mathcal{A}$ be an exact Frobenius category with enough projective-injectives.*

1. *The category $\mathrm{mod}\,\underline{\mathcal{B}}$ is an abelian Frobenius category with enough projective-injectives.*
2. *The syzygy functor $\Omega^1 : \underline{\mathcal{B}} \to \underline{\mathcal{B}}$ is an auto-equivalence and so are $\mathrm{Tor}_2^{\mathcal{B}}(\ ,\underline{\mathcal{B}})$ on $\mathrm{mod}\,\underline{\mathcal{B}}$ and the syzygy functor on $\underline{\mathrm{mod}\,\underline{\mathcal{B}}}$.*
3. *Each functor of the form $\mathrm{Ext}^i_{\mathcal{A}}(\ ,A)$ for A in $\mathcal{B}$ is representable as $(\ ,\underline{\Omega^{-i}A})$ on $\mathrm{mod}\,\underline{\mathcal{B}}$, and the complete complex attached at the beginning of this section to a finitely presented functor F on $\underline{\mathcal{B}}$ is a complete resolution.*

4. *The functor $F \mapsto F^\vee$ is a duality between* $\operatorname{mod}\underline{\mathcal{B}}$ *and* $\underline{\mathcal{B}}\operatorname{mod}$ *and applying this duality to a complete resolution of F returns a complete resolution of* $\Omega^{-1}_{\underline{\mathcal{B}}\operatorname{mod}}(F^\vee)$.

PROOF. (1) The preceding two lemmata show that the representable functors are projective-injectives in $\operatorname{mod}\underline{\mathcal{B}}$ and that the representable functors $\operatorname{Ext}^1_{\mathcal{A}}(\ ,A)$ for A in $\mathcal{B}$ form enough injectives.

(2) The syzygy functor on a Frobenius category modulo its (enough) projective-injectives is an auto-equivalence, and so is then $\operatorname{Tor}_2^{\mathcal{B}}(\ ,\underline{\mathcal{B}})$ in view of (4.3).

(3) Choosing an admissible monomorphism from an object A into a $\mathcal{B}$-injective I_A defines through the exact sequence $0 \to A \to I_A \to \Omega^{-1}A \to 0$ a representative of the first "co-syzygy object" $\Omega^{-1}A$ of A and $(\ ,\underline{\Omega^{-1}A}) \cong \operatorname{Ext}^1_{\mathcal{A}}(\ ,A)$.

By induction, each $\operatorname{Ext}^i_{\mathcal{A}}(\ ,A)$ for A in $\mathcal{B}$ is seen to be representable on $\operatorname{mod}\underline{\mathcal{B}}$ through $(\ ,\underline{\Omega^{-i}A})$ whence the complete complex attached to F is a complex of projectives in $\operatorname{mod}\underline{\mathcal{B}}$. As each representable functor in $\operatorname{mod}\underline{\mathcal{B}}$ is injective, dualizing the complete complex into such a functor returns an exact complex of abelian groups.

(4) Considering the complex

$$\cdots(\ ,\underline{C}) \xrightarrow{(\ ,\beta)} (\ ,\underline{B}) \xrightarrow{(\ ,\alpha)} (\ ,\underline{A}) \to \operatorname{Ext}^1_{\mathcal{A}}(\ ,C) \to \operatorname{Ext}^1_{\mathcal{A}}(\ ,B) \to \operatorname{Ext}^1_{\mathcal{A}}(\ ,A)\cdots$$

as a complete resolution of F, the term $(\ ,\underline{A})$ is usually placed in complex degree zero, and applying the duality $(\ )^\vee$ to the complex returns

$$\cdots(\underline{A},\ ) \xrightarrow{(\alpha,\ )} (\underline{B},\ ) \xrightarrow{(\beta,\ )} (\underline{C},\ ) \to (\underline{\Omega^1 A},\ ) \to (\underline{\Omega^1 B},\ ) \to (\underline{\Omega^1 C},\ )\cdots$$

as $(\ ,\underline{X})^\vee = (\underline{\Omega^1 X},\ )$ by (4.7). As $F^\vee$ appears as the image of $(\underline{C},\ )$ in $(\underline{\Omega^1 A},\ )$ by its construction in (4.6), and as $(\ ,\underline{A})^\vee = (\underline{\Omega^1 A},\ )$ sits in complex degree zero, the result follows. □

Now we tie in the duality above with the naive duality $F \mapsto F^*$ from (3.3) that exchanges finitely presented modules with finitely co-represented ones.

If $\operatorname{mod}\underline{\mathcal{B}}$ is an abelian Frobenius category as above, then every object has a complete resolution and finitely presented or finitely co-represented are the same notions. Furthermore, $\overline{\mathcal{B}} = \underline{\mathcal{B}}$ so that $(\ )^*$ can be regarded as a duality between finitely presented right and left $\underline{\mathcal{B}}$-modules.

PROPOSITION 5.6. *If $\mathcal{B}$ is an exact Frobenius category with enough projective-injectives, then*

$$\operatorname{Tor}_2^{\mathcal{B}}(F,\underline{\mathcal{B}}) \cong (F^\vee)^*$$

for every F in $\operatorname{mod}\mathcal{B}$ *and $F \mapsto (F^*)^\vee$ is an inverse to* $\operatorname{Tor}_2^{\mathcal{B}}(\ ,\underline{\mathcal{B}})$.

PROOF. With notation as in (4.6), $F^\vee$ has finite presentation

$$(\underline{B},\ ) \xrightarrow{(\beta,\ )} (\underline{C},\ ) \to F^\vee \to 0$$

as left $\underline{\mathcal{B}}$-module, thus $(F^\vee)^*$ has finite co-presentation

$$0 \to (F^\vee)^* \to (\ ,\underline{C}) \xrightarrow{(\ ,\beta)} (\ ,\underline{B})\ .$$

The result follows from (4.3). □

If one views the equivalence $(\ )^{*\vee}$ as the *relative Nakayama equivalence* $\mathcal{N}_{\underline{\mathcal{B}}/\mathcal{B}}$ with respect to the morphism $\mathcal{B} \to \underline{\mathcal{B}}$ of homological codimension 2, then the result says that $\mathrm{Tor}_2^{\mathcal{B}}(\ ,\underline{\mathcal{B}})$ represents $\mathcal{N}^{-1}_{\underline{\mathcal{B}}/\mathcal{B}}$. To see how this categorical notion compares to the actual Nakayama automorphism in case there are almost split sequences, cf. [**7**].

6. Finite Representation Type

6.1. Let $\mathcal{B}$ be any additive category closed under direct summands. Following M.Auslander [**2**], we say that $\mathcal{B}$ is of *finite representation type* if there is an object M in $\mathcal{B}$ such that $\mathcal{B} = \mathrm{add}(M)$, that is, every object in $\mathcal{B}$ is direct summand of $M^{\oplus n}$ for some n. Such an object M is called a *representation generator for* $\mathcal{B}$.

Note that finite representation type in this sense does not imply that $\mathcal{B}$ admits only finitely many isomorphism classes of indecomposable objects. In analogy to [**3**, Sect.10] we say that $\mathcal{B}$ is of *finite type*, if there are only finitely many isomorphism classes of indecomposable objects and every object is a finite direct sum of indecomposable ones. If $M_1, \dots, M_m$ runs through all isomorphism types of indecomposables, then $M = M_1 \oplus \cdots \oplus M_m$ is a representation generator.

In case the reader finds this confusing, consider the category of those vectorspaces over a field k whose dimension is denumerable. This category has $k^{(\mathbb{N})}$ as a representation generator and only one isomorphism class of indecomposables, but it is not of finite type.

In case $\mathcal{B}$ is a Krull-Schmidt category, "finite representation type" and "finite type" are the same notions. By a recent result of R. Wiegand [**36**, Thm.1.1.], if M is a finitely generated module over a commutative semilocal noetherian ring R, then $\mathrm{add}(M)$ is of finite type, but in general it is not a Krull-Schmidt category.

6.2. If M is a representation generator for $\mathcal{B}$, the endomorphism ring $\Lambda = \mathrm{End}_{\mathcal{B}}(M)$ is the *Auslander algebra* of M and $\Gamma = \mathrm{End}_{\underline{\mathcal{B}}}(\underline{M})$, also written as $\Gamma = \underline{\mathrm{End}}_{\mathcal{B}}(M)$ is the associated *stable Auslander algebra.*

If M, M' are different choices of representation generators, then the respective Auslander algebras Λ, Λ' are Morita equivalent, and if $\Lambda' \cong \mathrm{End}_{\Lambda}(P)^{op}$ for some finitely generated and generating projective (left) Λ-module P, then $\Gamma' \cong \mathrm{End}_{\Gamma}(\Gamma \otimes_{\Lambda} P)^{op}$ so that the "same" Morita equivalence works for Γ, Γ' as well. In this sense, the Morita equivalence class of the *stabilization homomorphism* $\Lambda \to \Gamma$ is an invariant of the additive category $\mathcal{B}$ of finite representation type.

Now we translate the results obtained so far to obtain

6.3. *Proof of Theorem* (1.1). Let M be a representation generator of $\mathcal{B}$, a representation finite additive category that is closed under direct summands and embedded as a full additive subcategory with plenty of projectives into an abelian category $\mathcal{A}$. With $\Lambda = \mathrm{End}_{\mathcal{B}}(M)$ and $\Gamma = \underline{\mathrm{End}}_{\mathcal{B}}(M)$, evaluation in M defines by [**2**, Chap.III.2] equivalences of categories

$$ev_M : \mathrm{mod}\,\mathcal{B} \xrightarrow{\cong} \mathrm{mod}\,\Lambda \quad , \quad G \mapsto G(M)\,,$$

$$ev_{\underline{M}} : \mathrm{mod}\,\underline{\mathcal{B}} \xrightarrow{\cong} \mathrm{mod}\,\Gamma \quad , \quad F \mapsto F(\underline{M})\,.$$

As $\mathcal{B}$ has plenty of projectives, $\mathrm{mod}\,\underline{\mathcal{B}}$ is abelian by (4.1.2) and so is then $\mathrm{mod}\,\Gamma$, which means that Γ is right coherent, see [**2**, Chap.III.2]. Theorem (4.1) shows then as well that every finitely presented right Γ-module admits a projective resolution of length 2 as Λ-module and that is the claim in (1.1.1).

Theorem (4.3) yields the vanishing of Tors as claimed in (1.1.2). To see this more clearly, observe that the forgetful functor $p_*\colon \operatorname{mod}\underline{\mathcal{B}} \to \operatorname{mod}\mathcal{B}$ corresponds to the forgetful functor $\operatorname{mod}\Lambda \to \operatorname{mod}\Gamma$ under the equivalences above. So then correspond the (left) derived functors of the left adjoint to the respective forgetful functor. These were called $\operatorname{Tor}^{\mathcal{B}}(\ ,\underline{\mathcal{B}})$ on the functor side, whereas they are the usual Tors on the module side,

$$ev_{\underline{M}} \operatorname{Tor}^{\mathcal{B}}(G,\underline{\mathcal{B}}) \cong \operatorname{Tor}^{\Lambda}(ev_M(G),\Gamma)\,.$$

To obtain (1.1.3), observe first that $\operatorname{Tor}_2^{\Lambda}(\ ,\Gamma)$ is an exact functor by (1.1.2). As any Tor-functor commutes with arbitrary direct sums, it follows that $\operatorname{Tor}_2^{\Lambda}(\ ,\Gamma) \cong (\ )\otimes_\Gamma \operatorname{Tor}_2^{\Lambda}(\Gamma,\Gamma)$ as functors on $\operatorname{mod}\Gamma$ by the Eilenberg-Watts Theorem, cf. [**11**, Chap.II,Thm.2.3]. Left exactness of $\operatorname{Tor}_2^{\Lambda}(\ ,\Gamma)$ is then equivalent to left flatness of $L = \operatorname{Tor}_2^{\Lambda}(\Gamma,\Gamma)$. Applying $(\ )\otimes_\Lambda \Gamma$ to a projective resolution

$$0 \to P_2 \to P_1 \to P_0 \to \Gamma \to 0$$

of Γ through finite projective right Λ-modules yields the exact sequence of right Γ-modules

$$0 \to \operatorname{Tor}_2^{\Lambda}(\Gamma,\Gamma) \to P_2\otimes_\Lambda \Gamma \to P_1\otimes_\Lambda \Gamma \to P_0\otimes_\Lambda \Gamma \to \Gamma \to 0$$

as $\operatorname{Tor}_1^{\Lambda}(\Gamma,\Gamma) = 0$ and $\operatorname{Tor}_0^{\Lambda}(\Gamma,\Gamma) = \Gamma\otimes_\Lambda\Gamma \to \Gamma$ is an isomorphism. In this last exact sequence, all terms but the leftmost one are known already to be finite projective, whence this exact sequence splits and exhibits $\operatorname{Tor}_2^{\Lambda}(\Gamma,\Gamma)$ as a finite projective right Γ-module as well. That $(\ )\otimes_\Gamma L$ represents Ω^3_Γ follows from theorem (4.3).

Finally assume that $\mathcal{B}$ is exact Frobenius with enough projective-injectives. As these notions are selfdual, $\mathcal{B}^{op}$ is in particular also a representation finite category with plenty of projectives, M is again a representation generator and $\operatorname{End}_{\mathcal{B}^{op}}(M) = \Lambda^{op}$, $\underline{\operatorname{End}}_{\mathcal{B}^{op}}(M) = \Gamma^{op}$. Thus claims (1.1.1-3) hold with left and right interchanged. Furthermore, $\operatorname{Tor}_2^{\mathcal{B}}(\ ,\underline{\mathcal{B}})$ is an auto-equivalence on $\operatorname{mod}\underline{\mathcal{B}}$ and so is then $\operatorname{Tor}_2^{\Lambda}(\ ,\Gamma) \cong (\ )\otimes_\Gamma L$. It follows that L is an invertible Γ-bimodule. The final assertions on form and property of L^{-1} will be established in the proposition (6.5) below. □

Before we describe all the tensor powers of $L = \operatorname{Tor}_2^{\Lambda}(\Gamma,\Gamma)$, together with there inverses in the Frobenius case, let us make the following comment.

REMARK 6.4. The proof above shows that in Theorem (1.1), statement (1.1.2) implies (1.1.3) and that this implication is just a ring theoretic statement not requiring any detour through functor categories. In a similar vein, one may view (1.1.2) as a ring theoretic consequence of (1.1.1): just remark, as in [**12**, (87)], that for a surjective ring homomorphism $\Lambda \to \Gamma$ with kernel I one has $\operatorname{Tor}_1^{\Lambda}(\Gamma,\Gamma) = I/I^2$ and this vanishes iff I is idempotent. In the case at hand,

$$I = \{\varphi \in \operatorname{End}_{\mathcal{B}}(M) \mid \varphi \text{ factors through a } \mathcal{B}\text{-projective}\} \subseteq \Lambda$$

is clearly an idempotent ideal as by assumption any $\mathcal{B}$-projective P is a direct summand of $M^{\oplus n}$ for some n.

PROPOSITION 6.5. *With notations as before, the tensor powers of L can be described as*

$$L^{\otimes i} \cong \underline{\operatorname{Hom}}_{\mathcal{B}}(M,\Omega^i M) \quad \textit{for} \quad i \geq 0\,.$$

If $\mathcal{B}$ is exact Frobenius, then the inverse of L as invertible Γ-bimodule is given by
$$L^{-1} \cong \operatorname{Ext}^2_\Lambda(\Gamma, \Lambda)\,.$$
The negative powers $L^{-i} = \operatorname{Hom}_\Gamma(L^{\otimes i}, \Gamma)$ satisfy
$$L^{-i} \cong \underline{\operatorname{Hom}}_{\mathcal{B}}(M, \Omega^{-i}M) \cong \underline{\operatorname{Hom}}_{\mathcal{B}}(\Omega^i M, M) \quad \textit{for} \quad i \geq 0\,.$$
One has a functorial isomorphism
$$\operatorname{Hom}_\Gamma(\ \,, L^{-1}) \cong \operatorname{Ext}^2_\Lambda(\ \,, \Lambda)$$
of dualities on the category of coherent Γ-modules.

PROOF. From (4.4) we get $\operatorname{Tor}_2^{\mathcal{B}}((\ \,,\underline{M}), \underline{\mathcal{B}}) \cong (\ \,, \underline{\Omega^1 M})$, and evaluation in M yields $L = \operatorname{Tor}_2^\Lambda(\Gamma, \Gamma) \cong \underline{\operatorname{Hom}}_{\mathcal{B}}(M, \Omega^1 M)$. Furthermore, by induction,
$$\begin{aligned} L^{\otimes i} &\cong L^{\otimes(i-1)} \otimes_\Gamma L \cong \operatorname{Tor}_2^\Lambda(L^{\otimes(i-1)}, \Gamma) \\ &\cong ev_{\underline{M}} \operatorname{Tor}_2^{\mathcal{B}}((\ \,, \underline{\Omega^{i-1}M}), \underline{\mathcal{B}}) \cong ev_{\underline{M}}(\ \,, \underline{\Omega^1(\Omega^{i-1}M)}) \quad \text{by (4.4)} \\ &\cong \underline{\operatorname{Hom}}_{\mathcal{B}}(M, \Omega^i M)\ . \end{aligned}$$
In case $\mathcal{B}$ is exact Frobenius, we use (5.6), noting that under evaluation in M the duality $(\ )^*$ on $\operatorname{mod}\underline{\mathcal{B}}$ becomes $\operatorname{Hom}_\Gamma(\ \,, \Gamma)$, whereas the duality $(\ )^\vee$ becomes $\operatorname{Ext}^2_\Lambda(\ \,, \Lambda)$. If F is a right $\underline{\mathcal{B}}$-module, set $N = F(M)$ to translate the isomorphism
$$\operatorname{Tor}_2^{\mathcal{B}}(F, \underline{\mathcal{B}})^* \cong F^\vee$$
into
$$\operatorname{Hom}_\Gamma(\operatorname{Tor}_2^\Lambda(N, \Gamma), \Gamma) \cong \operatorname{Ext}^2_\Lambda(N, \Lambda)\ . \tag{2}$$
Taking $N = \Gamma$ gives the desired description of L^{-1}, and the lefthand side of (2) can be expanded to
$$\operatorname{Hom}_\Gamma(\operatorname{Tor}_2^\Lambda(N, \Gamma), \Gamma) \cong \operatorname{Hom}_\Gamma(N \otimes_\Gamma L, \Gamma) \cong \operatorname{Hom}_\Gamma(N, L^{-1})$$
which implies the claimed isomorphism of dualities on the category of coherent Γ-modules. The assertion on the form of L^{-i} follows as Ω^{-i} is an inverse to Ω^i on $\underline{\mathcal{B}}$. □

Now we turn to the

6.6. *Proof of Corollary* (1.2). The first claim follows from (4.8), as the assumption implies that $\Gamma\operatorname{mod}$ and $\operatorname{mod}\Gamma$ are noetherian abelian categories, dual to each other. Thus both categories are artinian and as they are also Frobenius, the ring Γ is selfinjective artinian.

For (1.2.2), note first that the claim is invariant under Morita equivalence, thus independent of the particular choice of representation generator. It suffices thus to take $M = M_1 \oplus \cdots \oplus M_t$ where the objects M_i run exactly once through all isomorphism classes of indecomposable objects in $\mathcal{B}$. As $\mathcal{B}$ is exact Frobenius with enough projective-injectives, Ω^1 is an equivalence on $\underline{\mathcal{B}}$ and a first syzygy of a nonprojective indecomposable M_i is of the form $M_{\tau(i)} \oplus P$ for a unique indecomposable nonprojective $M_{\tau(i)}$. Assuming that the index set $\{1, \ldots, s\}$ with $s < t$ corresponds to the nonprojective indecomposables, τ is a permutation of this set and we get a right Γ-module isomorphism
$$\tilde{\tau}\colon L = \underline{\operatorname{Hom}}_\Gamma\Big(\bigoplus_{i=1}^s M_i, \bigoplus_{i=1}^s M_{\tau(i)}\Big) \to \Gamma = \underline{\operatorname{Hom}}_\Gamma\Big(\bigoplus_{i=1}^s M_i, \bigoplus_{i=1}^s M_i\Big)\ .$$

This means that $L \cong {}_1\Gamma_\gamma$ for some automorphism γ of Γ. If τ is of order n, then $\tilde{\tau}^n$ induces automorphisms μ_i of the local algebras $\underline{\mathrm{End}}_{\mathcal{B}}(M_i)$ and so each of them is conjugate to a quasi-unipotent automorphism by the Skolem-Noether theorem. The claim follows. To guard against misconceptions, observe that $\tilde{\tau}^n$ is generally not the identity with respect to the left Γ-module structure, it expresses rather the (non identity) isomorphisms $\Omega^n_{\mathcal{B}} M_i \cong M_i$.

It remains to establish (1.2.3). In view of (1.1), the third syzygy functor Ω^3_Γ is represented by $(\ )\otimes_\Gamma L$ where L is invertible. Applying Morita equivalence, it follows that R is an artinian Gorenstein ring whose 3'rd syzygy functor is represented by an invertible R-bimodule L'. Such invertible bimodules are isomorphic to R as one-sided modules, cf. [**11**, III.Prop (7.5)] and so tensor product with them preserves lengths. Localizing at one of the finitely many maximal ideals $\mathfrak{m} \subset R$, it follows that the ranks of the free modules in a minimal resolution of the residue class field $k = R/\mathfrak{m}$ over $R_\mathfrak{m}$ are bounded. But this implies that $R_\mathfrak{m}$ is a quotient of a discrete valuation ring modulo a finite power of its maximal ideal, see for example [**10**]. It follows that Ω^{2i}_R is represented by the identity functor as resolutions over (products of) hypersurface rings are periodic of period 2 by a result of D. Eisenbud [**17**]; cf. also [**10**] or [**6**, Proof of Prop.3.2]. Thus already Ω^1_R is represented by an invertible module. Looking again at a residue class field $k = R/\mathfrak{m}$ of R, it follows that the length of the first syzygy in a minimal resolution is at most 1, that is, $V/\mathfrak{m}^\nu$ is of length either 1 or 2. □

7. Periodic Hochschild (Co-)Homology

If $K \to A$ is an algebra over a commutative ring K, we denote $A^{ev} = A \otimes_K A^{op}$ its enveloping K-algebra and $\mathrm{HH}(A/K,\ )$ or $\mathrm{HH}(A,\ )$ the corresponding Hochschild (co-)homology groups. Recall that $\mathrm{HH}^*(A/K, M) \cong \mathrm{Ext}^*_{A^{ev}}(A, M)$ for any A-bimodule M as soon as A is projective as K-module.

7.1. *Proof of Theorem* (1.5). If $\Lambda \to \Gamma$ is a pseudoflat ring epimorphism where Λ is a k-algebra with $\mathrm{HH}\dim_k \Lambda = 2$ and k a field, let

$$P_\bullet \equiv \quad 0 \to P_2 \to P_1 \to P_0 \to \Lambda \to 0$$

be a projective resolution of Λ over its enveloping k-algebra Λ^{ev}. As

$$\mathrm{Tor}^{\Lambda^{ev}}(\Gamma^{ev}, \Lambda) \cong \mathrm{Tor}^\Lambda(\Gamma, \Gamma)$$

by [**13**, Thm.IX.2.8.], and $\mathrm{Tor}_1^\Lambda(\Gamma,\Gamma) = 0$, $\mathrm{Tor}_0^\Lambda(\Gamma,\Gamma) \cong \Gamma$, tensoring the resolution on both sides with Γ over Λ yields an exact sequence of Γ-bimodules

$$0 \to \mathrm{Tor}_2^\Lambda(\Gamma,\Gamma) \to Q_2 \to Q_1 \to Q_0 \to \Gamma \to 0\,,$$

where $Q_i = \Gamma \otimes_\Lambda P_i \otimes_\Lambda \Gamma$. As all terms to the right of $L = \mathrm{Tor}_2^\Lambda(\Gamma,\Gamma)$ in this sequence are Γ-projective on either side, the same holds for L. It follows that each $Q_i \otimes_\Gamma L$ is again a projective Γ-bimodule and $\Gamma \otimes_\Lambda P_\bullet \otimes_\Lambda L^{\otimes i}$ yields the exact sequence

$$0 \to L^{\otimes(i+1)} \to Q_2 \otimes_\Gamma L^{\otimes i} \to Q_1 \otimes_\Gamma L^{\otimes i} \to Q_0 \otimes_\Gamma L^{\otimes i} \to L^{\otimes i} \to 0\,.$$

Stringing these sequences together resolves Γ through a quasiperiodic projective Γ^{ev}-resolution of period 3 with periodicity factor L as desired.

Now assume that $L^{\otimes n} \cong \Gamma$ as Γ-bimodule. This means that the $3n$'th syzygy of Γ over Γ^{ev} is again Γ and the corresponding segment of the projective resolution gives a class z in $\mathrm{HH}^{3n}(\Gamma/k, \Gamma)$. Using M.Gerstenhaber's theorem [**19**] that Hochschild cohomology of associative algebras is graded commutative, it follows that $z^2 = 0$ if $3n$ is odd and the characteristic of k is not 2. But z^2 in $\mathrm{HH}^{6n}(\Gamma/k,\Gamma) = \mathrm{Ext}^{6n}_{\Gamma^{ev}}(\Gamma,\Gamma)$ is the class of the extension of Γ through its 6'th syzygy module given by the corresponding segment of the projective resolution. By definition, over any ring A and for any A-module M, the tautological class in $\mathrm{Ext}^i_A(M,\Omega^i_A M)$ is trivial iff $\mathrm{projdim}_A M < i$, and the syzygy module $\Omega^i_A M$ is then A-projective. Finally note that Γ is projective over its enveloping k-algebra iff $k \to \Gamma$ is a separable algebra. □

Now we turn to the generalized Ringel-Schofield theorem (1.9), using heavily [**14**] that implicitly contains a proof of the following.

LEMMA 7.2. *For any finite graph Q and any commutative ring K, the preprojective algebra $\Pi^0(KQ)$ is graded through the lenghts of paths, and each homogeneous component is a finite free K-module whose rank is independent of K.*

PROOF. For this proof, unadorned tensor products are over K. To begin with, orient all the edges in Q to obtain a quiver, and denote KQ its path algebra over K. Assigning degree 0 to the vertices and degree 1 to the arrows, this becomes a positively graded K-algebra. We abbreviate $\Pi_K = \Pi^0(KQ)$ and assume first that K is a field. According to [**14**, Thm.3.1 & Lemma 1.4], the preprojective algebra Π_K allows the description $\Pi_K \cong \mathrm{T}_{KQ}\,\mathrm{HH}^1(KQ, KQ\otimes KQ)$ where the term on the right is the tensor algebra over the path algebra KQ on the KQ-bimodule given by the first Hochschild cohomology group with values in the bimodule $KQ\otimes KQ$. The minimal graded projective resolution of KQ over its enveloping algebra is given by

$$(3)\qquad 0 \to \bigoplus_{a\colon i\to j \text{ in } Q} KQe_j \otimes e_i KQ \xrightarrow{\partial} \bigoplus_i KQe_i \otimes e_i KQ \xrightarrow{mult} KQ \to 0$$

with $\partial((e_j\otimes e_i)_a) = (e_j\otimes a)_j - (a\otimes e_i)_i$.

Now $\mathrm{HH}^*(KQ, KQ\otimes KQ) = \mathrm{Ext}^*_{KQ^{ev}}(KQ, KQ^{ev})$ and it is a simple exercise to show that only isolated vertices i_s of Q contribute to Hochschild degree zero

$$\mathrm{Hom}_{KQ^{ev}}(KQ, KQ^{ev}) = \bigoplus_{i_s} e_{i_s} KQ e_{i_s} = \bigoplus_{i_s} e_{i_s} K e_{i_s}\,.$$

Dualizing the exact sequence (3) yields the exact sequence of KQ-bimodules

$$0 \to \bigoplus_{i'} KQe_{i'} \otimes e_{i'} KQ \to \bigoplus_{a\colon i\to j \text{ in } Q} KQe_i \otimes e_j KQ \to \mathrm{HH}^1(KQ, KQ\otimes KQ) \to 0$$

where i' in the left sum runs over the nonisolated vertices. We used here the identifications of KQ-bimodules

$$\mathrm{Hom}_{KQ^{ev}}(KQe_j\otimes e_i KQ, KQ^{ev}) \cong e_j KQ \otimes KQe_i \cong KQe_i \otimes e_j KQ$$

pointed out in (loc.cit).

With $P_0 = \bigoplus_{a\colon i\to j} KQe_i \otimes e_j KQ$ and $P_1 = \bigoplus_{i'} KQe_{i'}\otimes e_{i'}KQ$, it follows that the algebra Π_K is the homomorphic image of $\mathrm{T}_{KQ}\,P_0$ modulo the ideal generated by P_1. Now $\mathrm{T}_{KQ}\,P_0 = K\overline{Q}$, the path algebra of the preprojective quiver associated to Q, and the (minimal graded) projective $K\overline{Q}$-resolution of Π_K is

$$(4)\qquad 0 \to \bigoplus_{i'} K\overline{Q}e_{i'} \otimes e_{i'} K\overline{Q} \xrightarrow{1\otimes\partial^*\otimes 1} K\overline{Q} \to \Pi_K \to 0$$

as P_1 generates the ideal defining Π_K and $\mathrm{T}_{KQ} P_0$ is of global dimension 1.

Each term in this exact sequence is graded through path length and each homogeneous component is a finite K-module. The corresponding dimensions are independent of the field for the projective $K\overline{Q}$-modules in this presentation, as these dimensions are given in terms of the combinatorics of $\overline{Q}$. It follows that the dimensions of the graded components of Π_K are independent of the field as well.

Each bigraded component of the $\mathbb{Z}$-algebra $\Pi_{\mathbb{Z}}$ is then finite free, of rank equal to the dimension of the corresponding component over any field. Now for an arbitrary commutative ring K one has $\Pi_K = \Pi_{\mathbb{Z}} \otimes_{\mathbb{Z}} K$ and the claim follows. $\square$

We conclude with the proof of the generalized Ringel-Schofield theorem.

7.3. *Proof of Theorem* (1.9). For the moment let Q be any finite graph as in the previous lemma, Π its preprojective algebra over some fixed commutative ring K, and $K\overline{Q}$ the path algebra of the preprojective quiver. As any epimorphism of K-rings; see e.g. [**12**, (97)]; the surjective K-algebra homomorphism $K\overline{Q} \to \Pi$ gives rise to an exact sequence

$$0 \to \mathrm{Tor}_1^K(\Pi,\Pi) \to \mathrm{Tor}_1^{K\overline{Q}}(\Pi,\Pi) \to \Pi \otimes_{K\overline{Q}} \Omega^1_{K\overline{Q}/K} \otimes_{K\overline{Q}} \Pi \to \Omega^1_{\Pi/K} \to 0\,,$$

where this time Ω^1 represents the universal module of differentials.

The preceding lemma says that the leftmost term $\mathrm{Tor}_1^K(\Pi,\Pi)$ vanishes. Now $\Omega^1_{\Pi/K}$ is a first syzygy module (sic!) of Π over its enveloping K-algebra and the Π-bimodule $\Pi \otimes_{K\overline{Q}} \Omega^1_{K\overline{Q}/K} \otimes_{K\overline{Q}} \Pi$ is projective as $\Omega^1_{K\overline{Q}/K}$ is projective as $K\overline{Q}^{ev}$-module. It follows that $\mathrm{Tor}_1^{K\overline{Q}}(\Pi,\Pi)$ is a projective Π-bimodule iff $\mathrm{HH}\dim_K \Pi \le 2$.

The presentation (7.2.(4)) gives a natural surjection

$$\Pi \otimes_{KQ} P_1 \otimes_{KQ} \Pi \cong \bigoplus_{i'} \Pi e_{i'} \otimes e_{i'}\Pi \to \mathrm{Tor}_1^{K\overline{Q}}(\Pi,\Pi)$$

which is a graded projective cover. Furthermore, applying (7.2.(3)) to the quiver $\overline{Q}$, it follows that $\Pi \otimes_{K\overline{Q}} \Omega^1_{K\overline{Q}/K} \otimes_{K\overline{Q}} \Pi \cong \oplus_{b:\, i\to j \text{ in } \overline{Q}} \Pi e_j \otimes e_i \Pi$.

Taken together, we get a minimal Π^{ev}-presentation of $\Omega^1_{\Pi/K}$,

$$\bigoplus_{i'} \Pi e_{i'} \otimes e_{i'}\Pi \to \bigoplus_{b:\, i\to j \text{ in } \overline{Q}} \Pi e_j \otimes e_i \Pi \to \Omega^1_{\Pi/K} \to 0$$

and we need to know when this is already all there is to the resolution. Each module in the presentation is graded and each homogeneous component is a finite free K-module, whose finite rank is independent of K. It follows that the presentation is a minimal resolution as soon as that holds over *some* ring K, for example a field of characteristic zero.

This finishes the proof for the extended Coxeter-Dynkin diagrams, as we know that at least over an algebraically closed field of characteristic zero their preprojective algebras are Morita equivalent to algebras of Hochschild dimension 2, and Hochschild dimension is Morita invariant [**29**]. The claim for the deformed preprojective algebras of extended Coxeter-Dynkin diagrams follows immediately: It is established in [**15**, Lemma 1.1] that the family $\Pi^\lambda(K\tilde{\Sigma})$ is flat in λ and Hochschild dimension is upper semi-continuous in flat families. The claim for the ordinary Coxeter-Dynkin diagrams follows as a "deformed" preprojective algebra $\Pi^\lambda(K\Sigma)$ is a pseudoflat quotient of the corresponding algebra $\Pi^\lambda(K\tilde{\Sigma})$, the kernel being the idempotent ideal generated by the idempotent corresponding to the extra vertex in $\tilde{\Sigma}$. (We put the term "deformed" in quotation marks as for ordinary Coxeter-Dynkin diagrams the family $\Pi^\lambda(K\Sigma)$ is certainly not flat in λ. In fact, most of the time these are zero rings as $\Pi^\lambda(K\tilde{\Sigma})$ has generally no nontrivial artinian quotients, see [**15**, Cor. 6.3].) □

Note that the proof exhibits the minimal graded resolution of the preprojective algebra of (a disjoint union of) extended Coxeter-Dynkin diagrams that was originally found by Ringel-Schofield [**34**]: With notations as above it is of the form

$$0 \to \bigoplus_{i'} \Pi e_{i'} \otimes e_{i'}\Pi \to \bigoplus_{b:\, i\to j \text{ in } \overline{Q}} \Pi e_j \otimes e_i \Pi \to \bigoplus_i \Pi e_i \otimes e_i \Pi \to \Pi \to 0\,.$$

For a Coxeter-Dynkin diagram of type A_n, K.Erdmann and N.Snashall [**18**] give an explicit description of the Hochschild cohomology of the preprojective algebra. In general, the Hochschild cohomology of the (deformed) preprojective algebras of a Coxeter-Dynkin diagram should of course be "explained" through the geometry of the corresponding simple surface singularities.

It would be particularly interesting to know whether the deformed preprojective algebras are again (stable) Auslander algebras for the category of reflexive modules over the deformed singularity. If so, it would confirm for this case that finite representation type is open in flat families, an open question in the representation theory of modules over higher dimensional rings, cf. [**27**]. An affirmative answer would also allow to connect the work in [**15**] with the classification theory in [**32**].

References

[1] M. Auslander: *Coherent Functors*, in "Proceedings of the Conference on Categorical Algebra, La Jolla 1965,"pp.189–231, Springer Verlag, New York 1966

[2] M. Auslander: *Representation dimension of artin algebras*, Queen Mary College Mathematics Notes, Queen Mary College, London 1971, 179p.

[3] M. Auslander: *Isolated singularities and almost split sequences*, in: Proc. ICRA IV, Springer Lect. Notes in Math. **1178** (1986), 194–241.

[4] M. Auslander: *Rational Singularities and almost split sequences*, Trans. Amer. Math. Soc. **293** (1986), 511–531.

[5] M. Auslander and I. Reiten: *Stable equivalence for artin algebras*, in: Conf. on Orders, Group Rings and Related Topics, Springer Lect. Notes in Math. **353** (1973), 8–71.

[6] M. Auslander and I. Reiten: *Stable Equivalence of Dualizing R-Varieties*, Adv. in Math. **12** (1974), 306–366.

[7] M. Auslander and I. Reiten: DTr-*periodic modules and functors*, in: Bautista, Raymundo (ed.) et al., "Representation theory of algebras. Seventh international conference, August 22-26, 1994, Cocoyoc, Mexico." CMS Conf. Proc. **18**, 39–50, American Mathematical Society, Providence 1996.

[8] M. Auslander, I. Reiten, and S. Smalø: *Representation theory of Artin algebras*, Cambridge Studies in Advanced Mathematics **36**, University Press, Cambridge 1995.

[9] M. Auslander and S. Smalø: *Preprojective modules over artin algebras*, J. Algebra **66** (1980), 61–122.

[10] L. L. Avramov, *Infinite free resolutions*, Six lectures on commutative algebra, Barcelona 1996 (J. Elias, J. Giral, M.-R. Miró-Roig, S. Zarzuela, eds.), Progr. Math. **166**, 1–118, Birkhäuser, Basel 1998

[11] H. Bass: *Algebraic K-Theory*, W.A.Benjamin, Inc. , New York 1968

[12] G. M. Bergman and W. Dicks: *Universal derivations and universal ring constructions.* Pac. J. Math. **79** (1979), 293–337.

[13] H. Cartan and S. Eilenberg: *Homological Algebra*, Seventh Printing 1973, Princeton University Press, Princeton 1956

[14] W. Crawley-Boevey: *Preprojective algebras, differential operators and a Conze embedding for deformations of Kleinian Singularities*, Univ. of Leeds, Dept. of Pure Math. Preprint Series no. **4** (1998), 19p.

[15] W. Crawley-Boevey and M. P. Holland: *Noncommutative Deformations of Kleinian Singularities*, Duke Math. J. **92** (1998), 605–635.

[16] E. Dieterich: *The Auslander-Reiten quiver of an isolated singularity*, in: Singularities, representation of algebras, and vector bundles, Proc., Lambrecht 1985, (ed. by G.-M. Greuel, G. Trautmann), Springer Lect. Notes Math. **1273** (1987), 244–264.

[17] D. Eisenbud: *Homological algebra on a complete intersection, with an application to group representations*, Trans. Amer. Math. Soc. **260** (1980), 35–64.

[18] K. Erdmann and N. Snashall: *On Hochschild cohomology of preprojective algebras, I,II.* J. Algebra **205** (1998), 391–412, 413–434.

[19] M. Gerstenhaber: *The Cohomology Structure of an Associative Ring*, Ann. of Math. **78**(2) (1963), 267–288.

[20] D. Happel: *Hochschild cohomology of finite-dimensional algebras.* Séminaire d'algèbre P. Dubreil et M.-P. Malliavin, Proc., Paris/Fr. 1987/88, Springer Lect. Notes Math. **1404** (1989), 108–126.

[21] D. Happel, U. Preiser and C. M. Ringel: *Vinberg's characterization of Coxeter-Dynkin diagrams using subadditive functions with application to periodic modules.* in: Representation theory II, Proc. 2nd int. Conf., Ottawa 1979, Springer Lect. Notes Math. **832** (1980), 280–294.

[22] D. Happel, U. Preiser and C. M. Ringel: *Binary polyhedral groups and Euclidean diagrams.* Manuscr. Math. **31** (1980), 317–329.

[23] R. Hartshorne: *Residues and Duality*, Springer Lect. Notes Math. **20**, Springer Verlag, New York 1966

[24] A. Heller: *Homological algebra in abelian categories*, Ann. of Math. **68** (1958), 484–525.

[25] J. Herzog: *Ringe mit nur endlich vielen Isomorphieklassen von maximalen, unzerlegbaren Cohen-Macaulay Moduln.* Math. Ann. **233** (1978), 21–34.

[26] P. Hilton: *Homotopy Theory and Duality*, Notes on Mathematics and its Applications, Gordon and Breach, New York 1965

[27] H. Knörrer: *Torsionsfreie Moduln bei Deformation von Kurvensingularitäten*, in: Singularities, representation of algebras, and vector bundles, Proc., Lambrecht 1985, (ed. by G.-M. Greuel, G. Trautmann), Springer Lect. Notes Math. **1273** (1987), 150–155.

[28] H. Knörrer: *Cohen-Macaulay modules on hypersurface singularities. I.* Invent. Math. **88** (1987), 153–164.

[29] J. L. Loday: *Cyclic Homology.* Grundlehren der math. Wissenschaften **301**, Springer Verlag, New York 1992

[30] M. Lorenz: *On the Homology of Graded Algebras*, Comm. in Algebra, **20**(2) (1992), 489–507.

[31] D. Quillen: *Higher Algebraic K-Theory: I*, in *"Algebraic K-Theory I(Batelle Institute Conf. 1972)"*, 85–147, Springer Lect. Notes in Math. **341**, Springer-Verlag Berlin 1973

[32] I. Reiten and M. Van den Bergh: *Two-dimensional tame and maximal orders of finite representation type*, Mem. Amer. Math. Soc., **80** (1989), no. 408, viii+72 pp.

[33] C. Riedtmann: *Algebren, Darstellungsköcher, Überlagerungen und zurück.* Comment. Math. Helv. **55** (1980), 199–224.

[34] C. M. Ringel and A. Schofield: *Wild Algebras with periodic Auslander-Reiten Translate*, manuscript, 17 pp.

[35] **SGA 4**. M. Artin, A. Grothendieck, and J. L. Verdier: *Théorie des Topos et Cohomologie Etale des Schémas*, Séminaire de Géométrie Algébrique du Bois Marie 1963/64, SGA 4, Springer Lect. Notes in Math. **269**, **270**, **305**, Springer Verlag, New York 1972-73

[36] R. Wiegand: *Local Rings of Finite Cohen-Macaulay Type*, J. Algebra **203** (1998), 156–168.

Department of Mathematics, University of Toronto, 100 St.George St., Toronto, Ont. M5S 3G3, Canada

E-mail address: ragnar@math.utoronto.ca

Contemporary Mathematics
Volume **229**, 1998

The syzygy theorem for monomial algebras

M. C. R. Butler

ABSTRACT. This note contains a refinement and generalisation of the remarkable recent proof in [**2, 3**] that the second syzygy module of any module over a finite-dimensional monomial algebra is a direct sum of cyclic modules, these being isomorphic to principal one-sided ideals generated by monomials.

1. Introduction

Let Λ be a *monomial* algebra of finite dimension over a field k, that is, the quotient algebra of the path algebra of a finite quiver by an admissable ideal generated by paths in the quiver. It is shown in [**2**] and [**3**] that the first syzygy module of any submodule of a projective Λ-module and, hence, the second syzygy module of an arbitrary Λ- module, is a direct sum of copies of principal one-sided ideals generated by monomials. In this note we refine the original proof of this remarkable syzygy theorem in two ways. Firstly, we show that there is a rather natural class of modules whose first syzygies are direct sums of copies of principal ideals, this class including many examples, such as the semisimple modules, which need not be submodules of projectives. Secondly, we show that the condition of finite-dimensionality on the algebra can be usefully relaxed.

Notation and terminology We fix throughout a field k and a quiver Q with finite vertex set Q_0 and arbitrary arrow set Q_1. The path algebra $\Gamma = kQ$ of Q over k has a canonical k-basis which will be identified with the set of all paths, also called *monomials*, in the quiver. The set of monomials of a fixed length $l \geq 0$ will be denoted by Q_l, and their union by Q_*. (Of course the Q_l's are empty for all sufficiently large l if Q is an acyclic quiver.) We write S for the subspace of Γ with basis Q_0 of primitive orthogonal idempotents, so that $S \cong k^{|Q_0|}$ as a k-algebra, and J for the augmentation ideal of Γ, that is, the ideal generated by Q_1. Obviously, for each $l \geq 0$, the ideal J^l is generated both as a right ideal and as a left ideal by the set Q_l.

Throughout the note, the terms *monomial right ideal, monomial ideal, monomial subspace, I, of* Γ signify that the set $I \cap Q_*$ is a k- basis of the object I. For any monomial ideal, I, the algebra Γ/I will be called a *monomial algebra.* Unless otherwise stated, modules are assumed to be right modules.

1991 *Mathematics Subject Classification.* 16D40, 16D70, 16G20, 16S99.

2. Statement of the syzygy theorem

We first describe the class of modules whose first syzygies are the main concern of this note. Let M be any Γ- module and $l \geq 0$. Since Q_l generates J^l as a left ideal, the submodule MJ^l of M is spanned by its S-submodules Mp, where p runs through Q_l.

Definition The right Γ-module M will be called *a path separated module* if, for each $l \geq 0$, the decomposition of S-modules

$$MJ^l = \sum_{p \in Q_l} Mp$$

is a direct sum decomposition.

Clearly Γ itself is a path separated module because, for all $l \geq 0$, we have $J^l = \bigoplus_{p \in Q_l} \Gamma p$. However the following proposition shows that the class includes all submodules of projective modules over arbitrary monomial quotient algebras of Γ.

PROPOSITION 2.1. *(a) A cyclic module of the form Γ/I, where I is a monomial right ideal of Γ, is a path separated Γ- module. (b) The class of path separated Γ-modules is closed under the formation of submodules and of arbitrary direct sums and direct products.*

PROOF. We only prove (a), (b) being an obvious formal consequence of the definition. Let $M = \Gamma/I$ and $l \geq 0$. Any element $\bar{x} \in MJ^l = \sum_{p \in Q_l} Mp$ can be represented by a finite sum $x = \sum_{p \in Q_l} x_p p$, in which each $x_p \in \Gamma$ is itself a finite sum, $x_p = \sum_{q \in Q_*} \lambda_{p,q} q$, with coefficients $\lambda_{p,q} \in k$. Let $\bar{x} = 0$, so that $x \in I$. Now I is a monomial right ideal, and any two monomials $q_1 p_1$ and $q_2 p_2$ in which the factors p_1 and p_2 have the same length are equal if and only if $p_1 = p_2$ and $q_1 = q_2$. Hence $x \in I$ requires that each term $\lambda_{p,q} qp$ in x has to be in I. In particular, for each p, $x_p p \in I$, which shows that the sum $\sum_{p \in Q_l} Mp$ is direct. This proves (a). $\square$

We also need to know that our Γ-modules possess minimal projective resolutions. This will be achieved by assuming that each Γ-module, M, is either *J-bounded* in the sense that $MJ^m = 0$ for some natural number m, or is *J-graded* in the sense that it is the direct sum of right S-submodules $M(0), M(1), M(2), \ldots$ such that $M(n)Q_1 \subset M(n+1)$ for all $n \geq 0$.

Note that any module over a quotient algebra of Γ by an *admissable* ideal is automatically J-bounded as a Γ- module.

We can now state the main theorem of the paper.

THEOREM 2.2. *Let $\Lambda = \Gamma/I$ be a monomial algebra. Let M be a right Λ-module which is path separated as a Γ-module, and is either J-bounded or J-graded. Then the first syzygy module of M as a Λ- module is a direct sum of cyclic modules isomorphic to right ideals in Λ generated by monomials (modulo I).*

3. Projective covers of Γ-modules

The proof of the theorem begins, as does the proof of the Syzygy Theorem in [**2, 3**], with a careful analysis of the structure of a projective resolution of M as a Γ- module, the resolution being J- graded when M is so. So in this section we analyse a fixed Γ- module epimorphism,

$$\theta : P \longrightarrow M,$$

in which P is projective and

either M is J-bounded,

or M and P are J- graded and θ is homogeneous and of degree 0.

As a projective Γ- module, P is known to be isomorphic to $(P/PJ)\otimes_S\Gamma$ (for a proof, see [**1**], Appendix). Therefore, since $S \cong k^{|Q_0|}$, we may choose $A \subset P$ such that there is a direct decomposition of Γ,

$$P = \bigoplus_{a\in A} a\Gamma,$$

into indecomposable summands in which, for each $a \in A$, there is an idempotent $e = e(a) \in Q_0$ such that the map $e\gamma \mapsto a\gamma$ is an isomorphism of $e\Gamma$ onto $a\Gamma$. *We write* $P = A\Gamma$ *to describe this situation, and refer to* A *as a* Γ- *basis of* P. Clearly the Γ- basis A determines a k-basis $\{ap : a \in A, p \in e(a)Q_*\}$ of P, which we denote by AQ_* and refer to as *a monomial basis of* P; of course it depends on the choice of A. Further, Γ is hereditary, so any submodule of P is projective and must have a Γ-basis. In particular, for each l, AQ_l is just such a basis of PJ^l. (In the J- graded case, we always assume in what follows that the elements of A have been chosen, as is possible, to be homogeneous.)

Notation For any subset X of a module, we write Xk for the k-subspace spanned by X.

LEMMA 3.1. *Let* $\theta : P \longrightarrow M$ *be as above, and let* A *be a* Γ- *basis of* P. *Then:*

(1) There exist subsets A_0 *of* A *and* A_{l+1} *of* A_lQ_1 *such that, for each* $l \geq 0$, *the restriction* $\theta_l : A_l\Gamma \to MJ^l$ *of* θ *is a minimal projective cover.*

(2) Write $A_* = \cup_{l\geq 0}A_l$. *Then the restriction of* θ *to the subspace* A_*k *of* P *is a* k- *linear isomorphism onto* M.

(3) Write $B_0 = A\setminus A_0$, $B_{l+1} = A_lQ_1 \setminus A_{l+1}$ *for* $l \geq 0$, *and* $B_* = \cup_{l\geq 0}B_l$. *Then there are* Γ- *module decompositions* $P = A_0\Gamma \oplus B_0\Gamma$ *and* $A_lJ = A_{l+1}\Gamma \oplus B_{l+1}\Gamma$ *for each* $l \geq 0$, *and a vector space decomposition* $P = A_*k \oplus B_*\Gamma$.

(4) $\ker(\theta) \cong B_*\Gamma$.

PROOF. Note first that the assumptions on M are inherited by its submodules MJ^l. Then well-known classical results (summarised for this context in [**1**], Section 2) show that, if W_l is any S- module complement to MJ^{l+1} in MJ^l, then the map $W_l\otimes_S\Gamma \to MJ^l$ is a minimal projective cover of MJ^l (in the graded case, the W_l's must also be suitably graded). The proof of (1) is by an inductive construction. Since θ maps A to a set of module generators of M, $\theta(A)$ must span an S- module complement of MJ in M. Now, for each $a \in A, aS = ak$. Hence there is a subset A_0 of A which is mapped bijectively by θ to a k basis of an S- module complement $W_0 = \theta(A_0)k$ of MJ in M. Then the result quoted above shows that $\theta_0 : A_0\Gamma \to M$ is a minimal projective cover. This proves (1) with $l = 0$. For the case $l = 1$, we note that $A_0\Gamma = A_0k \oplus A_0J$ as a vector space, and θ_0 maps the projective module $A_0J = A_0Q_1\Gamma$ onto MJ. So replacing A by A_0Q_1 in the argument for the case $l = 0$, we deduce that A_0Q_1 contains a subset A_1 mapped bijectively by θ to a k-basis of an S- module complement of MJ^2 in MJ, and hence that $\theta_1 : A_1\Gamma \to MJ$ is a minimal projective cover. This is (1) for $l = 1$, and the argument can obviously be repeated to prove the result for each $l \geq 1$.

We prove (2) and (3) simultaneously, again proceeding inductively. Obviously $P = A\Gamma = A_0\Gamma \oplus B_0\Gamma$ because A is the disjoint union of A_0 and B_0. Also then $P = A_0k\oplus A_0J\oplus B_0\Gamma$ as a vector space, and then the construction of A_1 and choice of B_1 show that $A_0J = A_1\Gamma \oplus B_1\Gamma$ as a Γ- module. Similarly, for any $l \geq 0$, we obtain decompositions

$$P = (\bigoplus_{i=0}^{l} A_ik) \oplus A_lJ \oplus (\bigoplus_{i=0}^{l} B_i\Gamma)$$

as a vector space, and

$$A_lJ = A_{l+1}\Gamma \oplus B_{l+1}\Gamma$$

as a Γ- module. The first of these sequences of formulae shows that the vector subspace $P' = A_*k + B_*\Gamma$ of P is the direct sum of its two summands. To see why $P' = P$, we note that by construction, $A = A_0 \cup B_0$, and quite generally, for each $l \geq 0$,

$$AQ_l = A_l \cup (\bigcup_{i+j=l} B_iQ_j).$$

Since each monomial on the right hand side is in one or other of the two summands of P', it follows that P' contains the entire canonical basis AQ_* of P, and so $P = P'$, as (3) asserts. The choice of the A_l's shows that, for each l, the sets $\theta(A_0), \dots, \theta(A_l)$ are disjoint and their union is a k- basis of a complement of MJ^{l+1} in M. Our assumptions on M imply then that $\theta(A_*)$ is a k- basis of M, and so from (1), θ maps A_*k isomorphically to M. This completes the proof of (2) and (3).

The proof of (4) copies the analogous proof in [**2, 3**]. Each element of the set B_* is a monomial $b \in B_l$ for some $l \geq 0$, and so has a unique representation in the form

$$b = a(b)p(b) = be(b),$$

where $a(b) \in A, p(b) \in Q_l$, and $e(b)$ is the unique idempotent in Q_0 such that $p(b)e(b) = p(b)$. Now $\theta(b) \in MJ^le(b) = \theta(A_l\Gamma e(b))$, so we can choose some $u(b) \in A_l\Gamma e(b)$ such that

$$c(b) = b - u(b) = c(b)e(b) \in \ker(\theta).$$

Therefore

$$c(b)\Gamma \cong e(b)\Gamma \cong b\Gamma \text{ and } A_l\Gamma \oplus b\Gamma = A_l\Gamma \oplus c(b)\Gamma.$$

Let $C_l = \{c(b) : b \in B_l\}$. Then $C_l\Gamma \subset \ker(\theta)$,

$$A_{l-1}J \;=\; A_l\Gamma \oplus B_l\Gamma \;=\; A_l\Gamma \oplus C_l\Gamma,$$

and $C_l\Gamma \cong B_l\Gamma$. On writing C_* for the union of the C_l's, we have $C_*\Gamma \subset \ker(\theta)$ and $P = A_*k \oplus C_*\Gamma$. Since θ maps A_*k isomorphically onto M, we deduce finally the formula

$$\ker(\theta) = C_*\Gamma \cong B_*\Gamma$$

needed to prove (4). □

When the module M is path separated, the process just used to prove (4) has a crucial refinement which we set out in the following corollary.

COROLLARY 3.2. *Assume the module M is also path separated. Then for each $b = a(b)p(b) \in B_l \subset B_*$, as above, the corresponding elements $c(b)$ of the Γ- basis C_* of $\ker(\theta)$ and $u(b) \in A_l\Gamma$ may be chosen to be elements of $Pp(b)$.*

PROOF. In the notation of the preceding proof, the element $u(b) \in A_l\Gamma$ can be expressed as a sum, $u(b) = \sum_{p\in Q_l} u_p$, in which each u_p is a k- linear combination of monomials in $A_lQ_* \cap AQ_*p$, so in particular, $u_p \in A_l\Gamma \cap Pp$. Now the sum $\sum_{p\in Q_l} Mp$ is a direct sum, $\sum_{p\in Q_l} \theta(u_p) = \theta(b) = \theta(a(b)p(b)) \in Mp(b)$, and $\theta(u_p) \in$

Mp for each $p \in Q_l$. Hence $\theta(b) = \theta(u_{p(b)})$, which shows that the initial choice of $u(b)$ may be made from $A_l\Gamma \cap Pp(b)$. □

There is also an important corollary when M is annihilated by a monomial ideal. Since Γ is hereditary, any of its ideals are projective as right modules and so possess Γ- bases. However, a monomial ideal I actually possesses a *canonical* Γ-basis as a projective right module which consists entirely of monomials, namely the set $(I \setminus IJ) \cap Q_*$.

COROLLARY 3.3. *Assume the module M is annihilated by the monomial ideal I in Γ. Let R denote the canonical monomial Γ- basis of I as a projective right module. Then $AR \subset B_*Q_*$, $PI = \sum_{b\in B_*} PI \cap b\Gamma$ and, for each $l \geq 0$,*

$$(A_l\Gamma + PI) \cap (B_l\Gamma + PI) = PI.$$

PROOF. We have $I = R\Gamma$ and $MI = 0$. Hence $PI = AR\Gamma \subset \ker(\theta)$. However, $AR \subset AQ_* = A_* \cup B_*Q_*$, and θ is injective on A_*. Hence $AR \subset B_*Q_*$, and so each summand $ar\Gamma$ of PI (where $(a, r) \in A \times R$) is a submodule of a unique summand $b\Gamma$ of $B_*\Gamma$. This shows that $PI = \sum_{b\in B_*} PI \cap b\Gamma$. Lastly, let $l \geq 0$ and consider an element $x \in A_l\Gamma \cap (B_l\Gamma + PI)$. Applying Lemma 3.1 to $A_l\Gamma$ instead of to $A\Gamma$, one can express x in the form $x = x_A + x_B$, with $x_A \in \sum_{m\geq l} A_m k$ and $x_B \in \sum_{m>l} B_m\Gamma$. Since $PI \subset B_*\Gamma$ we have $x_A \in A_*k \cap B_*\Gamma = 0$, and so $x = x_B$. There is a $y \in B_l\Gamma$ such that

$$y - x \in PI \cap \sum\nolimits_{m\geq l} B_m\Gamma = \sum\nolimits_{m\geq l} PI \cap B_m\Gamma,$$

and since the sum on the right is direct, we see that x and y are both in PI. This proves the formula $(A_l\Gamma + PI) \cap (B_l\Gamma + PI) = PI$. □

4. Proof of Theorem 2.2

Let M and I satisfy the conditions of the theorem. We choose any Γ- module epimorphism $\theta : P \to M$ with P a projective Γ -module, as in the previous sections, and let A and R be Γ- bases of P and I. We may then choose the sets A_l, B_l to have all the properties listed in Lemma 3.1 and Corollary 3.3. Since M is path separated, we may also use Corollary 3.2 to express the kernel $K = \ker(\theta)$ of θ as a direct sum,

$$K = \bigoplus\nolimits_{b\in B_*} c(b)\Gamma,$$

in which, given any $b \in B_l$, we have $b = a(b)p(b)$ for some $a(b) \in A$ and $p(b) \in Q_l$, and $c(b) - b = u(b) \in Pp(b) \cap A_l\Gamma$. Now $PI \subset K$, and θ clearly induces an epimorphism of the projective Λ- module $\bar{P} = P/PI = \oplus_{a\in A}\bar{a}\Lambda$ onto M, with kernel - a first syzygy of M as a Λ-module - given by $\bar{K} = K/PI$. Let $x \mapsto \bar{x}$ denote the canonical map from P to $\bar{P}$. Then it is clear that to complete the proof of the theorem, it will be sufficient to prove,

firstly, that the sum

$$\bar{K} = \sum_{b\in B_*} \bar{c}(b)\Lambda$$

is a direct sum and,

secondly, for each $b \in B_*$, that $\bar{c}(b)\Lambda$ is isomorphic to the right ideal in Λ generated by the image of the monomial $p(b)$.

Suppose, to obtain a contradiction of the first assertion, that the above sum is not direct. Then PI contains a sum of terms of the form $c(b)\gamma_b$, for various $b \in B_*$ and $\gamma_b \in \Gamma$, no one of which is itself in PI. Using the formulae $c(b) = b - u(b)$, we can re-arrange this as a congruence

$$\sum b'\gamma_{b'} + PI = \sum u(b')\gamma_{b'} - \sum c(b'')\gamma_{b''} + PI,$$

in which each $b' \in B_l$, for one l, and each $b'' \in B_m$ for some $m \geq l+1$. Hence the terms on the left side are all in $B_l\Gamma + PI$, whereas those on the right side are all in $A_l\Gamma + PI$. By Corollary 3.3, both sides are in PI, and so then are each of the individual terms $b'\gamma_{b'} \in PI$; but that implies that $p(b')\gamma_{b'} \in I$, which in turn shows, firstly, that $u(b')\gamma_{b'} \in PI$, and so finally, that $c(b')\gamma_{b'} \in PI$, contrary to the assumption that none of the terms $c(b)\gamma_b$ are in PI. This shows that the sum above is direct. Finally, we show that

$$\bar{c}(b)\Lambda = (c(b)\Gamma + PI)/PI \cong (p(b)\Gamma + I)/I.$$

Here, the right hand side is isomorphic to Γ/ρ, where $\rho = \{\gamma \in \Gamma : p(b)\gamma \in I\}$. Since $b = a(b)p(b)$ and $a(b)$ is an element of a Γ- basis of P, we also have $\rho = \{\gamma \in \Gamma : b\gamma \in PI\}$. Also Corollary 3.2 shows that $u(b) \in Pp(b)$, so that $u(b)\rho \subset PI$. Hence, the formula $c(b) = b - u(b)$ and Corollary 3.3 show that $c(b)\gamma \in PI$ if and only if $\gamma \in \rho$. Therefore, the left side of the formula above is also isomorphic to Γ/ρ, as we needed to show.

References

[1] M. C. R. Butler and A. D. King, *Minimal resolutions of algebras,* J. Algebra, to appear.

[2] Birge Zimmermann Huisgen, *Predicting syzygies over monomial relations algebras,* Manuscripta Math. 70, 157-182(1991).

[3] Birge Zimmermann Huisgen, *Homological domino effects and the first Finitistic Dimension Conjecture,* Invent. math. 108, 369 - 383 (1993).

Department of Mathematical Sciences, The University of Liverpool, Liverpool, L69 3BX, UK

E-mail address: mcrb@liv.ac.uk

Contemporary Mathematics
Volume **229**, 1998

ALGEBRAS WHOSE DERIVED CATEGORY IS TAME

J. A. DE LA PEÑA

Let A be a finite dimensional algebra over an algebraically closed field k. In recent years the consideration of the derived category $D^b(A)$ of the category $\operatorname{mod} A$ of (left-)modules over A has played an important role in Representation Theory (see for example [22]).

For a few algebras A, the description of the indecomposable objects of $D^b(A)$ (and the corresponding Auslander-Reiten quiver $\Gamma_{D^b(A)}$) is known. For example, for $A = kQ$ an hereditary algebra of finite or tame type (that is, A is the path algebra of a quiver Q of Dynkin or extended Dynkin type), the description of $\Gamma_{D^b(A)}$ is well-known (see [22]); in the case in which A is a tubular algebra, $\Gamma_{D^b(A)}$ is described in [23]. It is worth recalling that the category of coherent sheaves on the projective space $\mathbb{P}^n$ is derived equivalent to the module category mod_{A_n} of a certain finite dimensional algebra A_n, [11]. Moreover, for a tubular algebra A, there is a weighted projective line $\mathbb{X}$ such that $D^b(\operatorname{coh} \mathbb{X})$ is triangle equivalent to $D^b(A)$, which establishes a link between Atiyah's classification of vector bundles on elliptic curves and Ringel's classification of indecomposable modules over A, see [20].

A useful tool for describing the derived category of an algebra A is the *repetitive category* $\hat{A}$ (see [22] and section 1). If A has finite global dimension, then $D^b(A)$ is triangle equivalent to $\underline{\operatorname{mod}}\,\hat{A}$ the quotient of $\operatorname{mod} \hat{A}$ by the maps factorizing through projective $\hat{A}$-modules. We shall say that A is *derived-tame* if $\operatorname{gl.dim} A < \infty$ and the category $\hat{A}$ is tame. We recall that a k-category C is said to be tame if for each dimension $d \in \mathbb{N}$ and every object x of C, almost all the indecomposable modules $M: C \to \operatorname{mod} k$ with $\sum\limits_{c \in C} \dim_k M(c) = d$ and $M(x) \neq 0$ occur in a finite number of one-parameter families. We show in section 1 that the class of derived-tame algebras is closed under derived-equivalence. Important examples of derived-tame algebras are hereditary tame algebras and tubular algebras as shown in [4, 22, 23] and pg-critical algebras (that is, simply connected tame algebras which are minimal not of polynomial growth, see [26]) as will be shown in section 1.

If $\operatorname{gl.dim} A < \infty$, the *homological bilinear form* is given, for the classes $\mathbf{dim}\, X = x$, $\mathbf{dim}\, Y = y$ of the modules X, Y in the Grothendieck group $K_0(A)$ of A, by $\langle x, y \rangle_A = \sum\limits_{i=0}^{\infty} (-1)^i \dim_k \operatorname{Ext}_A^i(X, Y)$. The corresponding quadratic form, denoted by χ_A, is called the *Euler form* of A. Work has been done to characterize algebras

1991 Mathematics Subject Classification: 16G10, 16G60, 16D90

A whose derived category $D^b(A)$ is triangle equivalent to $D^b(C)$ for C a tame hereditary or a tubular algebra (see [3, 4, 5, 7]); in these cases, the Euler form χ_A is non-negative.

We recall that a basic algebra of the form $A = kQ/I$ (see [19] and (1.1)) is a *tree algebra* if the underlying graph of Q is a tree. A tree algebra $A = kQ/I$ always has $\operatorname{gl.dim} A < \infty$. It was shown in [17] that a tree algebra A with $\operatorname{gl.dim} A \leq 2$ and χ_A non-negative, is tame. Some of these algebras are known to be derived-tame (see section 2).

In section 2 of this work we introduce the *well-covered* algebras. This class includes algebras whose Euler form is non-negative, tame tilted algebras and others. Our *main result* states that any tree algebra which is well-covered and derived-tame has a non-negative Euler form. This is a special case of the following *conjecture:* a tree algebra A is derived-tame if and only if χ_A is non-negative.

We thank Ch. Geiss for some useful discussions and the referee for detailed comments. We acknowledge support form DGAPA, UNAM and CONACyT.

1. Derived categories and tame algebras

1.1. Let A be a basic, finite-dimensional k-algebra over a fixed algebraically closed field k. It is well-known that A may be written as $A = kQ/I$ where Q is a finite quiver and I is an admissible ideal of the path algebra kQ. We shall consider A as a k-category whose objects are the vertices Q_0 of Q and in which the space of maps $A(x,y)$ from x to y is $e_y A e_x$, where e_x denotes the primitive idempotent associated to the vertex x. See [19].

The *repetitive category* $\hat{A}$ is the k-category with objects $Q_0 \times \mathbb{Z}$ (which are denoted by $s[i]$, for $s \in Q_0$ and $i \in \mathbb{Z}$) and the only possibly non-zero morphism spaces are $\hat{A}(r[i], s[i]) = A(r,s) \times \{i\}$ and $\hat{A}(r[i], s[i+1]) = DA(s,r) \times \{i\}$, where $D = \operatorname{Hom}_k(-,k)$ denotes the usual duality. Clearly, the group $\mathbb{Z}$ acts freely on $\hat{A}$ in such a way that the quotient $\hat{A}/\mathbb{Z}$ is isomorphic to the *trivial extension* $T(A) = A \ltimes DA$ whose underlying vector space is $A \oplus DA$ and the multiplication is given by $(a,q)\cdot(a',q') = (aa', aq' + qa')$. See [22].

1.2. Let $A = kQ/I$ be as above. We recall that the *one-point extension* $A[M]$ of the algebra A by a module M is the category with objects $Q_0 \cup \{s\}$ and morphism spaces $A[M](s,x) = M(x)$, for $x \neq s$, $A[M](s,s) = k$ and such that A is a full convex (= path closed) subcategory. See [33].

Let i be a sink in Q. The *reflection* $S_i^+ A$ of A at i is the quotient of the one-point extension $A[I_i]$ by the two-sided ideal generated by e_i, where I_i denotes the injective envelope of the simple S_i at i. Dually, we consider the reflection $S_j^- A$ of A at a source j of Q. It was shown in [36] that $S_i^+ A$ and A are *tilting-cotilting equivalent*, that is, there is a sequence of algebras $A = A_0, A_1, \ldots, A_m = S_i^+ A$ and a sequence of modules ${}_{A_i}T^i$ $(0 \leq i < m)$ such that $A_{i+1} = \operatorname{End}_{A_i} T^i$ and T^i is either a tilting or a cotilting module. In particular, this implies that $S_i^+ A$ and A are derived equivalent (that is, have triangle equivalent derived categories).

Assume $A = kQ/I$ where Q is connected and without oriented cycles and let B be a finite dimensional algebra. Then by [24], $\hat{A}$ and $\hat{B}$ are isomorphic if and only if A and B are *reflection-equivalent*, that is, there is a sequence $A = A_0, A_1, \ldots, A_m =$

B where $A_{i+1} = S^-_{\sigma(i)} A_i$ or $S^+_{\tau(i)} A_i$ for some source $\sigma(i)$ or some sink $\tau(i)$ of A_i, $0 \leq i < m$.

1.3. An algebra A is said to be *tame* if for every $d \in \mathbb{N}$ there is a family $M_1, \ldots, M_{s(d)}$ of $A - k[t]$-bimodules which are finitely generated and free as right $k[t]$-modules and such that almost every indecomposable A-module X of dimension d is of the form $M_i \otimes_{k[t]} S_\lambda$ for some $1 \leq i \leq s(d)$ and $\lambda \in k$, where $S_\lambda = k[t]/(t-\lambda)$ is a simple $k[t]$-module. In this case, we denote by $\mu(d)$ the least value possible for the number $s(d)$ of bimodules in the family of M_i above. We say that A is of *polynomial growth* if $\mu(d) \leq d^m$ for some $m \in \mathbb{N}$ and all $d \in \mathbb{N}$.

Let B be a locally finite-dimensional k-category, that is, $\dim_k B(x,y)$ is finite for every pair of objects x, y of B. Then we say that B is *tame* if for every finite set of objects F, the full subcategory $B(F)$ of B with objects F is a tame algebra.

Clearly, if A is derived-tame, then A is tame. The following remark makes derived-tameness an interesting concept.

Proposition. *Let A be a derived-tame algebra and B be an algebra such that $D^b(B)$ is equivalent to $D^b(A)$ as a triangulated category. Then B is derived-tame.*

PROOF: As observed in [22, chapter III], $\text{gl.dim}\, B < \infty$. Hence $\underline{\text{mod}}\, \hat{A}$ and $\underline{\text{mod}}\, \hat{B}$ are triangle equivalent. Hence, if $\hat{A}$ is tame, then $\hat{B}$ is tame [25]. □

Remark: In the situation of the Proposition, the equivalence $F\colon \underline{\text{mod}}\, \hat{A} \to \underline{\text{mod}}\, \hat{B}$ is induced by a tensor product [32]. Thus the methods of [2] could be applied to show the result.

1.4. We recall that, in [35], a basic algebra of the form $A = kQ/I$ is called *strongly simply connected* if Q has no oriented cycles and for every convex (= path closed) subquiver Q' of Q, the induced algebra $B = kQ'/(I \cap kQ')$ has vanishing first Hochschild cohomology group. The class of strongly simply connected tame algebras is much studied (see for example [31]). Following [26], we say that a strongly simply connected algebra A is pg-*critical* if it is minimal in the class of tame algebras which are not of polynomial growth.

Proposition. *Let A be a pg-critical algebra, then A is derived-tame.*

PROOF: We note (see for example [26]) that A is derived equivalent to one to the algebras $A' = kQ'/I'$ described by one of the following diagrams:

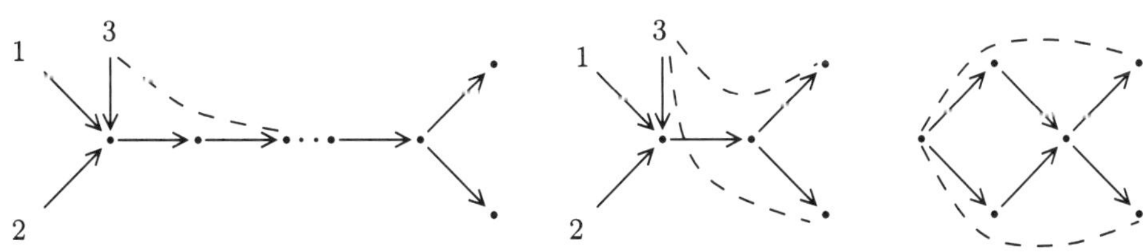

depending on the number n of vertices of Q'. As usual, a dotted line joining two vertices i and j means that the path between i and j belongs to the ideal I'. We consider the case $n = 8$, the other cases being similar.

By (1.3), it is enough to show that $\hat{A}'$ is tame. We may assume that $A = A'$.

The algebra $S_3^- S_2^- S_1^- A$ is tilting equivalent to the algebra B below

B: 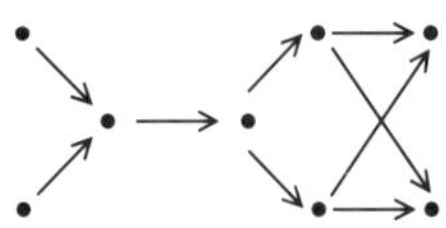 $\widehat{B}$: ... 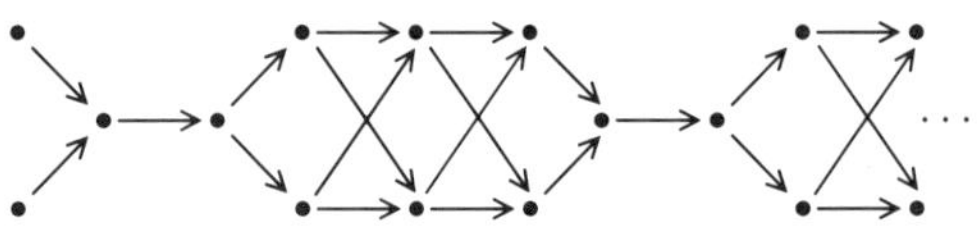

where the two squares of B are commutative. Then $\hat{B}$ has the quiver indicated above with the relations which make all squares commutative and $\hat{B}(a[i], b[i+1]) = 0$, for all $i \in \mathbb{Z}$ and all paths $a \to \cdots \to b$ in B.

The tameness of the category $\hat{B}$ is easy to verify (by looking at it either as a 'clan' [14] or as a full subcategory of one of the tame categories considered in [21]).

By (1.2), A is tilting-cotilting equivalent to B and $\underline{\mathrm{mod}}\,\hat{A} \simeq \underline{\mathrm{mod}}\,\hat{B}$ as triangulated categories by [36]. Now (1.3) implies that $\hat{A}$ is tame. □

As a somewhat more complicated example, we consider the algebra A_0 given by the following quiver with relations

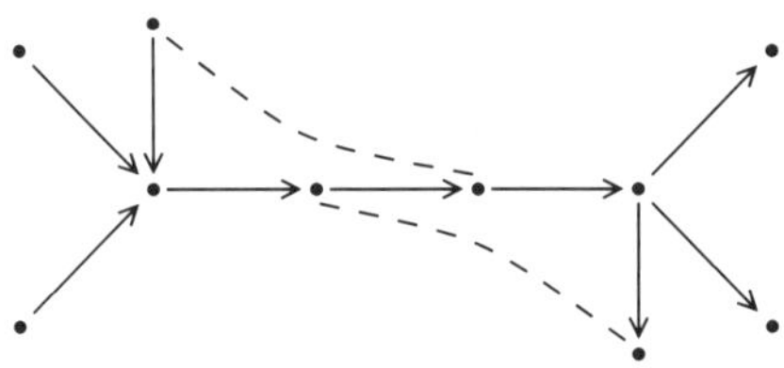

By applying some reflections, the algebra A_0 is shown to be derived-equivalent to the following algebra B with all squares commutative.

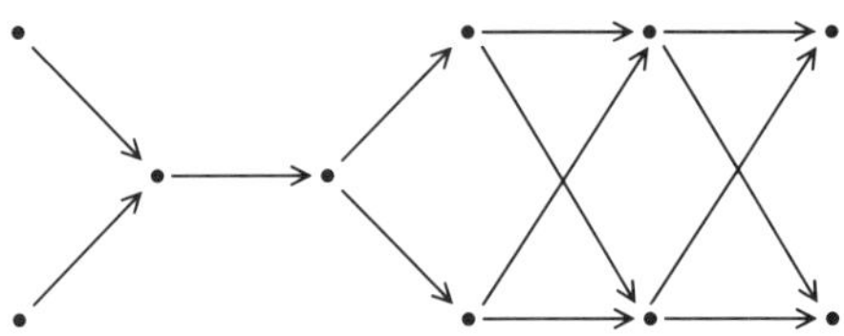

As before, $\hat{B}$ (and hence $\hat{A}$) is tame.

1.5. Let $F\colon D^b(A) \to D^b(B)$ be an equivalence of triangulated categories. Then there is an induced isomorphism $f\colon K_0(A) \to K_0(B)$ such that $f(\mathbf{dim}\, X^\cdot) = \mathbf{dim}\, FX^\cdot$, for any object $X^\cdot \in D^b(A)$, where $\mathbf{dim}\, X^\cdot = \sum_{i \in \mathbb{Z}} (-1)^i \mathbf{dim}\, X^i \in K_0(A)$. Moreover, f is an isometry. In particular, χ_A is non-negative if and only if χ_B is and, in this case, $\operatorname{corank} \chi_A = \operatorname{corank} \chi_B$.

Consider a reflection $B = S_i^- A$; then there is an isometry $f\colon K_0(A) \to K_0(B)$ whose restriction $K_0(A/(i)) \to K_0(B/(i))$ is the identity and $f(e_i) = -e_i - \mathbf{dim}\,\mathrm{rad}\,P_i$, where P_i is the indecomposable projective A-module corresponding to the vertex i.

We reformulate a result in [3] whose proof will be useful in later arguments.

Proposition. *Let $A = kQ/I$ be a strongly simply connected algebra. Then the following are equivalent:*

(a) *The category $\hat{A}$ is locally representation-finite.*

(b) *The form χ_A is positive.*

(c) *A is iterated tilted of Dynkin type.*

PROOF: (a) $\Rightarrow$ (c): For every admissible sequence of vertices $i_1, \dots, i_t$ in Q, the algebra $S_{i_t}^+ \dots S_{i_1}^+ A$ is representation-finite (as a full subcategory of $\hat{A}$). It follows from results in [1] that A is iterated tilted of Dynkin type.

(c) $\Rightarrow$ (a) is well-known (see [22]) and (c) $\Rightarrow$ (b) is clear.

(b) $\Rightarrow$ (c): Since A is strongly simply connected and χ_A is positive then A is representation-finite [13]. Consider any algebra of the form $B = S_{i_t}^+ \dots S_{i_1}^+ A$; we shall prove by induction on t that B is representation-finite (and strongly simply connected). Indeed, the case $t = 0$ is known; we assume that $t > 0$ and $B' = S_{i_{t-1}}^+ \dots S_{i_1}^+ A$ is strongly simply connected and representation-finite. Then by [3, (1.4)], $B = S_{i_t}^+ B'$ satisfies the separation condition. Since χ_B is positive, then again [13] implies that B is representation-finite. Then, by [12], B is also strongly simply connected. Again [1] yields the result. □

1.6. The following results are the motivation for the problem considered in this paper.

Theorem. *Let A be a strongly simply connected algebra.*

(1) [3, 7] *A is derived equivalent to a hereditary algebra of type $\tilde{\mathbb{D}}_m$ or $\tilde{\mathbb{E}}_p$ ($p = 6, 7$ or 8) if and only if χ_A is non-negative of corank 1.*

(2) [7] *A is derived equivalent to a tubular algebra if and only if χ_A is non-negative of corank 2 and $\chi_A^{-1}(1) \cap (\chi_A^{-1}(0))^\perp = \emptyset$, where $V^\perp = \{w \in K_0(A)\colon \langle v, w\rangle = 0$, for all $v \in V\}$.*

(3) [8] *Let A be a tree algebra such that χ_A is non-negative of corank 2. Then A is derived equivalent to either a tubular algebra or a pg-critical algebra. In particular, A is derived-tame.* □

Remarks: (a) Statement (1) of the Theorem above was shown in [3] in case A is representation-finite. In [7] the representation-infinite case was considered.

(b) For the example A_0 in (1.4), the form χ_{A_0} is non-negative with corank $\chi_{A_0} =$ 3.

1.7. Let A be an algebra with $\operatorname{gl.dim} A < \infty$. Let $S_1, \dots, S_n$ be a complete set of simple modules for A. Then for any module X we have

$$\chi_A(\mathbf{dim}\, X) = \sum_{s=0}^{\infty} (-1)^s \left[\sum_{i,j=1}^{n} \dim_k X(i) \dim_k X(j) \dim_k \operatorname{Ext}_A^s(S_i, S_j) \right].$$

Proposition. *Let A be a strongly simply connected algebra. If χ_A is weakly non-negative, then q_A is also weakly non-negative.*

PROOF: Assume q_A is not weakly non-negative. Since A is strongly simply connected, by [28] we get a convex full subcategory B of A which is hypercritical. In particular, B is concealed from a minimal wild algebra. Therefore $\operatorname{gl.dim} B \leq 2$ and $\chi_B = q_B$ is not weakly non-negative. Since B is convex in A, then χ_B is the restriction of χ_A to B, which contradicts the weak non-negativity of χ_A. □

Notation: For a quadratic form $q: \mathbb{Z}^n \to \mathbb{Z}$, we consider the symmetric bilinear form $q(v, w) = q(v + w) - q(v) - q(w)$. If q is non-negative and $q(v) = 0$ it follows that $q(v, -) = 0$. Observe that for the Euler form χ_A, we have $\chi_A(v, w) = \langle v, w \rangle_A + \langle w, v \rangle_A$.

2. Algebras with non-negative Euler form

2.1. An integral quadratic form $q: \mathbb{Z}^n \to \mathbb{Z}$ of shape

$$q(v) = \sum_{i=1}^{n} v(i)^2 + \sum_{i<j} q_{ij} v(i) v(j)$$

is called a *unit form.* Such a form q is said to be *balanced* if, for every vector $v \in \mathbb{Z}^n$ with $q(v) = 0$, the condition $q(v, -) = 0$ is satisfied. The following are some simple facts.

(a) If q is balanced, then $q^{-1}(0)$ is a vector space.

(b) If q is non-negative, then q is balanced. Indeed, consider the canonical basis $e_1, \dots, e_n$. If $q(v) = 0$, then

$$0 \leq q(2v \pm e_i) = \pm 2q(v, e_i) + 1$$

and so $q(v, e_i) = 0$.

(c) The form $x^2 + y^2 - 3xy$ is balanced but it is indefinite.

The following result will be essential in our considerations.

Theorem. [9] *Let $q: \mathbb{Z}^n \to \mathbb{Z}$ be a unit form. Then q is non-negative if and only if both the following conditions are satisfied:*

(a) $-2 \leq q_{ij} \leq 2$ *for every* $i < j$*;*

(b) q *is balanced.* □

2.2. Let $A = kQ/I$ be a tree algebra. Then χ_A is a unit form. Moreover, if $\chi_A(v) = \sum_{i=1}^{n} v(i)^2 + \sum_{i<j} \chi_{ij} v(i)v(j)$, then $-1 \leq \chi_{ij} \leq 1$.

Corollary. *Let A be a tree algebra. Then χ_A is non-negative if and only if χ_A is balanced.* □

2.3. The following result exhibits the situation that we shall encounter in the cases when the Euler form is indefinite.

Proposition. *Let $A = kQ/I$ be a tree algebra. Suppose that A contains a convex subcategory C which is derived equivalent to a tame hereditary algebra $k\Delta$. Let $0 \neq v \in K_0(C)$ with $\chi_C(v) = 0$ be such that $\chi_A(v, e_s) \neq 0$ for some vertex s of Q.*

Then A is not derived-tame.

PROOF: We shall assume that $\langle e_s, v\rangle_A \neq 0$ and consider $R = \operatorname{rad} P_s$ the radical of the indecomposable projective P_s. We observe that we may suppose that $\operatorname{supp} R \cap \operatorname{supp} v \neq \emptyset$. Indeed, since χ_C is non-negative and $\chi_C(v, e_s) = \chi_A(v, e_s) \neq 0$, then $v(s) = 0$ and $0 \neq \langle e_s, v\rangle_A = \langle \mathbf{dim}\, P_s - \mathbf{dim}\, R, v\rangle_A = -\sum_x \dim_k R(x)\langle e_x, v\rangle_A$. Hence there exists $x \in \operatorname{supp} R$ with $\langle e_x, v\rangle_A \neq 0$. Either $v(x) \neq 0$ and we are done or x is a proper successor of s with $0 \neq \langle e_x, v\rangle_A = \chi_A(v, e_x)$ and we may sustitute s by x. Since Q has no oriented cycles, this process can be repeated only finitely many times until we reach a vertex s with the desired property. Moreover, we choose s such that there are no successors x of s with $x \notin C$ and $\langle e_x, v\rangle_A \neq 0$.

Let R_0 be the restriction of R to C. Consider the algebra $B = C[R_0]$ which is a full subcategory of A. Clearly, $\hat{B}$ is a full subcategory of $\hat{A}$; hence to show that A is not derived-tame, it is enough to show that B is not derived-tame. Let t be the source of B such that $\operatorname{rad} P_t = R_0$. Then using the way that s is chosen and the convexity of C, we get

$$\begin{aligned} 0 \;\neq\; & \langle \mathbf{dim}\, R, v\rangle_A = \sum_x \dim_k R(x)\langle e_x, v\rangle_A = \\ = \; & \sum_{x\in C} \dim_k R_0(x)\langle e_x, v\rangle_A = \langle \mathbf{dim}\, R_0, v\rangle_C. \end{aligned}$$

Therefore, $\chi_B(v, e_t) = -\langle \mathbf{dim}\, R_0, v\rangle_C \neq 0$ and χ_B is indefinite by (2.1 b).

Let $F\colon D^b(C) \to D^b(k\Delta)$ be a triangular equivalence. Since A is a tree, then R_0 is an indecomposable C-module. By the description of $D^b(k\Delta)$ (see [22]), we know that (up to a shift) $FR_0 \in \operatorname{mod} k\Delta$. Hence by [6], we get a triangular equivalence $F'\colon D^b(C[R_0]) \to D^b(k\Delta[FR_0])$ which implies that $\chi_{k\Delta[FR_0]}$ is indefinite. The main result in [27] implies that either the one-point extension $k\Delta[FR_0]$ is wild or the one-point coextension $[FR_0]k\Delta$ is wild. By (1.3), $B = C[R_0]$ is not derived-tame. This completes the proof. □

2.4. We say that an algebra $A = kQ/I$ is *well-covered* if there exist convex subcategories $C_1, \ldots, C_s$ of A satisfying:

(i) for each $i = 1, \ldots, s$, the algebra C_i is derived equivalent to a tame hereditary algebra $k\Delta_i$ (and hence, there is a vector $v_i \in K_0(C_i)$ such that $\chi_{C_i}^{-1}(0) = \mathbb{Z}v_i$).

(ii) for each vector $v \in K_0(A)$ such that $\chi_A(v) = 0$, there exist rational numbers $\lambda_1, \ldots, \lambda_s$ such that $v = \sum_{i=1}^{s} \lambda_i v_i$.

Examples of well-covered algebras are the following:

(a) Algebras which are derived-equivalent to tame concealed algebras, to tubular algebras or to pg-critical algebras. See [15, 30].

(b) Tame tilted algebras as shown in [29]. For such an algebra A, the form χ_A is weakly non-negative, but not necessarily non-negative.

(c) The algebra A given by the following quiver with relations

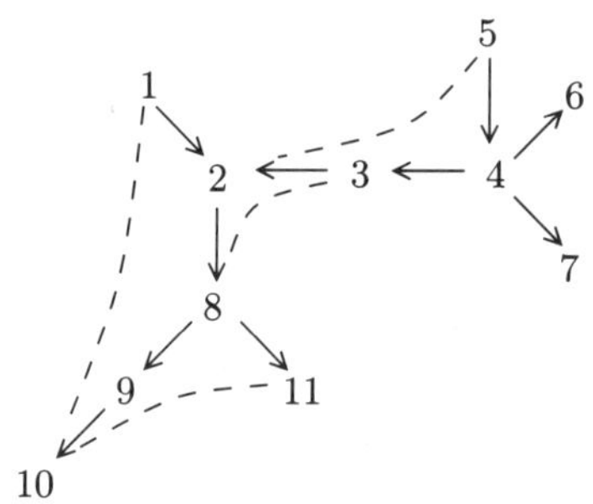

The full subcategories C_1, C_2 and C_3 of A defined by the vertices $\{3,4,5,6,7\}$, $\{1,2,3,4,6,7,8,9,10\}$ and $\{1,2,8,9,10,11\}$ respectively, satisfy the conditions (i) and (ii) in the definition above. Using (2.2) we check that χ_A is non-negative of corank 3.

Remarks: (a) Let C be a tree algebra with χ_C non-negative and such that $\chi_C^{-1}(0) \neq \emptyset$. Assume that for every convex subcategory D of C we have $\chi_C^{-1}(0) = \emptyset$. Then C is derived equivalent to a tame hereditary algebra.

(b) Let A be a tree algebra with χ_A non-negative and which does not contain any convex subcategory C derived equivalent to a tame hereditary algebra. Then χ_A is positive.

PROOF: (a): By (1.6), we just need to show that χ_C is of corank 1.

Assume that w_1 and w_2 are two linearly independent elements of $\chi_C^{-1}(0)$, then the minimality of C implies that $w_1(i) \neq 0 \neq w_2(i)$ for every extremal vertex i of C. Then we choose a vertex $j \neq y$ extremal in C and form a linear combination $w = aw_1 + bw_2$ with $w(j) = 0$. Then the algebra with support $C \setminus \{y\}$ provides a contradiction against the choice of C.

(b): If χ_A is not positive, then there exists a vector $0 \neq v \in K_0(A)$ with $\chi_A(v) = 0$. Choose C as a minimal convex subcategory of A with $\chi_C^{-1}(0) \neq 0$. Then (a) yields a contradiction. □

2.5. Proposition. *Let $A = kQ/I$ be a tree algebra such that χ_A is non-negative. Then A is well-covered.*

PROOF: Let $C_1, \ldots, C_s$ be all convex subcategories of A satisfying property (i). We shall prove (ii) by induction on the number of vertices of Q. By (2.4), if $s = 0$, then

χ_A is positive and there is nothing to prove. Assume $s > 0$ and choose x a vertex in $\bigcup_{i=1}^{s} \operatorname{supp} C_i$ with the property that $\left(\bigcup_{i=1}^{s} \operatorname{supp} C_i\right) \setminus \{x\}$ is contained in exactly one connected component Q' of $Q \setminus \{x\}$ (this is possible because Q is a tree).

Let $0 \neq v \in K_0(A)$ be such that $\chi_A(v) = 0$. Observe that $\operatorname{supp} v \subset Q' \cup \{x\}$: otherwise, choose $y \in (\operatorname{supp} v) \setminus (Q' \cup \{x\})$ and let C be a minimal convex subcategory of A containing y and such that $\chi_C(w) = 0$ for some vector $0 \neq w \in K_0(C)$. Then by (2.4), C is derived-equivalent to a tame hereditary algebra. This contradicts the choice of $C_1, \dots, C_s$.

If $v(x) = 0$, then we may consider the convex subcategory B of A with support Q'. Since χ_B is non-negative, by induction hypothesis, $v = \sum_{i \in J} \lambda_i v_i$ for some $\lambda_i \in \mathbb{Q}$ $(i \in J)$, where $i \in J$ if $\operatorname{supp} C_i \subset Q'$ and $\chi_{C_i}^{-1}(0) = \mathbb{Z} v_i$.

If $v(x) \neq 0$, we may assume that $v_1(x) \neq 0$ and form a linear combination $w = av + bv_1$ such that $w(x) = 0$. As above we get $w = \sum_{i \in J} \lambda_i v_i$, which shows the result. $\square$

2.6. The main result of this section is the following.

Theorem. *Let $A = kQ/I$ be a tree algebra which is well-covered. If A is derived-tame, then χ_A is non-negative.*

Proof: In view of (2.2), we need only to prove that χ_A is balanced. Let $v \in K_0(A)$ be such that $\chi_A(v) = 0$ and assume that $\chi_A(v, e_s) \neq 0$ for some s in Q. Since A is well-covered, there exist convex subcategories $C_1, \dots, C_t$ of A such that C_i is derived-equivalent to a tame hereditary algebra $k\Delta_i$ and $\chi_{C_i}^{-1}(0) = \mathbb{Z} v_i$, $i = 1, \dots, t$. Moreover, $v = \sum_{i=1}^{t} \lambda_i v_i$ for some rational numbers $\lambda_1, \dots, \lambda_t$.

Since $\chi_A(v, e_s) \neq 0$, there is some $j \in \{1, \dots, t\}$ with $\chi_A(v_j, e_s) \neq 0$. Then (2.3) implies that A is not derived-tame, a contradiction which establishes the result. $\square$

2.7. *Remarks:* (a) Example (2.4 c) is derived-tame.

(b) In joint work with Ch. Geiss we consider special cases of the converse of Theorem (2.6).

(c) Possibly the *conjecture* stated in the introduction holds for the more general class of strongly simply connected algebras. There are examples showing that it does not hold in general.

2.8. Finally, we especialize to our situation a result from [9].

Corollary. *Let A be a well-covered tree algebra which is derived-tame. Then all the eigenvalues of the Coxeter matrix ϕ_A have modulus one.*

Proof: The Euler form $\chi_A(v) = v^t C^{-1} v$ is a unit form which is non-negative (here C is the Cartan matrix of A [25].) Then $\phi_A = -C^{-t}C$ has eigenvalues on $\mathbb{S}^1$ by [9]. $\square$

References

1. I. Assem, D. Happel and O. Roldán. *Representation-finite trivial extension algebras.* J. Pure Appl. Algebra 33 (1984) 235–242.

2. I. Assem and J.A. de la Peña. *On the tameness of trivial extensions of algebras.* Fundamenta Mathematicae 149 (1996) 171–181.

3. I. Assem and A. Skowroński. *Quadratic forms and iterated tilted algebras.* J. Algebra 128 (1990) 55–85.

4. I. Assem and A. Skowroński. *On tame repetitive algebras.* Fundamenta Math. 142 (1993), 59–84.

5. M. Barot. *Representation-finite derived tubular algebras.* Preprint. México (1996).

6. M. Barot and H. Lenzing. *Extensions of derived equivalences.* Preprint. México (1998).

7. M. Barot and J. A. de la Peña. *Derived tubular strongly simply connected algebras.* To appear in Proceedings Am. Math. Soc.

8. M. Barot and J. A. de la Peña. *Algebras whose Euler form is non-negative.* To appear in Colloq. Math.

9. M. Barot and J. A. de la Peña. *The Dynkin type of a non-negative unit form.* Preprint. México (1998).

10. R. Bautista, F. Larrión and L. Salmerón. *On simply connected algebras.* J. London Math. Soc. (2) 27 (1983), 212–220.

11. A. A. Beilinson. *Coherent sheaves on* $\mathbb{P}^n$ *and problems of linear algebra.* Func. Anal. and Appl. Vol. 12 (1978) 214–216.

12. O. Bretscher and P. Gabriel. *The standard form of a representation-finite algebra.* Bull. Soc. Math. France 111 (1983), 21–40.

13. K. Bongartz. *Algebras and quadratic forms.* J. London Math. Soc. (2) 28 (1983) 461–469.

14. W.W. Crawley-Boevey. *Functorial filtrations II: Clans and the Gelfand problem.* J. London Math. Soc. 40 (1989) 9–30.

15. A. Dean and J. A. de la Peña. *Algorithms for weakly non-negative quadratic forms.* Linear Alg. Appl. 235 (1996) 35–46.

16. P. Dräxler. *Completely separating algebras.* J. Algebra 165 (1994) 550–565.

17. P. Dräxler and J.A. de la Peña. *Tree algebras with non-negative Tits form.* Preprint. México (1996).

18. P. Dräxler and A. Skowroński. *Biextensions by indecomposable modules of derived regular length* 2. Preprint. Bielefeld (1997).

19. P. Gabriel and A. V. Roiter. *Representations of finite-dimensional algebras.* Algebra VIII, Encyclopaedia of Math. Sc. Vol 73. Springer (1992).

20. W. Geigle and H. Lenzing. *A class of weighted projective curves arising in representation theory of finite dimensional algebras.* In Singularities, representation of algebras and vector bundles. Springer LNM 1273 (1987) 265–297.

21. Ch. Geiss and J.A. de la Peña. *An interesting family of algebras.* Arch. Math. Vol. 60 (1993) 25–35.

22. D. Happel. *Triangulated categories in the Representation Theory of finite dimensional algebras.* London Math. Soc. LN 119 (1988).

23. D. Happel and C. M. Ringel. *The derived category of a tubular algebra.* In Representation Theory I. LNM 1177. Springer (1986) 156–180.

24. D. Hughes and J. Waschbüsch. *Trivial extensions of tilted algebras.* Proc. London Math. Soc. 46 (3) (1983) 347–364.

25. H. Krause. *Stable module categories and their representation type.* Preprint Bielefeld (1996).

26. R. Nörenberg and A. Skowroński. *Tame minimal non-polynomial growth strongly simply connected algebras.* Proc. ICRA VII. CMS–AMS Conference Proceedings Vol. 18 (1996) 519–538.

27. J. A. de la Peña. *On the representation type of one-point extensions of tame concealed algebras.* Manuscr. Math. 61 (1988) 183–194.

28. J. A. de la Peña. *Algebras with hypercritical Tits form.* In Topics in Algebra. Banach Center Publications, Vol. 26, Part I, PWN - Polish Scientific Publishers, Warsaw (1990) 353–369.

29. J. A. de la Peña. *Tame algebras with sincere directing modules.* J. Algebra 161 (1993) 171–185.

30. J. A. de la Peña. *On the corank of the Tits form of a tame algebra.* J. Pure Appl. Alg. 107 (1996) 89–105.

31. J. A. de la Peña and A. Skowroński. *Geometric and homological characterizations of polynomial growth simply connected algebras.* Inventiones Math. 126 (1996) 287–296.

32. J. Rickard. *Derived equivalences as derived functors.* J. London Math. Soc. 43 (1991) 37–48.

33. C. M. Ringel. *Tame algebras.* Representation Theory I. Proc. ICRA II (Ottawa, 1979), Springer LNM 831 (1980) 137–287

34. C. M. Ringel. *Tame algebras and integral quadratic forms.* LNM 1099 Springer (1984).

35. A. Skowroński. *Simply connected algebras and Hochschild cohomologies.* Canadian Math. Soc. Conf. Proceedings 14 (1993) 431–447.

36. T. Wakamatsu. *Stable equivalence between universal covers of trivial extensions.* Tsukuba J. Math. 9 (1985) 299–316.

Instituto de Matemáticas, UNAM, México 04510 D. F., México

e-mail address: jap@penelope.matem.unam.mx

Contemporary Mathematics
Volume **229**, 1998

Circular Biextensions of Tame Concealed Algebras

Peter Dräxler

ABSTRACT. The notion of a circular biextension of a tame concealed algebra is introduced. It is completely determined when such an algebra is of tame representation type.

1. Introduction and main result

Let k be an algebraically closed field. Associated with the extended Dynkin diagrams $\widetilde{\mathbb{A}}_n$ ($n \geq 1$), $\widetilde{\mathbb{D}}_n$ ($n \geq 4$), $\widetilde{\mathbb{E}}_6$, $\widetilde{\mathbb{E}}_7$, $\widetilde{\mathbb{E}}_8$ there are the tame concealed algebras. They are defined as the endomorphism algebras of basic preprojective tilting modules of hereditary algebras $k\vec{\Delta}$ where the underlying graph Δ of the quiver $\vec{\Delta}$ is one of the extended Dynkin diagrams. If Δ is different from $\widetilde{\mathbb{A}}_n$, then all quivers $\vec{\Delta}$ having Δ as underlying graph yield the same tame concealed algebras B. On the other hand, for $\Delta = \widetilde{\mathbb{A}}_n$ one obtains various classes depending on the number p of clockwise and q of counterclockwise arrows in $\vec{\Delta}$. Hence we define the *extended Dynkin type* of a tame concealed algebra B as Δ if B is the endomorphism algebra of a basic preprojective tilting module over $k\vec{\Delta}$ where Δ is one of the diagrams $\widetilde{\mathbb{D}}_n$, $\widetilde{\mathbb{E}}_6$, $\widetilde{\mathbb{E}}_7$, $\widetilde{\mathbb{E}}_8$. Moreover, the extended Dynkin type is defined as $\widetilde{\mathbb{A}}_{p,q}$ if B is the endomorphism algebra of a basic preprojective tilting module over $k\vec{\Delta}$ where Δ is $\widetilde{\mathbb{A}}_n$ and in $\vec{\Delta}$ there are p clockwise and q counterclockwise arrows. If we require $p \geq q$, then the extended Dynkin type is uniquely determined by B.

The tame concealed algebras furnish the simplest class of tame algebras which are not of finite representation type. Recall that a finite-dimensional k-algebra is said to be tame if for any non-negative integer d the indecomposable (left) A-modules of dimension d lie up to isomorphism on finitely many 1-parameter families.

Besides a preprojective and a preinjective component the Auslander-Reiten quiver of a tame concealed algebra B consists of a family $(\mathcal{T}_\lambda)_{\lambda \subset \mathbb{P}}$ of standard stable tubes $\mathcal{T}_\lambda$ where the index set $\mathbb{P} = \mathbb{P}^1(k)$ is the projective line over k. Each of the tubes $\mathcal{T}_\lambda$ is an abelian length category having only finitely many simple objects. The number $n(\lambda)$ of these simple objects is called the rank of the tube $\mathcal{T}_\lambda$. Almost each of the tubes $\mathcal{T}_\lambda$ is homogeneous, i.e. $n(\lambda) = 1$.

One-point extensions and more general branch enlargements by simple modules in tubes $\mathcal{T}_\lambda$ led to many examples of tame algebras (see e.g. [**Ri2**]) and even to the complete classification of strongly simply connected algebras which are tame of polynomial growth (see [**Sk**]). It is well-known that one-point extensions by modules

1991 *Mathematics Subject Classification.* Primary 16G20, 16G60.

of length 2 in tubes $\mathcal{T}_\lambda$ lead into the area of tame algebras of non-polynomial growth whereas one-point extensions by modules of length ≥ 3 always are wild (see [**Ri1**]). Multiple one-point extensions by modules of length 2 (even more general derived cases) were investigated in [**DrS**]. In some proofs in [**DrS**] a new class of tame algebras arose. The aim of this paper is to introduce this class of algebras systematically and to analyze the occurring representation types.

Let us introduce the necessary notation. The objects of length l in a tube $\mathcal{T}_\lambda$ form an orbit under the Auslander-Reiten translation τ of B. Let us fix a module R^λ of length 2 in $\mathcal{T}_\lambda$. For all $i = 1, \dots, n(\lambda)$ we put $R_i^\lambda := \tau^i R^\lambda$. Since $R_{n(\lambda)}^\lambda = R^\lambda$, we can assume that the index set for i is $\mathbb{Z}/n(\lambda)$. For each $i \in \mathbb{Z}/n(\lambda)$ there is a non-zero homomorphism $f_i : R_{i+1}^\lambda \to R_i^\lambda$ which is unique up to a multiple in k and satisfies $f_i f_{i+1} = 0$ for all $i \in \mathbb{Z}/n(\lambda)$.

Let us define for $n \in \mathbb{N}$ the finite-dimensional algebra C_n as the algebra kQ_n/I where Q_n is the quiver in Figure 1 whose vertex set is identified with $\mathbb{Z}/n$. The ideal I of the path algebra kQ_n is generated by the paths $X_{i+1}X_i$ for all $i \in \mathbb{Z}/n$.

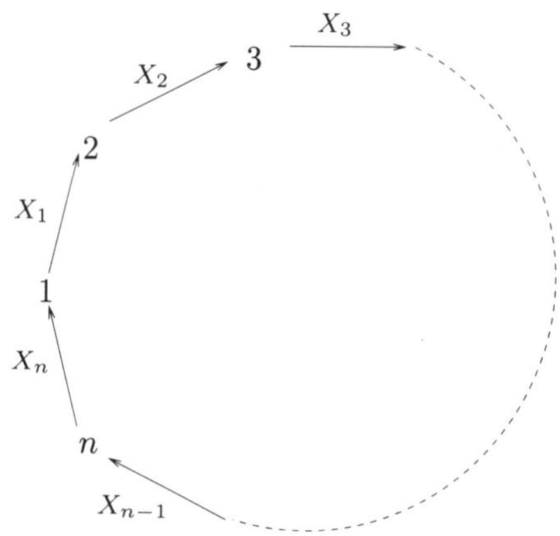

FIGURE 1. Circular quiver

For $\lambda \in \mathbb{P}$ we obtain a contravariant representation $R(\lambda)$ of $C_{n(\lambda)}$ in $B-$mod by mapping each vertex i of $Q_{n(\lambda)}$ to R_i^λ and each arrow X_i to f_i. But a representation of $C_{n(\lambda)}$ in $B-$mod is nothing but a B-$C_{n(\lambda)}$-bimodule. More general, for any finite subset $T = \{\lambda_1, \dots, \lambda_t\}$ of $\mathbb{P}$ we define the algebra $C(T)$ as the product $C_{n(\lambda_1)} \times \cdots \times C_{n(\lambda_t)}$ and the B-$C(T)$-bimodule $R(T)$ as $\oplus_{i=1}^t R(\lambda_i)$. Given a second finite subset S of $\mathbb{P}$ the *circular biextension* $\boxed{\text{S}}B\boxed{\text{T}}$ of B by the sets S, T is defined to be the matrix algebra

$$\begin{pmatrix} C(T) & 0 & 0 \\ R(T) & B & 0 \\ \mathrm{D}\,R(S)) \otimes_B R(T) & \mathrm{D}\,R(S) & C(S) \end{pmatrix}$$

where the multiplication is given by the usual matrix multiplication and $\mathrm{D} = \mathrm{Hom}_k(-, k)$ is the usual duality. We denote by $\boxed{\text{S}}B$ (resp. $B\boxed{\text{T}}$) the truncated versions where T (resp. S) is the empty set. The algebra $B\boxed{\text{T}}$ will be called *circular extension* of B by T. Dually $\boxed{\text{S}}B$ will be called *circular coextension* of B by S. Now we can formulate the first result of our paper:

THEOREM (A). *Let B be a tame concealed algebra and S, T be two finite subsets of $\mathbb{P}$. Then the circular biextension $\boxed{S}B\boxed{T}$ is tame in the following cases:*

(i) *The extended Dynkin type of B is $\widetilde{\mathbb{D}}_n$ and $T = \{\lambda\}$, $S = \{\mu\}$ such that $n(\lambda) = n(\mu) = n - 2$.*

(ii) *The extended Dynkin type of B is $\widetilde{\mathbb{A}}_{p,q}$ and $T = \{\lambda_1, \lambda_2\}$, $S = \{\mu_1, \mu_2\}$ such that $n(\lambda_1) = n(\mu_1) = p$ and $n(\lambda_2) = n(\mu_2) = q$.*

(iii) *The extended Dynkin type of B is $\widetilde{\mathbb{A}}_{2,2}$ and $T = \{\lambda\}$, $S = \{\mu\}$ such that $n(\lambda) = n(\mu) = 1$.*

(iv) *The extended Dynkin type of B is $\widetilde{\mathbb{A}}_{2,2}$ and $T = \{\lambda\}$, $S = \{\mu_1, \mu_2\}$ such that $n(\lambda) = 1$, $n(\mu_1) = n(\mu_2) = 2$ or $T = \{\lambda_1, \lambda_2\}$, $S = \{\mu\}$ such that $n(\lambda_1) = n(\lambda_2) = 2$, $n(\mu) = 1$.*

Clearly, one obtains tame biextensions $\boxed{S'}B\boxed{T'}$ if S' is a subset of S and T' a subset of T for sets T, S as in Theorem A. The question, whether the list of tame circular biextensins of tame concealed algebras obtained in this way is complete, will be answered positively by our second result:

THEOREM (B). *Suppose that the circular extension $B\boxed{T}$ of the tame concealed algebra B is tame where T is a non-empty finite subset of $\mathbb{P}$. Then one of the following three assertions is valid:*

(i) *The extended Dynkin type of B is $\widetilde{\mathbb{D}}_n$ and $T = \{\lambda\}$ such that $n(\lambda) = n - 2$.*

(ii) *The extended Dynkin type of B is $\widetilde{\mathbb{A}}_{p,q}$ and T is a subset of $\{\lambda_1, \lambda_2\}$ where $n(\lambda_1) = p$ and $n(\lambda_2) = q$.*

(iii) *The extended Dynkin type of B is $\widetilde{\mathbb{A}}_{2,2}$ and $T = \{\lambda\}$ such that $n(\lambda) = 1$.*

Let us finish the introduction by an example of a tame circular biextension $\boxed{S}B\boxed{T}$ where B is of extended Dynkin type $\widetilde{\mathbb{D}}_6$. It is given by the quiver in Figure 2 bound by the relations $\alpha_1\sigma_1 + \alpha_2\sigma_2 + \mu_1 X_1$, $\tau_1\gamma_1 + \tau_2\gamma_2 + Y_3\nu_3$, $\gamma_i\mu_3 + \gamma_i\beta_2\beta_1\alpha_1\sigma_1 X_4$ $(i = 1, 2)$, $\nu_1\alpha_i + Y_4\tau_1\gamma_1\beta_2\beta_1\alpha_i$ $(i = 1, 2)$, $\beta_i\mu_i + \mu_{i+1}X_{i+1}$ $(i = 1, 2)$, $Y_i\nu_i + \nu_{i+1}\beta_i$ $(i = 1, 2)$, $X_{i+1}X_i$ $(i = 1, \dots, 4)$, $Y_{i+1}Y_i$ $(i = 1, \dots, 4)$.

2. The associated bimodule problem

We recall some notation from [**GR**]. An aggregate $\mathcal{A}$ is a skeletally small k-additive category with splitting idempotents and finite-dimensional morphism spaces. The category $A-$mod of finite-dimensional left A-modules over a finite-dimensional algebra A is an aggregate.

A bimodule M over the two aggregates $\mathcal{A}$, $\mathcal{B}$ is a k-linear bifunctor $\mathcal{A} \times \mathcal{B} \to k-$mod which is contravariant in the first and covariant in the second argument (see [**Dd**]). A representation of M is a triple (a, m, b) where a is an object of $\mathcal{A}$, b is an object of $\mathcal{B}$ and m is an element of $M(a, b)$. A morphism from a representation (a, m, b) to another representation (a', m', b') of M is a pair (f, g) such that $f : a \to a'$ in $\mathcal{A}$, $g : b \to b'$ in $\mathcal{B}$ and $M(\mathrm{id}_a, g)(m) = M(f, \mathrm{id}_b)(m')$. Defining the composition of morphisms of representations of M in the obvious way, we obtain the category $\operatorname{rep} M$ which is again an aggregate.

Let us return to algebras $\boxed{S}B\boxed{T}$ where B is tame concealed and S, T are finite subsets of $\mathbb{P}$. It is well-known that the modules over such a biextension algebra can be identified with quintuples $(U, \gamma, X, \delta, V)$ where $U \in C(T)-$mod, $X \in B-$mod, $V \in C(S)-$mod, $\gamma \in \operatorname{Hom}_{C(T)}(U, \operatorname{Hom}_B(R(T), X)$, and $\delta \in \operatorname{Hom}_{C(S)}(\mathrm{D}\, R(S) \otimes_B$

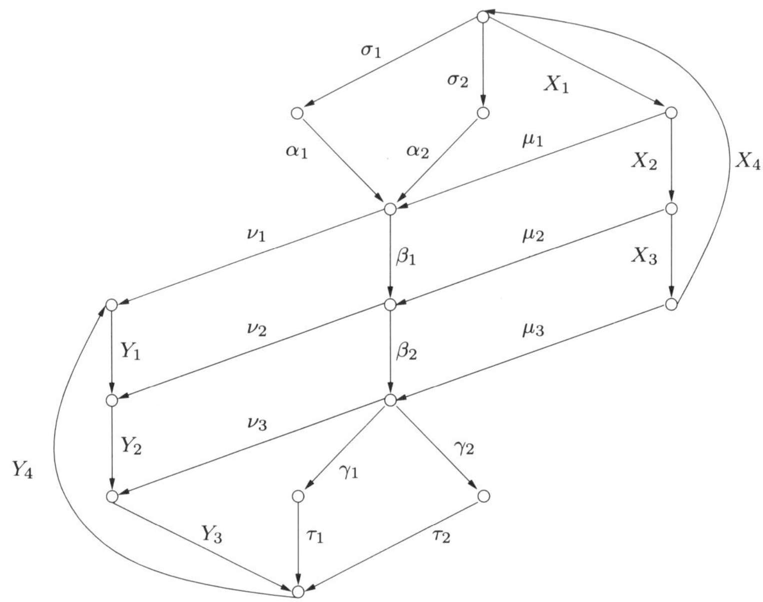

FIGURE 2. Example 1

$X, V)$. By mapping the quintuple $(U, \gamma, X, \delta, V)$ to the triple $(U \oplus V, \gamma \oplus \mathrm{D}\,\delta, X)$, we obtain an equivalence $\boxed{\mathrm{S}}\,B\,\boxed{\mathrm{T}}$ mod $\to \operatorname{rep} M(S, B, T)$ where $M(S, B, T)$ is the bimodule $(C(T) \times C(S))$–mod $\times$ B–mod $\to$ k–mod which maps $U \oplus V$ in $(C(T) \times C(S))$–mod and X in B–mod to the vector space

$$\operatorname{Hom}_{C(T)}(U, \operatorname{Hom}_B(R(T), X) \oplus \mathrm{D}\operatorname{Hom}_{C(S)}(\mathrm{D}\,R(S) \otimes_B X, V).$$

We want to characterize the tameness of $\boxed{\mathrm{S}}\,B\,\boxed{\mathrm{T}}$ via $M(S, B, T)$ and invoke [**Dr**][Theorem 3.3] which can be generalized immediately from vector space categories to our bimodule situation because the category $(C(T) \times C(S))$–mod has up to isomorphism only finitely many indecomposable objects. Let us define the category $\mathcal{B}$ as the factor category of B–mod by the ideal given by the intersection of the kernel of the functors $\operatorname{Hom}_B(R(T), -) : B\text{–mod} \to k\text{–mod}$ and the kernel of the functor $\mathrm{D}\,R(S) \otimes_B - : B\text{–mod} \to k\text{–mod}$. We denote by $\mathcal{M}(S, B, T)$ the induced bimodule which acts on $(C(T) \times C(S))$–mod $\times\, \mathcal{B}$.

LEMMA. *Let B be a tame concealed algebra and S, T be finite subsets of $\mathbb{P}$. Then the circular biextension algebra $\boxed{S}\,B\,\boxed{T}$ is tame if and only if the bimodule $\mathcal{M}(S, B, T)$ is tame.*

PROOF. Following [**Dr**][Theorem 3.3] the tameness of $\boxed{\mathrm{S}}\,B\,\boxed{\mathrm{T}}$ is characterized by three conditions. Fortunately, two of them are obviously satisfied. Namely, first the algebra B is tame and next the finite collection of tubes $\mathcal{T}_\lambda$ such that $\lambda \in S \cup T$ together with the preprojective and the preinjective component contain only finitely many indecomposable B-modules whose dimension is bounded by a fixed integer. The third condition is just the tameness of $\mathcal{M}(S, B, T)$. $\square$

Let B and B' be two tame concealed algebras of the same extended Dynkin type. Then there exists a preprojective tilting module X in B–mod whose endomorphism algebra is B'. That X is preprojective implies that the associated tilting functor $F = \operatorname{Hom}_B(X, -)$ induces an equivalence of cofinite subcategories of

B–mod and B'–mod. In particular, the respective families of tubes are mapped to each other. Therefore we will identify the two families. Using this convention the lemma above allows us to prove:

PROPOSITION. *Let B and B' two tame concealed algebras of the same extended Dynkin type and S, T be two finite subsets of $\mathbb{P}$. Then the algebras $\boxed{S}B\boxed{T}$ and $\boxed{S}B'\boxed{T}$ have the same representation type.*

PROOF. By [**Ha**] the tilting functor F extends to a triangle equivalence $D^b(B) \to D^b(B')$ of derived categories of bounded complexes which we will consider as an identification. Via the canonical embedding we obtain B–mod and B'–mod as full subcategories of $D^b(B)$. We observe the isomorphism $\mathrm{D}\,R(S) \otimes_B - \cong \mathrm{D}\,\mathrm{Hom}_B(-, R(S)) : B\text{–mod} \to k\text{–mod}$ and call $\mathcal{D}$ the factor category of $D^b(B)$ by the ideal given by the intersection of the kernel of the functor $\mathrm{Hom}_{D^b(B)}(R(T), -) : D^b(B) \to k\text{–mod}$ and the kernel of the functor $\mathrm{D}\,\mathrm{Hom}_{D^b(B)}(-, R(S)) : D^b(B) \to k\text{–mod}$. Thus $\mathcal{B}$ and $\mathcal{B}'$ become full subcategories of $\mathcal{D}$.

To establish our claim it is enough to find an autoequivalence of $\mathcal{D}$ which maps $\mathcal{B}$ into $\mathcal{B}'$. We denote by $\mathcal{D}_0$ the full subcategory of $\mathcal{B}$ whose objects are direct sums of indecomposables in the tubes $\mathcal{T}_\lambda$. The full subcategory $\mathcal{D}_\leq$ (resp. $\mathcal{D}_\geq$) of $\mathcal{D}$ has as objects direct sums of indecomposables which are predecessors (resp. successors) of objects in $\mathcal{D}_0$. Clearly, $\mathcal{D}_\leq \cup \mathcal{D}_\geq$ additively generates $\mathcal{D}$ and $\mathcal{D}_\leq \cap \mathcal{D}_\geq = \mathcal{D}_0$. Moreover, since the family $(\mathcal{T}_\lambda)_{\lambda \in \mathcal{P}}$ is separating in $\mathrm{D}^b(B)$, there is no non-zero map from $\mathcal{D}_\leq \setminus \mathcal{D}_0$ to $\mathcal{D}_\geq \setminus \mathcal{D}_0$. This shows that we can extend an autoequivalence $\phi_\leq$ of $\mathcal{D}_\leq$ and an autoequivalence $\phi_\geq$ of $\mathcal{D}_\geq$ to an autoequivalence ϕ of $\mathcal{D}$ provided that $\phi_\leq$ and $\phi_\geq$ coincide on $\mathcal{D}_0$.

To construct our desired autoequivalences we proceed as follows. Let N be a common multiple of all the ranks $n(\lambda)$ of the tubes $\mathcal{T}_\lambda$. It τ is the Auslander-Reiten translation in $D^b(B)$ then τ^{-N} induces an autoequivalence $\phi_\leq$ of $\mathcal{D}_\leq$ which is the identity on $\mathcal{D}_0$. Dually τ^N acts as an autoequivalence $\phi_\geq$ on $\mathcal{D}_\geq$ which leaves $\mathcal{D}_0$ invariant. Clearly, a sufficiently large power of the induced autoequivalence ϕ of $\mathcal{D}$ will map $\mathcal{B}$ into $\mathcal{B}'$ and also $\mathcal{B}'$ into $\mathcal{B}$. □

3. Proof of Theorem A

Let B be a tame concealed algebra and S, T be finite subsets of $\mathbb{P}$. If $S \cap T = \emptyset$, then there does not exist any indecomposable B-module X satisfying $\mathrm{Hom}_B(R(T), X) \neq 0$ and $\mathrm{Hom}_B(X, R(S)) \neq 0$. Consequently, every indecomposable representation of $M(S, B, T)$ is either a representation of $M(\emptyset, B, T)$ or a representation of $M(S, B, \emptyset)$. Thus we proved the following 'splitting lemma':

LEMMA. *If B is a tame concealed algebra and S, T are finite subsets of $\mathbb{P}$ such that $S \cap T = \emptyset$, then any indecomposable module over $\boxed{S}B\boxed{T}$ is either a module over $\boxed{S}B$ or a module over $B\boxed{T}$. In particular, $\boxed{S}B\boxed{T}$ is tame provided the algebras $\boxed{S}B$ and $B\boxed{T}$ are tame.*

We are now going to proof the tameness of $\boxed{\mathrm{S}}B\boxed{\mathrm{T}}$ in the four cases of Theorem A and start with case (i). Thus the extended Dynkin type of B is $\widetilde{\mathbb{D}}_n$ for some $n \geq 4$ and $T = \{\lambda\}$, $S = \{\mu\}$ such that $n(\lambda) = n(\mu) = n - 2$. If $n > 4$ this implies $S = T$. Only for $n = 4$ the case $S \neq T$ is possible. But then the tameness of $\boxed{\mathrm{S}}B\boxed{\mathrm{T}}$ follows from the tameness of $\boxed{\mathrm{S}}B$ and $B\boxed{\mathrm{T}}$ by the above splitting lemma. On the

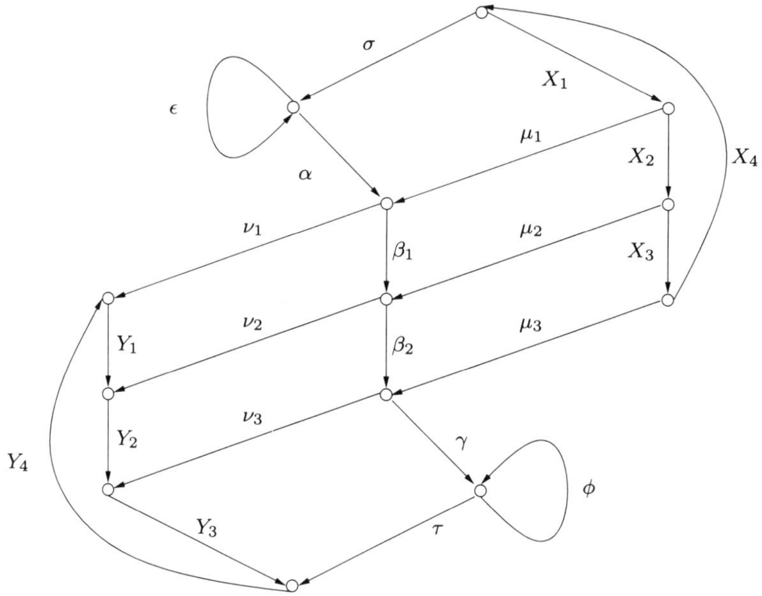

FIGURE 3. Clannish algebra 1

other hand the tameness of $\boxed{\mathrm{S}}B$ (resp. $B\boxed{\mathrm{T}}$) follows from the tameness of $\boxed{\mathrm{S}}B\boxed{\mathrm{S}}$ (resp. $\boxed{\mathrm{T}}B\boxed{\mathrm{T}}$). Thus we only have to deal with the case $S = T$.

By the Proposition from section 2 we can chose $B = k\vec{\Delta}$ hereditary such that the underlying graph Δ is $\widetilde{\mathbb{D}}_n$. We can even choose a convenient orientation and select the one from [**DR**] because this enables us to use the tables in this paper. Then $\boxed{\mathrm{S}}B\boxed{\mathrm{T}}$ has exactly the shape as in the example at the end of section one (for the case $n = 6$).

By replacing all X_i (resp. Y_i) in the relations by tX_i (resp. tY_i) we obtain a family of algebras $(A_t)_{t\in k}$ with the same dimension. Since $A_t \cong \boxed{\mathrm{S}}B\boxed{\mathrm{T}}$ for all $t \neq 0$, we obtain that A_0 is a degeneration of $\boxed{\mathrm{S}}B\boxed{\mathrm{T}}$. Thus by [**Ge**] the tameness of A_0 implies the tameness of $\boxed{\mathrm{S}}B\boxed{\mathrm{T}}$. But A_0 is a clannish algebra whose tameness is known by [**C-B**]. It is given by the quiver in Figure 3 together with the relations $\nu_1\alpha$, $\gamma\mu_3$, $\alpha\sigma$, $\tau\gamma$, $\beta_i\mu_i$ $(i = 1,2)$, $\nu_{i+1}\beta_i$ $(i = 1,2)$, $X_{i+1}X_i$ $(i = 1,\dots,4)$, $Y_{i+1}Y_i$ $(i = 1,\dots,4)$, $\epsilon(\epsilon - 1)$, $\phi(\phi - 1)$.

Because the proof of tameness in case (ii) is very similar, we skip it and mention only some special features. First, it is not too surprising that the degeneration algebra A_0 where we finally arrive at in this case is actually special biserial. Moreover, there are some more special cases to handle. Namely, for $q \geq 2$ one knows again that $S = T$. But for $p > 1$ and $q = 1$ the intersection $S \cap T$ may have 1 or 2 elements. Both cases have to be settled separately, but the arguments are parallel. For $p = q = 1$, which means that B is the Kronecker algebra, in addition $S \cap T = \emptyset$ is possible. Fortunately, this is settled by the splitting lemma and the other two cases work as for $p > 1$.

Let us give again a complete proof of the tameness of $\boxed{\mathrm{S}}B\boxed{\mathrm{T}}$ in case (iii). Since the extended Dynkin type of B is $\widetilde{\mathbb{A}}_{2,2}$, we may choose B to be of the shape $k\vec{\Delta}$ where the underlying graph Δ is $\widetilde{\mathbb{A}}_3$ and $\vec{\Delta}$ has two sources and two sinks.

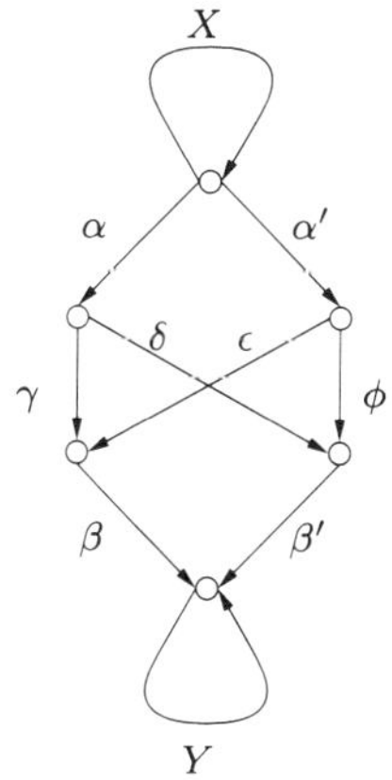

FIGURE 4. Example 2

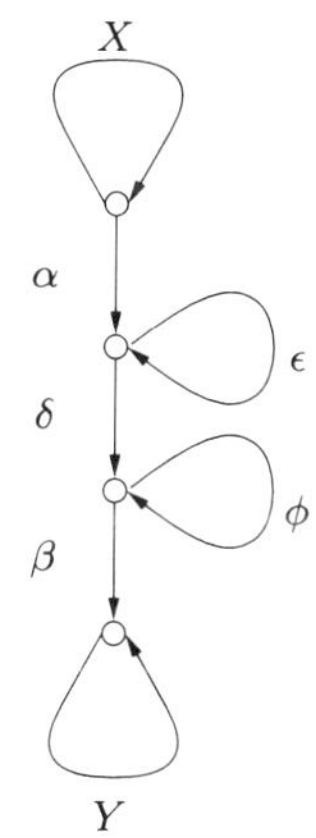

FIGURE 5. Clannish algebra 2

By the splitting lemma we may assume that $S = T$. After possibly applying an automorphism of B, we arrive at the algebra $\boxed{\text{S}}B\boxed{\text{T}}$ given by the quiver of Figure 4 with the relations $\epsilon\alpha + \phi\alpha'$, $\gamma\alpha X + \epsilon\alpha' + \gamma\alpha$, $\beta'\phi + \beta\epsilon$, $Y\beta\gamma + \beta'\delta + \beta\gamma$, X^2, Y^2.

Replacing X by tX and Y by tY we again obtain a family $(A_t)_{t\in k}$ of algebras of the same dimension where $A_t \cong \boxed{\text{S}}B\boxed{\text{T}}$ for all $t \neq 0$. Since A_0 is again a clannish algebra given by the quiver in Figure 5 and the relations $\delta\alpha$, $\beta\delta$, X^2, Y^2, $\epsilon(\epsilon - 1)$, $\phi(\phi - 1)$, tameness follows by [**C B**] as in case (i).

Finally, in case (iv) the tameness of $\boxed{\text{S}}B\boxed{\text{T}}$ follows from the cases (ii) and (iii) via the splitting lemma.

4. Proof of Theorem B

The circular extension $B\boxed{\text{T}}$ is wild provided one finds a subset T' of T such that $B\boxed{\text{T'}}$ is wild. The reason is that the second algebra appears as a convex subcategory of the first one. Therefore, in order to prove Theorem B, it is sufficient to prove the wildness of the circular extension $B\boxed{\text{T}}$ where B is a tame concealed algebra and T is a subset of $\mathbb{P}$ in the following cases:

(a) The extended Dynkin type of B is $\widetilde{\mathbb{E}}_n$ $(n = 6, 7, 8)$ and $T = \{\lambda\}$ for λ in $\mathbb{P}$.

(b) The extended Dynkin type of B is $\widetilde{\mathbb{D}}_n$ $(n \geq 4)$ and $T = \{\lambda\}$ for λ in $\mathbb{P}$ such that $n(\lambda) < n - 2$.

(c) The extended Dynkin type of B is $\widetilde{\mathbb{D}}_4$ and $T = \{\lambda, \mu\}$ for λ, μ in $\mathbb{P}$ such that $n(\lambda) = n(\mu) = 2$.

(d) The extended Dynkin type of B is $\widetilde{\mathbb{A}}_{p,q}$ $(p \geq 3,\ q \geq 2)$ and $T = \{\lambda\}$ for λ in $\mathbb{P}$ such that $n(\lambda) = 1$.

(e) The extended Dynkin type of B is $\widetilde{\mathbb{A}}_{p,1}$ $(p \geq 2)$ and $T = \{\lambda, \mu\}$ for λ, μ in $\mathbb{P}$ such that $n(\lambda) = n(\mu) = 1$.

(f) The extended Dynkin type of B is $\widetilde{\mathbb{A}}_{1,1}$ and $T = \{\lambda, \mu, \nu\}$ for λ, μ, ν in $\mathbb{P}$ such that $n(\lambda) = n(\mu) = n(\nu) = 1$.

(g) The extended Dynkin type of B is $\widetilde{\mathbb{A}}_{2,2}$ and $T = \{\lambda, \mu\}$ for λ, μ in $\mathbb{P}$ such that $n(\lambda) = 2$ and $n(\mu) = 1$.

Before going through these various cases, we need some auxiliary results. Note that for λ in $\mathbb{P}$ the module R^λ is of length 2 in the tube $\mathcal{T}_\lambda$ which is an abelian length category. Hence it has a socle which we denote by V^λ and a top which we denote by W^λ. Consequently, $R^\lambda / V^\lambda \cong W^\lambda$ holds and $V^\lambda = W^\lambda$ if $n(\lambda) = 1$.

LEMMA. *Let B be a tame concealed algebra and λ in $\mathbb{P}$ such that $T = \{\lambda\}$. Then the circular extension $B\boxed{T}$ is wild if one of the following assertions holds:*

(i) $n(\lambda) > 1$ *and the one-point extension* $B[R^\lambda]$ *is wild.*

(ii) $n(\lambda) = 1$ *and the one-point extension* $(B \times k)[W^\lambda \oplus k]$ *is wild.*

PROOF. For $n(\lambda) > 1$ the algebra $B[R^\lambda]$ is a subcategory of $B\boxed{\mathrm{T}}$. This proves (i). To see (ii), we observe that $\begin{pmatrix} 0 & 0 \\ V^\lambda & 0 \end{pmatrix}$ is an ideal of $B\boxed{\mathrm{T}}$ such that the associated factor algebra is isomorphic to $\begin{pmatrix} C(T) & 0 \\ W^\lambda & B \end{pmatrix}$ where X_1 acts trivially on W^λ. In the canonical $\mathbb{Z}$-covering of this algebra obtained by stretching the loop X_1 to a line, the algebra $(B \times k)[W^\lambda \oplus k]$ appears as convex subcategory. By [**DoS**] the wildness of $(B \times k)[W^\lambda \oplus k]$ implies the wildness of $B\boxed{\mathrm{T}}$. □

Note, that $(B \times k)[W^\lambda \oplus k]$ is a branch extension of B by the simple regular module W^λ where the added branch is of length 2 (see [**Ri2**]). Hence the algebra $(B \times k)[W^\lambda \oplus k]$ is wild provided its tubular type is neither a Dynkin nor an extended Dynkin diagram (see [**Pe**]).

Using the lemma, cases (a) and (b) are settled easily. Namely, for $n(\lambda) > 1$ the wildness of $B[R^\lambda]$ is known by [**Ri1**] and for $n(\lambda) = 1$ the tubular type of $(B \times k)[W^\lambda \oplus k]$ is neither Dynkin nor extended Dynkin. In case (c) the two-point extension $B[R^\lambda, R^\mu]$ is a wild subcategory of $B\boxed{\mathrm{T}}$.

To solve the remaining four cases dealing with B of extended Dynkin type $\widetilde{\mathbb{A}}_{p,q}$, we recall that by the Proposition in section 2 we can choose $B = k\vec{\Delta}$ where $\vec{\Delta}$ is the quiver from Figure 6.

In the cases (d), (e), (f) we form the factor algebras of $B\boxed{\mathrm{T}}$ by the idempotents associated with z and possibly some predecessors of z, obtaining one of the algebras whose quivers appear in Figure 7 whereas the relations are X^2 in case (d), X^2, Y^2 in case (e) and X^2, Y^2, Z^2 in case (f). It is easy to see that in the universal Galois covering of each of these algebras there appears a convex subcategory which is a hereditary algebra associated with a hyperextended Dynkin diagram. Consequently,

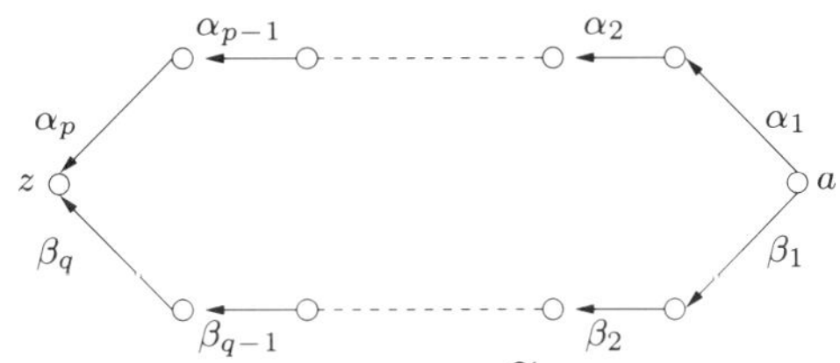

FIGURE 6. $\widetilde{\mathbb{A}}_{p,q}$

the coverings and therefore (see [**DoS**]) also the factor algebras are wild. Case (g) is eventually settled by factoring out from $B\boxed{T}$ the ideal generated by the idempotents belonging to z and one suitable predecessor. The factor algebra is the same as for case (e).

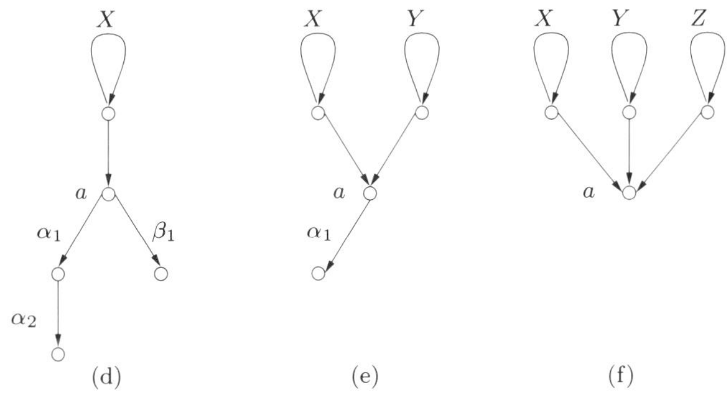

FIGURE 7. Factor algebras

References

[C-B] W. Crawley-Boevey, *Functorial filtrations II: Clans and the Gelfand problem,* J. London Math. Soc. 40 (1989), 9-30.

[DR] V. Dlab, C.M. Ringel, *Indecomposable representations of graphs and algebras,* Mem. Amer. Math. Soc. 173 (1973).

[DoS] P. Dowbor, A. Skowroński: *On the representation type of locally bounded categories,* Tsukuba J. Math. 10 (1986), 63-72.

[Dr] P. Dräxler, *On the density of fiber sum functors,* Math. Z. 216 (1994), 645-656.

[DrS] P. Dräxler, A. Skowroński, *Tame biextensions by indecompsable modules of derived regular length 2,* Preprint 97-081 SFB 343, Bielefeld.

[Dd] Ju. A. Drozd, *Tame and wild matrix problems,* Lecture Notes in Math. 832 (1980), 242-258.

[GR] P. Gabriel, A.V. Roiter, *Representations of finite dimensional algebras,* Encyclopedia of the Mathematical Sciences, Vol. 73, Algebra VIII, A.I. Kostrikin and I.V. Shafarevich (Eds.), Berlin, Heidelberg, New York, 1992.

[Ge] C. Geiß, *On degenerations of tame and wild algebras,* Arch. Math. 64 (1995), 11-16.

[Ha] D. Happel, *Triangulated categories in the representation theory of finite dimensional algebras,* LMS lecture notes series 119 (1988).

[HV] D. Happel, D. Vossieck, *Minimal algebras of infinite representation type with preprojective component,* Manuscr. Math. 42 (1983), 221-243.

[NS] R. Nörenberg, A. Skowroński, *Tame minimal non-polynomial growth strongly simply connected algebras,* Colloq. Math. 73 (1997), 301-330.

[Pe] J.A. de la Peña, *On the representation type of one-point extensions of tame concealed algebras,* Manuscr. Math. 61 (1988), 183-194.

[Ri1] C.M. Ringel, *Tame algebras,* in: Representation Theory I, Lecture Notes in Math. 831 (1980), 134-287.

[Ri2] C.M. Ringel, *Tame algebras and integral quadratic forms,* Lecture Notes in Math. 1099 (1984).

[Sk] A. Skowroński, *Simply connected algebras of polynomial growth,* to appear in Compositio Math.

FAKULTÄT FÜR MATHEMATIK, UNIVERSITÄT BIELEFELD, POBOX 100131, D-33501 BIELEFELD, GERMANY

E-mail address: draexler@mathematik.uni-bielefeld.de

Contemporary Mathematics
Volume **229**, 1998

ON THE DISTRIBUTION OF AR-COMPONENTS OF RESTRICTED LIE ALGEBRAS

ROLF FARNSTEINER

ABSTRACT. In this paper we study the distribution of indecomposable modules of the reduced enveloping algebra $u(L, \chi)$ associated to a finite dimensional restricted Lie algebra $(L, [p])$. Each component of the stable Auslander-Reiten quiver possesses only finitely modules of a given dimension. We combine this fact with geometric techniques in order to produce families of components of type $\mathbb{Z}[A_\infty]$.

0. INTRODUCTION AND PRELIMINARIES

Let F be an algebraically closed field, Λ a finite dimensional F-algebra of infinite representation type. If Λ is tame, it follows from Brauer Thrall II and the work of Crawley-Boevey [5, Corollary E] that there exists a number $d \in \mathbb{N}$, such that for each multiple ℓd there are infinitely many non-isomorphic indecomposable Λ-modules of dimension ℓd. In case Λ is wild, this conclusion continues to hold (with d even) since there is a representation embedding from the module category of the Kronecker algebra $F[X, Y]/(X^2, Y^2)$ into the module category of Λ. A related question concerning the distribution of the dimensions within a given Auslander-Reiten component was raised by Ringel [30]: Given an AR-component Θ and $d > 0$, is the set

$$\Theta(d) := \{[M] \in \Theta \ ; \ dim_F M \leq d\}$$

finite? While the answer is affirmative for some classes of algebras (cf. for instance [5, Corollary F]), there exist symmetric algebras which admit infinite $\Theta(d)$ (cf. [24]).

In this note we shall study questions related to the distribution of AR-components for reduced enveloping algebras of restricted Lie algebras. Our results, some of which will be applied in a forthcoming paper [12], illustrate the utility of the geometric methods that are based on support varieties.

In the sequel we shall consider finite dimensional restricted Lie algebras $(L, [p])$, defined over an algebraically closed field F of characteristic $p > 0$. The universal enveloping algebra of L will be denoted $\mathcal{U}(L)$. Given a linear form $\chi \in L^*$, the *χ-reduced enveloping algebra* of L is defined by

$$u(L, \chi) := \mathcal{U}(L)/(\{x^p - x^{[p]} - \chi(x)^p 1 \ ; \ x \in L\}).$$

1991 *Mathematics Subject Classification.* Primary 16G70, 17B50 ; Secondary 20G15.

Supported by N.S.A. Grant MDA904-96-1-0040 and by the D.F.G.

We refer the interested reader to [32] concerning basic facts of restricted Lie algebras and their enveloping algebras. A few properties of $u(L,\chi)$ that are relevant for our purposes are summarized in the following list (cf. [6, 8, 13]).

- The algebra $u(L,\chi)$ is finite dimensional, more precisely $dim_F u(L,\chi) = p^{dim_F L}$.
- $u(L,\chi)$ is a Frobenius algebra whose Nakayama automorphism μ is uniquely determined by

$$\mu(x) = x - tr(ad\, x)1 \quad \forall\, x \in L.$$

- The restricted enveloping algebra $u(L) := u(L,0)$ is a Hopf algebra.
- Let Θ be a component of the stable Auslander-Reiten quiver $\Gamma_s(L,\chi)$ of $u(L,\chi)$. Then Θ is either a finite or infinite tube of rank $k \in \{1,p\}$, or

$$\Theta \cong \mathbb{Z}[A_\infty],\ \mathbb{Z}[A_\infty^\infty],\ \mathbb{Z}[D_\infty],\ \mathbb{Z}[\tilde{A}_n],\ \mathbb{Z}[\tilde{A}_{12}],\ \mathbb{Z}[\tilde{D}_n],\ \mathbb{Z}[\tilde{E}_\ell] \quad (6 \leq \ell \leq 8).$$

In the following all $u(L,\chi)$-modules are assumed to be finite dimensional. We also set $\Gamma_s(L) := \Gamma_s(L,0)$.

Let M be a $u(L,\chi)$-module. In view of the work of Friedlander and Parshall (cf. [15, 16, 17]) and Jantzen [22] we define the *cohomological support variety* of M via

$$\mathcal{V}_L(M) := \{x \in L\ ;\ x^{[p]} = 0 \textit{ and } M|_{\langle x\rangle} \textit{ is not free}\}.$$

Here $M|_{\langle x\rangle}$ denotes the restriction of M to the subalgebra of $u(L,\chi)$ that is generated by x.

It is well-known (cf. [15, 16, 17]) that $\mathcal{V}_L(M)$ is a conical variety whose dimension coincides with the complexity $c_{u(L,\chi)}(M)$. In particular, M is projective if and only if $\mathcal{V}_L(M) = \{0\}$. Moreover, if $\Theta \subset \Gamma_s(L,\chi)$ is a component, then we have $\mathcal{V}_L(M) = \mathcal{V}_L(N) \quad \forall\ [M],[N] \in \Theta$. Accordingly, we may define

$$\mathcal{V}_L(\Theta) := \mathcal{V}_L(M) \qquad [M] \in \Theta.$$

After providing an affirmative answer to Ringel's question (Theorem 1.2), we turn in §2 to algebraic Lie algebras, i.e., those where $L = Lie(G)$ is the Lie algebra of an algebraic group G. In this setting the stabilizer G_χ of χ (with respect to the coadjoint action) operates on $\Gamma_s(L,\chi)$ via automorphisms. If G_χ is reductive (e.g., if G is reductive and χ is semisimple), then the components invariant relative to this action are shown to consist of modules that are stable under the twist by the adjoint representation. In the same vein, we employ support varieties to see that representation infinite enveloping algebras possess infinitely many indecomposables for each multiple of a suitable p-power. This result, as well as those of the final section extend recent work of [26] in various directions. In particular, we show that each restricted simple Lie algebra $L \not\cong s\ell(2)$ admits infinitely many stable AR-components of type $\mathbb{Z}[A_\infty]$ containing modules of certain p-power dimensions.

This paper was written during the author's stay at the Sonderforschungsbereich 343. He would like to take this opportunity to express his gratitude for the support he has received.

1. The Distribution of Indecomposable Modules

Throughout this section we fix a finite dimensional restricted Lie algebra $(L, [p])$ as well as a linear form $\chi \in L^*$. The following subsidiary result is the key step in proving the finiteness of $\Theta(d)$.

Lemma 1.1. *Let $\Theta \subset \Gamma_s(L,\chi)$ be a component, $\mathcal{T} \subset \Theta$ a τ-orbit. Then $\mathcal{T} \cap \Theta(d)$ is finite for every $d > 0$.*

Proof. Assume there is $d > 0$ such that $\mathcal{T} \cap \Theta(d)$ is infinite. Then Θ is a nonperiodic component. By assumption there exists a vertex $[M] \in \Theta(d)$ such that

$$dim_F \Omega^{2j}(M) \leq d$$

for j belonging to some infinite set $J \subset \mathbb{Z}$. Without loss of generality we may assume that $0 \in J$. Consequently,

$$dim_F Ext^{2j}_{u(L,\chi)}(M,M) \leq dim_F Hom_{u(L,\chi)}(\Omega^{2j}(M), M) \leq d^2 \quad \forall\, j \in J \cap \mathbb{N},$$

while

$$dim_F Ext^{-2j}_{u(L,\chi)}(M,M) \leq dim_F Hom_{u(L,\chi)}(M, \Omega^{2j}(M)) \leq d^2 \;\; \forall\, j \in J \cap -\mathbb{N}.$$

Hence there are infinitely many $n \in \mathbb{N}$ such that

$$dim_F Ext^{2n}_{u(L,\chi)}(M,M) \leq d^2.$$

Since Θ is not periodic, it follows from [2, (3.4)] that $dim\mathcal{V}_L(M) = c_{u(L,\chi)}(M) \geq 2$. By work of Jantzen [22] and Friedlander-Parshall [15], the variety $\mathcal{V}_L(M)$ is the zero locus of the ideal $ker\,\Phi_M$, where

$$\Phi_M : S(L^*) \longrightarrow Ext^{ev}_{u(L,\chi)}(M,M)^{(-1)}$$

is a graded algebra homomorphism from the symmetric algebra of L^* into the even part of the Yoneda algebra of M. The superscript (-1) indicates the twisting of the underlying F-space by the Frobenius automorphism of F (cf. [16, (1.4)] for details). The Noether Normalization Theorem provides $X, Y \in L^*$ such that

$$dim_F Ext^{2n}_{u(L,\chi)}(M,M) \geq dim_F S^n[FX \oplus FY] \geq n+1 \quad \forall\, n \geq 0,$$

a contradiction. □

Theorem 1.2. *Let $\Theta \subset \Gamma_s(L,\chi)$ be a component. Then $\Theta(d)$ is finite for every $d > 0$.*

Proof. Suppose first that Θ is periodic. Then Θ is either finite or an infinite tube. From the structure of the latter (cf. [8, (5.3)]) we readily obtain $|\Theta(d)| \leq pd$.

Let Θ be nonperiodic. According to (1.1) the set $\Theta(d)$ is finite whenever Θ has only finitely many τ-orbits. If Θ has infinitely many τ-orbits, then our introductory remarks leave us with $\Theta \cong \mathbb{Z}[A_\infty]$, $\mathbb{Z}[D_\infty]$, or $\Theta \cong \mathbb{Z}[A_\infty^\infty]$.

Suppose that $\Theta \cong \mathbb{Z}[A_\infty]$, so that $\mathbb{Z} \times \mathbb{N}$ is its set of vertices. By definition the arrows and the Auslander-Reiten translation of Θ are given by

$$(i,a) \to (i,a+1) \ ; \ (i,a) \to (i+1,a-1)$$

and

$$\tau(i,a) = (i-1,a),$$

respectively. If Θ is not regular, then [11, (4.1)] shows that there exists precisely one principal indecomposable $u(L,\chi)$-module P that is attached to Θ. By the same token, $[Rad(P)]$ is located at an end of Θ, and we may set

$$[Rad(P)] = (0,1).$$

Induction on n now shows that

$$(1) \qquad dim_F(j,n) = \sum_{k=j}^{n+j-1} dim_F(k,1) - \sum_{k=j}^{n-2+j} \delta_{k,0} dim_F P \quad \forall\, j \in \mathbb{Z},$$

where we set $P = (0)$ in case Θ is regular.

Now consider the set

$$\Upsilon := \{n \in \mathbb{N} \ ; \ \exists\, j \in \mathbb{Z} \ : \ (j,n) \in \Theta(d)\}.$$

Suppose Υ is infinite and choose for each $n \in \Upsilon$ an element $j(n) \in \mathbb{Z}$ such that $(j(n),n) \in \Theta(d)$. If $\{j(n) \ ; \ n \in \Upsilon\}$ is infinite, then (1) implies that the τ-orbit of $(0,1)$ contains infinitely many elements of $\Theta(d + dim_F P)$. This contradicts (1.1). Alternatively, the set $\{n + j(n) - 1 \ ; \ n \in \Upsilon\}$ is infinite, and we arrive at the same contradiction.

Consequently, Υ is finite and there exist τ-orbits $\mathcal{T}_1, \ldots, \mathcal{T}_r$ such that $\Theta(d) = \bigcup_{i=1}^r (\mathcal{T}_i \cap \Theta(d))$. Thanks to (1.1), the set $\Theta(d)$ is finite.

We next consider $\Theta \cong \mathbb{Z}[A_\infty^\infty]$ and label its vertices by $\mathbb{Z} \times \mathbb{Z}$. By definition there are arrows

$$(i,a) \to (i,a+1) \ ; \ (i,a) \to (i+1,a-1).$$

Let $\{P_1, \ldots, P_r\}$ be the set of principal indecomposable modules that are attached to Θ and put $t := \sum_{i=1}^r dim_F P_i$. Then we have

$$dim_F(i,a) + dim_F(0,0) = dim_F(i,-i) + dim_F(0,a+i) + q(i,a),$$

where $-t \leq q(i,a) \leq t$ depends on the number of projective meshes contained in the "parallelogram" $\{(0,0),\ (0,a+i),\ (i,-i),\ (i,a)\}$. Consequently,

$$(i,-i) \in \Theta(d + dim_F(0,0) + t)$$

whenever $(i,a) \in \Theta(d)$. The arguments of [25, p.131] now show that, for any $\ell > 0$ the set $\Theta(\ell) \cap \{(i,-i) \ ; \ i \in \mathbb{Z}\}$ is finite. Thus, $\Theta(d)$ is contained in a union of finitely many τ -orbits and we may apply (1.1) to see that $\Theta(d)$ is finite.

Finally, let $\Theta \cong \mathbb{Z}[D_\infty]$. The set of vertices is $\mathbb{Z} \times (\mathbb{N}_0 \cup \{-1\})$. The subset $\mathbb{Z} \times \mathbb{N}_0$ is a subquiver isomorphic to $\mathbb{Z}[A_\infty]$. To obtain Θ we add arrows

$$(i,-1) \to (i,1) \quad ; \quad (i,1) \to (i+1,-1) \quad \forall\ i \in \mathbb{Z}.$$

Induction on n shows that

$$dim_F(j,n) = \begin{cases} dim_F(j,0) + dim_F(j+n,0) + q(j,n) & n \equiv 1 \ \ mod(2) \\ dim_F(j,-1) + dim_F(j+n,0) + q(j,n) & n \equiv 0 \ \ mod(2) \end{cases}$$

for all $n \geq 1$. As before $-t \leq q(j,n) \leq t$ depends on the number of projective meshes of the wing $\{(j,n),(j,-1),(j+n+1,-1)\}$. The proof may now be completed by using the arguments of the case $\Theta = \mathbb{Z}[A_\infty]$. □

Corollary 1.3. *Let $\mathcal{B} \subset u(L,\chi)$ be a block. Then $\mathcal{B}$ has finite representation type if and only if its Auslander-Reiten quiver $\Gamma(\mathcal{B})$ is connected.*

Proof. Necessity follows from a well-known theorem of Auslander (cf. [1, (VI.1.4)]). The reverse direction is a direct consequence of Brauer-Thrall II and (1.2). By way of illustration, we sketch a second argument that is based on the geometric approach.

Suppose that $\Gamma(\mathcal{B})$ is connected. If $\Gamma_s(\mathcal{B})$ is not connected, then $\mathcal{B}$ is readily seen to be a Nakayama algebra. Alternatively, let S be a simple $\mathcal{B}$-module. Given $x \in \mathcal{V}_L(S) \setminus \{0\}$, a suitable modification of [15, (2.2)] implies that there exists an indecomposable $\mathcal{B}$-module M such that $\mathcal{V}_L(M) = Fx$. Since $[M]$ and $[S]$ belong to the same stable component, we obtain $\mathcal{V}_L(S) = Fx$. Thanks to [8, (3.2)] this implies that $\mathcal{B}$ is a Nakayama algebra. □

2. Indecomposables of Algebraic Lie Algebras

In the following we shall be primarily interested in the situation where L is the Lie algebra of a (connected) algebraic group G. In that case a natural action of G on the set of stable Auslander-Reiten components provides an additional tool.

We begin with a subsidiary result that is valid for arbitrary restricted Lie algebras $(L,[p])$ admitting a *triangular decomposition*

$$L = L^- \oplus L_0 \oplus L^+.$$

By definition, the constituents are p-subalgebras satisfying $[L_0, L^\pm] \subset L^\pm$. We endow a $u(L_0)$-module M with the structure of a $u(L_0 \oplus L^+)$-module by letting L^+ act trivially and define

$$M{\uparrow}^L := u(L) \otimes_{u(L_0 \oplus L^+)} M.$$

A restricted Lie algebra $(L, [p])$ is said to be *p-nilpotent* if some power of its p-map is trivial. The Lie algebras of unipotent algebraic groups are p-nilpotent.

Lemma 2.1. *Suppose that L^- is p-nilpotent and let M and N be $u(L_0)$-modules.*

(1) If M is indecomposable, then $M{\uparrow}^L$ is indecomposable.
(2) If $M{\uparrow}^L \cong N{\uparrow}^L$, then $M \cong N$.
(3) $\mathcal{V}_{L_0}(M) = \mathcal{V}_{L_0}(M{\uparrow}^L) \subset \mathcal{V}_L(M{\uparrow}^L) \subset \mathcal{V}_{L_0\oplus L^+}(F)$.

Proof. (1). This follows directly from [7, (4.1)].

(2). According to [7, (3.2)] there exists an automorphism ω of $u(L_0)$ such that the $u(L_0)$-module $(M{\uparrow}^L)^{L^-}$ of L^--invariants of $M{\uparrow}^L$ is isomorphic to the $u(L_0)$-module $M^{(\omega)}$ obtained from M by twisting the action of $u(L_0)$. This readily yields the assertion.

(3). We apply [9, (3.4)] to see that the support variety of $M{\uparrow}^L$ is contained in $\mathcal{V}_{L_0\oplus L^+}(F)$. Let us denote by $u(L)_{ad}$ the L-module with action $x \cdot u := xu - ux \quad \forall\, x \in L, u \in u(L)$. Since $[L_0, L^-] \subset L^-$, it follows that $u(L^-)$ and its augmentation ideal $u(L^-)^\dagger$ are L_0-submodules of $u(L)_{ad}$. If $u(L^-)$ is endowed with this structure, then the map

$$u(L^-) \otimes_F M \longrightarrow M{\uparrow}^L \quad ; \quad u \otimes m \mapsto u \otimes m$$

is an isomorphism of L_0-modules. From the decomposition $u(L^-) = F \oplus u(L^-)^\dagger$, we infer that the L_0-module M is a direct summand of $M{\uparrow}^L$. Consequently, $\mathcal{V}_{L_0}(M) \subset \mathcal{V}_{L_0}(M{\uparrow}^L)$. For the reverse inclusion we combine the above isomorphism with [15, (2.1)] to obtain

$$\mathcal{V}_{L_0}(M{\uparrow}^L) = \mathcal{V}_{L_0}(u(L^-) \otimes_F M) = \mathcal{V}_{L_0}(u(L^-)) \cap \mathcal{V}_{L_0}(M) \subset \mathcal{V}_{L_0}(M).$$

□

Remark 1. Properties (1) and (2) can be rephrased by saying that $M \mapsto M{\uparrow}^L$ is a representation embedding from the module category of $u(L_0)$ to the module category of $u(L)$.

Let $(L, [p])$ be a restricted Lie algebra with group $Aut_p(L)$ of automorphisms (of the restricted Lie algebra). Given a homomorphism $\pi : G \longrightarrow Aut_p(L)$ of linear algebraic groups, and a linear form $\chi \in L^*$ we consider the stabilizer

$$G_\chi := \{g \in G \; ; \; \chi \circ \pi(g^{-1}) = \chi\}$$

as well as its connected component G_χ°. For a $u(L, \chi)$-module M and $g \in G_\chi$, we let $M^{(g)}$ denote the $u(L, \chi)$-module with underlying F-space M and twisted action

$$x \cdot m := \pi(g^{-1})(x)\, m \quad \forall\, x \in L,\ m \in M.$$

There results an operation

$$G_\chi \times \Gamma_s(L, \chi) \longrightarrow \Gamma_s(L, \chi) \quad ; \quad (g, [M]) \mapsto [M^{(g)}]$$

of G_χ on $\Gamma_s(L,\chi)$. As noted in [22, (2.2)], we have

$$\mathcal{V}_L(M^{(g)}) = \pi(g)(\mathcal{V}_L(M)) \quad \forall\, g \in G_\chi. \tag{2}$$

Let $g \in G_\chi$. The equivalence $M \mapsto M^{(g)}$ induces an automorphism $[M] \mapsto [M^{(g)}]$ of the quiver $\Gamma_s(L,\chi)$. Accordingly, G_χ operates on the set $\mathcal{C}_s(L,\chi)$ of components of $\Gamma_s(L,\chi)$, and for every component $\Theta \subset \Gamma_s(L,\chi)$ and $g \in G_\chi$ we either have

$$\Theta^{(g)} = \Theta$$

or

$$\Theta^{(g)} \cap \Theta = \emptyset.$$

Owing to (2) the stabilizer

$$G_\Theta := \{g \in G_\chi \ ; \ \Theta^{(g)} = \Theta\}$$

of a component $\Theta \subset \Gamma_s(L,\chi)$ is contained in the closed subgroup

$$G_{\mathcal{V}_L(\Theta)} := \{g \in G_\chi \ ; \ \pi(g)(\mathcal{V}_L(\Theta)) = \mathcal{V}_L(\Theta)\}.$$

In particular, if the orbit $\{\Theta^{(g)} \ ; \ g \in G_\chi\}$ is finite, then $G_{\mathcal{V}_L(\Theta)}$ is a closed subgroup of finite index. Consequently, $G_\chi^\circ \subset G_{\mathcal{V}_L(\Theta)}$, so that $\mathcal{V}(\Theta)$ is G_χ°-invariant.

Let G be a connected algebraic group with Lie algebra L, $\chi \in L^*$ a linear form. Considering the adjoint representation $Ad : G \longrightarrow Aut_p(L)$, we call a component $\Theta \subset \Gamma_s(L,\chi)$ *invariant* if $\Theta^{(g)} = \Theta \quad \forall\, g \in G_\chi^\circ$. A $u(L,\chi)$-module M is referred to as *Ad-stable* if $M \cong M^{(g)}$ for every $g \in G_\chi^\circ$. If M is a rational G-module, then the corresponding $u(L)$-module is *Ad*-stable. Moreover, since G_χ° is connected, every simple $u(L,\chi)$-module is *Ad*-stable.

Evidently, every component containing an Ad-stable vertex is stable. The converse is the content of our first application of (1.2), which has turned out to be useful in the study of invariant components of reductive groups (cf. [12]).

Proposition 2.2. *Let G be a connected algebraic group with Lie algebra L. Suppose that $\chi \in L^*$ is given such that G_χ is reductive. If $\Theta \subset \Gamma_s(L,\chi)$ an invariant component, then every vertex of Θ is defined by an Ad-stable module.*

Proof. Let $\Theta \subset \Gamma_s(L,\chi)$ be an invariant component and suppose that $[M] \in \Theta$. By assumption we have

$$\mathcal{O} := \{[M^{(g)}] \ ; \ g \in G_\chi^\circ\} \subset \Theta(dim_F M),$$

and (1.2) implies that $\mathcal{O}$ is finite. There results a homomorphism

$$\varrho : G_\chi^\circ \longrightarrow Sym(\mathcal{O})$$

of abstract groups sending G°_χ to the symmetric group of $\mathcal{O}$. Let $G_i \subset G^\circ_\chi$ be an almost simple constituent. Then $ker\,\varrho|_{G_i}$ is a normal subgroup of G_i of finite index that contains the center of G_i. Thanks to [20, Corollary 29.5] we obtain $ker\,\varrho|_{G_i} = G_i$. Since the reductive group G°_χ is the product of its almost simple constituents and its center, ϱ is the trivial homomorphism and $M^{(g)} \cong M \quad \forall\, g \in G^\circ_\chi$. □

Remark 2. Let G be a connected, reductive group of characteristic $p > 3$, $\chi \in L^*$ a linear form. We can decompose

$$\chi = \chi_s + \chi_n$$

into its semisimple and nilpotent parts (cf. [23, Thm.4] and [29, (3.1)]). Since the stabilizer G_{χ_s} of the semisimple part χ_s is reductive (cf. [23, (3.1)]), (2.2) shows that every vertex of an invariant component $\Theta \subset \Gamma_s(L, \chi_s)$ is Ad-stable.

We continue by considering the distribution of indecomposable modules within $\Gamma_s(L, \chi)$. For a natural number d, we let $\mathcal{C}_s(L, \chi, d)$ and $\mathcal{C}^{per}_s(L, \chi, d)$ be the subsets of $\mathcal{C}_s(L, \chi)$ containing a module of dimension d, and a periodic module of dimension d, respectively.

Let $\Theta \subset \Gamma_s(L, \chi)$ be a periodic component, $[M_1] \in \Theta$ a vertex of minimal dimension d. According to the remark of [8, p.62] there exists, for any $\ell > 0$, a vertex $[M_\ell] \in \Theta$ with $dim_F M_\ell = \ell\, d$. In case $L = s\ell(2)$ and $p \geq 3$, the indecomposable modules of length 2 are periodic and of dimension p. For $p = 2$ the principal block of $u(s\ell(2))$ is isomorphic to the Kronecker algebra $F[X, Y]/(X^2, Y^2)$. Hence there exist nonprojective indecomposables of dimension ℓp in all cases. Since the principal indecomposable $u(s\ell(2))$-modules are known to have dimension p or $2p$, it readily follows from Pollack's work [28, Thm.4] and the representation theory of the Kronecker algebra that $u(s\ell(2))$ admits nonprojective indecomposable modules in each dimension. We shall use these observations repeatedly in the sequel.

Example 1. Let G be connected, reductive with maximal torus $T \neq G$ and Lie algebra L. Let $B \supset T$ be a Borel subgroup, Δ the corresponding set of simple roots. We choose a proper subset $\emptyset \neq I \subset \Delta$ as well as a standard parabolic subgroup $P_I \supset B$. The Levi decomposition of $Lie(P_I)$ induces a triangular decomposition

$$L = L^- \oplus L_0 \oplus L^+,$$

where L^+ and L^- are p-nilpotent subalgebras and L_0 is a Levi factor of $K := Lie(P_I) = L_0 \oplus L^+$.

If I is a singleton, contained in a connected component of Δ, then $L_0 \cong Q \oplus s\ell(2)$, $Q \subset C(L_0)$, and $\mathcal{V}_K(F)$ does not contain any G-invariant subsets $\neq \{0\}$. Hence (1) and (3) of (2.1) imply that a nonprojective indecomposable $u(s\ell(2))$-module M defines infinitely many components containing a module $(M\!\uparrow^L)^{(g)}$ of dimension $p^{dim_F L^-}(dim_F M)$. In particular, there are infinitely many nonisomorphic indecomposables of dimension ℓp^r. (More precisely, we can choose $r := max\{|\Phi^+| - 1, 1\}$, where Φ^+ is the set of positive roots.) This result, which readily extends to Lie algebras of Cartan type (cf. §3 below), is the main theorem of [26] (cf. [26, (4.8)]).

The following subsidiary result deals with the complementary case of a solvable Lie algebra. We fix a toral Lie algebra T and linear forms $\alpha \neq \beta \in T^*$ such that $\gamma(t^{[p]}) = \gamma(t)^p \ \ \forall\, t \in T, \gamma \in \{\alpha, \beta\}$. Let $L(\alpha, \beta)$ denote the restricted Lie algebra with underlying F-vector space $T \oplus Fx_\alpha \oplus Fx_\beta$, Lie bracket

$$[t, x_\gamma] = \gamma(t)x_\gamma \quad \gamma \in \{\alpha, \beta\}, \quad [x_\alpha, x_\beta] = 0,$$

and p-map given by $x_\gamma^{[p]} = 0 \ \ \gamma \in \{\alpha, \beta\}$.

Lemma 2.3. *Let M be a $u(L(\alpha, \beta))$-module such that*
(a) $x_\alpha, x_\beta \notin \mathcal{V}_{L(\alpha,\beta)}(M)$, and
(b) $dim_F Top(M) = 1$.
Then we have $dim_F M = p^2$.

Proof. Since the top of M is simple, M is generated by a weight vector m_λ. Accordingly, we have

$$M = \sum_{0 \leq r,s \leq p-1} Fx_\alpha^r x_\beta^s m_\lambda, \tag{3}$$

so that $dim_F M \leq p^2$. Let $\varrho : u(L(\alpha, \beta)) \longrightarrow End_F(M)$ be the representation of M. From (3) we obtain $\varrho(x_\alpha)^{p-1}(M) = \sum_{s=0}^{p-1} Fx_\alpha^{p-1} x_\beta^s m_\lambda$. Thus, if $d = dim_F \varrho(x_\alpha)^{p-1}(M)$, then

$$\varrho(x_\alpha)^{p-1}(M) = \bigoplus_{s=0}^{d-1} Fx_\alpha^{p-1} x_\beta^s m_\lambda.$$

Condition (a) implies

$$dim_F M = p\, dim_F im \varrho(x_\gamma)^{p-1}(M) \qquad \gamma = \alpha, \beta,$$

so that

$$M = \bigoplus_{r=0}^{p-1} \bigoplus_{s=0}^{d-1} Fx_\alpha^r x_\beta^s m_\lambda. \tag{4}$$

By interchanging the rôles of x_α and x_β we obtain

$$M = \bigoplus_{r=0}^{d-1} \bigoplus_{s=0}^{p-1} Fx_\alpha^r x_\beta^s m_\lambda. \tag{5}$$

Let $M = \bigoplus_{\zeta \in \Lambda} M_\zeta$ denote the weight space decomposition of M relative to the torus T. From (4) and (5) it follows that

$$\Lambda = \{\lambda + r\alpha + s\beta \ ; \ (r, s) \in \mathbb{N}_{p-1} \times \mathbb{N}_{d-1}\}$$

as well as

$$\Lambda = \{\lambda + r\alpha + s\beta \ ; \ (r,s) \in \mathbb{N}_{d-1} \times \mathbb{N}_{p-1}\}.$$

Here we have set $\mathbb{N}_q := \{0, 1, \dots, q\}$.

If $\{\alpha, \beta\}$ is linearly independent over the Galois field $GF(p)$, then we readily obtain $d = p$ and $dim_F M = p^2$. Alternatively, we may assume that $\beta = k\alpha$ for $k \in \{0, 2, 3, \dots, p-1\}$. Since $dim \mathcal{V}_{L(\alpha,\beta)}(M) \leq 1$, it follows from the result dual to [10, (4.4)] that $Soc(M)$ is simple. Writing $Soc(M) = F_\zeta$ we obtain from (4) and (5)

$$((d-1)k - 1)\alpha + \lambda = \zeta = ((d-1) - k)\alpha + \lambda,$$

whence $kd = d$. Since $k \neq 1$, we obtain $d = p$, as desired. □

Theorem 2.4. *Let G be a connected algebraic group with Lie algebra L. If $u(L)$ has infinite representation type, then there exists $r > 0$ such that $\mathcal{C}_s(L, \ell p^r)$ is infinite for all $\ell > 0$. Moreover, if G is reductive, then there exists $d > 0$ such that $\mathcal{C}_s^{per}(L, \ell d)$ is infinite for all $\ell > 0$.*

Proof. In view of (1.2) it suffices to show the existence of infinitely many indecomposable modules of the desired dimension.

Let $R_u(G) \subset G$ be the unipotent radical of G. According to [19, (11.11)], $Lie(R_u(G)) = rad_p(L)$ is the p-nilpotent radical of L. We consider $\hat{G} := G/R_u(G)$ as well as its Lie algebra $\hat{L} = L/rad_p(L)$. If $\hat{G}$ is not a torus, then the example preceding (2.3) yields our result. Alternatively, G is solvable, so that G is a semidirect product of a torus and $R_u(G)$ (cf. [20, (19.3)]). Thus,

$$L = T \oplus rad_p(L),$$

where $T \subset L$ is a maximal torus. Note that $K := [rad_p(L), rad_p(L)] + \langle rad_p(L)^{[p]} \rangle$ is a p-ideal of L.

If $dim_F H^1(u(rad_p(L)), F) = 1$, then the local algebra $u(rad_p(L))$ is a Nakayama algebra and $rad_p(L) = (Fx)_p$ is cyclic (cf. [28, §4] or [8, p.58]). Thus, by a result due to Pfautsch and Voigt [27, Thm.3] (cf. also [8, (4.3)] and [14, (2.4)]), the algebra $u(L)$ has finite representation type, a contradiction. An application of [18, (2.1)] now shows that $dim_F rad_p(L)/K = dim_F H^1(u(rad_p(L)), F) \geq 2$. We consider the restricted Lie algebra $\mathcal{L} := L/K$

$$\mathcal{L} = T \oplus \tilde{V} \ ; \ \tilde{V}^{[p]} = \{0\} \ , \ [\tilde{V}, \tilde{V}] = (0), \ dim_F \tilde{V} = n \geq 2$$

and decompose $\tilde{V}$ into its root spaces relative to T:

$$\tilde{V} = \bigoplus_{\alpha \in R} \tilde{V}_\alpha.$$

If there is $\alpha_0 \in R$ with $dim_F \tilde{V}_{\alpha_0} \geq 2$, then we consider the factor algebra

$$\mathcal{L}_1 := \mathcal{L}/(\sum_{\alpha \neq \alpha_0} \tilde{V}_\alpha) \cong T \oplus \tilde{V}_{\alpha_0}.$$

Given $v \in \tilde{V}_{\alpha_0} \setminus \{0\}$, the module

$$M_v := u(\mathcal{L}_1) \otimes_{u(T \oplus Fv)} F$$

is indecomposable with support variety Fv. The assertion now follows from the structure of periodic components of $\Gamma_s(\mathcal{L}_1)$.

Hence we may assume that $dim_F \tilde{V}_\alpha = 1$ for every $\alpha \in R$. In that case $\mathcal{L}$ has an algebra $L(\alpha, \beta)$ as a quotient, and it suffices to establish the result for such an algebra. Letting $P(\gamma) := u(L(\alpha, \beta)) \otimes_{u(T)} F_\gamma$ denote the principal indecomposable $u(L(\alpha, \beta))$-module with top F_γ (cf. [7, (4.3)]), we see that

$$(0) \longrightarrow \Omega^2(F) \longrightarrow P(\alpha) \oplus P(\beta) \longrightarrow P(0) \longrightarrow F \longrightarrow (0)$$

are the initial terms of a minimal projective resolution of F. In particular, we have

$$dim_F \Omega^2(F) = p^2 + 1.$$

Now let $v \neq 0$ be an element of $V := Fx_\alpha \oplus Fx_\beta$. Choose a linear form $f \in L(\alpha, \beta)^*$ such that $(ker\, f) \cap V = Fv$. According to [16, (4.2)], there exists a linear map $\zeta : \Omega^2(F) \longrightarrow F$ corresponding to f such that such that the support variety of $\mathcal{K}_\zeta := ker \zeta$ equals Fv. Note that $dim_F \mathcal{K}_\zeta = p^2$. From the above resolution we obtain

$$\Omega^2(F) \subset P(\alpha) \oplus P(\beta),$$

so that $dim_F Soc(\mathcal{K}_\zeta) \leq 2$. If the socle had dimension 1, then $\mathcal{K}_\zeta$ would embed into a principal indecomposable module $P(\lambda)$. Since $dim_F \mathcal{K}_\zeta = p^2 = dim_F P(\lambda)$, this contradicts the fact that $\mathcal{K}_\zeta$ is not projective. As $\mathcal{K}_\zeta$ is periodic, the arguments of [10, (4.4)] show that

$$dim_F Top(\mathcal{K}_\zeta) = dim_F Soc(\mathcal{K}_\zeta) = 2.$$

Thus, if $\mathcal{K}_\zeta$ is decomposable, then

$$\mathcal{K}_\zeta = M_1 \oplus M_2,$$

where each M_i has a simple top and support variety Fv. It now follows from (2.3) that $\mathcal{K}_\zeta$ is indecomposable, whenever $v \notin Fx_\alpha \cup Fx_\beta$. Thus, there are infinitely many periodic components containing an indecomposable module of dimension p^2, so that our assertion follows.

Finally, we assume G to be reductive. Since L is not a torus, there is $x \in \mathcal{V}_L(F) \setminus \{0\}$. Owing to [15, (2.2)] there exists an Auslander-Reiten component $\Theta \subset \Gamma_s(L)$ with support variety $\mathcal{V}_L(\Theta) = Fx$. If Fx is stable under $Ad(G)$, then Fx is a p-nilpotent ideal of L. Since Lie algebras of reductive groups do not possess non-trivial p-nilpotent ideals [19, (11.8)], $\mathcal{V}_L(\Theta)$ is not invariant and $\{\Theta^{(g)}\ ;\ g \in G\}$ is infinite. □

Remark 3. Theorem 2.4 continues to hold if L is a completely solvable Lie algebra. In that case the derived algebra $[L, L]$ of L is nilpotent. If $T(L)$ denotes the largest toral ideal of L, then the factor algebra $\tilde{L} := L/T(L)$ admits a decomposition $\tilde{L} = \tilde{T} \oplus rad_p(\tilde{L})$. Moreover, $u(\tilde{L})$ is representation infinite whenever $u(L)$ has this property. Consequently, the above arguments apply.

3. AR-Components for Lie Algebras of Classical and Cartan Type

In this section we apply geometric methods in order to see that Auslander-Reiten components of type $\mathbb{Z}[A_\infty]$ usually occur most often. The exception is given by the algebra $u(s\ell(2))$: Here the Auslander-Reiten quiver has infinitely many periodic components of type $\mathbb{Z}[A_\infty]/(\tau)$, containing all modules of even length, and $p-1$ components of type $\mathbb{Z}[\tilde{A}_{12}]$ containing all modules of odd length.

We denote by $A_\infty(L,\chi,d)$ the set of those components of $\mathcal{C}_s(L,\chi,d)$ that are isomorphic to $\mathbb{Z}[A_\infty]$. As usual, we will suppress $\chi = 0$ in our notation. Throughout this section we assume that $p \geq 5$.

Theorem 3.1. *Let G be a connected, reductive algebraic group of semisimple rank ≥ 2. Then there exists $r > 0$ such that $A_\infty(L,\ell p^r)$ is infinite for every $\ell \geq 1$.*

Proof. Let $T \subset G$ be a maximal torus, Φ the root system corresponding to T, $B \supset T$ a Borel subgroup with corresponding simple system Δ. We proceed in several steps and first assume G to be almost simple. By assumption Δ contains at least two elements. If Δ is not of type A_2, there exists a singleton $I := \{\alpha\} \subset \Delta$ such that there either are $\beta_1 \neq \beta_2 \in \Delta$ such that $\alpha + \beta_i$ is a root $1 \leq i \leq 2$, or there is $\beta_1 \in \Delta$ such that $\alpha + i\beta_1$ is a root $1 \leq i \leq 2$. Consider the standard parabolic subgroup P_I, denote its opposite group by P_I^-, and let U_I^+ and U_I^- be the unipotent radicals of P_I and P_I^-, respectively. There results a triangular decomposition

$$L = L^- \oplus L_0 \oplus L^+,$$

where $L_0 = C(L_0) \oplus s\ell(2)$ is the Levi subalgebra of $Lie(P_I)$ and $L^\pm = Lie(U_I^\pm)$. We consider the root space decomposition

$$L = Lie(T) \oplus \bigoplus_{\gamma\in\Phi} L_\gamma$$

of L relative to T. We have $L_\gamma = Fe_\gamma \ \forall\, \gamma \in \Phi$ and $L_0 = Lie(T) \oplus Fe_\alpha \oplus Fe_{-\alpha}$ (cf. [20, (26.2)]). Let Φ^+ be the set of positive roots. The group U_I^+ is generated by the root groups U_β, $\beta \in \Phi^+ \setminus \{\alpha\}$. Given $x_\beta \in U_\beta$, general theory (cf. [20, §27]) shows that $Ad(x_\beta)(e_\gamma) \in \sum_{k\in\mathbb{N}_0} L_{\gamma+k\beta}$. Since $Ad(x_\beta)$ is unipotent, this implies

$$Ad(x_\beta)(e_\gamma) \equiv e_\gamma \quad mod(L^+)$$

for $\gamma \in \{\alpha, -\alpha\}$. By the same token, we have

$$Ad(x_\beta)(t) \equiv t \quad mod(L^+) \quad \forall\, t \in Lie(T).$$

Now let M be a $u(L_0)$-module. It follows from the above that the $u(L_0 \oplus L^+)$-module M with trivial L^+-structure is $Ad(U_I^+)$-stable. Direct computation then shows that $M\uparrow^L$ also enjoys this property. In view of (2.1), we therefore have

$$U_I^+ \cdot \mathcal{V}_{L_0}(M) \subset \mathcal{V}_L(M\uparrow^L).$$

Let H be a closed subgroup of G, x and element of L. We consider the centralizers $C_H(x) := \{h \in H \; ; \; Ad(h)(x) = x\}$ and $C_{Lie(H)}(x) := \{y \in Lie(H) \; ; \; [y,x] = 0\}$. General theory yields

$$\begin{aligned} dim(H \cdot x) &= dimH - dimC_H(x) = dim_F Lie(H) - dim_F Lie(C_H(x)) \\ &\geq dim_F Lie(H) - dim_F C_{Lie(H)}(x). \end{aligned}$$

Let M_ℓ be an ℓp-dimensional periodic $u(s\ell(2))$-module with support variety $\mathcal{V}_{L_0}(M_\ell) = Fe_\alpha$. (Take for instance a vertex of quasilength ℓ of the periodic component containing a non-projective Verma module.) We give M_ℓ the structure of an L_0-module by letting $C(L_0)$ act trivially and consider $M_\ell\uparrow^L$. According to the above, we have

$$B \cdot e_\alpha \subset \mathcal{V}_L(M_\ell\uparrow^L).$$

Let $\gamma := \beta_2$ for $|\Delta| \geq 3$ and $\gamma := \alpha + \beta_1$, otherwise. Owing to [19, (3.2)] there is an element $t_0 \in Lie(T)$ such that the differential $d(\alpha)$ satisfies $d(\alpha)(t_0) \neq 0$. Note that $C_{Lie(B)}(e_\alpha)$ is T-invariant. Using standard properties of classical Lie algebras (cf. [31, (II.3)]), we thus obtain

$$(Ft_0 \oplus Fe_{\beta_1} \oplus Fe_\gamma) \cap C_{Lie(B)}(e_\alpha) = (0),$$

so that

$$dim\mathcal{V}_L(M_\ell\uparrow^L) \geq dim(B \cdot e_\alpha) \geq 3.$$

We proceed by considering the case where $\Delta = \{\alpha, \beta\}$ is of type A_2. Then $L \cong s\ell(3)$ and we may assume that $G \cong SL(3)$. Consider the element $x_0 := e_\alpha + e_\beta \in \mathcal{V}_L(F)$. We propose to show that $x_0 \in \mathcal{V}_L(M_\ell\uparrow^L)$ and note that this is equivalent to $Ext^1_{u(Fx_0)}(F, M_\ell\uparrow^L) \neq (0)$. Consider the induced module $A_0 := u(L) \otimes_{u(Fx_0)} F$. Since $u(L) : u(Fx_0)$ is a Frobenius extension of first kind, [13, (1.6)] provides isomorphisms

$$A_0^* \cong Hom_{u(Fx_0)}(u(L), F) \cong A_0$$

of $u(L)$-modules. Let $\omega : u(Lie(P_I)) \longrightarrow u(Lie(P_I))$ be the automorphism that is given by $\omega(x) = x - tr((ad_{L/Lie(P_I)})(x))1 \quad \forall \; x \in Lie(P_I)$. Then $u(L) : u(Lie(P_I))$ is an ω-Frobenius extension (cf. [13, p.158]), and there result isomorphisms

$$\begin{aligned} Ext^1_{u(Fx_0)}(F, M_\ell\uparrow^L) &\cong Ext^1_{u(L)}(A_0, M_\ell\uparrow^L) \cong Ext^1_{u(Lie(P_I))}(A_0, M_\ell^{(\omega)}) \\ &\cong H^1(u(Lie(P_I)), A_0^* \otimes_F M_\ell^{(\omega)}) \\ &\cong H^1(u(Lie(P_I)), A_0 \otimes_F M_\ell^{(\omega)}). \end{aligned}$$

Here the superscript "(ω)" indicates the twisting of the module structure by ω. Since $\mathcal{V}_{L^+}(A_0) \subset L^+ \cap Fx_0 = \{0\}$, the module $A_0 \otimes_F M_\ell^{(\omega)}$ is $u(L^+)$-projective. Consequently, the Grothendieck spectral sequence

$$H^p(u(L_0), H^q(u(L^+), A_0 \otimes_F M_\ell^{(\omega)})) \Rightarrow H^n(u(Lie(P_I)), A_0 \otimes_F M_\ell^{(\omega)})$$

collapses, and we obtain in particular an isomorphism

$$H^1(u(Lie(P_I)), A_0 \otimes_F M_\ell^{(\omega)}) \cong H^1(u(L_0), (A_0 \otimes_F M_\ell^{(\omega)})^{L^+}).$$

Since L^+ operates trivially on $M_\ell^{(\omega)}$, the foregoing isomorphisms combine to

$$Ext^1_{u(Fx_0)}(F, M_\ell{\uparrow}^L) \cong H^1(u(L_0), A_0^{L^+} \otimes_F M_\ell^{(\omega)}).$$

The $u(L^+)$-projectivity of A_0 readily yields $dim_F A_0^{L^+} = p^5$. Direct computation then shows that

$$A_0^{L^+} = \{\epsilon_0 u \otimes 1 \ ; \ u \in u(L_0 \oplus L^-)\},$$

where $\epsilon_0 = e_\beta^{p-1} e_{\alpha+\beta}^{p-1}$ is an integral of the Hopf algebra $u(L^+)$. Let $\nu : u(L_0) \longrightarrow u(L_0)$ be the automorphism of $u(L_0)$ given by

$$\nu(x) = x - tr(ad_{L^+}(x))1 \quad \forall\, x \in L_0.$$

Using [13, (1.2)] it is not hard to see that

$$a\,\epsilon_0 = \epsilon_0\,\nu(a) \qquad \forall\, a \in u(L_0).$$

Since L^- is an ideal of $L_0 \oplus L^-$, we thus have a decomposition

$$A_0^{L^+} = V \oplus W$$

of $u(L_0)$-modules, with $V := \{\epsilon_0 u \otimes 1 \ ; \ u \in u(L_0)\}$ and $W := \{\epsilon_0 u \otimes 1 \ ; \ u \in u(L_0 \oplus L^-)L^-\}$. The canonical map

$$u(L_0) \times F \longrightarrow V \ \ ; \ \ (u, r) \mapsto \epsilon_0 \nu(u) \otimes r$$

induces an isomorphism

$$u(L_0) \otimes_{u(Fe_\alpha)} F \longrightarrow V$$

of $u(L_0)$-modules. Observing the self-duality of $u(L_0) \otimes_{u(Fe_\alpha)} F$ we therefore obtain

$$\begin{aligned} dim_F H^1(u(L_0), A_0^{L^+} \otimes_F M_\ell^{(\omega)}) &\geq dim_F H^1(u(L_0), V \otimes_F M_\ell^{(\omega)}) \\ &= dim_F H^1(u(L_0), V^* \otimes_F M_\ell^{(\omega)}) \\ &= dim_F Ext^1_{u(L_0)}(V, M_\ell^{(\omega)}) \\ &= dim_F Ext^1_{u(Fe_\alpha)}(F, M_\ell) \neq 0, \end{aligned}$$

where the last inequality follows from the fact that $e_\alpha \in \mathcal{V}_{L_0}(M_\ell)$. Thus, $Ext^1_{u(Fx_0)}(F, M_\ell{\uparrow}^L) \neq (0)$ and $x_0 \in \mathcal{V}_L(M_\ell{\uparrow}^L)$. Consequently, we have

$$dim \mathcal{V}_L(M_\ell{\uparrow}^L) \geq dim(B \cdot x_0) = 3$$

in this case as well.

It now follows from [11, (2.1)] that $M_\ell{\uparrow}^L$ belongs to a component of type $\mathbb{Z}[A_\infty]$. Since $\mathcal{V}_{Lie(P_I)}(F)$ does not contain any G-invariant subsets we have established our assertion in case G is almost simple.

Now let G be semisimple with almost simple components $G_1, \ldots, G_r$. Then we have

$$L = \bigoplus_{i=1}^{r} Lie(G_i) \qquad (r \geq 2),$$

and the first part of the proof yields the assertion provided there exists a component of rank ≥ 2. Alternatively, $s\ell(2)$ is a direct summand of L. Writing $L = s\ell(2) \oplus K$, we give every $s\ell(2)$-module M the structure of an L-module by letting K act trivially. Thus,

$$dim\mathcal{V}_L(M) = dim\mathcal{V}_{s\ell(2)}(M) + dim\mathcal{V}_K(F) \geq dim\mathcal{V}_{s\ell(2)}(M) + 2,$$

and the above arguments show that the periodic $s\ell(2)$-modules give rise to $u(L)$-modules belonging to components of type $\mathbb{Z}[A_\infty]$.

If G is reductive, then we consider the semisimple group $\mathcal{G} := G/Rad(G)$, whose Lie algebra we denote by $\mathcal{L}$. As $Rad(G) = Z(G)^\circ$ is a torus, it readily follows that the canonical projection

$$\pi : L \longrightarrow \mathcal{L}$$

induces an isomorphism $\pi : \mathcal{V}_L(F) \longrightarrow \mathcal{V}_{\mathcal{L}}(F)$. Thus, if M is a $u(\mathcal{L})$-module, then its pull-back $\tilde{M}$ has support variety

$$\mathcal{V}_L(\tilde{M}) \cong \mathcal{V}_{\mathcal{L}}(M).$$

As $\mathcal{G}$ has rank ≥ 2, our result follows. □

Remark 4. Let $L = Lie(G)$, where G is connected and reductive. Suppose there is $d > 0$ such that $\mathcal{C}_s(L, d)$ is finite. Since $G/Z(G)$ operates on the finite set $\mathcal{C}_s(L, d)$, it follows from the structure of semisimple groups that G operates trivially on $C_s(L, d)$. Hence every component $\Theta \in \mathcal{C}_s(L, d)$ is invariant. It can be shown (cf. [12]) that such components are either isomorphic to $\mathbb{Z}[A_\infty]$ or $\mathbb{Z}[\tilde{A}_{12}]$, with the latter occurring only if Θ belongs to a tame block of $u(L)$.

Let G be reductive with maximal torus $T \subset G$. As before we consider a triangular decomposition

$$L = L^- \oplus L_0 \oplus L^+ \;\; ; \;\; L_0 = Lie(T)$$

of L. Following Kac and Weisfeiler [23] we refer to a linear form $\chi \in L^*$ as *semisimple* if there exists $g \in G$ such that $g{\cdot}\chi \in (L^- {\oplus} L^+)^\perp$. It is well-known (cf. [29, (3.1)]) that the stabilizer G_χ of a semisimple linear form $\chi \in L^*$ is reductive and that $Lie(G_\chi)$ coincides with the infinitesimal stabilizer

$$L_\chi := \{x \in L \; ; \; \chi([x, L]) = (0)\}.$$

Corollary 3.2. *Suppose that $p > 5$ and let G be a semisimple, simply connected algebraic group. If $\chi \in L^*$ is semisimple such that G_χ has semisimple rank ≥ 2, then there exists $r > 0$ such that $A_\infty(L, \chi, \ell p^r)$ is infinite for every $\ell > 0$.*

Proof. Owing to [17, (3.2)] there exists a parabolic subalgebra $L_\chi \subset P \subset L$ such that the induction functor $X \mapsto u(L,\chi) \otimes_{u(P,\chi|_P)} X$ is an equivalence between the module categories of $u(L_\chi, \chi|_{L_\chi})$ and $u(L,\chi)$. Since $dim_F u(L,\chi) \otimes_{u(P,\chi|_P)} X = p^{dim_F L/P} dim_F X$, it suffices to verify the result for the former algebra.

We put $K := [L_\chi, L_\chi]$. Then K is a p-subalgebra of L_χ such that $\chi(K) = (0)$, and there exists a central torus $T \subset L_\chi$ such that $L_\chi = T \oplus K$. Consequently,

$$u(L_\chi, \chi|_{L_\chi}) \cong u(T, \chi|_T) \otimes_F u(K) \cong u(T) \otimes_F u(K) \cong u(L_\chi).$$

The desired result now follows directly from (3.1). □

Remark 5. If G_χ is a torus, that is, if χ is regular semisimple, then $u(L,\chi)$ is semisimple. If G_χ has semisimple rank 1, then the arguments of the proof of (3.2) show that

$$u(L_\chi) \cong \bigoplus_{i=1}^{r} u(s\ell(2)).$$

As a result, $\Gamma_s(L,\chi)$ does not contain components of type $\mathbb{Z}[A_\infty]$ in that case.

Aside from the classical simple Lie algebras of types $A - G$ there exist four series of restricted simple Lie algebras, labelled $(W(n))_{n\geq 1}$, $(S(n))_{n\geq 3}$, $(H(2r))_{r\geq 1}$ and $K(2r+1)_{r\geq 1}$, that are not associated to algebraic groups. These are referred to as the restricted Lie algebras of *Cartan type.* The Block-Wilson classification theorem (cf. [3]) insures that every restricted simple Lie algebra of characteristic $p > 7$ is either classical or of Cartan type.

Corollary 3.3. *Let $L \ncong W(1)$ be a restricted Lie algebra of Cartan type. Then there exists $r > 0$ such that $A_\infty(L, \ell p^r)$ is infinite for every $\ell \geq 1$.*

Proof. Since L is of Cartan type, L possesses a $\mathbb{Z}$-grading $L = \bigoplus_{i=-r}^{s} L_i$ such that $L_i^{[p]} \subset L_{ip}$. This grading induces a triangular decomposition

$$L = L^- \oplus L_0 \oplus L^+$$

such that $L^+ := \sum_{i>0} L_i$ and $L^- := \sum_{i<0} L_i$ are p-nilpotent. Moreover, we have $W(n)_0 \cong g\ell(n)$, $S(n)_0 \cong s\ell(n)$, $H(2r)_0 \cong sp(2r)$, and $K(2r+1)_0 \cong sp(2r) \oplus F$. In particular, L_0 is the Lie algebra of a reductive group G_0 (cf. [32, Chapter IV] for details). If $L \ncong W(2), H(2), K(3)$, then G_0 has semisimple rank ≥ 2. According to (3.1) there exists $s > 0$ such that for every $\ell \geq 1$ there is an infinite set $\mathcal{X}_\ell$ of isoclasses of indecomposable $u(L_0)$-modules of dimension ℓp^s with $dim \mathcal{V}_{L_0}(X) \geq 3 \;\; \forall \; [X] \in \mathcal{X}_\ell$. We may now apply (2.1) to see that $\{[X\uparrow^L] \; ; \; [X] \in \mathcal{X}_\ell\}$ is an infinite set of isoclasses of indecomposable $u(L)$-modules of dimension $\ell p^{s+dim_F L^-}$. By the same token each consituent has a support variety of dimension ≥ 3. The result now follows from a consecutive application of [11, (2.1)] and (1.2).

It remains to consider the cases $L \cong W(2)$, $H(2)$, $K(3)$. We put $L_{(s-1)} := L_{s-1} \oplus L_s$ and note that $L_{(s-1)} \subset \mathcal{V}_L(F)$. Moreover, $dim_F H(2)_{(s-1)} = 5$, while $dim_F K(3)_{(s-1)} = 3 = dim_F W(2)_{(s-1)}$. Since $L_0 \cong s\ell(2) \oplus T$, where $T \subset L_0$ is a central torus, there exists for every $x \in \mathcal{V}_{L_0}(F) \setminus \{0\}$ and every $\ell \geq 1$ an indecomposable $u(L_0)$-module $M(x,\ell)$ of dimension ℓp and with support variety Fx. An application of the Cartan-Weyl formula shows that the annihilator of an element $y \in L_{(s-1)}$ in $M(x,\ell)\uparrow^L$ has dimension $> \frac{1}{p} dim_F M(x,\ell)\uparrow^L$. Consequently, $L_{(s-1)} \subset \mathcal{V}_L(M(x,\ell)\uparrow^L)$, so that $[M(x,\ell)\uparrow^L]$ belongs to a component $\Theta(x,\ell)$ of type $\mathbb{Z}[A_\infty]$. According to (2.1) we have $\mathcal{V}_L(\Theta(x,\ell)) \cap L_0 = Fx$, implying that $A_\infty(L, \ell p^{dim_F L^- +1})$ is infinite. □

Remark 6. Consider the p-dimensional Witt algebra $W(1) := \bigoplus_{i=-1}^{p-2} Fe_i$ whose multiplication and p-map are given by

$$[e_i, e_j] = (j-i)\, e_{i+j} \quad ; \quad e_i^{[p]} = \delta_{i,0}\, e_0,$$

respectively. (For $p=3$ we have $W(1) \cong s\ell(2)$.) Note that $W(1)$ has a triangular decomposition with $W(1)^- := Fe_{-1}$, $W(1)_0 := Fe_0$, and $W(1)^+ := \sum_{i\geq 1} Fe_i$. The subalgebra $Fe_{-1} \oplus Fe_0 \oplus Fe_1$ is isomorphic to $s\ell(2)$. We set $B := W(1)_0 \oplus W(1)^-$.

According to [4] the algebra $u(W(1))$ has p simple modules $S_1, \ldots, S_p$. Aside from the trivial module there exists one module of dimension $p-1$ and $p-2$ modules of dimension p. For $i \in \{0, \ldots, p-1\}$ we let F_i denote the one-dimensional B-module with trivial $W(1)^-$-action and $e_0 \cdot 1 = i$. In virtue of (2.1) the modules $M_i := u(W(1)) \otimes_{u(B)} F_i$ are indecomposable and pairwise nonisomophic. Moreover, [9, (3.4)] implies $\mathcal{V}_{W(1)}(M_i) \subset Fe_{-1}$. If each of the M_i is projective, then they form a full set of representatives of the principal indecomposable $u(W(1))$-modules and we obtain

$$p^p = dim_F u(W(1)) = p^{p-2} + (p-1)p^{p-2} + (p-2)p^{p-1},$$

a contradiction. Hence there is an index i such that $\mathcal{V}_{W(1)}(M_i) = Fe_{-1}$. The Auslander-Reiten component of M_i contains modules of dimension ℓp^{p-2} for every $\ell \geq 1$. Since the space Fe_{-1} is not invariant under the action of the connected component of the automorphism group of $W(1)$ (cf. [21, Thm.6]), there are infinitely many indecomposables in each dimension ℓp^{p-2}. (Note that this result is also valid for $p=3$).

Consider $\mathcal{V} := \{x + y \in W(1) \; ; \; x \in \mathcal{V}_{s\ell(2)}(F),\ y \in Fe_2\}$. Then $\mathcal{V} \cong \mathcal{V}_{s\ell(2)}(F) \times F$ is an irreducible, conical variety of dimension 3. According to [15, (2.2)] there exists a $u(W(1))$-module M with $\mathcal{V}_{W(1)}(M) = \mathcal{V}$. Since $\mathcal{V}$ is irreducible, at least one indecomposable summand N of M has the same property. By [11, (2.1)] the component Θ of N is isomorphic to $\mathbb{Z}[A_\infty]$. Since $\mathcal{V}$ is not invariant under $Aut_p(W(1))^\circ$, there are infinitely many such components. Let $[X] \in \Theta$ be a quasi-simple vertex of minimal dimension. As the dimension of every principal indecomposable $u(W(1))$-module is divisible by p^{p-2}, it readily follows from (1) that

$$dim_F Y \equiv q\ell(Y) dim_F X \quad mod(p^{p-2}) \quad \forall\ [Y] \in \Theta.$$

Here $q\ell(Y)$ denotes the quasilength of $[Y]$. Since the support variety $\mathcal{V}_L(\Theta) = \mathcal{V}$ intersects the p-nilpotent algebra $W(1)_{(3)} = \sum_{i\geq 3} Fe_i$ trivially, it follows that X is free over the local algebra $u(W(1)_{(3)})$. Consequently, $dim_F X = \ell_0 p^{p-4}$. For each $r > 0$, let $[Y_r]$ be a vertex of quasilength rp^2. The above discussion ensures the existence of $\ell_r \geq 0$ such that

$$dim_F Y_r = (r\ell_0 + \ell_r)p^{p-2}.$$

Owing to (1.2) the dimensions of the Y_r are not bounded. Consequently, $A_\infty(W(1), \ell p^{p-2})$ is infinite for infinitely many $\ell > 0$.

References

[1] M. Auslander, I. Reiten and S. Smalø. *Representation Theory of Artin Algebras.* Cambridge Studies in Advanced Math. **36.** Cambridge University Press, 1995

[2] D. Benson and J. Carlson. *Complexity and Multiple Complexes.* Math. Z. **195** (1987), 221-238

[3] R. Block and R. Wilson. *Classification of Restricted Simple Lie Algebras.* J. Algebra **114** (1988), 115-259

[4] H. Chang. *Über Wittsche Lie Ringe.* Abh. Math. Sem. Univ. Hamburg **14** (1941), 151-184

[5] W. Crawley-Boevey. *On Tame Algebras and Bocses.* Proc. London Math. Soc. **56** (1988), 451-483

[6] K. Erdmann. *The Auslander-Reiten Quiver of Restricted Enveloping Algebras.* CMS Conf. Proc. **18** (1996), 201-214

[7] R. Farnsteiner. *Representations of Lie algebras with Triangular Decomposition.* Symposia Gaussiana **2** (1995), 275-286

[8] R. Farnsteiner. *Periodicity and Representation Type of Modular Lie Algebras.* J. reine angew. Math. **464** (1995), 47-65

[9] R. Farnsteiner. *Representations of Blocks Associated to Induced Modules of Restricted Lie Algebras.* Math. Nachr. **179** (1996), 57-88

[10] R. Farnsteiner. *On Auslander-Reiten Quivers of Enveloping Algebras of Restricted Lie Algebras.* Math. Nachr. (to appear)

[11] R. Farnsteiner. *On Support Varieties of Auslander-Reiten Components.* Indag. Math. (to appear)

[12] R. Farnsteiner. *Auslander-Reiten Components for Lie Algebras of Reductive Groups.* Manuscript in Preparation.

[13] R. Farnsteiner and H. Strade. *Shapiro's Lemma and its Consequences in the Cohomology Theory of Modular Lie Algebras.* Math. Z. **206** (1991), 153-168

[14] J. Feldvoss and H. Strade. *Restricted Lie Algebras with Bounded Cohomology and Related Classes of Algebras.* Manuscripta math. **74** (1992), 47-67

[15] E. Friedlander and B. Parshall. *Support Varieties for Restricted Lie Algebras.* Invent. math. **86** (1986), 553-562

[16] E. Friedlander and B. Parshall. *Geometry of p-Unipotent Lie Algebras.* J. Algebra **109** (1987), 25-45

[17] E. Friedlander and B. Parshall. *Modular Representation Theory of Lie Algebras.* Amer. J. Math. **110** (1988), 1055-1098

[18] G. Hochschild. *Cohomology of Restricted Lie Algebras.* Amer. J. Math. **76** (1954), 555-580

[19] J. Humphreys. *Algebraic Groups and Modular Lie Algebras.* Mem. Amer. Math. Soc. **71** (1967)

[20] J. Humphreys. *Linear Algebraic Groups.* Graduate Texts in Mathematics **21.** Springer-Verlag, 1981

[21] N. Jacobson. *Classes of Restricted Lie Algebras of Characteristic p, II.* Duke Math. J. **10** (1943), 107-121

[22] J. Jantzen. *Kohomologie von p-Lie-Algebren und nilpotente Elemente.* Abh. Math. Sem. Univ. Hamburg **56** (1986), 191-219

[23] V. Kac and B. Weisfeiler. *Coadjoint Action of a Semisimple Algebraic Group and the Center of the Enveloping Algebra in Characteristic p.* Indag. Math. **38** (1976), 136-151

[24] S. Liu and R. Schulz. *The Existence of a Bounded Infinite DTr-Orbit.* Proc. Amer. Math. Soc. **122** (1994), 1003-1005

[25] E. Marmolejo and C. Ringel. *Modules of Bounded Length in Auslander-Reiten Components.* Arch. Math. **50** (1988), 128-133

[26] D. Nakano and R. Pollack. *On the Construction of Indecomposable Modules over Restricted Enveloping Algebras.* J. Pure Appl. Algebra **107** (1996), 61-73

[27] W. Pfautsch and D. Voigt. *The Representation-finite Algebraic Groups of Dimension zero.* C. R. Acad. Sci. Paris **306** (1988), 685-689

[28] R. Pollack. *Restricted Lie Algebras of Bounded Type.* Bull. Amer. Math. Soc. **74** (1968), 326-331

[29] A. Premet. *Irreducible Representations of Lie Algebras of Reductive Groups and the Kac-Weisfeiler Conjecture.* Invent. math. **121** (1995), 79-117

[30] C. Ringel. *Representation Theory of Finite Dimensional Algebras.* London Math. Soc. Lecture Notes **116** (1986), 7-79

[31] G. Seligman. *Modular Lie Algebras.* Ergebnisse der Mathematik **40** Springer-Verlag, 1967

[32] H. Strade and R. Farnsteiner. *Modular Lie Algebras and their Representations.* Marcel Dekker Monographs **116** New York, 1988

FAKULTÄT FÜR MATHEMATIK, UNIVERSITÄT BIELEFELD, 33501 BIELEFELD, GERMANY

Current address: Department of Mathematics, University of Wisconsin, Milwaukee, WI 53201

E-mail address: rolf@uwm.edu

Contemporary Mathematics
Volume **229**, 1998

COMPATIBLE DEFORMATIONS

MURRAY GERSTENHABER AND ANTHONY GIAQUINTO

ABSTRACT. Two deformations of the same algebra A are *compatible* if one can interpolate a continuous (in some sense) family of deformations between them. When A has finite dimension this is equivalent to having the deformed algebras lie on a common irreducible component of the variety of algebras of dimension that of A. If the infinitesimals of the deformations are $\alpha, \beta \in H^2(A, A)$ then a necessary condition for compatibility is the vanishing of $[\alpha, \beta] \in H^3(A, A)$. Analogous assertions hold for complex manifolds and bialgebras. In general, there are higher obstructions. As an application we show that while the quantum plane has a deformation to the Weyl algebra, the analog of the quantum plane in characteristic 2 can not be deformed to a separable algebra. When one deformation is a rigid algebra there are stronger compatibility conditions. Using these one can show that the Donald–Flanigan conjecture fails for the quaternion group (but some more subtle conjecture probably holds).

If two deformations, A_t, A'_t of the same algebra A can be joined by a continuous (in a sense to be made more precise in Section 1) family of deformations (a 'variety of algebras'), then the deformations will be called *compatible*. There is a natural obstruction to compatibility, namely, the (Gerstenhaber) bracket of the infinitesimals of the deformations. In the case of a finite dimensional algebra A over a field this can be used to show that two deformations of A lie on different components of the variety of algebras of the given dimension. It follows, in particular, that neither deformed algebra can be deformed to the other. When one of the deformations, say A_t, is rigid then there are stronger conditions for compatibility due to the fact that A_t must also be a deformation of A'_t, so every cocycle of A_t must be a "lifting" of one of A'_t. In particular, if some derivation of A_t is not a lifting of one of A'_t then the deformations are not compatible. This condition is often computationally more manageable than the vanishing of the bracket of the infinitesimals.

This paper grew out of an attempt to decide if the Donald–Flanigan conjecture holds for the eight-element quaternion group $\mathcal{Q} = \{\pm 1, \pm i, \pm j, \pm ij\}$, where $ij = -ji$ and $i^2 = j^2 = k^2 = -1$. The conjecture asserts that if p is a prime dividing the order of a finite group G, then the group algebra $\mathbb{F}_p G$ (which always has a non-zero radical) can be deformed to a separable algebra. The conjecture fails for $\mathcal{Q}$, as we show here, for a curious reason. Filter the group algebra $\mathbb{F}_2\mathcal{Q}$ by powers of its radical and let B be the associated graded algebra. The group algebra is then a

1991 *Mathematics Subject Classification.* Primary 16S80.

Key words and phrases. deformations, varieties of algebras, Donald-Flanigan conjecture.

Both authors thank Birge Huisgen-Zimmermann for many helpful suggestions regarding this paper deformation of B (and is, in fact, a "jump deformation"). Now if the the Donald–Flanigan conjecture held for $\mathcal{Q}$ then the group algebra would be deformable either to the direct sum $S_1 := M_2(K) \oplus M_2(K)$ of two copies of the 2×2 matrix algebra over the (algebraically closed) coefficient field K (which must always be enlarged to contain the deformation parameter), or to $S_2 := M_2(K) \oplus K^4$, the direct sum of M_2 plus four copies of K. But the associated graded algebra B can be deformed both to S_1 and to S_2, and neither of these deformations is compatible with the deformation of B to the original group algebra. We show this incompatibility by exhibiting derivations of S_1 and S_2 which are not are not liftings of derivations of the group algebra $\mathbb{F}_2\mathcal{Q}$. While this can probably be done directly, it is difficult because there is no a priori identification of the underlying vector spaces of these algebras and therefore there is no good criterion to tell when a derivation can not be a lifted one. Using the graded algebra B tells us what the identification of the underlying vector spaces of $\mathbb{F}_2\mathcal{Q}$ and S_1 (or S_2) must be, for they are both naturally identified with B.

Compatibility of deformations is meaningful also in the categories of complex analytic manifolds and of bialgebras, and one finds analogous obstructions. (In the complex case, although exhibiting the obstruction is just as trivial as for algebras, it is worth noting that one really does not need an independent construction since all the infinitesimal aspects of the analytic deformation theory have been algebraized. For bialgebras, however, we do not yet have a full graded homotopy Lie algebra structure on the cochains, but all we really need to understand is the obstruction to an infinitesimal deformation.)

1. Components of the variety of algebras. Suppose that A is an algebra over a ring k and that we have two deformations of A, with multiplications $*$ and $*'$ of the form

$$a * b = ab + tF_1(a,b) + t^2F_2(a,b) + \dots, \qquad a *' b = ab + uG_1(a,b) + u^2G_2(a,b) + \dots.$$

Recall that the condition for associativity of $*$ is that for all $n > 0$ we have $\sum_{i=1}^{n-1} F_i \circ F_{n-i} = \delta F_n$. (Here δ is the Hochschild coboundary, and if F, G are any 2-cochains of A with coefficients in itself, then $F \circ G(a,b,c) = F(G(a,b),c) - F(a, G(b,c))$). In particular, $F_1 \circ F_1$ is a coboundary. Now if F is a 2-cocycle of A (with coefficients in itself), whose cohomology class we denote by α, then the class of $F \circ F$ (which is always a 3-cocycle) depends only on that of F and may be denoted $\alpha \circ \alpha$. It is the obstruction to finding an F_2 such that the multiplication $ab + tF(a,b) + t^2F_2(a,b)$ is associative modulo t^3 (the coefficient ring now being $k[t]/t^3$, where k is the original coefficient ring). Now denote the deformed algebras defined by $*$ and $*'$ by A_t and A_u, respectively.

Definition. Deformations A_t and A_u of an algebra A are *compatible* if there is a deformation with two parameters, t and u, of the form $a * b = ab + \Phi_1(a,b;t,u) + \Phi_2(a,b;t,u) + \dots$, where $\Phi_1 = tF_1 + uG_1$ and each Φ_i is a homogeneous polynomial of total degree i in t and u which reduces to F_i when $u = 0$ and to G_i when $t = 0$.

This is, in effect, an algebraic way of asserting that there is a continuous family of deformations connecting the two original ones, i.e., that they lie on the same component in the variety of algebras. In this case, in addition to $F = F_1$ and

$G = G_1$ being 2-cocycles, any linear combination of F and G must also be a 2-cocycle. It follows, in particular, that the primary obstruction to $F + G$, viewed as an infinitesimal deformation of A, must vanish. This obstruction is the cohomology class of $(F+G)\circ(F+G)$. Now $(F+G)\circ(F+G) = F\circ F + G\circ G + F\circ G + G\circ F = F\circ F + G\circ G + [F,G]$ where $[F,G] = F\circ G + G\circ F$ is the Gerstenhaber bracket, cf [G1]. Since we started with two full deformations it follows that the classes of $F\circ F$ and $G\circ G$ already vanish, for we have $F\circ F = \delta F_2, G\circ G = \delta G_2$. So we have

Theorem 1. *If two deformations of an algebra A are compatible, and if the infinitesimals of these deformations are, respectively, the classes α, β of the 2-cocycles F and G, then $[\alpha, \beta] = 0$, i.e., $[F, G]$ must be a coboundary.* □

It follows that the class of $[F, G]$ may be viewed as an obstruction to the compatibility of these deformations.

The main ambiguity in what has been said is the meaning of a 'variety of algebras'. This is straightforward when A is a finite-dimensional algebra, say of dimension N, over an algebraically closed field k. For if we choose a basis for A, then it can be represented by its structure constants (relative to this basis) as a point in affine N^3 space over k. (Of course, many points may represent the same algebra; the general linear group $Gl(N,k)$ operates on the space and the isomorphism classes of algebras are its orbits.) The space of structure constants $\mathcal{V}$ is an algebraic variety which is, in general, a union of various irreducible components $\mathcal{C}$. Having chosen a basis for A, a deformation of A like A_t above defines a new set of structure constants which are now functions of t and may be viewed as defining a generic point of some irreducible subvariety of $\mathcal{V}$. This subvariety must then be entirely contained in one or more of the irreducible components of $\mathcal{V}$. (It may be contained in more than one.) The theorem then gives a necessary condition that there exist a single component of $\mathcal{V}$ containing the subvarieties defined by the two deformations A_t, A_u. Conversely, if such a component exists then the deformations are compatible. In the infinite dimensional case, or over an arbitrary coefficient ring k, it may be that deformations provide the best way to define irreducible components of the 'variety of algebras'.

The ideas above hold not only for associative algebras, but also for Lie algebras and commutative associative algebras, except that one must use, respectively, the Hochschild, Chevalley–Eilenberg and Harrison cohomologies. They carry over immediately to some categories such as complex analytic manifolds and bialgebras, where the cohomology, if not always well understood is at least well defined. They probably extend to most situations where deformations are meaningful, although one may have to use more exotic means, such as triples or operads, to define the appropriate cohomology. (The basic operation in the definition of an operad, namely composition, arose first in algebraic deformation theory, [G1, G2, G4]. The formal definition due to May comes from homotopy theory, which in some aspects is not far removed from algebraic deformation theory.)

2. Higher obstructions. There exist, in general, higher obstructions to compatibility of deformations. These are similar to those one meets when trying to construct a deformation of a single algebra, except that in place of a single obstruction at each stage there are $n-1$ obstructions at the nth stage. For suppose that we have deformations

$$* = \pi + tF_1^0 + t^2F_2^0 + t^3F_3^0 + \dots, \qquad *' = \pi + uF_1^1 + u^2F_2^1 + \dots,$$

so that in our previous notation $F_1^0 = F, F_1^1 = G$ and π denotes the original multiplication. By definition, these deformations are compatible if there exists a deformation of the form

$$*'' = \pi + (tF^0 + uF^1) + (t^2F_2^0 + tuF_2^1 + u^2F_1^2) + \dots = \sum_{n=0}^{\infty} \sum_{i+j=n} t^i u^j F_n^j,$$

where $F_0^0 = \pi$. This is equivalent to saying that for any fixed n we have

$$\sum_{\substack{i+j=n \\ i,j>0}} \sum_{\substack{k+l=m \\ k,l\geq 0}} F_i^k \circ F_j^l = \delta F_n^m, \quad m = 0, 1, \dots, n.$$

So if the construction has survived to level $n-1$ then the left sides of the preceding $n+1$ equations, which are guaranteed to be 3-cocycles, must all be coboundaries. This is automatically true of the first and last because $*$ and $*'$ were deformations, so there are $n-1$ obstructions at level n.

As with the deformations of a single algebra, we generally have the following problem: The obstruction classes at each level depend on all previous choices, which are determined only up to cocycles, and not merely on the original infinitesimals. Making the wrong choice could produce a non-zero obstruction where a cleverer one might succeed. Unfortunately, the computations rapidly become unwieldy.

3. The complex analytic case. The analogue of Theorem 1 holds for complex analytic manifolds without separate proof because 1) all of the infinitesimal concepts in that deformation theory can be captured purely algebraically using the deformation theory of presheaves of algebras, [GS] and 2) the Cohomology Comparison Theorem of that paper in turn reduces that theory to the theory of a single algebra. Compact complex analytic manifolds behave very much like finite dimensional algebras. The deformations of a single such manifold $\mathcal{M}$ form, by Kuranishi's Theorem, a finite dimensional variety which may have several irreducible components. Let T denote the sheaf of germs of holomorphic tangent vector fields on $\mathcal{M}$ and $\bigwedge T$ be its exterior algebra. The graded Lie algebra structure on $\bigwedge T$ induces a graded Lie structure on the total cohomology $H^*(\mathcal{M}, \bigwedge T)$ analogous to the Gerstenhaber bracket for the total Hochschild cohomology $H^*(A, A)$ of an associative algebra A with coefficients in A itself. Infinitesimal deformations of the complex structure are now elements of $H^1(\mathcal{M}, T) = H^1(\mathcal{M}, \bigwedge^1 T)$, these being in an algebraic sense, of total degree 2. If we have two full deformations of the complex structure, with infnitesimals the classes of cocycles F and G, respectively, then the class of $[F, G]$, which lies in $H^2(\mathcal{M}, T)$, may be viewed as an obstruction to having both deformations lie on a common component containing $\mathcal{M}$ of the variety of complex manifolds.

4. Differential deformations. If A is an algebra with differential operators then one can define 'differential cochains' to be ones which are cup products of differential operators. These are typically closed under the Hochschild coboundary, so one can define 'differential Hochschild cohomology'. Since differential operators are also normally closed under composition, one can in these favorable cases define 'differential deformations'. One can then also define differential compatibility. This raises the question of whether two deformations which are compatible are differentially compatible. This will, of course, be the case if all cochains are differential.

The operators may be purely algebraic. One can, for example, always take as differential operators the polynomials in the derivations of A into itself. As a special case, on the polynomial ring $A = k[x_1, \dots, x_n]$ in n variables over an arbitrary coefficient ring k we have the differential operators $\partial_i = \partial/\partial x_i$. In characteristic zero, all cochains are differential if one is willing to allow formal infinite sums of monomials in the ∂_i and the variables. Moreover, inclusion of the complex of differential cochains into the Hochschild complex induces an isomorphism on cohomology. The cohomology of a polynomial ring A in characteristic zero is given by the theorem of Hochschild-Kostant-Rosenberg (HKR) which asserts, in this case, that the cohomology of A is just the exterior algebra on the derivations of A (linear combinations of the ∂_i with coeficients in A) where the cup product replaces the exterior product. In finite characteristics there are more cohomology classes. For example, in characteristic zero the class of $F = \partial_x \smile \partial_x$ vanishes, since there $F = -\frac{1}{2}\delta(\partial_x^2)$ but in characteristic 2 it is not zero. Similarly, the class of $\partial_x^2 \smile \partial_x + \partial_x \smile \partial_x^2$ vanishes in characteristic zero but not in characteristic 3. However, in characteristic p the cochains of a truncated polynomial algebra $k[x_1, \dots, x_n]/(x_1^p, \dots, x_n^p)$ are all still differential (and now one only needs finite sums). Note that the HKR theorem implies, in particular, that if $A = \mathbb{C}[x_1, \dots, x_n]$, then $H^m(A, A) = 0$ for $m > n$, but this no longer holds in positive characteristics.

5. The quantum plane in characteristic zero and the Weyl algebra. The complex polynomial ring $A = \mathbb{C}[x, y]$ in two variables has two natural deformations. For the first, set $a * b = ab + \sum_{i=1}^{\infty}(t^i/i!)\partial_x a \partial_y b$. (Denoting the original multiplication by $\pi : A \otimes A \to A$, tacitly extended to be bilinear over $\mathbb{C}[[t]]$, this can be written $* = \pi \exp(\partial_x \otimes \partial_y)$. Although the series is infinite, the deformed product of two polynomials is a polynomial in t.) With this we have $x * y = xy + t, y * x = xy$, so $x * y - y * x = t$. The deformed algebra is thus just the smallest Weyl algebra. It is simple, but being infinite dimensional, it is not separable, and its cohomology does not vanish identically. (A separable algebra over an algebraically closed field is just a finite sum of matrix algebras or 'multimatrix algebra'; its cohomology in positive dimensions with coefficients in any bimodule is identically zero.) However, like any deformation to a separable algebra, the deformation from the polynomial algebra to the Weyl algebra is a jump deformation: The value of t is not important as all non-zero complex values of t give isomorphic algebras. The infinitesimal of the deformation is the class of $F = \partial_x \smile \partial_y$.

For the second deformation, set $*' = \exp(x\partial_x \otimes y\partial_y)$. Here we have $x *' y = e^t xy$ and $y *' x = xy$. This is the standard quantum plane with $q = e^t$. Different complex values of t in general give non-isomorphic algebras. The infinitesimal of the deformation is the class of $G = x\partial_x \smile y\partial_y$. Surprisingly, but for trivial reasons, we have

Proposition. *The deformations of $A = \mathbb{C}[x, y]$ to the Weyl algebra and to the quantum plane are compatible.*

Proof. By the HKR Theorem, $H^3(A, A) = 0$, so all obstructions vanish. □

In some sense, therefore, the Weyl algebra and the standard quantum plane lie on a common variety of algebras. In fact, the same proof shows that all deformations of $\mathbb{C}[x, y]$ form a single variety of algebras. (The next section shows that the analogous assertion does not hold for the finite analogs in positive characteristics.) The Weyl algebra is rigid since its cohomology vanishes in positive dimensions when the coefficient module is the algebra itself. So it is the quantum plane which deforms to the Weyl algebra. What naturally happens is that the quantum plane deforms to a quantum Weyl algebra with relations $x * y - qy * x = 1$ by means of a certain universal deformation formula. When working formally, this quantum algebra is isomorphic to the standard Weyl algebra, and so this deformation explicitly illustrates the result of the previous proposition. These results hold only formally though. If q is treated as a complex number, then the quantum Weyl algebra is never isomorphic to the usual Weyl algebra (unless $q = 1$). The formal isomorphism between them has zero radius of convergence, [GZ]

6. Deformations of a truncated polynomial algebra in positive characteristic. Suppose now that the characteristic is p and let $A = k[x, y]/(x^p, y^p)$. This has deformations analogous to the Weyl algebra and to the quantum plane, as well as others. The simplest is to a direct sum of p^2 copies of the rational function field $k(t)$. (The coefficient ring must be enlarged by adjoining the deformation parameter.) For note that $k[x]/x^p$ can be deformed to $k(t)[x]/(x^p - t^{p-1}x)$, which is just a direct sum of p copies of $k(t)$, so $k[x]/x^p \otimes k[y]/y^p$ deforms to a direct sum of p^2 copies. For the analog of the deformation to the Weyl algebra, it is simplest to define the deformation by setting $x * y = xy + t, y * x = xy$. It is then not difficult but here not necessary to write a general deformed product $a * b$ as a polynomial in t. All we will need later is that the infinitesimal F of the deformation again has the form $F = \partial_x \smile \partial_y$. We again have $x * y - y * x = t$. Now, however, we have an isomorphism to the algebra M_p of $p \times p$ matrices: Send x to the matrix $X = (e_{21}+e_{32}+\cdots+e_{p,p-1})$ and y to the matrix $Y = t(e_{12}+2e_{23}+\cdots+(p-1)e_{p-1,p})$. These generate all of M_p. The Weyl algebra may be viewed as one form of limit of matrix algebras; letting $p \to \infty$, its generators can be represented by two infinite dimensional matrices. Unlike the Weyl algebra, M_p is separable. Similarly, it is easiest to define the deformation of the truncated polynomial algebra A in characteristic p to the finite 'quantum plane' simply by setting $x *' y = qy *' x$ with $q = 1 + t$. The infinitesimal of this deformation is then $G = x\partial_x \smile y\partial_y$. We no longer have an HKR Theorem. In fact, using only the primary obstruction, one can show that in characteristic 2 the deformations $*$ and $*'$ are not compatible.

Recall that $F \circ G(a, b, c) = F(G(a, b, c) - F(a, G(b, c))$and $[F, G] = F \circ G + G \circ F$. In our case, a simple computation independent of the coefficient ring gives

$$[F, G] = [\partial_x \smile \partial_y, x\partial_x \smile y\partial_y] = \partial_x \smile (y\partial_y - x\partial_x) \smile \partial_y. \tag{1}$$

Note that separable algebras are rigid because their cohomology is trivial; in particular they have no non-trivial infinitesimal deformations. A separable or more generally rigid algebra A (over an algebraically closed field) lies on a unique irreducble component of the variety of structure constants. This component generally will have subvarieties representing algebras which are not rigid but which deform to algebras isomorphic to A; we have just seen such an example. But the points of this component representing algebras isomorphic to A form a Zariski open subset.

Proposition. *In characteristic* 2, *the deformations from the truncated polynomial algebra* $A = k[x,y]/(x^2,y^2)$ *to separable algebras and to the quantum plane are all mutually incompatible. In particular, one can not deform the quantum plane in characteristic* 2 *to a separable algebra.*

Proof. Since A has dimension four, the only multimatrix algebras to which it could deform would be either the direct sum of four copies of $k(t)$ or $M_2 = M_2(k(t))$. Clearly, deformations to distinct rigid algebras are incompatible, and a deformation to a commutative rigid algebra can not be compatible with any deformation to a non-commutative algebra. Therefore, we only have to show the incompatibility of the deformations of A to M_2 and to the quantum plane. We show that the obstruction cocycle $[F,G]$ is not a coboundary.

Suppose that $[F,G] = \delta f$ for some 2-cochain f. The 2-cocycle $a\partial_x \smile \partial_x + b\partial_x \smile \partial_y + c\partial_y \smile \partial_x + d\partial_y \smile \partial_y$ takes the values a,b,c,d on $(x,x),(x,y),(y,x),(y,y)$ respectively, so adding to f a suitable 2-cocycle of the given form (which does not change its coboundary), we may assume that f vanishes on these arguments. Further, $[F,G]$ vanishes whenever its first argument is y, and in particular on (y,x,x), so we have $\delta f(y,x,x) = 0$. The left side is just $f(yx,x)$, which therefore vanishes, and we have similarly $f(xy,y) = f(x,xy) = f(y,xy) = 0$. Finally, $\delta f(y,x,xy) = 0$. In view of the preceding, the left side is just $f(xy,xy)$. So all values of f vanish, which is impossible. □

In characteristics greater than two, the primary obstruction to compatibility (of the deformations of the truncated polynomial algebra to the quantum plane and to M_p) becomes a coboundary, namely, of $\frac{1}{2}(x\partial_x^2 \smile \partial_y - y\partial_x \smile \partial_y^2)$. In characteristic 3 the obstruction to constructing the tu^2 term in the notation of §2 becomes nontrivial, so the deformations are still incompatible. (One finds that its vanishing would force $(\partial_x^2 \smile \partial_x + \partial_x \smile \partial_x^2) \smile \partial_y$ to be a coboundary, which is not the case in characteristic 3.) One expects that in characteristic p all obstructions vanish up to that coming from the term in tu^{p-1}, but that one then meets a non-trivial obstruction (no matter how the choices of coboundary are made). This would be consistent with the fact that in the limit, when one gets to characteristic zero, there no longer is any obstruction, and the deformations become at least formally compatible.

7. Deformation to a rigid algebra. The obstructions to compatibility so far defined are generally difficult to use because the computations become involved, but when one deformation yields a rigid algebra there are simpler criteria. To avoid the ambiguities of the infinite dimensional case, assume that A is a finite dimensional algebra over some field k. When we have just one deformation A_t of A with multiplication $* = \pi + tF_1 + t^2F_2 + \dots$, a cocycle $G \in Z^n(A,A)$ is 'liftable' to A_t if the latter has an n-cocycle of the form $G_t = G + tG_1 + t^2 + \dots, G_i \in C^n(A,A)$; otherwise, G is obstructed. A lifting, if it exists, is unique, and lifted cocycles span

$Z^n(A_t, A_t)$. The coboundary of a cochain is liftable to the coboundary of the same cochain with respect to the deformed multiplication, but cocycles not originally coboundaries may lift to coboundaries. The latter are called jump cocycles and $H^n(A_t, A_t)$ = (liftable n-cocycles)/(jump n-cocycles), a result originally due to Griffiths in the analytic case.

Let $\dim A = N$, let $\mathcal{V}$ denote the variety (in N^3 space) of N dimensional algebras, and suppose now that A_t is rigid. In particular, A_t may be separable, in which case any lifting must be to a coboundary. Then A_t lies on a unique irreducible component $\mathcal{C}$ of $\mathcal{V}$, and there is a Zariski open subset of $\mathcal{V}$ consisting of algebras which (possibly after some extension of the coefficient field) are isomorphic to A_t. It follows that if A_u is another deformation of A which is compatible with A_t, then A_t must in fact also be a deformation of A_u (again, after suitable extension of coefficient fields). Consequently, $Z^n(A_t, A_t)$ must be spanned by liftings of n-cocycles of A_u, so this is a criterion for compatibility. When $n = 1$ this says that every derivation of A_t must be a lifting of a (unique) derivation of A_u. Denoting the multiplications in A_t, A_u by $*, *'$, respectively, it follows that for any $a \in A$ the lifting to A_t of the inner derivation of A_u sending x to $a *' x - x *' a$ is the inner derivation sending x to $a * x - x * a$.

8. The Donald–Flanigan conjecture. Let G be a finite group whose order we denote by $\#G$. If k is a (commutative, unital) ring in which $\#G$ is invertible, then Maschke's Theorem asserts that the group ring kG is separable. When k is an algebraically closed field this says that kG is a direct sum of matrix algebras of various dimensions. If, however, p is a prime which divides $\#G$ then $\mathbb{F}_pG$ has a non-zero radical. (The sum of all the group elements is a central element of $\mathbb{F}_p$ and has square equal to zero.) The Donald–Flanigan conjecture is a "modular form" of Maschke's Theorem: it asserts that if k is a field of characteristic p dividing $\#G$ then kG can be deformed to a separable algebra. It has further been conjectured that, passing to the algebraic closure of the coefficient field, the deformed algebra becomes a direct sum of matrix algebras which can be paired off with those arising for the complex group algebra $\mathbb{C}G$ such that the corresponding matrix blocks have the same sizes. Such a deformation is called a "strong" solution to the Donald–Flanigan problem for the group G, while a deformation which is separable but which does not decompose as $\mathbb{C}G$ does is called a "weak" one. There may be both a strong and a weak solution (or conceivably several weak solutions) for a single group G. In fact, we have just seen this. For if we denote by C_n the cyclic group of n elements, then $\mathbb{F}_2[C_2 \times C_2]$ is the truncated polynomial algebra in two variables, which can be deformed both to a direct sum of four copies of the (algebraically closed) coefficient field, the strong solution, and to an algebra of 2×2 matrices, a weak solution. (As noted before, the "coefficient field" has actually been enlarged to contain the deformation parameter.) For a bibliography on the Donald–Flanigan conjecture, cf. [GSps]

The ideas of the preceding section will now be used to show that the conjecture fails for the 8-element quaternion group $\mathcal{Q}$. When working in characteristic 2, it is convenient to denote the central group element "-1" by z; the group then consists of $\{1, i, j, ij, z, zi, zj, zij\}$ where $i^2 = j^2 = z$ and $ji = zij$. Let k be a field of characteristic 2 and $A := k\mathcal{Q}$ be the group algebra. The radical R of A, which consists of all linear combinations of group elements in which the sum of the coefficients is the zero element of k, is generated by $1+i$ and $1+j$. It is nilpotent of index 5, i.e., we have $R^4 \neq 0, R^5 = 0$. Filtering A by the powers of R, the associated graded algebra B is generated by two elements, a, b of degree 1 (representing the classes of $1+i, 1+j$, respectively). An elementary computation will show that that the defining relations are $a^2 = b^2 = ab + ba, a^4 = 0$. Writing $B = B_0 + B_1 + B_2 + B_3 + B_4$, the spanning sets of these homogeneous summands are $\{1\}, \{a, b\}, \{a^2, ab\}, \{a^3, a^2b\}, \{a^3b\}$, respectively. The center of B is spanned by 1 and the multiples of a^2.

A basic fact about any filtered algebra A that it is a deformation of its associated graded algebra B. In fact, since here the algebras are finite dimensional there is a polynomial (rather than a power series) $*' = \pi + uF_1 + \cdots + u^rF_r$ such that our original group algebra A is just B_1 (i.e., obtained from the above by setting $u = 1$). Moreover, all the algebras obtained by specializing u to any non-zero element of the coefficient field are canonically isomorphic. Now let K be the algebraic closure of $k(t)$ with whatever deformation parameters are needed adjoined, and denote the direct sum of four copies of K by K^4. We will see that the associated graded algebra B can also be deformed both to $M_2(K) \oplus M_2(K)$ and to $M_2(K) \oplus K^4$. In view of the dimension and the fact that A is not commutative, these are the only separable k- algebras to which A could possibly deform. (A deformation to the first would be a weak solution to the Donald–Flanigan conjecture and one to the second would be a strong solution, if either existed.) But both of these deformations of B to separable algebras will be seen to have derivations which are not liftings of derivations of A, so the deformation from B to the original group algebra A is not compatible with any deformation from B to a separable algebra. It will follow that the Donald–Flanigan conjecture does not hold for Q.

To exhibit the deformations of B to separable algebras, define derivations ∂_a, ∂_b of B by setting $\partial_a a = 1, \partial_a b = 0, \partial_b a = 0, \partial_b b = 1$; these are well defined because they preserve the defining relations. These derivations commute and both have square zero, so setting $* = \pi + t\partial_a \smile \partial_b$ already defines a deformation. (This special case of a general proposition can easily be verified directly.) Note that ∂_a and ∂_b are still derivations of the deformed algebra with multiplication $*$. Now observe that modulo the central ideal R' of square zero consisting of the multiples of a^2 we have just a 2×2 matrix algebra, so by Wederburn's theorem $(A, *)$ is just $M_2 \oplus R'$. Moreover, as a module over the separable component M_2 the radical R' is just a copy of M_2, so enlarging coefficients, $(A, *)$ is isomorphic to $M_2(K) \otimes K[a^2]$. Now we can deform further by setting $a^4 + ta^2 = 0$. This creates a new central idempotent, $e := t^{-1}a^2$, and the resulting algebra is just $M_2(K) \oplus M_2(K)$. The derivations ∂_a, ∂_b have been preserved because a^2 behaves like a constant relative to them. Denote this algebra by S_1.

Now perform first the deformation in which a^4 becomes ta^2; it is well-defined because it was so above; there one could have replaced ∂_a, ∂_b by any constant multiples of themselves which could then be specialized to zero. The resulting algebra is now isomorphic to a direct sum of two copies of $k[x, y]/(x^2, y^2)$; the first is generated by ae and be, the second by $a(1+e), b(1+e)$. The first copy can be deformed to a 2×2 matrix algebra by setting $ae * be + be * ae = te$, and the second to the direct sum of four copies of the coefficient field by setting

$(a(1+e))^2 = ta(1+e), (b(1+e))^2 = tb(1+e)$. This gives the second deformation of B to a separable algebra, S_2. Writing $a = ae + a(1+e)$ and similarly for b, one sees that in S_2 we have $a*b+b*a = te = a^2$. Note that in B we had $ab+ba = a^2$ and that in S_2 we have $a*a = a^2$. (The last equation is not a tautology; on the right the square is taken with respect to the multiplication in B and on the left it is in S_2.)

It is evident now that the group algebra A can not be deformed to S_2: The operation of $\operatorname{ad} b$ on a is precisely the same in S_2 as in B; it carries a to a^2. But A and B have the same underlying vector space, and the elements labeled a, b in B are the elements $1+i, 1+j$ in A, respectively. So here the operation of $\operatorname{ad} b$ on a differs from the original (we get $ij + ji \neq (1+i)^2$) and no further deformation can undo it. In S_1 we have $a*b+b*a = a^2+t, a*a = a^2$ so the operation has changed, but only by a multiple of the unit element, and the argument is essentially the same. (Note that without loss of generality we may assume that the unit element remains the unit element in all deformations.) So the Donald-Flanigan conjecture fails for the quaternion group.

This leaves us with the paradox that the prediction from the Donald–Flanigan conjecture given in [GG] and proven in [FJL] is true even though the conjecture, as stated, is not. The curious way in which the conjecture fails here makes us suspect that a somewhat subtler assertion leading to the same conclusion does hold, but at present we have no idea what it might be.

References

[FJL] P. Fleischmann, I. Janiszczak,and W. Lempken, *Finite groups have non-Schur centralizers*, Manuscripta Mathematika **80** (1993), 213–224.

[G1] M. Gerstenhaber, *The cohomology structure of an associative ring*, Annals of Math. **78** (1963), 267–288.

[G2] ______, *On the deformation of rings and algebras*, Annals of Math. **79** (1964), 59–103.

[G3] ______, *On the deformation of rings and algebras. II*, Annals of Math. **84** (1966), 1–19.

[G4] ______, *On the deformation of rings and algebras. III*, Annals of Math. **88** (1968), 1–34.

[GG] M. Gerstenhaber and D. J. Green, *A group-theoretic consequence of the Donald–Flanigan conjecture*, Jour. of Algebra **166** (1994), 356–363.

[GS] M. Gerstenhaber and S. D. Schack, *Algebraic cohomology and deformation theory*, in Deformation Theory of Algebras and Structures and Applications (M. Hazewinkel and M. Gerstenhaber, eds.), Kluwer, Dordrecht, 1988, pp. 11–264.

[GSps] M. Gerstenhaber and Mary E. Schaps, *Hecke algebras, $U_q sl_n$, and the Donald–Flanigan conjecture for S_n*, Trans. Amer. Math. Soc. **349** (1997), 3353–3371.

[GZ] A. Giaquinto and J.J. Zhang, *Quantum Weyl Algebras*, Jour. of Algebra **176** (1995), 861–881.

Dept. of Mathematics, University of Pennsylvania, Philadelphia, PA 19104-6395
E-mail address: mgersten@math.upenn.edu

Dept. of Mathematics, Loyola University Chicago, Chicago, Il 60626
E-mail address: tonyg@math.luc.edu

Contemporary Mathematics
Volume **229**, 1998

Directing objects in hereditary categories

Dieter Happel and Idun Reiten

ABSTRACT. If $\mathcal{H}$ is a hereditary abelian category with tilting object, then it is shown that $\mathcal{H}$ is derived equivalent to a module category over a hereditary artin algebra provided $\mathcal{H}$ contains a directing object.

Tilting objects in hereditary abelian categories have been introduced in [**HRS1**] in order to provide a common framework for two parallel classes of artin algebras which were considered quite intensively in the last years. These are tilted algebras (compare for example [**HRi1**]) on one hand and canonical algebras or variations thereof (compare for example [**LS**]) on the other hand. Note that we have to impose some finiteness conditions on the categories we study. For a complete list of those conditions we refer to section one.

These developments led to the more general concept of quasitilted algebras which by definition are just the endomorphism algebras of tilting objects in a hereditary abelian category. They also can be characterized homologically as was shown in [**HRS1**]. In fact, an artin algebra Λ is quasitilted if and only if the global dimension of Λ is at most two and each indecomposable Λ-module X satisfies either $\text{proj.dim}_\Lambda X \leq 1$ or $\text{inj.dim}_\Lambda X \leq 1$. For further details we refer to [**HRS1**], [**HRe1**] and [**H**].

The two classes mentioned above account for the two most prominent examples of hereditary abelian categories containing a tilting object. The first class corresponds to mod H, the category of finitely generated left modules over a hereditary artin algebra. The second class corresponds to coh $\mathbb{X}$, the category of coherent sheaves over a weighted projective line in the sense of [**GL**]. It is conjectured that these are the only examples up to derived equivalence.

The aim of this article is to give a solution to this conjecture in the following case. For the precise definitions we refer to the later sections. A different case will be considered in a forthcoming paper [**HRe2**].

Theorem: *Let $\mathcal{H}$ be a connected hereditary abelian category containing a tilting object. If $\mathcal{H}$ contains an indecomposable directing object, then $\mathcal{H}$ is derived equivalent to mod H for a hereditary artin algebra H.*

1991 *Mathematics Subject Classification.* 16G10, 16G70, 16E10.

These results were obtained during a visit of the first author at NTNU under a grant of the Norwegian Research Council. He would like to thank the second author for her kind hospitality.

In the first section we have collected the precise definitions and some of the general results used in the later part. Moreover we show here that a hereditary abelian category $\mathcal{H}'$ which is derived equivalent to a hereditary abelian category $\mathcal{H}$ containing a tilting object will also contain a tilting object. This is then generalized to hereditary abelian categories containing a tilting complex (for the precise definition see section 1). The second section is devoted to the proof of the Theorem above.

1. Preliminaries

Let R be a commutative artin ring and let $\mathcal{H}$ be a hereditary, abelian category (i.e. $\operatorname{Ext}^2_{\mathcal{H}}(X,Y) = 0$ for all $X, Y \in \mathcal{H}$ and $\mathcal{H}$ is abelian). We will assume throughout that for all $X, Y \in \mathcal{H}$ we have that $\operatorname{Hom}_{\mathcal{H}}(X,Y)$ and $\operatorname{Ext}^1_{\mathcal{H}}(X,Y)$ are R-modules which are of finite length over R and we have that the composition is bilinear over R. Note that these assumptions imply that the theorem of Krull-Remak-Schmidt holds in $\mathcal{H}$. For an R-module M we denote by $l_R(M)$ the length of M over R. These assumptions on $\mathcal{H}$ will be fixed throughout.

Given $X \in \mathcal{H}$ we denote by $\operatorname{add} X$ the full subcategory of $\mathcal{H}$ with objects the finite direct sums of direct summands of X. Further we denote by $\operatorname{Fac} X$ the full subcategory of $\mathcal{H}$ with objects the epimorphic images of objects in $\operatorname{add} X$ and by $\mathcal{E}(X)$ the full subcategory of $\mathcal{H}$ with objects Y satisfying $\operatorname{Ext}^1_{\mathcal{H}}(X,Y) = 0$.

We say that $T \in \mathcal{H}$ is a tilting object if $\operatorname{Fac} T = \mathcal{E}(T)$. We point out that this was not the original definition in [**HRS1**]. It was shown in [**H**] that it is equivalent to the original definition. It is sometimes easier to verify in concrete situations.

The existence of a tilting object in $\mathcal{H}$ has some strong consequences. We will recall some of those important to us from [**HRS1**].

Let $\mathcal{H}$ be a hereditary abelian category with tilting object T. Then $\mathcal{H}$ has almost split sequences. So for each indecomposable nonprojective $Z \in \mathcal{H}$ there is a nonsplit short exact sequence

$$0 \longrightarrow \tau_{\mathcal{H}} Z \xrightarrow{\mu} E \xrightarrow{\pi} Z \longrightarrow 0$$

such that $\tau_{\mathcal{H}} Z$ is indecomposable and for each $f \in \operatorname{Hom}_{\mathcal{H}}(W,Z)$ which is not a split epi there is $g \in \operatorname{Hom}_{\mathcal{H}}(W,E)$ such that $f = \pi g$.

Moreover, if $K_0(\mathcal{H})$ denotes the Grothendieck group of $\mathcal{H}$, then $K_0(\mathcal{H})$ is free abelian of finite rank and is endowed with a bilinear form $\langle -,-\rangle_{\mathcal{H}}$ which is non-degenerate. The bilinear form is defined for $X, Y \in \mathcal{H}$ by

$$\langle X,Y\rangle_{\mathcal{H}} = l_R(\operatorname{Hom}_{\mathcal{H}}(X,Y)) - l_R(\operatorname{Ext}^1_{\mathcal{H}}(X,Y))$$

and then extended to $K_0(\mathcal{H})$ by bilinearity.

This has the following consequence as noted in [**H**]. Note that the assumption on the non-degeneracy of the form is automatically satisfied if $\mathcal{H}$ contains a tilting object.

LEMMA 1.1. *Let $\mathcal{H}$ be a hereditary abelian category. Let $T \in \mathcal{H}$ be an object satisfying $Ext^1_{\mathcal{H}}(T,T) = 0$. Assume that $\langle -,-\rangle_{\mathcal{H}}$ is non-degenerate on $K_0(\mathcal{H}) \simeq \mathbb{Z}^n$ for some $n > 0$. If $T = T_1 \oplus \cdots \oplus T_n$ with T_i indecomposable and $T_i \not\simeq T_j$ for $i \neq j$, then T is a tilting object.*

We also mention the following consequence which will be needed later.

COROLLARY 1.2. *Let $\mathcal{H}$ be a hereditary abelian category containing a tilting object. We assume that $rk K_0(\mathcal{H}) = n$. Let $X_1, \dots X_m$ be pairwise nonisomorphic indecomposable objects in $\mathcal{H}$ and let $X = \oplus_{i=1}^m X_i$. If $m > n$, then $Ext^1_{\mathcal{H}}(X, X) \neq 0$.*

Proof: If $\mathrm{Ext}^1_{\mathcal{H}}(X, X) = 0$, then we may assume that $X_1 \dots X_m$ are ordered such that $\mathrm{Hom}_{\mathcal{H}}(X_i, X_j) = 0$ for $i < j$ (compare for example [**HRi1**]). Let $X' = \oplus_{i=1}^n X_i$. By 1.1 we have that X' is a tilting object. Now $\langle X', X_{n+1}\rangle_{\mathcal{H}} = 0$, in contrast to 2.4 of [**H**].

For a hereditary abelian category $\mathcal{H}$ we denote by $D^b(\mathcal{H})$ its bounded derived category. We recall some of the relevant features. We have an embedding $\mathcal{H} \to D^b(\mathcal{H})$ by sending objects to the corresponding stalk complex concentrated in degree zero. We denote by $\mathcal{H}[i]$ for $i \in \mathbb{Z}$ the image of $\mathcal{H} = \mathcal{H}[0]$ under the i-th power of the shift functor. It is well known and easy to see that each indecomposable object $X^\bullet \in D^b(\mathcal{H})$ is isomorphic to an indecomposable object in $\mathcal{H}[i]$ for some $i \in \mathbb{Z}$. Moreover for $X, Y \in \mathcal{H}$ and $i, j \in \mathbb{Z}$ we have that

$$\mathrm{Hom}_{D^b(\mathcal{H})}(X[i], Y[j]) \simeq \mathrm{Ext}^{j-i}_{\mathcal{H}}(X, Y).$$

We will also need our joint result presented in [**H**]. Let $\mathcal{H}$ be a connected hereditary abelian category with tilting object T. Then $\mathcal{H} \simeq \mathrm{mod}\, H$ for some hereditary artin algebra H if and only if $\mathcal{H}$ contains a nonzero projective object.

We also will use the following essential properties of a connected hereditary abelian category with tilting object, which is not equivalent to $\mathrm{mod}\, H$ for a hereditary artin algebra H. In this case the translation $\tau_{\mathcal{H}}$ introduced above is an equivalence of categories and for $X, Y \in \mathcal{H}$ we have that $\mathrm{Ext}^1_{\mathcal{H}}(X, Y) \simeq D\mathrm{Hom}_{\mathcal{H}}(Y, \tau_{\mathcal{H}} X)$, where D denotes the usual duality (compare for example [**HRS1**]).

Next we show that the property of containing a tilting object is an invariant of the derived equivalence class.

THEOREM 1.3. *Let $\mathcal{H}$ be a connected hereditary abelian category with tilting object T. Then any hereditary abelian category $\mathcal{H}'$ which is derived equivalent to $\mathcal{H}$ contains a tilting object.*

Proof: Since $\mathcal{H}$ is connected and $\mathcal{H}'$ is derived equivalent to $\mathcal{H}$ then trivially $\mathcal{H}'$ is also connected. Note that $\mathcal{H}'$ automatically satisfies the general assumptions given in the beginning of this section. If $\mathcal{H}' \simeq \mathrm{mod}\, H'$ for some hereditary artin algebra H' there is nothing to show, since in this case ${}_{H'}H'$ is a tilting object. So we may assume that $\tau_{\mathcal{H}'}$ is an equivalence on $\mathcal{H}'$. Also it is easily seen that the assertion holds if $\mathcal{H} = \mathrm{mod}\, H$ for some hereditary artin algebra H. So we may also assume that $\tau_{\mathcal{H}}$ is an equivalence. We may assume that $T \in \mathcal{H}$ is multiplicity-free, so $T = T_1 \oplus \cdots \oplus T_n$ with T_i indecomposable and $T_i \not\simeq T_j$ for $i \neq j$. Let $F : D^b(\mathcal{H}) \to D^b(\mathcal{H}')$ be a derived equivalence. Since F is a triangle equivalence we have that F commutes with τ. Applying the shift functor if necessary, we may assume that $F(T[0]) \subset \bigcup_{j=0}^{r} \mathcal{H}'[j]$ for some $r \geq 0$ and that $F(T[0]) \cap \mathcal{H}'[0] \neq 0 \neq F(T[0]) \cap \mathcal{H}'[r]$. This induces a direct sum decomposition of T as follows. For $0 \leq j \leq r$ let $T^j = \oplus T_i$ where the direct sum is taken over all those T_i with $F(T_i[0]) \in \mathcal{H}'[j]$. Since $\mathrm{End}_{\mathcal{H}} T$ is a connected artin algebra we have $T^j \neq 0$ for all $0 \leq j \leq r$. Let ${}_{\mathcal{H}'}T' = \bigoplus_{j=0}^{r} \tau^j_{\mathcal{H}'} F(T^j)[-j]$. Since $F(T^j) \in \mathcal{H}'[j]$ and $\tau_{\mathcal{H}'}$ is an equivalence we infer that ${}_{\mathcal{H}'}T' \in \mathcal{H}'[0]$. We claim that ${}_{\mathcal{H}'}T'$ is a tilting object in $\mathcal{H}'$. Since $\mathcal{H}$ and $\mathcal{H}'$ are derived equivalent we see that $\mathcal{H}'$ satisfies the assumptions of Lemma 1.1. Thus it is enough to show that $Ext^1_{\mathcal{H}'}(T', T') = 0$.

For this let $0 \le i, j \le r$. Then

$$\begin{aligned}
&\mathrm{Ext}^1_{\mathcal{H}'}(\tau^i_{\mathcal{H}'}F(T^i)[-i], \tau^j_{\mathcal{H}'}F(T^j)[-j]) \\
&\qquad \simeq \mathrm{Hom}_{D^b(\mathcal{H}')}(\tau^i_{\mathcal{H}'}F(T^i), \tau^j_{\mathcal{H}'}F(T^j)[-j+i+1]) \\
&\qquad \simeq \mathrm{Hom}_{D^b(\mathcal{H})}(\tau^i_{\mathcal{H}}T^i, \tau^j_{\mathcal{H}}T^j[-j+i+1]) \\
&\qquad \simeq \mathrm{Ext}^{-j+i+1}_{\mathcal{H}}(\tau^i_{\mathcal{H}}T^i, \tau^j_{\mathcal{H}}T^j)
\end{aligned}$$

Note that we have used that $\tau_{\mathcal{H}}$ is an equivalence and that F commutes with τ and the shift functor. The last term is possibly nonzero only in the following two cases. If $-j+i+1 = 0$, then $i = j-1$. So

$$\mathrm{Hom}_{\mathcal{H}}(\tau^{j-1}T^{j-1}, \tau^j T^j) \simeq \mathrm{Hom}_{\mathcal{H}}(T^{j-1}, \tau T^j) \simeq DExt^1_{\mathcal{H}}(T^j, T^{j-1}) = 0,$$

since T is a tilting object in $\mathcal{H}$. If $-j+i+1 = 1$, then $j = i$. So $\mathrm{Ext}^1_{\mathcal{H}}(\tau^i_{\mathcal{H}}T^i, \tau^i_{\mathcal{H}}T^i) \simeq \mathrm{Ext}^1_{\mathcal{H}}(T^i, T^i) = 0$, since T is a tilting object in $\mathcal{H}$.

In the remaining part of this section we want to generalize this to tilting complexes. This will give an alternative proof of the Theorem above. However we think that it is worthwhile to have both proofs. We will also include an example showing that the construction in the proof above will not generalize. For this we first have to recall some of the relevant definitions.

An object $T^\bullet \in D^b(\mathcal{H})$ is called a tilting complex if $\mathrm{Hom}_{D^b(\mathcal{H})}(T^\bullet, T^\bullet[i]) = 0$ for all $i \neq 0$ and $\mathrm{add}\, T^\bullet$ generates $D^b(\mathcal{H})$ as a triangulated category.

It follows from [**Rc**] that for a tilting complex $T^\bullet \in D^b(\mathcal{H})$ the right derived functor of $F = \mathrm{Hom}(T^\bullet, -)$ is a triangle equivalence $D^b(\mathcal{H}) \to D^b(\mathrm{End}\, T^\bullet)$. This has the following consequence.

LEMMA 1.4. *An object $T \in \mathcal{H} \subset D^b(\mathcal{H})$ is a tilting complex if and only if T is a tilting object.*

Proof: Let T be a tilting object. For any $X \in \mathcal{H}$ there exists an exact sequence $0 \to X \to E \to T' \to 0$ with $E \in \mathrm{Fac}\, T$ and $T' \in \mathrm{add}\, T$. Therefore $\mathrm{add}\, T$ generates $D^b(\mathcal{H})$ as a triangulated category. Thus T is a tilting complex.

Conversely, assume that T is a tilting complex. We have to show that $\mathrm{Fac}\, T = \mathcal{E}(T)$. For this let $0 \neq X \in \mathcal{E}(T)$. Let F be the triangle equivalence mentioned above F:$D^b(\mathcal{H}) \to D^b(\mathrm{End}\, T)$. Then $0 \neq F(X) = \mathrm{Hom}(T, X)$. It is easy to see that a minimal right $\mathrm{add}\, T$-approximation of X is surjective, hence $X \in \mathrm{Fac}\, T$.

We recall from [**HRS2**] the definition of a piecewise hereditary algebra. An artin algebra Λ is called piecewise hereditary of type $\mathcal{H}$, if there exists a hereditary abelian category $\mathcal{H}$ and a triangle equivalence $D^b(\Lambda) \to D^b(\mathcal{H})$. The type of Λ is of course only defined up to triangle equivalence. The following two statements are important but straightforward observations.

LEMMA 1.5. *Let Λ be a piecewise hereditary algebra of type $\mathcal{H}$ and let F be a triangle equivalence from $D^b(\Lambda)$ to $D^b(\mathcal{H})$. Then $F({}_\Lambda\Lambda)$ is a tilting complex.*

LEMMA 1.6. *Let $T^\bullet = \oplus_{i=0}^r T_i[i]$ be in $D^b(\mathcal{H})$ with $T_0, \ldots, T_r \in \mathcal{H}$. Then $Hom(T^\bullet, T^\bullet[j]) = 0$ for $j \neq 0$ if and only if $Hom_{\mathcal{H}}(T_s, T_t) = 0$ for $s \neq t$ and $Ext^1_{\mathcal{H}}(T_s, T_t) = 0$ for $t \neq s+1$.*

Before we present the next result, let us consider the following example. Let H be the path algebra over a field k of the following quiver:

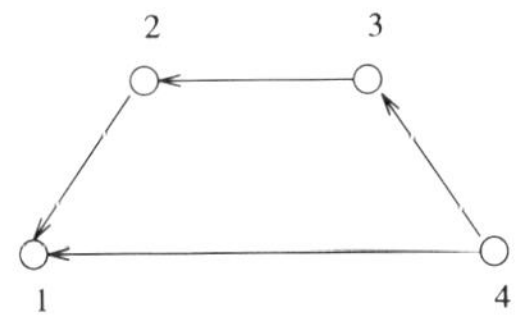

Let $\mathcal{H} = \operatorname{mod} H$. We choose $T_0 = P(1) \oplus P(4) \oplus \tau_H^2 S(3)$ and $T_1 = S(3)$, where $P(i)$ is the projective cover of the simple H-module $S(i)$ attached to the vertex i. Then it is easy to check that $T_0 \oplus T_1[1]$ is a tilting complex. However $T_0 \oplus \tau_H S(3)$ is not a tilting object in $\mathcal{H}$. This shows that the construction of the tilting object as in 1.3 does not work in general. We point out that we may choose a hereditary category $\mathcal{H}'$ triangle equivalent to $\mathcal{H}$, say $G : D^b(\mathcal{H}) \to D^b(\mathcal{H}')$ is a triangle equivalence, such that $\mathcal{H}'$ is not a module category, where $G(T_0) \in \mathcal{H}'[0]$ and $G(T_1[1]) \in \mathcal{H}'[1]$. Then $G(T_0) \oplus G(T_1)[1]$ is a tilting complex in $D^b(\mathcal{H}')$ where the construction in 1.3 does not yield a tilting object, even though $\tau_{\mathcal{H}'}$ is an equivalence.

THEOREM 1.7. *Assume that $D^b(\mathcal{H})$ contains a tilting complex. Then $\mathcal{H}$ contains a tilting object.*

Proof: If $D^b(\mathcal{H})$ contains a tilting complex then clearly there is a tilting complex $T^\bullet = \oplus_{i=0}^r T_i[i]$ in $D^b(\mathcal{H})$ with $T_0, \dots, T_r \in \mathcal{H}$ with $T_0 \neq 0 \neq T_r$. Amongst all such tilting complexes choose one with r minimal. If $r = 0$, then $T^\bullet \in \mathcal{H}$, so is a tilting object by 1.4. So assume that $r \geq 1$. Consider the universal extension

$$(*) \qquad 0 \to T_r \to E \to T_{r-1}^s \to 0$$

for some $s \geq 0$, so that the connecting homomorphism

$$\operatorname{Hom}_{\mathcal{H}}(T_{r-1}, T_{r-1}^s) \to \operatorname{Ext}^1_{\mathcal{H}}(T_{r-1}, T_r)$$

is surjective. Consider the complex $S^\bullet = \oplus_{i=0}^{r-1} T_i[i] \oplus E[r-1]$. We claim that $S^\bullet$ is a tilting complex. This gives a contradiction to the assumption that $r \geq 1$. In fact it follows from $(*)$ that $\operatorname{add} S^\bullet$ generates $D^b(\mathcal{H})$ as a triangulated category since $\operatorname{add} T^\bullet$ generates $D^b(\mathcal{H})$ as a triangulated category. Using Lemma 1.6 and $(*)$ it is straightforward to check that also $\operatorname{Hom}(S^\bullet, S^\bullet[j]) = 0$ for $j \neq 0$.

Note that the proof of the Theorem above actually shows a stronger statement.

PROPOSITION 1.8. *If $T^\bullet = \oplus_{i=0}^r T_i[i]$ is a tilting complex in $D^b(\mathcal{H})$ for some hereditary abelian category $\mathcal{H}$ with $T_0, \dots, T_r \in \mathcal{H}$ with $T_0 \neq 0 \neq T_r$, then there is a tilting object $T = T_0 \oplus Z$ and a tilting object $T' = Z' \oplus T_r$ in $\mathcal{H}$ for some $Z, Z' \in \mathcal{H}$.*

We refer to [**HRe2**] for an application of this result to piecewise hereditary algebras.

2. Directing objects

The aim of this section is to give a proof of the Theorem from the introduction. We keep the notation from the previous section. First we will have to recall some additional terminology which is quite familiar in representation theory of artin algebras. We also point out that the proof of the Theorem follows closely the strategy of the existing proofs that a module category over an artin algebra containing a complete slice is a tilted algebra (see for example [**Ri**]).

Let $\mathcal{H}$ be a hereditary abelian category containing a tilting object. We will assume throughout the rest of the paper that $\mathcal{H}$ does not contain nonzero projective objects. As noted before, $\mathcal{H}$ has almost split sequences. So we can consider the AR-quiver $\Gamma_{\mathcal{H}}$ of $\mathcal{H}$. This is a locally finite translation quiver. If X in $\mathcal{H}$ is indecomposable we denote by $\mathcal{C}_X$ the connected component of $\Gamma_{\mathcal{H}}$ containing X. A component $\mathcal{C}$ of $\Gamma_{\mathcal{H}}$ is called generalized standard (compare [**Sk**] or [**Li**]) if $\operatorname{rad}^{\infty}(X,Y) = 0$ for all indecomposable $X, Y \in \mathcal{C}$. If $X, Y \in \mathcal{H}$ are indecomposable then a path from X to Y in $\mathcal{H}$ is a sequence $(X_0, X_1, \cdots, X_t)$ of indecomposable objects $X_i \in \mathcal{H}$ for $0 \leq i \leq t$ with $X = X_0$ and $Y = X_t$ such that rad $(X_i, X_{i+1}) \neq 0$ for $0 \leq i < t$. If $(X_0, X_1, \cdots, X_t)$ is a path in $\mathcal{H}$ we say that the path has length t. A path in $\mathcal{H}$ from X to X of length at least one is called a cycle. We say that an indecomposable X in $\mathcal{H}$ is directing if it does not lie on a cycle. Finally we recall that a path $(X_0, X_1, \cdots, X_t)$ in $\mathcal{H}$ is sectional if $X_i \not\simeq \tau_{\mathcal{H}} X_{i+2}$ for $0 \leq i \leq t-2$ and there exists an irreducible map $X_i \to X_{i+1}$ for $0 \leq i < t$.

LEMMA 2.1. *Let $\mathcal{H}$ be a hereditary abelian category containing a tilting object. If $\mathcal{H}$ contains an indecomposable directing object X, then each $Y \in \mathcal{C}_X$ is directing.*

Proof: Let $Y \in \mathcal{C}_X$ and assume that there is an irreducible map $Y \to X$. We will show that Y is directing. The assertion follows from this and its dual.

Assume to the contrary that Y lies on a cycle $Y \xrightarrow{f_1} E_1 \xrightarrow{f_2} \cdots E_r \xrightarrow{f_{r+1}} Y$. Since $\operatorname{End}_{\mathcal{H}}(Y)$ is of finite length over R we may assume that $f_{r+1} \cdots f_1 = 0$. So we may choose $1 \leq s \leq r$ such that $f = f_s \cdots f_1 \neq 0$ but $f_{s+1} f = 0$. Then there exists an indecomposable object $Z \in \mathcal{H}$ with $\operatorname{Hom}_{\mathcal{H}}(Y, \tau_{\mathcal{H}} Z) \neq 0$ and $\operatorname{Hom}_{\mathcal{H}}(Z, E_{s+1}) \neq 0$ (compare [**HRi2**], note that the proof given there carries over to this situation). Since there is an irreducible map $Y \to X$, there is an irreducible map $\tau_{\mathcal{H}} X \to Y$. In particular we obtain a path from $\tau_{\mathcal{H}} X$ to $\tau_{\mathcal{H}} Z$ and hence a path from X to Z, since $\tau_{\mathcal{H}}$ is an equivalence. Since $\operatorname{Hom}_{\mathcal{H}}(Z, E_{s+1}) \neq 0$ there is a path $Z \to E_{s+1} \to \cdots \to Y \to X$. This contradicts the fact that X is directing.

PROPOSITION 2.2. *Let $\mathcal{H}$ be a hereditary abelian category containing a tilting object. Let $X \in \mathcal{H}$ be indecomposable directing. Then $\mathcal{C}_X$ is generalized standard. Moreover the length of any sectional path is bounded and there exists a finite quiver $\vec{\Delta}$ such that $\mathcal{C}_X \simeq \mathbb{Z}\vec{\Delta}$.*

Proof: Assume to the contrary that the component $\mathcal{C}_X$ is not generalized standard. Then there exist $U, V \in \mathcal{C}_X$ and a non-zero map $f : U \to V$ with $f \in \operatorname{rad}^{\infty}(U, V)$. Since $\mathcal{C}_X$ is a stable translation quiver without oriented cycles we infer that $\mathcal{C}_X \simeq \mathbb{Z}\vec{\Delta}$ for some possibly infinite quiver $\vec{\Delta}$ by [**Li**], [**Zh**]. Thus there exists $i \in \mathbb{Z}$ and a path of irreducible maps from $\tau_{\mathcal{H}}^i V$ to U. Since V is directing by 2.1 we have that $i > 0$. In particular, we obtain a path $\tau_{\mathcal{H}}^i V = V_0 \xrightarrow{g_0} V_1 \xrightarrow{g_1} \cdots \xrightarrow{g_{t-1}} V_t = V$ with $g_j \in \operatorname{rad}^{\infty}(V_j, V_{j+1})$ for some $0 \leq j < t$ and $V_s \in \mathcal{C}_X$ for $0 \leq s \leq t$. For this V, we now choose i minimal such that there exists a path with these properties. Since $g_j \in \operatorname{rad}^{\infty}(V_j, V_{j+1})$ there exists for each $m > 0$ a path

$$V_j = V_{j,0} \xrightarrow{\alpha} V_{j,1} \longrightarrow \cdots \longrightarrow V_{j,m} = V_{j+1}$$

with $V_{j,s} \in \mathcal{C}_X$ for $0 \le s \le m$ and $\alpha \in \operatorname{rad}^\infty(V_{j,0}, V_{j,1})$. Let $m = n+1$ where $n = \operatorname{rk} K_0(\mathcal{H})$. We consider the object $W = \bigoplus_{s=1}^{m} V_{j,s}$. By 1.2 we have that

$$0 \neq \operatorname{Ext}^1_{\mathcal{H}}(W, W) \simeq D\operatorname{Hom}_{\mathcal{H}}(W, \tau_{\mathcal{H}} W).$$

Thus there exist $s, s' \ge 1$ with $\operatorname{Hom}_{\mathcal{H}}(V_{j,s}, \tau_{\mathcal{H}} V_{j,s'}) \neq 0$. So we obtain a path

$$\tau^i_{\mathcal{H}} V = V_0 \to V_1 \to \cdots \to V_{j,0} \xrightarrow{\alpha} V_{j,1} \cdots \to V_{j,s} \to \tau_{\mathcal{H}} V_{j,s'} \to$$

$$\to \cdots \to \tau_{\mathcal{H}} V_{j,m} \to \cdots \to \tau_{\mathcal{H}} V_t = \tau_{\mathcal{H}} V.$$

If we apply $\tau^-_{\mathcal{H}}$ to this path we obtain a path from $\tau^{i-1}_{\mathcal{H}} V$ to V. Since $\tau^-_{\mathcal{H}} \alpha \in \operatorname{rad}^\infty$ this contradicts the minimality of i. Hence $\mathcal{C}_X$ is generalized standard.

Since $\mathcal{C}_X$ is generalized standard we have that the direct sum Z of indecomposable objects on a sectional path satisfies $\operatorname{Ext}^1_{\mathcal{H}}(Z, Z) = 0$. By 1.2 we infer that the length of any sectional path is bounded. In particular we see that $\vec{\Delta}$ is finite. This finishes the proof of the Proposition.

Let $\mathcal{H}$ be a hereditary abelian category containing a tilting object and let $X \in \mathcal{H}$ be indecomposable directing. We consider $\mathcal{S}(\to X) = \{Y \in \mathcal{C}_X \mid Y \text{ has a sectional path to } X\}$. Since $\mathcal{C}_X$ is generalized standard we infer that $\operatorname{Ext}^1_{\mathcal{H}}(Z, Z) = 0$ for any Z which is a direct sum of indecomposables in $\mathcal{S}(\to X)$. By 2.2 there are only finitely many indecomposables in $\mathcal{S}(\to X)$. So we may consider $S = \bigoplus_{Y \in \mathcal{S}(\to X)} Y$. The rest of this section is devoted to showing that S is a tilting object in $\mathcal{H}$. For this we consider the following pair $(\mathcal{T}, \mathcal{F})$ of subcategories in $\mathcal{H}$, where $\mathcal{F} = \operatorname{Sub} S$ is the full subcategory of objects cogenerated by S and $\mathcal{T} = \{Y \in \mathcal{H} \mid \operatorname{Hom}_{\mathcal{H}}(Y, S) = 0\}$. Clearly this is a torsion pair provided we can find for each $Y \in \mathcal{H}$ a short exact sequence $0 \to t(Y) \to Y \to Y/t(Y) \to 0$ with $t(Y) \in \mathcal{T}$ and $Y/t(Y) \in \mathcal{F}$. This short exact sequence will be called the canonical exact sequence for Y.

This canonical exact sequence is obtained a follows. Let $Y \in \mathcal{H}$ and let $Y \xrightarrow{f} S'$ be the minimal left add S-approximation. Then $\operatorname{im} f \in \operatorname{Sub} S$. Thus $\operatorname{Ext}^1_{\mathcal{H}}(\operatorname{im} f, S) = 0$, since $\operatorname{Ext}^1_{\mathcal{H}}(S, S) = 0$ and $\mathcal{H}$ is hereditary. But then $\ker f \in \mathcal{T}$, since $\operatorname{Hom}_{\mathcal{H}}(f, S)$ is surjective and $\operatorname{Ext}^1_{\mathcal{H}}(\operatorname{im} f, S) = 0$. Thus

$$0 \to \ker f \to Y \to \operatorname{im} f \to 0$$

is the canonical exact sequence for Y. This shows that $(\mathcal{T}, \mathcal{F})$ is a torsion pair.

In the following preliminary assertions we have stated for convenience the dual properties explicitly. However we omit the obvious proofs of the dual statements.

LEMMA 2.3. *Using the notation as above we have that $\tau_{\mathcal{H}} S \in \operatorname{Sub} S$. In particular $(\mathcal{T}, \mathcal{F})$ is a split torsion pair. Dually we have that $\tau^-_{\mathcal{H}} S \in \operatorname{Fac} S$.*

Proof: To show the first part we proceed by induction on the maximal length $l(Y)$ of a sectional path from Y to X for $Y \in \mathcal{S}(\to X)$ indecomposable. If $l(Y) = 0$, then $Y \simeq X$. Then consider the almost split sequence $0 \to \tau_{\mathcal{H}} X \to E \to X \to 0$. By definition of $\mathcal{S}(\to X)$ we have that $E \in \operatorname{add} S$, hence $\tau_{\mathcal{H}} X \in \operatorname{Sub} S$. Now let $Y \in \mathcal{S}(\to X)$ be indecomposable with $l(Y) > 0$. If $Y \to Y'$ is irreducible with $Y' \in \mathcal{S}(\to X)$ we have $l(Y') < l(Y)$, so by induction we have that $\tau_{\mathcal{H}} Y' \in \operatorname{Sub} S$. So consider the almost split sequence $0 \to \tau_{\mathcal{H}} Y \to F \to Y \to 0$. Then we have that

$F = (\tau_{\mathcal{H}} Y')^r \oplus F'$ with r maximal. But then $F' \in \mathcal{S}(\to X)$, hence $F \in \operatorname{Sub} S$ and so $\tau_{\mathcal{H}} Y \in \operatorname{Sub} S$.

For the remaining part, choose $A \in \mathcal{F}$ and $B \in \mathcal{T}$. Since $A \in \mathcal{F} = \operatorname{Sub} S$ we have an injective map $A \to S'$ for some $S' \in \operatorname{add} S$. Since $\tau_{\mathcal{H}}$ is an equivalence we have an injective map $\tau A \to \tau S'$. By the first part we infer that $\tau S' \in \operatorname{Sub} S$, hence $\tau A \in \operatorname{Sub} S$. Thus $\operatorname{Ext}^1_{\mathcal{H}}(A, B) \simeq D \operatorname{Hom}_{\mathcal{H}}(B, \tau A) = 0$, so the torsion pair splits.

LEMMA 2.4. *Using the notation above there exists for each $A \in \mathcal{F}$ a short exact sequence $0 \to A \to S' \to S'' \to 0$ with $S', S'' \in \operatorname{add} S$. Dually there exists for each $B \in \operatorname{Fac} S$ a short exact sequence $0 \to S'' \to S' \to B \to 0$ with $S', S'' \in \operatorname{add} S$.*

Proof: Let $A \in \mathcal{F} = \operatorname{Sub} S$. So there exists a short exact sequence

$$0 \to A \xrightarrow{\mu} S' \xrightarrow{\pi} Y \to 0$$

with $S' \in \operatorname{add} S$ and μ the minimal left $\operatorname{add} S$-approximation of A. We claim that $Y \in \operatorname{add} S$.

Since $\operatorname{Ext}^1_{\mathcal{H}}(S, S) = 0$ and μ is the minimal left $\operatorname{add} S$-approximation we have that $\operatorname{Ext}^1_{\mathcal{H}}(Y, S) = 0$. Let Y_i be an indecomposable summand of Y. Note that $Y_i \in \operatorname{Fac} S$.

We now distinguish the following two cases.

First assume that $\operatorname{Hom}_{\mathcal{H}}(Y_i, S) \neq 0$. Then $Y_i \in \mathcal{F}$ by 2.3 and therefore there exists an injective map $Y_i \to S''$ for some $S'' \in \operatorname{add} S$. Since $\mathcal{H}$ is hereditary, there exists a short exact sequence $0 \to S' \to Y_i \oplus E \to S'' \to 0$ for some $E \in \mathcal{H}$ (compare [**HRS1**]). Since $\operatorname{Ext}^1_{\mathcal{H}}(S, S) = 0$ we infer that $Y_i \in \operatorname{add} S$.

Next we consider the case that $\operatorname{Hom}_{\mathcal{H}}(Y_i, S) = 0$. We have noted above that $\operatorname{Ext}^1_{\mathcal{H}}(Y_i, S) = 0$. We claim that then also $\operatorname{Hom}_{\mathcal{H}}(Y_i, \tau S) = \operatorname{Ext}^1_{\mathcal{H}}(Y_i, \tau S) = 0$.

To see this let $Z \in \mathcal{S}(\to X)$ be indecomposable. Again we proceed by induction on the maximal length $l(Z)$ of a sectional path from Z to X. If $l(Z) = 0$, then $Z \simeq X$. We consider the almost split sequence $0 \to \tau X \to E \to X \to 0$. Then $E \in \mathcal{S}(\to X)$. Applying $\operatorname{Hom}_{\mathcal{H}}(Y_i, -)$ to this sequence then shows $\operatorname{Hom}_{\mathcal{H}}(Y_i, \tau X) = 0 = \operatorname{Ext}^1_{\mathcal{H}}(Y_i, \tau X)$. Now let $Z \in \mathcal{S}(\to X)$ be indecomposable with $l(Z) > 0$. If $Z \to Z'$ is irreducible with $Z' \in \mathcal{S}(\to X)$ we have $l(Z') < l(Z)$, so by induction we have that $\operatorname{Hom}_{\mathcal{H}}(Y_i, \tau Z') = 0 = \operatorname{Ext}^1_{\mathcal{H}}(Y_i, \tau Z')$. Then consider the almost split sequence $0 \to \tau Z \to F \to Z \to 0$. Then $F = (\tau Z')^r \oplus F'$ with $F' \in \operatorname{add} S$. Applying $\operatorname{Hom}_{\mathcal{H}}(Y_i, -)$ to this sequence then shows $\operatorname{Hom}_{\mathcal{H}}(Y_i, \tau Z) = 0 = \operatorname{Ext}^1_{\mathcal{H}}(Y_i, \tau Z)$.

Therefore $0 = \operatorname{Ext}^1_{\mathcal{H}}(Y_i, \tau S) \simeq D\operatorname{Hom}_{\mathcal{H}}(\tau S, \tau Y_i) \simeq D\operatorname{Hom}_{\mathcal{H}}(S, Y_i)$, contradicting the fact that $Y_i \in \operatorname{Fac} S$.

LEMMA 2.5. *Using the notation above and letting $A \in \mathcal{F}$ and $B \in \mathcal{H}$ be indecomposable with $\operatorname{Hom}_{\mathcal{H}}(A, B) \neq 0$, then $B \in \operatorname{Sub} S \cup \operatorname{Fac} S$. Dually we have for each $B \in \operatorname{Fac} S$ and each indecomposable $A \in \mathcal{H}$ with $\operatorname{Hom}_{\mathcal{H}}(A, B) \neq 0$, that $A \in \operatorname{Sub} S \cup \operatorname{Fac} S$.*

Proof: Let $0 \neq f : A \to B$. We first show that the assertion holds if f is either mono or epi.

If f is epi we consider the exact sequence $0 \to K \to A \xrightarrow{f} B \to 0$. Since $A \in \mathcal{F} = \operatorname{Sub} S$ there exists by 2.4 a short exact sequence $0 \to A \to S' \to S'' \to 0$ with $S', S'' \in \operatorname{add} S$. This yields the following commutative diagram:

$$\begin{array}{ccccccccc}
 & & & & & & 0 & & \\
 & & & & & & \downarrow & & \\
 & & 0 & & & & B & & \\
 & & \downarrow & & & & \downarrow & & \\
0 & \longrightarrow & K & \longrightarrow & S' & \longrightarrow & Z & \longrightarrow & 0 \\
 & & \downarrow & & \| & & \downarrow & & \\
0 & \longrightarrow & A & \longrightarrow & S' & \longrightarrow & S'' & \longrightarrow & 0 \\
 & & \downarrow & & & & \downarrow & & \\
 & & B & & & & 0 & & \\
 & & \downarrow & & & & & & \\
 & & 0 & & & & & &
\end{array}$$

If $0 \neq \mathrm{Ext}^1_{\mathcal{H}}(S,B) \simeq D\mathrm{Hom}_{\mathcal{H}}(B,\tau S)$ we have $B \in \mathcal{F}$ by 2.3. If $\mathrm{Ext}^1_{\mathcal{H}}(S,B) = 0$, then $Z \simeq B \oplus S''$, so $B \in \mathrm{Fac}\, S$.

If f is mono we consider the exact sequence $(*)$ $0 \to A \to B \to Q \to 0$. As above we have that $B \in \mathrm{Fac}\, S$ in case $\mathrm{Hom}_{\mathcal{H}}(S,B) \neq 0$ and that $B \in \mathrm{Sub}\, S$ in case $\mathrm{Ext}^1_{\mathcal{H}}(S,B) \neq 0$. So we may assume that $\mathrm{Hom}_{\mathcal{H}}(S,B) = 0 = \mathrm{Ext}^1_{\mathcal{H}}(S,B)$. Applying $\mathrm{Hom}_{\mathcal{H}}(S,-)$ to the sequence $(*)$ then shows that $\mathrm{Hom}_{\mathcal{H}}(S,Q) \simeq \mathrm{Ext}^1_{\mathcal{H}}(S,A)$. If $\mathrm{Ext}^1_{\mathcal{H}}(S,A) = 0$, then $A \in \mathrm{add}\, S$ by 2.4, so that $\mathrm{Hom}_{\mathcal{H}}(S,B) \neq 0$. Hence we can assume that $\mathrm{Ext}^1_{\mathcal{H}}(S,A) \neq 0$, and so $\mathrm{Hom}_{\mathcal{H}}(S,Q) \neq 0$. Write $Q = Q' \oplus Q''$, where $\mathrm{Hom}_{\mathcal{H}}(S,Q'') = 0$ and Q'' is maximal with this property. Then Q' is nonzero and contained in $\mathrm{Fac}\, S$. Consider the following commutative diagram where $\mu : Q' \to Q$ is the canonical split inclusion and the first horizontal sequence is induced from the second one.

$$\begin{array}{cccccccccc}
 & & & & & 0 & & 0 & & \\
 & & & & & \downarrow & & \downarrow & & \\
 & 0 & \longrightarrow & A & \longrightarrow & B' & \longrightarrow & Q' & \longrightarrow & 0 \\
 & & & \| & & \downarrow & & \downarrow{\scriptstyle \mu} & & \\
(*) & 0 & \longrightarrow & A & \longrightarrow & B & \longrightarrow & Q & \longrightarrow & 0 \\
 & & & & & \downarrow & & \downarrow & & \\
 & & & & & Q'' & = & Q'' & & \\
 & & & & & \downarrow & & \downarrow & & \\
 & & & & & 0 & & 0 & &
\end{array}$$

Since $Q' \in \mathrm{Fac}\, S$, there exists a short exact sequence $0 \to S'' \to S' \xrightarrow{\pi} Q' \to 0$ with $S', S'' \in \mathrm{add}\, S$ by 2.4. This yields the following commutative diagram where the middle horizontal exact sequence is the pullback along π.

$$\begin{array}{ccccccccc}
 & & & & 0 & & 0 & & \\
 & & & & \downarrow & & \downarrow & & \\
 & & & & S'' & = & S'' & & \\
 & & & & \downarrow & & \downarrow & & \\
0 & \longrightarrow & A & \longrightarrow & L & \longrightarrow & S' & \longrightarrow & 0 \\
 & & \| & & \downarrow & & \downarrow{\scriptstyle \pi} & & \\
0 & \longrightarrow & A & \longrightarrow & B' & \longrightarrow & Q' & \longrightarrow & 0 \\
 & & & & \downarrow & & \downarrow & & \\
 & & & & 0 & & 0 & &
\end{array}$$

Now $\mathrm{Hom}_{\mathcal{H}}(S, B) = 0$ implies $\mathrm{Hom}_{\mathcal{H}}(S, B') = 0$. In particular, the pullback does not split. Also we see that $L \in \mathrm{Sub}\, S$. Since $\mathrm{Hom}_{\mathcal{H}}(S, B') = 0$ and $L \in \mathcal{F} = \mathrm{Sub}\, S$ we get by the first part of the proof, applied to the epimorphism $L \to B'$, that $\mathrm{Ext}^1_{\mathcal{H}}(S, B') \neq 0$. Applying $\mathrm{Hom}_{\mathcal{H}}(S, -)$ to the sequence $0 \to B' \to B \to Q'' \to 0$ from the diagram $(*)$ yields an exact sequence

$$\mathrm{Hom}_{\mathcal{H}}(S, Q'') \to \mathrm{Ext}^1_{\mathcal{H}}(S, B') \to \mathrm{Ext}^1_{\mathcal{H}}(S, B)$$

By assumption, the outer terms vanish, hence so does $\mathrm{Ext}^1_{\mathcal{H}}(S, B')$, a contradiction.

Now, let $0 \neq f : A \to B$ be an arbitrary map with $A \in \mathcal{F} = \mathrm{Sub}\, S$. Consider the canonical factorization $A \xrightarrow{\pi} \mathrm{im}\, f \xrightarrow{\mu} B$. Let Z be an indecomposable summand of $\mathrm{im}\, f$. Since there is an epimorphism $A \to Z$ we have by the previous considerations that $Z \in \mathcal{F} \cup \mathrm{Fac}\, S$. If $Z \in \mathrm{Fac}\, S$ and since B is a successor of Z we infer that $B \in \mathrm{Fac}\, S$. If $Z \in \mathcal{F}$ we have by the first part of the proof, applied to the monomorphism $Z \to B$, that $B \in \mathcal{F} \cup \mathrm{Fac}\, S$.

THEOREM 2.6. *Let $\mathcal{H}$ be a connected hereditary abelian category with tilting object T. Assume that $\mathcal{H}$ contains an indecomposable directing object X. Then $\mathcal{H}$ is derived equivalent to $\mathrm{mod}\, H$ for a hereditary artin algebra H.*

Proof: We consider $S = \bigoplus_{Y \in (\mathcal{S} \to X)} Y$. We have seen that $\mathrm{Ext}^1_{\mathcal{H}}(S, S) = 0$.

We first show that S is a tilting object in $\mathcal{H}$. By 2.5 we have that every successor of $\mathrm{Sub}\, S$ belongs to $\mathrm{Sub}\, S \cup \mathrm{Fac}\, S$ and that every predecessor of $\mathrm{Fac}\, S$ also belongs to $\mathrm{Sub}\, S \cup \mathrm{Fac}\, S$. In particular we see that $\mathrm{Sub}\, S \cup \mathrm{Fac}\, S$ is closed under predecessors and successors. Since $\mathcal{H}$ is connected, we infer that each indecomposable object in $\mathcal{H}$ belongs to $\mathrm{Sub}\, S \cup \mathrm{Fac}\, S$. We claim that $\mathrm{Fac}\, S = \mathcal{E}(S)$. Since $\mathcal{H}$ is hereditary we always have that $\mathrm{Fac}\, S \subset \mathcal{E}(S)$. Conversely, let $A \in \mathcal{E}(S)$ be indecomposable. We may assume that $A \in \mathrm{Sub}\, S$. By 2.4 there exists a short exact sequence

$$0 \to A \to S' \to S'' \to 0$$

with $S', S'' \in \mathrm{add}\, S$. By assumption this sequence splits, hence $A \in \mathrm{add}\, S \subset \mathrm{Fac}\, S$. So S is a tilting object.

It remains to show that $H = \mathrm{End}_{\mathcal{H}} S$ is a hereditary artin algebra. By definition, H is a quasitilted algebra. So $\mathrm{gl.dim}\, H \leq 2$. Thus it is enough to show that $\mathrm{inj.dim}_H P \leq 1$ for each indecomposable projective H-module. Let P be an indecomposable projective H-module. Then $\text{inj.dim}_H P \leq 1$ if and only if $\text{Hom}_H(\tau_H^- P, {}_H H) = 0$ (for example see [**Ri**]). Let S_i be an indecomposable direct summand of S such that $P \simeq \text{Hom}_{\mathcal{H}}(S, S_i)$. By 2.3 we infer that $\tau_{\mathcal{H}}^- S_i \in \text{Fac}\, S$. In particular, we have that $\tau_H^- P = \text{Hom}_{\mathcal{H}}(S, \tau_{\mathcal{H}}^- S_i)$. Since $0 = \text{Ext}^1_{\mathcal{H}}(S,S) \simeq D\text{Hom}_{\mathcal{H}}(\tau_{\mathcal{H}}^- S, S)$, we see that $\text{Hom}_{\mathcal{H}}(\tau_{\mathcal{H}}^- S_i, S) = 0$. But then by tilting theory (see [**HRS1**]) we have:

$$0 = \text{Hom}_{\mathcal{H}}(\tau_{\mathcal{H}}^- S_i, S) = \text{Hom}_H(\text{Hom}_{\mathcal{H}}(S, \tau_{\mathcal{H}}^- S_i), \text{Hom}_{\mathcal{H}}(S,S)) = \text{Hom}_H(\tau_H^- P, {}_H H).$$

So $\text{inj.dim}_H P \leq 1$, or equivalently H is a hereditary artin algebra. So $\mathcal{H}$ is derived equivalent to mod H, which finishes the proof.

References

[GL] W. Geigle, H. Lenzing, *A class of weighted projective curves arising in representation theory of finite dimensional algebras*, In: Singularies, representations of algebras, and vector bundles, Springer Lecture Notes 1273 (1987) 265-297.

[H] D. Happel, *Quasitilted algebras*, Proceedings ICRA VIII, to appear.

[HRe1] D. Happel, I. Reiten, *An introduction to quasitilted algebras*, An. St. Univ. Ovidius Constantza, Vol. 4 (1996), 137-149.

[HRe2] D. Happel, I. Reiten, *On hereditary categories with tilting object*, preprint.

[HRi1] D. Happel, C. M. Ringel, *Tilted algebras*, Trans. Amer. Math. Soc., 274 (1982), 399-443.

[HRi2] D. Happel, C. M. Ringel, *Directing projectives*, Archiv der Mathematik 60 (1993), 237-246.

[HRS1] D. Happel, I. Reiten, S. O. Smalø, *Tilting in abelian categories and quasitilted algebras*, Mem. Amer. Math. Soc. 575 (1996).

[HRS2] D. Happel, I. Reiten, S. O. Smalø, *Piecewise hereditary algebras*, Arch. Math. 66 (1996), 182-186.

[LS] H. Lenzing, A. Skowronski, *Quasitilted algebras of canonical type*, Coll. Math. 71 (1996), 161-181.

[Li] S. Liu, *Semi-stable components of an Auslander-Reiten quiver*, J. London Math. Soc. (2) 47 (1993), 405-416.

[Rc] J. Rickard, *A Morita theory for derived categories*, J. London Math. Soc. 39 (1989), 436-456.

[Ri] C. M. Ringel, *Tame algebras and integral quadratic forms*, Springer Lecture Notes 1099 (1984).

[Sk] A. Skowronski, *Regular Auslander-Reiten components containing directing modules*, Proc. Amer. Math. Soc. 120 (1994), 19-26.

[Zh] Y. Zhang, *The structure of stable components*, Canad. J. Math. 43 (1991), 652-672.

Fakultät für Mathematik, Technische Universität Chemnitz, D-09107 Chemnitz, Germany

E-mail address: `happel@mathematik.tu-chemnitz.de`

Institutt for matematiske fag, NTNU, Lade, N 7034 Trondheim, Norway

E-mail address: `idunr@math.ntnu.no`

Contemporary Mathematics
Volume **229**, 1998

On subcategories associated with tilting modules

Dieter Happel and Luise Unger

ABSTRACT. Let Λ be an Artin algebra. With a tilting module T there are associated full subcategories $\mathcal{E}^i = \{X \mid \operatorname{Ext}^j(T,X) = 0 \text{ for } j \neq i\}$ of the category of finitely generated Λ-modules for $0 \leq i \leq \operatorname{pd}_\Lambda T$. We will study here homological finiteness conditions of these subcategories.

Let Λ be an Artin algebra over a commutative Artin ring R and let $\operatorname{mod}\Lambda$ be the category of finitely generated left Λ-modules. By D we denote the standard duality on $\operatorname{mod}\Lambda$. For $X \in \operatorname{mod}\Lambda$ we denote by $\operatorname{pd}_\Lambda X$ the *projective dimension* of X. We will say that a module X is *exceptional* if $\operatorname{pd}_\Lambda X < \infty$ and $\operatorname{Ext}^i_\Lambda(X,X) = 0$ for $i > 0$. An exceptional module T is called a *tilting module* if there exists an exact sequence

$$0 \to {}_\Lambda\Lambda \to T^0 \to T^1 \to \cdots \to T^m \to 0$$

with $T^j \in \operatorname{add} T$ for all j, where $\operatorname{add} T$ is the full subcategory of $\operatorname{mod}\Lambda$ whose objects are direct sums of direct summands of T.

With a tilting module T of projective dimension d there are associated full subcategories $\mathcal{E}^i = \{X \mid \operatorname{Ext}^j(T,X) = 0 \text{ for } j \neq i\}$ for $0 \leq i \leq d$.

The interest in these subcategories is twofold. On one hand in these subcategories lie precisely those Λ-modules which are mapped to stalk complexes under the canonical derived equivalence $D^b(\Lambda) \to D^b(\operatorname{End} T)$, where $D^b(\Lambda)$ denotes the bounded derived category of $\operatorname{mod}\Lambda$. For details and some elementary consequences we refer to section 2. On the other hand some of these subcategories correspond in special situations to subcategories which are studied in applications. One prominent example is shown in the work of Ringel [**R2**] in case Λ is a quasihereditary algebra. This and some other examples will be given in section 3. Our main results will deal with the connection to spherical modules in the sense of Auslander-Bridger [**AB**]. These will be presented in section 4.

We denote the composition of morphisms $f : X \to Y$ and $g : Y \to Z$ in a given category $\mathcal{K}$ by fg.

The notation and terminology introduced here will be fixed throughout this article. For unexplained representation-theoretic terminology we refer to [**R1**] or [**ARS**].

1991 *Mathematics Subject Classification.* 16G10, 16G70, 16E10.

Part of the material contained in this article was obtained during a visit of the first author at the University of Paderborn. He would like to thank the second author for the invitation.

1. Preliminaries

In this section we will briefly recall the basic definitions and results we will use in the main part of this article.

Let $\mathcal{C}$ be a full subcategory of $\operatorname{mod}\Lambda$. It is always assumed to be closed under direct sums, direct summands and isomorphisms. An object $C \in \mathcal{C}$ is called a *cocover* of $\mathcal{C}$ if for all $X \in \mathcal{C}$ there exists an injective map $\mu : X \to C'$ for some $C' \in \operatorname{add} C$. Recall that $\operatorname{add} C$ denotes the full subcategory of $\operatorname{mod}\Lambda$ consisting of those modules which are direct sums of direct summands of C. A cocover C is called a *minimal cocover* in case no proper direct summand of C is a cocover. It is called a *strong* cocover if there exists an injective map μ as above with $\operatorname{cok}\mu \in \mathcal{C}$. If $\mathcal{C}$ is closed under extensions it is shown in [**AS1**] that a minimal cocover C is $\mathcal{C}$*-injective* in $\mathcal{C}$, so satisfies $\operatorname{Ext}^1_\Lambda(X, C) = 0$ for all $X \in \mathcal{C}$.

The subcategory $\mathcal{C}$ is called *contravariantly finite* in $\operatorname{mod}\Lambda$, if every $X \in \operatorname{mod}\Lambda$ has a *right* $\mathcal{C}$*-approximation*, i.e. there is a morphism $F_X \to X$ with $F_X \in \mathcal{C}$ such that the induced morphism $\operatorname{Hom}_\Lambda(C, F_X) \to \operatorname{Hom}_\Lambda(C, X)$ is surjective for all $C \in \mathcal{C}$. If X admits a right $\mathcal{C}$-approximation then it admits also a *minimal* right $\mathcal{C}$-approximation. This is a right $\mathcal{C}$-approximation which in addition is also *right minimal* in the sense that its restriction to any nonzero summand is nonzero.

The subcategory $\mathcal{C}$ is called *resolving* if $\mathcal{C}$ is closed under extensions, kernels of surjective maps and contains ${}_\Lambda\Lambda$. Note that for a contravariantly finite subcategory which is resolving every right approximation is surjective [**AR2**]. Indeed, the projective cover has to factor over the right approximation.

If $\mathcal{C}$ is a resolving subcategory and $C \in \mathcal{C}$ is $\mathcal{C}$-injective, then $\operatorname{Ext}^i_\Lambda(X, C) = 0$ for all $X \in \mathcal{C}$ and all $i > 0$.

We refrain from defining also the dual notions such as covariantly finite and coresolving. Also we will not explicitely write down the dual assertions of our main results but will use them freely.

A subcategory $\mathcal{C}$ which is both contravariantly and covariantly finite is called *functorially finite.*

For the convenience of the reader we have collected here some of the fundamental identities from Auslander-Reiten theory which will be used in remaining sections. Proofs of those can be found in [**ARS**] or [**AR1**].

Given $X, Y \in \operatorname{mod}\Lambda$ we denote by $\underline{\operatorname{Hom}}(X, Y)$ (resp. $\overline{\operatorname{Hom}}(X, Y)$) the factor space of $\operatorname{Hom}(X, Y)$ by the subspace of those maps factoring through a projective (resp. injective) $\Lambda-$module. The corresponding categories are denoted by $\underline{\operatorname{mod}}\ \Lambda$ and $\overline{\operatorname{mod}}\ \Lambda$. By τ (resp. τ^-) we denote the Auslander-Reiten translations. Then τ is an equivalence from $\underline{\operatorname{mod}}\ \Lambda \to \overline{\operatorname{mod}}\ \Lambda$ with inverse given by τ^-. The following identities are very useful for $X, Y \in \operatorname{mod}\Lambda$.

$$\operatorname{Ext}^1_\Lambda(X, Y) = D\,\overline{\operatorname{Hom}}(Y, \tau X) = D\,\underline{\operatorname{Hom}}(\tau^- Y, X).$$

If $\operatorname{pd}_\Lambda X = 1$ we have the following additional information. First we have that $\operatorname{Hom}(D(\Lambda_\Lambda), \tau X) = 0$ and that $\operatorname{Ext}^1_\Lambda(X, Y) = \operatorname{Hom}_\Lambda(Y, \tau X)$. Secondly it holds that $\tau X = D\operatorname{Ext}^1_\Lambda(X, {}_\Lambda\Lambda)$.

For a $\Lambda-$module X we denote by $\Omega(X)$ the syzygy module of X, so the kernel of a projective cover of X, and by $\Omega^-(X)$ the cosyzygy module of X.

The following easy fact and its dual will be used frequently in the later sections.

LEMMA 1.1. *Let* $0 \to X \to I \to \Omega^-(X) \to 0$ *be exact with* I *an injective* $\Lambda-$*module. If* $Z \in \operatorname{mod}\Lambda$ *satisfies* $\operatorname{Ext}^1_\Lambda(Z,X) = 0$, *then* $\overline{\operatorname{Hom}}_\Lambda(Z,\Omega^-(X)) = 0$.

2. Elementary properties

We keep the basic notation from the previous sections. Here we will mainly recall one of the reasons studying the subcategories defined above and collect some of their main known properties. In addition we will also investigate the special case of tilting modules of projective dimension 2.

Let T be a tilting module with $\operatorname{pd}_\Lambda T = d$. We denote by Γ the endomorphism algebra $\operatorname{End}_\Lambda T$. With T there are associated the following subcategories of $\operatorname{mod}\Lambda$ and $\operatorname{mod}\Gamma$. For $0 \leq j \leq d$ define:

$$\mathcal{E}^j = \{X \mid \operatorname{Ext}^i_\Lambda(T,X) = 0 \text{ for } i \neq j\} \subset \operatorname{mod}\Lambda$$

$$\mathcal{T}_j = \{Y \mid \operatorname{Tor}^\Gamma_i(Y,T) = 0 \text{ for } i \neq j\} \subset \operatorname{mod}\Gamma$$

The subcategory $\mathcal{E}^0$ is also denoted by $T^\perp$.

The main result in tilting theory can be formulated as follows [**H**]. There exists a triangle equivalence $F : D^b(\Lambda) \to D^b(\Gamma)$ such that the restriction of F to $\mathcal{E}^j$ is given by $\operatorname{Ext}^j_\Lambda(T,\ )[-j]$, where $[\cdot]$ denotes the translation functor, and it induces an equivalence between $\mathcal{E}^j$ and $\mathcal{T}_j$.

PROPOSITION 2.1. *If* T *is a tilting module with* $\operatorname{pd}_\Lambda T = d$, *let* $0 \leq i,j \leq d$ *and* $X \in \mathcal{E}^i, Y \in \mathcal{E}^j$. *Then* $\operatorname{Ext}^t_\Lambda(X,Y) = 0$ *for* $t < j - i$.

Proof: Indeed, by the usual properties of triangle equivalences and derived categories, we have the following line of identifications. For this let $0 \leq i,j \leq d$ and $X \in \mathcal{E}^i, Y \in \mathcal{E}^j$. We set $X' = \operatorname{Ext}^i_\Lambda(T,X)$ and $Y' = \operatorname{Ext}^j_\Lambda(T,Y)$. Then

$$\begin{aligned}\operatorname{Ext}^t_\Lambda(X,Y) &= \operatorname{Hom}_{D^b(\Lambda)}(X,Y[t]) = \operatorname{Hom}_{D^b(\Gamma)}(F(X),F(Y)[t]) \\ &= \operatorname{Hom}_{D^b(\Gamma)}(X'[-i],Y'[t-j]) = \operatorname{Ext}^{t+i-j}_\Gamma(X',Y').\end{aligned}$$

Thus $\operatorname{Ext}^t_\Lambda(X,Y) = 0$ if $t + i - j < 0$, or equivalently $t < j - i$.

In the following we collect some easy or well-known properties of these subcategories.

PROPOSITION 2.2. *If* T *is a tilting module with* $\operatorname{pd}_\Lambda T = d$, *then the following assertions hold:*

(1) *For* $0 \leq i \leq d$ *the subcategory* $\mathcal{E}^i$ *is closed under extensions.*
(2) $\mathcal{E}^0 = T^\perp$ *is functorially finite and coresolving.*
(3) *The subcategory* $\mathcal{E}^d$ *is closed under kernels of epimorphisms.*

Proof: (1) and (3) follow easily from the long exact sequence in homology. (2) is contained in [**AR2**].

In the rest of this section we will deal with the special case of a tilting module T with $\operatorname{pd}_\Lambda T = d = 2$. The ideas behind the next lemma will again be used in section 4.

LEMMA 2.3. *If T is a tilting module with $pd_\Lambda T = 2$, then $\tau\Omega(T)$ is a cocover of $\mathcal{E}^2$.*

Proof: First we show that $\tau\Omega(T) \in \mathcal{E}^2$. Since $\operatorname{pd}_\Lambda T = 2$, we infer that $\operatorname{pd}_\Lambda \Omega(T) = 1$. The assertion now follows from identities quoted in section one. $D\operatorname{Ext}^1(T, \tau\Omega(T)) = \underline{\operatorname{Hom}}(\Omega(T), T)$. The last term vanishes, since $\operatorname{Ext}^1(T,T) = 0$. Thus $\operatorname{Ext}^1(T, \tau\Omega(T)) = 0$. Now $D\operatorname{Hom}(T, \tau\Omega(T)) = \operatorname{Ext}^1(\Omega(T), T) = \operatorname{Ext}^2(T,T)$. Thus $\operatorname{Hom}(T, \tau\Omega(T)) = 0$, or $\tau\Omega(T) \in \mathcal{E}^2$. Next we show that for $X \in \mathcal{E}^2$ we have that $\operatorname{pd}\tau^-\Omega^-(X) = 2$ and that $\tau^-\Omega^-(X) \in {}^\perp T$, where

$$ {}^\perp T = \{Y \in \operatorname{mod}\Lambda \mid \operatorname{Ext}^i(Y,T) = 0 \text{ for } i \neq 0\}. $$

For this let $0 \to X \to I^0 \to I^1 \to I^2$ be the start of a minimal injective resolution of X. Applying the functor $\nu^- = \operatorname{Hom}(D(\Lambda_\Lambda), -)$ to this and using the facts that $X \in \mathcal{E}^2$ and $D(\Lambda_\Lambda) \in \mathcal{E}^0$ yields an exact sequence

$$ (*) \quad 0 \to \nu^- I^o \to \nu^- I^1 \to \nu^- I^2 \to \tau^-\Omega^- X \to 0. $$

Note that this is a minimal projective resolution of $\tau^-\Omega^-(X)$. Moreover we deduce from this that $\Omega(\tau^-\Omega^-(X)) \simeq \tau^- X$. In particular we see that $\operatorname{pd}_\Lambda \tau^-\Omega^-(X) = 2$. We also infer that $\tau\Omega(\tau^-\Omega^-(X)) \simeq X$.

Now $D\operatorname{Ext}^1(\tau^-\Omega^-(X), T) = \overline{\operatorname{Hom}}(T, \Omega^-(X)) = 0$, since $\operatorname{Ext}^1(T, X) = 0$. Moreover we see that $D\operatorname{Ext}^2(\tau^-\Omega^-(X), T) = D\operatorname{Ext}^1(\tau^- X, T) = \overline{\operatorname{Hom}}(T, X) = 0$, since $\operatorname{Hom}(T, X) = 0$.

We denote by $\mathcal{P}^{<\infty}(\Lambda)$ the full subcategory of $\operatorname{mod}\Lambda$ consisting of the modules of finite projective dimension. The previous considerations then show that $\tau^-\Omega^-(X) \in {}^\perp T \cap \mathcal{P}^{<\infty}(\Lambda)$ for $X \in \mathcal{E}^2$. By [**HU2**] we infer that ${}^\perp T \cap \mathcal{P}^{<\infty}(\Lambda)$ is contravariantly finite, in particular contains a strong cocover which can be chosen as T. Thus there exists an exact sequence

$$ (**) \quad 0 \to \tau^-\Omega^-(X) \to T' \to Z \to 0 $$

with $T' \in \operatorname{add} T$. Following the results in section 3 of [**AR2**] we infer that $\operatorname{pd}_\Lambda Z = 2$. Applying the functor $D\operatorname{Ext}^2(-, {}_\Lambda\Lambda)$ to $(**)$ gives an injective map

$$ D\operatorname{Ext}^2(\tau^-\Omega^-(X), {}_\Lambda\Lambda) \to D\operatorname{Ext}^2(T', {}_\Lambda\Lambda). $$

By the previous considerations and the formulas in section one the first term is readily identified with X and the second term is in $\operatorname{add}\tau\Omega(T)$, hence $\tau\Omega(T)$ is a cocover for $\mathcal{E}^2$.

THEOREM 2.4. *If T is a tilting module with $pd_\Lambda T = 2$, then the subcategory $\mathcal{E}^2$ is covariantly finite.*

Proof: We consider the following two subcategories of $\operatorname{mod}\Lambda$ associated with the tilting module T.

$$ \mathcal{T} = \{X \mid \operatorname{Hom}_\Lambda(X, \mathcal{E}^2) = 0\} $$

$$ \mathcal{T}' = \{X \mid \operatorname{Ext}^2_\Lambda(T, X) = 0\}. $$

Both categories are closed under extensions and epimorphic images, hence can be considered as the torsion classes of two torsion pairs in $\operatorname{mod}\Lambda$. We claim that $\mathcal{T} = \mathcal{T}'$.

For this we first observe that for $X \in \operatorname{mod}\Lambda$ the following holds.

$$\begin{aligned} \operatorname{Ext}^2_\Lambda(T, X) &= \operatorname{Ext}^1_\Lambda(\Omega(T), X) = D\,\overline{\operatorname{Hom}}_\Lambda(X, \tau\Omega(T)) \\ &= D\operatorname{Hom}_\Lambda(X, \tau\Omega(T)) \end{aligned}$$

where the last equality holds, since $\operatorname{Hom}(D(\Lambda_\Lambda), \tau\Omega(T)) = 0$.

So we see that for $X \in \mathcal{T}$ we have that $X \in \mathcal{T}'$, and thus $\mathcal{T} \subset \mathcal{T}'$. Conversely let $X \notin \mathcal{T}$. Then there exists $Y \in \mathcal{E}^2$ such that $\operatorname{Hom}(X, Y) \neq 0$. By the previous Lemma we have that Y is cogenerated by $\tau\Omega(T)$. Therefore also $\operatorname{Hom}(X, \tau\Omega(T)) \neq 0$, hence $X \notin \mathcal{T}'$. Thus $\mathcal{T} = \mathcal{T}'$. We denote the torsion pair in $\operatorname{mod}\Lambda$ induced by $\mathcal{T}$ by $(\mathcal{T}, \mathcal{F})$.

Next we construct a left $\mathcal{E}^2$-approximation for X. If $X \in \mathcal{T}$ there is nothing to show. So we may assume that $\operatorname{Hom}(X, \mathcal{E}^2) \neq 0$. Construct the universal extension

$$0 \to X \xrightarrow{\ \mu\ } E \longrightarrow T^r \to 0$$

of X by T such that the connecting homomorphism $\operatorname{Hom}_\Lambda(T, T^r) \to \operatorname{Ext}^1_\Lambda(T, X)$ is surjective, and so $\operatorname{Ext}^1_\Lambda(T, E) = 0$. Moreover any map from $X \to Y$ with $Y \in \mathcal{E}^2$ factors through E. Next we consider the canonical exact sequence

$$(*)\quad 0 \to t(E) \xrightarrow{\ \alpha\ } E \xrightarrow{\ \beta\ } E/t(E) \to 0$$

with $t(E) \in \mathcal{T}$ and $E/t(E) \in \mathcal{F}$. Then any map from $E \to Y$ with $Y \in \mathcal{E}^2$ factors through β. So any map from $X \to Y$ with $Y \in \mathcal{E}^2$ factors through $\mu\beta$. So, if we show that $E/t(E) \in \mathcal{E}^2$, then $\mu\beta$ is an $\mathcal{E}^2$-approximation for X.

For this we apply the functor $\operatorname{Hom}_\Lambda(T, -)$ to the exact sequence $(*)$ and obtain an exact sequence

$$\cdots \to \operatorname{Ext}^1_\Lambda(T, E) \to \operatorname{Ext}^1_\Lambda(T, E/t(E)) \to \operatorname{Ext}^2_\Lambda(T, t(E)) \to \cdots.$$

By construction $\operatorname{Ext}^1_\Lambda(T, E) = 0$ and $\operatorname{Ext}^2_\Lambda(T, t(E)) = 0$, since $t(E) \in \mathcal{T} = \mathcal{T}'$. Thus $\operatorname{Ext}^1_\Lambda(T, E/t(E)) = 0$. Since $T \in \mathcal{T}$ and $E/t(E) \in \mathcal{F}$ we finally conclude that $E/t(E) \in \mathcal{E}^2$. This finishes the proof of the theorem.

3. Examples and special cases

In this section we will present several examples of classes of algebras and tilting modules. We will indicate the occurence of some of the subcategories introduced above in these special situations. In the final part we will show that in general these subcategories do not have the pleasant known properties as in the classical case which we will recall first.

3.1. The classical case. Here we recall some of the main properties of tilting modules of projective dimension one as established in [**HR**]. Let ${}_\Lambda T$ be a tilting module with $\mathrm{pd}_\Lambda T = 1$. Then the pair of subcategories $(\mathcal{E}^0, \mathcal{E}^1)$ is a torsion pair in $\mathrm{mod}\,\Lambda$. Note that in this case the subcategory $\mathcal{E}^0$ can be identified with the subcategory $\mathrm{Fac}\,T$ of modules generated by T. In this case both these subcategories are functorially finite, and hence have almost split sequences by a result in [**AS2**]. Since $(\mathcal{E}^0, \mathcal{E}^1)$ is a torsion pair in $\mathrm{mod}\,\Lambda$ any simple Λ-module S is either torsion or torsion-free. The following easy Lemma will show that this property will not hold in general. Note that there are examples of algebras of finite global dimension which violate the conclusion of the Lemma. We also point out that $T = D(\Lambda_\Lambda)$ is an example of a tilting module for an algebra Λ of finite global dimension and that in this case we have that $\mathrm{pd}_\Lambda T = \mathrm{gl.dim}\,\Lambda$. We denote by $K_0(\Lambda)$ the Grothendieck group of Λ. This is a free abelian group whose rank is denoted by $\mathrm{rk}\,K_0(\Lambda)$.

LEMMA 3.1. *Let T be a tilting module of projective dimension d. If $\mathcal{E}^0 \cup \cdots \cup \mathcal{E}^d$ contains all simple Λ-modules, then $d < \mathrm{rk}\,K_0(\Lambda)$.*

Proof: Let S be a simple Λ-module and let $P(S)$ denote the projective cover of S. Consider a minimal projective resolution of T

$$0 \to P^d \to \cdots \to P^1 \to P^0 \to T \to 0$$

If $S \in \mathcal{E}^i$ for some i, then $P(S)$ is a direct summand of P^i and $P(S)$ is not a direct summand of P^j for all $j \neq i$. Thus $d < \mathrm{rk}\,K_0(\Lambda)$.

3.2. Modules of finite projective dimension. Recall that we have denoted by $\mathcal{P}^{<\infty}(\Lambda)$ the full subcategory of $\mathrm{mod}\,\Lambda$ consisting of the modules of finite projective dimension. If $\mathcal{P}^{<\infty}(\Lambda)$ is contravariantly finite then it contains a tilting module T such that $\mathcal{P}^{<\infty}(\Lambda) = \{X \mid \mathrm{Ext}^i_\Lambda(X,T) = 0 \text{ for all } i \neq 0\} \cap \mathcal{P}^{<\infty}(\Lambda)$ (compare [**HU2**]). In particular we obtain a canonical tilting module T for a representation-finite algebra Λ.

3.3. Quasihereditary algebras. The following summarizes some of the results in [**R2**]. For details we refer to this article. Let Λ be a quasi-hereditary algebra. Let $\mathcal{F}(\Delta)$ (resp. $\mathcal{F}(\nabla)$) be the subcategories with Δ- (resp. ∇-) filtration. Then there exists a tilting module T with $\mathrm{add}\,T = \mathcal{F}(\Delta) \cap \mathcal{F}(\nabla)$. Moreover in this case $\mathcal{F}(\nabla) = \mathcal{E}^0$.

3.4. Specific examples. In two concrete examples we will compute the subcategories associated with a tilting module. We consider the path algebra H of the following quiver over an algebraically closed field k.

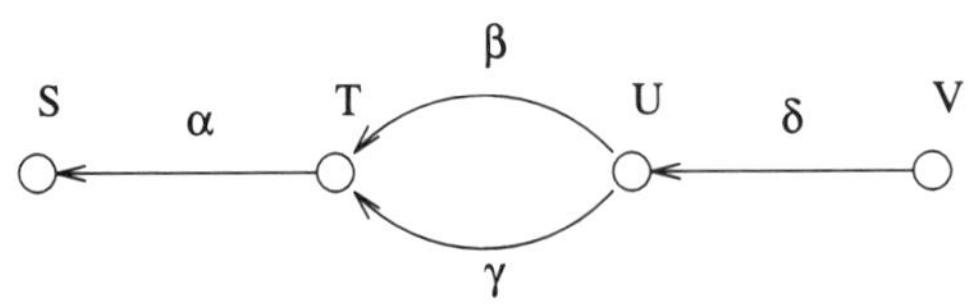

In the first example we choose the twosided ideal I_1 in H generated by $\delta\beta, \gamma\alpha$. Let $\Lambda = H/I_1$. Then it is easily seen that $\mathrm{gl.dim}\,\Lambda = 2$. So we may choose $D(\Lambda_\Lambda)$

as a tilting module of projective dimension 2. We claim that the subcategory $\mathcal{E}^1$ is neither covariantly nor contravariantly finite. Let us briefly recall the relevant structure of $\operatorname{mod}\Lambda$. There exists a unique preprojective component $\mathcal{P}$ and a unique preinjective component $\mathcal{Q}$ as well as an infinite family of pairwise orthogonal tubes $\mathcal{T}_i$ separating $\mathcal{P}$ from $\mathcal{Q}$. There are precisely two of these tubes which are not homogeneous. It is easy to see that all homogeneous tubes are contained in $\mathcal{E}^1$ and that no module from the preprojective component and no module from the preinjective component is contained in $\mathcal{E}^1$. But then it is straightforward to see that $\mathcal{E}^1$ is neither covariantly nor contravariantly finite.

In the second example we choose the twosided ideal I_2 in H generated by $\delta\beta, \gamma\alpha, \beta\alpha$. Let $\Lambda = H/I_2$. Then it is easily seen that $\operatorname{gl.dim}\Lambda = 3$. So we may choose $D(\Lambda_\Lambda)$ as a tilting module of projective dimension 3. Trivially we have that $\mathcal{E}^0 = \operatorname{add} D(\Lambda_\Lambda)$. Next we show that $\mathcal{E}^i = 0$ for $1 \leq i \leq 3$. For this let $X \in \mathcal{E}^i$ for $i > 0$ be indecomposable and not isomorphic to the simple projective module S corresponding to the vertex labelled with S. It is easy to see that then $\operatorname{Hom}(D(\Lambda_\Lambda), X) \neq 0$. For this note that $P(T) = I(S)$, where $I(S)$ is the injective envelope of the simple S. Thus we have to consider $X = S$. By computing a minimal injective resolution of S it is easy to see that $\operatorname{Ext}^2_\Lambda(I(U), S) \neq 0$ and that $\operatorname{Ext}^3_\Lambda(I(V), S) \neq 0$. This shows the assertion.

4. Spherical modules

Spherical modules have been introduced by Auslander and Bridger in [**AB**]. For each positive integer d we consider the subcategory $\mathcal{S}^d$ of spherical modules where

$$\mathcal{S}^d = \{X \in \operatorname{mod}\Lambda \mid \operatorname{Ext}^i_\Lambda(X, {}_\Lambda\Lambda) = 0 \text{ for } 0 < i < d, \operatorname{pd}_\Lambda X \leq d\}$$

Note that a spherical module is either projective or has projective dimension d. In this way we slightly deviate from the original definition in [**AB**], where a module X is called d-spherical if $\operatorname{Ext}^i_\Lambda(X, {}_\Lambda\Lambda) = 0$ for $0 < i < d$ and $\operatorname{pd}_\Lambda X = d$. This has the following technical advantage. In this case any minimal spherical approximation, if it exists, will be automatically surjective. On the other hand, we will study spherical modules in the stable category, and then there is no difference between these two definitions.

For a given tilting module T with $\operatorname{pd}_\Lambda T = d$ we consider

$$^{\perp}T \cap \mathcal{P}^{<\infty}(\Lambda) = \{X \in \operatorname{mod}\Lambda \mid \operatorname{Ext}^i_\Lambda(X, T) = 0 \text{ for } i > 0, \operatorname{pd}_\Lambda X < \infty\}$$

This is a contravariantly finite resolving subcategory of $\operatorname{mod}\Lambda$.

Finally we denote by

$$\mathcal{C}^d = ({}^{\perp}T \cap \mathcal{P}^{<\infty}(\Lambda)) \cap \mathcal{S}^d.$$

Note that in the situation of 3.2. we have that $\mathcal{C}^d = \mathcal{S}^d$.

In this section we will be mainly interested in the subcategory $\mathcal{C}^d$ and its relation to the subcategory $\mathcal{E}^d$. For this we first recall a result of Auslander and Reiten [**AR1**].

For each integer d we consider functors F, G where $F : \overline{\text{mod}}\ \Lambda \to \underline{\text{mod}}\ \Lambda$ with $F(X) = \tau^{-}\Omega^{-(d-1)}(X)$, and $G : \underline{\text{mod}}\ \Lambda \to \overline{\text{mod}}\ \Lambda$ with $G(Y) = \tau\Omega^{d-1}(Y)$. Then F, G is a pair of adjoint functors. We denote the adjunction isomorphism by $\eta_{X,Y}$, so

$$\eta_{X,Y} : \underline{\text{Hom}}(F(X), Y) \simeq \overline{\text{Hom}}(X, G(Y)).$$

Note that the subcategory $\mathcal{E}^d \subset \overline{\text{mod}}\ \Lambda$ by 2.1. We denote by $\underline{\mathcal{C}}^d$ the full subcategory of $\underline{\text{mod}}\ \Lambda$ with objects the modules in $\mathcal{C}^d$.

THEOREM 4.1. *Let T be a tilting module with $\text{pd}_\Lambda T = d$. The restrictions of F and G induce inverse equivalences between $\mathcal{E}^d$ and $\underline{\mathcal{C}}^d$.*

Proof: Given $X \in \mathcal{E}^d$ and $Y \in \mathcal{C}^d$, we first show that $F(X) \in \mathcal{C}^d$ and that $G(Y) \in \mathcal{E}^d$. For this we need some additional notation. We denote by $\nu^- = \text{Hom}_\Lambda(D(\Lambda_\Lambda), -)$ and by $\nu = D\text{Hom}_\Lambda(-, {}_\Lambda\Lambda)$ the Nakayama functors. Recall that this is a pair of inverse equivalences between the full subcategories of injective Λ-modules and projective Λ-modules.

Let $X \in \mathcal{E}^d$ and

$$(1)\ 0 \to X \xrightarrow{\mu} I^0 \xrightarrow{e^0} I^1 \to \cdots \to I^{d-1} \xrightarrow{e^{d-1}} I^d \to \cdots$$

be a minimal injective resolution of X.

Applying ν^- and using the fact that $X \in \mathcal{E}^d$ we obtain a long exact sequence where $\nu^- I^j$ is projective for $0 \leq j \leq d$.

$$(2)\ 0 \to \nu^- I^0 \xrightarrow{\nu^-(e^0)} \nu^- I^1 \to \cdots \to \nu^- I^{d-1} \xrightarrow{\nu^-(e^{d-1})} \nu^- I^d \xrightarrow{\pi} F(X) \to 0$$

It is easy to see that this in fact is a minimal projective resolution of $F(X)$. In particular we see that $\text{pd}_\Lambda F(X) = d$. Also we see that $\Omega^t(F(X)) = \tau^- \Omega^{-(d-1-t)}(X)$ for $0 \leq t \leq d-1$. Applying ν to (2) gives back (1), which is exact, hence $\text{Ext}^i_\Lambda(F(X), {}_\Lambda\Lambda) = 0$ for $0 < i < d$. Let $j > 0$. Then

$$\begin{aligned} \text{Ext}^j_\Lambda(F(X), T) &= \text{Ext}^1_\Lambda(\Omega^{j-1}(F(X)), T) \\ &= \text{Ext}^1_\Lambda(\tau^- \Omega^{-(d-j)}(X), T) = D\,\overline{\text{Hom}}_\Lambda(T, \Omega^{-(d-j)}(X)). \end{aligned}$$

If $d = j$ the last term is zero. So we may assume that $0 < j < d$. In this case $0 = \text{Ext}^{d-j}_\Lambda(T, X) = \text{Ext}^1(T, \Omega^{-(d-j)+1}(X))$, hence $D\,\overline{\text{Hom}}_\Lambda(T, \Omega^{-(d-j)}(X)) = 0$. Thus $F(X) \in \mathcal{C}^d$.

Next let $Y \in \mathcal{C}^d$ and let

$$(3)\ 0 \to P_d \to P_{d-1} \to \cdots P_1 \to P_0 \to Y \to 0$$

be a minimal projective resolution of Y. Applying ν to this gives the start of a minimal injective resolution of $G(Y)$

$$(4)\ 0 \to G(Y) \to \nu P_d \to \nu P_{d-1} \to \cdots \to \nu P_1 \to \nu P_0$$

Note that $\Omega^{-t}(G(Y)) = \tau\Omega^{d-1-t}(Y)$ for $0 < t < d$. Applying ν^- to (4) gives back (3), in particular we see that $\mathrm{Hom}_\Lambda(D(\Lambda_\Lambda), G(Y)) = 0$, and also we see that $\mathrm{Hom}_\Lambda(T, G(Y)) = \overline{\mathrm{Hom}}_\Lambda(T, G(Y))$. But

$$\overline{\mathrm{Hom}}_\Lambda(T, G(Y)) = D\mathrm{Ext}^1_\Lambda(\Omega^{(d-1)}(Y), T) = D\mathrm{Ext}^d_\Lambda(Y, T) = 0,$$

since $Y \in \mathcal{C}^d$. So let $0 < j < d$. Then

$$\begin{aligned}\mathrm{Ext}^j_\Lambda(T, G(Y)) &= \mathrm{Ext}^1_\Lambda(T, \Omega^{-(j-1)}(G(Y))) \\ &= \mathrm{Ext}^1_\Lambda(T, \tau\Omega^{d-1-(j-1)}(Y)) \\ &= D\,\underline{\mathrm{Hom}}(\Omega^{d-j}(Y), T).\end{aligned}$$

Now $0 = \mathrm{Ext}^{d-j}_\Lambda(Y, T) = \mathrm{Ext}^1_\Lambda(\Omega^{d-j-1}(Y), T)$ shows that $\underline{\mathrm{Hom}}(\Omega^{d-j}(Y), T) = 0$, hence $G(Y) \in \mathcal{E}^d$.

Note that the arguments given above also show that both F and G are dense. It is straightforward to see that both functors are full. We will show that F is faithful. The same property for G can be proved in a similar way.

For this let $f : X \to X'$ be a map in $\mathcal{E}^d$ with $F(f) = 0$ in $\underline{\mathcal{C}}^d$. Let (I^i, e^i) and (J^i, h^i) be minimal injective resolutions of X and X'. Let $\mu : X \to I^0$ and $\mu' : X' \to J^0$ be the injective envelopes. We denote by $f^i : I^i \to J^i$ a lifting of f. Now $F(f) : F(X) = \mathrm{cok}\,\nu^-(e^d) \to F(X') = \mathrm{cok}\,\nu^-(h^d)$ is the map induced by $\nu^-(f^i)$. We denote by $\pi : \nu^- I^d \to F(X)$ and $\pi' : \nu^- J^d \to F(X')$ the quotient maps. Since $F(f) = 0$ in $\underline{\mathcal{C}}^d$ we obtain a map $g^{d+1} : F(X) \to \nu^- J^d$ with $F(f) = g^{d+1}\pi'$. Using the usual arguments for constructing homotopies we obtain inductively maps $g^i : \nu^- I^i \to \nu^- J^{i-1}$ with $\nu^-(f^i) = g^i\nu^-(h^{i-1}) + \nu^-(e^i)g^{i+1}$ for all $1 \le i \le d$. It is easy to see that also $\nu^-(e^0)g^1 = \nu^-(f^0)$. Therefore we see that $e^0\nu^-(g^1) = f^0$. On the other hand we have that $\mu f^0 = f\mu'$, hence $f\mu' = \mu e^0\nu^-(g^1) = 0$, and therefore $f = 0$, since μ' is injective. In particular F is faithful.

This finishes the proof of the theorem.

COROLLARY 4.2. *If $\mathcal{C}^d$ is contravariantly finite, then the subcategory $\mathcal{E}^d$ is contravariantly finite.*

Proof: It is easy to see that the arguments in Proposition 5.9 from [**AR2**] can be adapted to our situation. Also note that it has been shown in [**AS1**] that a subcategory $\mathcal{C}$ of $\mathrm{mod}\,\Lambda$ is contravariantly finite if and only if $\underline{\mathcal{C}}$ is contravariantly finite in $\underline{\mathrm{mod}}\,\Lambda$.

In the rest of this section we study conditions under which the subcategory $\mathcal{C}^d$ is contravariantly finite. We start with an easy Lemma. For this let T be a tilting module with $\mathrm{pd}_\Lambda T = d$. We choose a minimal coresolution of ${}_\Lambda\Lambda$ as follows:

$$0 \to {}_\Lambda\Lambda \xrightarrow{e^0} T^0 \xrightarrow{e^1} T^1 \to \cdots \to T^{d-1} \xrightarrow{e^d} T^d \to 0$$

Note that is shown in [**H**] that the length of such a coresolution can be chosen in this way. We denote by $Q^i = \mathrm{cok}\,e^i$ for $0 \le i \le d$. Note that $Q^i \in {}^\perp T$ for all $0 \le i \le d$.

LEMMA 4.3. *Using the notation as above. If $\mathrm{Ext}^j_\Lambda(T^i, {}_\Lambda\Lambda) = 0$ for $0 \le i \le d-3$ and $i+2 \le j < d$, then $\mathrm{Ext}^1_\Lambda(Q^i, Q^i) = 0$ for $0 \le i \le d-3$ and $\mathrm{Ext}^j_\Lambda(Q^i, {}_\Lambda\Lambda) = 0$ for $0 \le i \le d-3$ and $i+2 \le j < d$. Moreover, if $X \in \mathcal{C}^d$, then $\mathrm{Ext}^1_\Lambda(X, Q^i) = 0$ for $0 \le i \le d-3$.*

Proof: We first show by induction on i that $\mathrm{Ext}^j_\Lambda(Q^i, {}_\Lambda\Lambda) = 0$ for $0 \le i \le d-3$ and $i+2 \le j < d$. If $i = 0$, we consider the short exact sequence

$$0 \to {}_\Lambda\Lambda \to T^0 \to Q^0 \to 0.$$

Applying $\mathrm{Hom}_\Lambda(-, {}_\Lambda\Lambda)$ to this sequence gives for $j \ge 2$ that $\mathrm{Ext}^j_\Lambda(Q^0, {}_\Lambda\Lambda) = \mathrm{Ext}^j_\Lambda(T^0, {}_\Lambda\Lambda)$, hence $\mathrm{Ext}^j_\Lambda(Q^0, {}_\Lambda\Lambda) = 0$ for $2 \le j < d$.

Now assume that $\mathrm{Ext}^j_\Lambda(Q^s, {}_\Lambda\Lambda) = 0$ for $0 \le s \le i$ and $s+2 \le j < d$ and consider the short exact sequence

$$0 \to Q^i \to T^{i+1} \to Q^{i+1} \to 0.$$

Applying $\mathrm{Hom}_\Lambda(-, {}_\Lambda\Lambda)$ to this sequence gives an exact sequence

$$\mathrm{Ext}^{j-1}_\Lambda(Q^i, {}_\Lambda\Lambda) \to \mathrm{Ext}^j_\Lambda(Q^{i+1}, {}_\Lambda\Lambda) \to \mathrm{Ext}^j_\Lambda(T^{i+1}, {}_\Lambda\Lambda).$$

Let $i+3 \le j < d$. By induction the first term vanishes and by assumption the last term vanishes, hence so does the middle term.

Next we compute $\mathrm{Ext}^1_\Lambda(Q^i, Q^i)$ for $0 \le i \le d-3$. For this set $Q^{-1} = {}_\Lambda\Lambda$ and consider for $i \ge j \ge 0$ the short exact sequences

$$(*)\ 0 \to Q^{j-1} \to T^j \to Q^j \to 0$$

Applying $\mathrm{Hom}_\Lambda(Q^i, -)$ successively to these sequences and using the fact that $Q^i \in {}^\perp T$ shows that $\mathrm{Ext}^1_\Lambda(Q^i, Q^i) = \mathrm{Ext}^{i+2}_\Lambda(Q^i, {}_\Lambda\Lambda)$. The last term vanishes by the first part of the Lemma.

Finally, let $X \in \mathcal{C}^d$. Let $0 \le i \le d-3$. We apply $\mathrm{Hom}_\Lambda(X, -)$ to the exact sequences $(*)$ and obtain, while using that $X \in {}^\perp T$, that $\mathrm{Ext}^1_\Lambda(X, Q^i) = \mathrm{Ext}^{i+2}_\Lambda(X, {}_\Lambda\Lambda)$. The last term vanishes, since X is spherical and $0 < i+2 < d$.

THEOREM 4.4. *Let T be a tilting module with $pd_\Lambda T = d$. If $Ext^j_\Lambda(T^i, {}_\Lambda\Lambda) = 0$ for $0 \le i \le d-3$ and $i+2 \le j < d$, then the subcategory $\mathcal{C}^d$ is contravariantly finite.*

Proof: We have noted before that the subcategory ${}^\perp T \cap \mathcal{P}^{<\infty}(\Lambda)$ is contravariantly finite. For $X \in \mathrm{mod}\,\Lambda$ we denote by $\pi_X : F_X \to X$ a minimal approximation. Suppose that we have for each $Z \in {}^\perp T \cap \mathcal{P}^{<\infty}(\Lambda)$ a map $\phi_Z : G_Z \to Z$ with $G_Z \in \mathcal{C}^d$ and with the property that for each $Y \in \mathcal{C}^d$ the induced map $\mathrm{Hom}_\Lambda(Y, \phi_Z)$ is surjective, then clearly $\phi_{F_X}\pi_X$ is a $\mathcal{C}^d$-approximation of X.

So let $Z \in {}^\perp T \cap \mathcal{P}^{<\infty}(\Lambda)$. We will now construct ϕ_Z. Again we set $Q^{-1} = {}_\Lambda\Lambda$. For $-1 \le j \le d-3$ we first construct recursively universal extensions

$$(*)\ 0 \to (Q^j)^{t_j} \longrightarrow E_j \xrightarrow{\phi_j} E_{j+1} \to 0$$

such that the connecting homomorphism $\mathrm{Hom}_\Lambda((Q^j)^{t_j}, Q^j) \to \mathrm{Ext}^1_\Lambda(E_{j+1}, Q^j)$ is surjective.

We only give the start of the recursion. Let $E_{d+2} = Z$ and choose an extension

$$0 \to (Q^{d-3})^{t_{d-3}} \longrightarrow E_{d-3} \xrightarrow{\phi_{d-3}} Z \to 0$$

such that the connecting homomorphism

$$\mathrm{Hom}_\Lambda((Q^{d-3})^{t_{d-3}}, Q^{d-3}) \to \mathrm{Ext}^1_\Lambda(Z, Q^{d-3})$$

is surjective.

Let $G_Z = E_{-1}$ and $\phi_Z = \phi_{-1}\phi_0 \dots \phi_{d-3}$. We claim that $\phi_Z : G_Z \to Z$ has the property required above.

Since ${}^\perp T \cap \mathcal{P}^{<\infty}(\Lambda)$ is closed under extensions we infer that $E_j \in {}^\perp T \cap \mathcal{P}^{<\infty}(\Lambda)$ for $-1 \le j \le d-3$. By the previous Lemma we have that $\mathrm{Ext}^1_\Lambda(Q^j, Q^j) = 0$ for $-1 \le j \le d-3$, hence $\mathrm{Ext}^1_\Lambda(E_j, Q^j) = 0$ by construction of these sequences. We apply $\mathrm{Hom}_\Lambda(E_j, -)$ to the short exact sequences

$$0 \to Q^{s-1} \to T^s \to Q^s \to 0$$

for $0 \le s \le j$ and obtain that $0 = \mathrm{Ext}^1_\Lambda(E_j, Q^j) = \mathrm{Ext}^{j+2}_\Lambda(E_j, {}_\Lambda\Lambda)$.
By 4.3 the sequences $(*)$ give recursively that $\mathrm{Ext}^t_\Lambda(E_s, {}_\Lambda\Lambda) = 0$ for $-1 \le s \le d-3$ and $s+2 \le t \le d-1$. In particular, we infer that $G_Z = E_{-1} \in \mathcal{C}^d$.

Finally we show the approximation property. For this let $Y \in \mathcal{C}^d$. By the previous Lemma we have that $\mathrm{Ext}^1_\Lambda(Y, Q^j) = 0$ for $-1 \le j \le d-3$. By considering again the sequences $(*)$ we infer that $\mathrm{Hom}_\Lambda(Y, \phi_j)$ is surjective, hence also $\mathrm{Hom}_\Lambda(Y, \phi_Z)$ is surjective, and therefore $\mathcal{C}^d$ is contravariantly finite.

This finishes the proof of the theorem.

COROLLARY 4.5. *Let T be a tilting module of projective dimension two. Then the subcategory $\mathcal{C}^2$ is contravariantly finite. In particular $\mathcal{E}^2$ is functorially finite, and hence has almost split sequences.*

Proof: If $d = 2$ there are no assumptions in the preceding Theorem. The remaining assertions follow from 4.2, 2.4 and [**AS1**].

The following is an immediate consequence of the Theorem above.

COROLLARY 4.6. *Let T be a tilting module of projective dimension d. If T is spherical, then the subcategory $\mathcal{C}^d$ is contravariantly finite.*

COROLLARY 4.7. *Let T be a tilting module of projective dimension d. If in a minimal coresolution $0 \to {}_\Lambda\Lambda \to T^0 \to \cdots \to T^d \to 0$ of ${}_\Lambda\Lambda$ we have that $\mathrm{pd}_\Lambda T^i \le i+1$ for $0 \le i \le d-3$, then the subcategory $\mathcal{C}^d$ is contravariantly finite. Moreover, in this case there exist indecomposable nonprojective spherical modules.*

Proof: The assumptions of the Theorem are trivially satisfied. For the remaining assertion let Z be an indecomposable module in ${}^\perp T \cap \mathcal{P}^{<\infty}(\Lambda)$ with $\mathrm{pd}_\Lambda Z = d$. Such a module exists, for we may choose $Z = T_0$ where T_0 is an indecomposable direct summand of T with $\mathrm{pd}_\Lambda T_0 = d$. Let $\phi_Z : G_Z \to Z$ be the $\mathcal{C}^d$-approximation constructed above. So G_Z is spherical. Since ${}_\Lambda\Lambda \in \mathcal{C}^d$ we infer that ϕ_Z is surjective. We denote by $K_Z = \ker \phi_Z$. Since ${}^\perp T \cap \mathcal{P}^{<\infty}(\Lambda)$ is resolving we infer that $K_Z \in {}^\perp T \cap \mathcal{P}^{<\infty}(\Lambda)$, in particular we have that $\mathrm{pd}_\Lambda K_Z \le d$. Since $\mathrm{pd}_\Lambda T^i \le i+1$ we infer that $\mathrm{Ext}^d_\Lambda(Q^i, {}_\Lambda\Lambda) = 0$ for $0 \le i \le d-3$.

But then $0 \ne \mathrm{Ext}^d_\Lambda(Z, {}_\Lambda\Lambda) = \mathrm{Ext}^d_\Lambda(G_Z, {}_\Lambda\Lambda)$ by applying $\mathrm{Hom}_\Lambda(-, {}_\Lambda\Lambda)$ to the sequences $(*)$ in 4.4. Thus $\mathrm{pd}_\Lambda G_Z = d$, hence G_Z has an indecomposable direct summand which has projective dimension d and is spherical.

If Λ is a Gorenstein algebra and $T = D(\Lambda_\Lambda)$ then the conditions above are related to the k–Gorenstein conditions in [**AR3**].

There is a class of examples where the assumptions of the last Corollary are always satisfied. Let S be a simple Λ–module and let $P(S)$ be its projective cover. Let $S(1) = S, S(2), \ldots, S(n)$ be a complete list of representatives from the isomorphism classes of simple Λ–modules. Let $P[S] = \oplus_{i=2}^n P(S(i))$. Then $P[S]$ is an almost complete tilting module (see [**HU3**] for details). Let X be an indecomposable complement to $P[S]$ with $\mathrm{pd}_\Lambda X = d$. Then it is shown in [**HU3**] that $T = P[S] \oplus X$ satisfies the assumptions of the last corollary. In fact, it is shown that there exists a long exact sequence

$$0 \to P(S) \to P^1 \to \cdots \to P^d \to X \to 0$$

with $P^i \in \operatorname{add} P[S]$ for $1 \leq i \leq d$.

References

[AB] M. Auslander, M. Bridger, *Stable module theory*, Mem. Amer. Math. Soc. 94(1969).

[AR1] M. Auslander, I. Reiten, *Representation theory of Artin algebras V, Methods for computing almost split sequences and irreducible morphisms*, Comm. in Alg. 5(1977), 519-554.

[AR2] M. Auslander, I. Reiten, *Applications of contravariantly finite subcategories*, Adv. Math. 86(1991), 111-152.

[AR3] M. Auslander, I. Reiten, *k-Gorenstein algebras and syzygy modules*, J. Pure and Appl. Alg. 1992(1994), 1-27.

[ARS] M. Auslander, I. Reiten, S. Smalø, *Representation theory of artin algebras*, Cambridge University Press 1995.

[AS1] M. Auslander, S. Smalø, *Preprojective modules over Artin algebras*, J. Algebra 66(1980), 61-122.

[AS2] M. Auslander, S. Smalø, *Almost split sequences in subcategories*, J. Algebra 69(1981), 426-454.

[H] D. Happel, *Triangulated categories in the representation theory of finite dimensional algebras*, LMS Lecture Note Series 119, Cambridge 1988.

[HR] D. Happel, C.M. Ringel, *Tilted algebras*, Trans. Amer. Math. Soc., 274(1982), 399-443.

[HU1] D. Happel, L. Unger, *Partial tilting modules and covariantly finite subcategories*, Comm. Algebra 22(1994), 1723-1727.

[HU2] D. Happel, L. Unger, *Modules of finite projective dimension and cocovers*, Math. Ann., 306(1996), 445-457.

[HU3] D. Happel, L. Unger, *Complements and the Generalized Nakayama Conjecture*, to appear.

[R1] C. M. Ringel, *Tame algebras and integral quadratic forms*, Springer Lecture Notes in Mathematics 1099, Heidelberg 1984.

[R2] C. M. Ringel, *The category of good modules with good filtrations over a quasi-hereditary algebra has almost split sequences*, Math. Zeit. 208(1991), 209-225.

Fakultät für Mathematik, Technische Universität Chemnitz, 09107 Chemnitz, Germany

E-mail address: happel@mathematik.tu-chemnitz.de

Fachbereich für Mathematik, Universität Paderborn, 33098 Paderborn, Germany

E-mail address: luise@uni-paderborn.de

Contemporary Mathematics
Volume **229**, 1998

Modules of the Highest Homological Dimension over a Gorenstein Ring

YASUO IWANAGA AND JUN-ICHI MIYACHI

Dedicated to Professor Kent R. Fuller on his 60th birthday

ABSTRACT. We will study modules of the highest injective, projective and flat dimension over a Goresntein ring. Let R be a Gorenstein ring of self-injective dimension n and $0 \to {}_RR \to E_0 \to \cdots \to E_n \to 0$ a minimal injective resolution. Then it is shown in [F-I] that the flat dimension and projective dimension of E_n is n, the highest dimension. In this note, we shall prove that if M is a left R-module of injective dimension n, then the last injective term $E^n(M)$ in a minimal injective resolution of M has projective and flat dimension n, and any indecomposable summand of $E^n(M)$ embeds in E_n. As a consequence, we obtain that if R is Auslander-Gorenstein, then $E^n(M)$ has essential socle.

1. Introduction

A Noether ring R is called **Gorenstein** if R has left and right finite self-injective dimensions. Further, a Noether ring R is called **Auslander-Gorenstein** if R is Gorenstein and in a minimal injective resolution $0 \to {}_RR \to E_0 \to E_1 \to \cdots$, each E_i has flat dimension at most i. This concept was introduced by Auslander as

1991 *Mathematics Subject Classification.* 16E10, 18G10, 16G60.

The first author was supported by Grant-in-Aid for Scientific Research from the Ministry of Education, No.08640021.

The second author was supported by Grant-in-Aid for Scientific Research from the Ministry of Education, No.09640014.

a non-commutative version of the Gorenstein condition for commutative rings, studied by Bass [Ba]. In the non-commutative case, we can also see the ubiquity of Auslander-Gorenstein rings in several articles, for example, [B-G], [G-L], [Le], [L-S], [S-Z].

We showed in [Iw1] that for a Gorenstein ring of self-injective dimension n, finiteness of the injective, projective and flat dimensions of a module are all equivalent, and all of these dimensions are at most n. This fact motivated our interest in modules with the highest injective, projective or flat dimension. Let R be a Gorenstein ring of self-injective dimension n and $0 \to {}_RR \to E_0 \to \cdots \to E_n \to 0$ a minimal injective resolution. Then it is shown in [F-I] that any direct summand of E_n has the highest projective and flat dimension n. In this note, we will study the relationship between more general modules of projective (or flat) dimension n and the module E_n.

Throughout this note, $\mathrm{id}(M)$, $\mathrm{pd}(M)$ and $\mathrm{fd}(M)$ stand for the injective, projective and flat dimension of a module M, respectively. Further, $0 \to M \to E^0(M) \to \cdots \to E^n(M) \to \cdots$ is a minimal injective resolution of M.

The results obtained in this note are the following.

Theorem 1. *Let R be a Gorenstein ring of self-injective dimension n and $0 \to {}_RR \to E_0 \to \cdots \to E_n \to 0$ a minimal injective resolution. If a left R-module M has injective dimension n, then any indecomposable direct summand E of $E^n(M)$ is isomorphic to a summand in E_n. As a consequence E has projective and flat dimension n.*

As a byproduct, [Mi2, Corollary 1.3] and [I-S2, Theorem 6] yield a generalization of [I-S2, Theorem 6] for Auslander-Gorenstein rings. Hoshino showed that an injective indecomposable module of flat dimension i over an Auslander-Gorenstein ring appears in i-th injective term of a minimal injective resolution of the ring ([Ho, Theorem 6.3]). Miyachi showed that any injective indecomposable module over a Gorenstein ring appears in some injective term of a minimal injective resolution of the ring ([Mi1, Corollary 4.7]).

Theorem 2. *If R is an Auslander-Gorenstein ring of self-injective dimension n, then any injective indecomposable left R-module of flat dimension n is isomorphic to a direct summand of E_n and is* *of the form $E(S)$ for a simple left module S. Thus if a left R-module M has injective dimension n, $E^n(M)$ has essential socle.*

The final result generalizes [I-S1, Theorem; I-S2, Theorem 2]. It appears interesting to study the distribution of injective indecomposables along the terms of a minimal injective resolution of a Gorenstein ring.

Proposition 3. *Let R be a Noether ring and $0 \to {}_RR \to E_0 \to \cdots \to E_n \to \cdots$ a minimal injective resolution of ${}_RR$.*

(1) If M is a left R-module with $0 < i = id(M) < \infty$, then E_0 and $E^i(M)$ have no isomorphic direct summands in common.

(2) If R has left self-injective dimension $n \geq 1$, then E_0 and E_n have no isomorphic direct summands in common.

2. The Proofs

Proof of Theorem 1.

By [Iw1, Theorem 2], M has projective dimension at most n. Thus let $0 \to P_n \to \cdots \to P_1 \to P_0 \to M \to 0$ be a projective resolution of M and consider an injective resolution of each P_i $(0 \leq i \leq n)$

$$0 \to P_i \to E^0(P_i) \to E^1(P_i) \to \ \cdots \ \to E^n(P_i) \to 0.$$

Then $E^j(P_i)$ for each j $(0 \leq j \leq n)$ is a direct summand of a direct sum of copies of E_j. Hence by [Mi2, Corollary 1.3], M has an injective resolution of the following form

$$\begin{aligned} 0 \to M \to Q \longrightarrow \bigoplus_{i=0}^{n-1} E^{i+1}(P_i) \longrightarrow \bigoplus_{i=0}^{n-2} E^{i+2}(P_i) \longrightarrow \cdots \\ \longrightarrow E^{n-1}(P_0) \oplus E^n(P_1) \longrightarrow E^n(P_0) \to 0. \end{aligned}$$

Here Q is a direct summand of $\bigoplus_{i=0}^{n} E^i(P_i)$. Then $E^n(M)$ is a direct summand of $E^n(P_0)$, and so a direct summand of direct sum of copies of E_n. Since any indecomposable summand E of $E^n(M)$ is uniform, E embeds in E_n. $\square$

Proof of Theorem 2.

Let E be an injective indecomposable left module of flat dimension n. By [Mi1, Corollary 4.7], E is isomorphic to a direct summand in E_n. Since $\mathrm{Soc}(E_n)$ is essential in E_n ([I-S 2, Theorem 6]), E is of the form $\mathrm{E}(S)$ for some simple module S.

By Theorem 1 and [F-I, Proposition 1.1], any direct summand of $E^n(M)$ has flat dimension n and so has essential socle. That is, the socle of $E^n(M)$ is essential. □

Proof of Proposition 3.

(1) Let U be any nonzero submodule of E_0 and $V = U \cap R \neq 0$. Then from the exact sequence

$$0 \to V \to R \to R/V \to 0,$$

we have an exact sequence

$$\mathrm{Ext}^i_R(R,\ M) \longrightarrow \mathrm{Ext}^i_R(V,\ M) \longrightarrow \mathrm{Ext}^{i+1}_R(R/V,\ M).$$

Here $\mathrm{Ext}^i_R(R,\ M) = 0$ from $i > 0$ and $\mathrm{Ext}^{i+1}_R(R/V,\ M) = 0$ from $\mathrm{id}(M) = i$. Hence we obtain $\mathrm{Ext}^i_R(V,\ M) = 0$ and thus we see that V is not monomorphic to $E^i(M)$.

(2) is obvious from (1). □

3. Examples

Let us conclude this note with a few examples. In the following examples, if R is a path algebra given by a quiver $\mathcal{Q}$ with set $\mathcal{Q}_0$ of vertices and $i \in \mathcal{Q}_0$, then $S(i)$ denotes the simple R-module corresponding to the vertex i and $E(i)$ its injective hull.

(1) Theorem 1 and Proposition 3 prompt us to raise the following question: *Let R be a Gorenstein ring of self-injective dimension n and E an injective indecomposable R-module of projective dimension n. Then does there exist an R-module M of injective dimension n such that E embeds in $E^n(M)$?* It's easy to see that the question is affirmative if R is Auslander-Gorenstein. However, the answer is negative for Gorenstein rings. For example, let R be

a finite dimensional algebra over any field given by the following quiver

$$\begin{array}{ccccccc} 1 & & & & & & \\ \alpha \searrow & & & & & & \\ & & 3 & \xrightarrow{\gamma} & 4 & \underset{\varepsilon}{\overset{\delta}{\rightleftarrows}} & 5 \\ \beta \nearrow & & & & & & \\ 2 & & & & & & \end{array}$$

with the relations $\gamma\alpha = \gamma\beta = \varepsilon\delta = \delta\varepsilon = 0$. Then R is a Gorenstein ring of self-injective dimension 2 and has infinite global dimension. $E(3)$ has projective dimension 2 but never appears in $\mathrm{E}^2(M)$ for any R-module M of injective dimension 2. Also we can see from this observation that an injective indecomposable module with the highest projective dimension does not necessarily embed in the last term of a minimal injective resolution of a Gorenstein ring.

Moreover, $E(1)$ and $E(2)$ are both direct summands of the last injective term E_2 in a minimal injective resolution $0 \to {}_RR \to E_0 \to E_1 \to E_2 \to 0$ but $\mathrm{Ext}^1_R(E(i),\ R) \neq 0$ $(i = 1, 2)$. Hence $E(1)$ and $E(2)$ are not holonomic. Here a finitely generated module X over a Gorenstein ring R of self-injective dimension n is called **holonomic** if $\mathrm{Ext}^i_R(X,\ R) = 0$ for all $i \neq n$.

Finally we can see in this example that all injective terms E_0, E_1 and E_2 have the highest projective and flat dimensions.

(2) In [Iw2], it is proved that any holonomic module over an Auslander-Gorenstein ring has finite composition length and embeds in a direct sum of finitely many copies of the last injective term in a minimal injective resolution of the ring. However a submodule of finite composition length in the last injective term is not necessarily holonomic.

For example, let R be a finite dimensional algebra over any field given by the following quiver

$$\begin{array}{ccccc} & & 2 & & \\ \alpha \nearrow & & & & \searrow \gamma \\ 1 & \overset{\mu}{\leftarrow} & 5 & \overset{\lambda}{\leftarrow} & 4 \\ \beta \searrow & & & & \nearrow \delta \\ & & 3 & & \end{array}$$

with the relations $\mu\lambda = \alpha\mu = \beta\mu = 0$ and $\gamma\alpha = \delta\beta$. Then R is Auslander-Gorenstein of self-injective dimension 4.

Consider the left R-module M of dimension vector $(0,1,1,1,0)$, then M is a submodule of the last injective term of a minimal injective resolution of ${}_RR$ but not holonomic. For, we can see $\mathrm{Ext}_R^1(M,\ R) \neq 0$, that is, the grade of M is one.

(3) We can see that if R is a Gorenstein ring of self-injective dimension n and S is a simple submodule of the last injective term E_n in a minimal injective resolution of ${}_RR$, then $\mathrm{pd}(S) = \mathrm{fd}(S) = n$ or ∞. Conversely, if S is a simple R-module of the highest projective dimension n, S appears in the socle of E_n. There is an example of a Gorenstein ring R with a simple module of infinite projective and flat dimension not appearing in E_n.

Let R be a finite dimensional algebra over any field given by the following quiver

$$\begin{array}{ccccc} 1 & & \xleftarrow{\delta} & & 3 \\ & \alpha \searrow & & \nearrow \gamma & \\ & & 2 & & \\ & & \circlearrowright \beta & & \end{array}$$

with the relations $\alpha\delta = \gamma\alpha = \beta^2 = 0$. Then R is Auslander-Gorenstein of self-injective dimension 3. We can see

$$\mathrm{pd}(S(1)) = 2, \quad \mathrm{pd}(S(2)) = \infty, \quad \mathrm{pd}(S(3)) = 3$$

and

$$E_0 = E(1)^{(4)}, \quad E_1 = E(2)^{(2)}, \quad E_2 = E(1), \quad E_3 = E(3).$$

Here, $M^{(t)}$ stands for a direct sum of t copies of a module M.

References

[A-R] M. Auslander and I. Reiten, *k-Gorenstein algebras and syzygy modules*, J. of Pure and Applied Algebra **92** (1994), 1-27.

[Ba] H. Bass, *On the ubiquity of Gorenstein rings*, Math. Z. **82** (1963), 8-28.

[B-G] K. A. Brown and K. R. Goodearl, *Homological aspects of Noetherian P.I. Hopf algebras, and irreducible modules of maximal dimension*, Preprint.

[F-I] K. R. Fuller and Y. Iwanaga, *On n-Gorenstein rings and Auslander rings of low injective dimension*, Proceedings of ICRA VI, CMS Proceedings Series **14** (1993), 175–183.

[G-L] K. R. Goodearl and T. H. Lenagan, *Catenarity in quantum algebras*, J. Pure and Applied Algebra **111** (1996), 123–142.

[Ho] M. Hoshino, *Noetherian rings of self-injective dimension two*, Comm. Algebra **21** (1993), 1071–1094.

[Iw1] Y. Iwanaga, *Rings with finite self-injective dimension II*, Tsukuba J. of Math. **4** (1980), 107–113.

[Iw2] Y. Iwanaga, *Duality over Auslander-Gorenstein rings*, Math. Scand. **81** (1997), 184–190.

[I-S1] Y. Iwanaga and H. Sato, *Minimal injective resolutions of Gorenstein rings*, Comm. in Algebra **18** (1990), 3835–3856.

[I-S2] Y. Iwanaga and H. Sato, *On Auslander's n-Gorenstein rings*, J. of Pure and Applied Algebra **106** (1996), 61–76.

[Le] T. Levasseur, *Some properties of non-commutative regular graded rings*, Glasgow Math. J. **34** (1992), 277–300.

[L-S] T. Levasseur and J. T. Stafford, *The quantum coordinate ring of the special linear group*, J. Pure and Applied Algebra **86** (1993), 181–186.

[Mi1] J. Miyachi, *Duality for derived categories and cotilting bimodules*, J. Algebra **185** (1996), 583–603.

[Mi2] J. Miyachi, *Injective resolutions of Noetherian rings and cogenerators*, Preprint, 1997.

[S-Z] J. T. Stafford and J. J. Zhang, *Homological properties of (graded) Noetherian PI rings*, J. of Algebra **168** (1994), 988–1026.

Faculty of Education, Shinshu University, Nishi-Nagano 6, Nagano 380, Japan

E-mail address: iwanaga@gipnc.shinshu-u.ac.jp

Department of Mathematics, Tokyo Gakugei University, Koganei-shi, Tokyo 184, Japan

E-mail address: miyachi@u-gakugei.ac.jp

Contemporary Mathematics
Volume **229**, 1998

DERIVED EQUIVALENCE OF GRAPH ALGEBRAS

MICHAEL KAUER

ABSTRACT. Here we show how the derived equivalences of blocks of group algebras of dihedral defect can be obtained by a quite mechanical process. To this end we consider a block of dihedral defect or more generally an algebra of dihedral type as a Brauer graph algebra. Brauer graph algebras are defined by means of a locally embedded graph (Brauer graph), which is a direct generalization of a Brauer tree. In analogy to a construction of GABRIEL-RIEDTMANN [Ga-Rie 79] applying to trees one is able to associate to a locally embedded graph a Brauer graph algebra via a quiver with relations.
We prove that every Brauer graph algebra is derived equivalent to a Brauer graph algebra where all edges of the graph are incident with one given vertex. This generalizes an analog result of RICKARD (cf. [Ri 89,2]) for trees.

1. GRAPHS, ALGEBRAS AND ORDERS – DEFINITIONS

To fix notation once and for all let $G = (\mathrm{V}, \mathrm{E})$ denote a finite connected **graph** (V the vertices, E the edges), possibly with loops and multiple edges. Let n_v be the valency of a vertex v (= number of edges incident with this vertex v where loops are counted twice). Edges with one incident vertex of valency one are called **end edges**. The **multiplicity** of a graph is a given valuation $m_v \in \mathbb{N}$ of the vertices.

A graph G is called **locally embedded** (or **Brauer graph**), if at each vertex v a cyclic ordering of the edges incident with this vertex is given, this means a numbering $\varepsilon(v) = (e_v(1), \ldots, e_v(i), \ldots, e_v(n_v))$ with $i \in \mathbb{Z}/n_v\mathbb{Z}$ modulo a cyclic permutation where the loops at v show up twice, the other edges at v exactly once.

Apart from the loops there shall be one further kind of edges, which are incident with only one vertex, the so called **truncated edges**, arising from end edges by deleting the vertex of valency one.

If the graph G is a tree without truncated edges and with at most one vertex of multiplicity bigger than 1, this concept gives the classical Brauer tree.

A **walk** of a graph is a path of oriented edges $(e_j)_j$ where e_{j+1} is the next edge to e_j in the local orientation at the vertex, to which e_j is pointing. A truncated edge is to be oriented outwards; these become "dead ends". This generalizes the so called "walk around the Brauer tree".

Example 1.1.

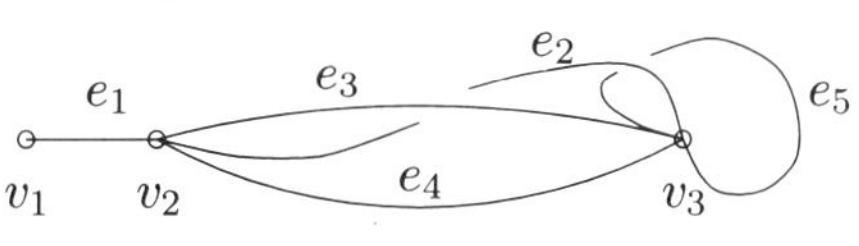

The local embeddings:
$\varepsilon(v_1) = (e_1)$,
$\varepsilon(v_2) = (e_1, e_3, e_2, e_4)$,
$\varepsilon(v_3) = (e_2, e_5, e_4, e_3, e_5)$

The walks:
$(e_1, e_3, e_5, e_4, e_1)$,
$(e_2, e_5, e_2, e_4, e_3)$

1991 *Mathematics Subject Classification.* Primary 16G30; Secondary 16E99.

I would like to thank the organizers for giving me the opportunity to present these results and K. W. Roggenkamp whose ideas have been essential in achieving these results.

In analogy to Brauer trees (cf. [Ga-Rie 79], 1.4) we construct to a locally embedded graph G a **quiver** $Q := (Q_0, Q_1)$. As vertices Q_0 of the quiver we take the edges E of the locally embedded graph G, so $Q_0 := \mathrm{E}$. For each vertex $v \in \mathrm{V}$ the tuple $\varepsilon(v) = (e_v(1), \dots, e_v(n_v))$ defines a cycle of arrows $e_v(i) \xrightarrow{\alpha_{v,i}} e_v(i+1)$ with $i \in \mathbb{Z}/n_v\mathbb{Z}$ (especially one has $e_v(n_v) \xrightarrow{\alpha_{v,0}} e_v(1)$). Because the cycles $c_{v,i} := \alpha_i \cdots \alpha_{i-1}$ correspond to a complete walk around the vertex v (starting at $e_v(i)$) they will be called **complete cycles** (we will see that a complete cycle may split in two shorter cycles). This way one gets a quiver Q with

1. Q is an union of cycles,
2. each vertex belongs to exactly 2 cycles.

Let now k be a field. In the path algebra kQ we define for $t = \pm 1$ the ideal I_t to be generated by the following elements:

$$I_t := \langle \begin{array}{rl} \alpha_{v,i}\alpha_{v',i'} & : \forall v, v' \in \mathrm{V}, i \in \mathbb{Z}/n_v\mathbb{Z}, i' \in \mathbb{Z}/n_{v'}\mathbb{Z}, (v', i') \neq (v, i+1), \\ c_{v,i}^{m_v} - t c_{v',i'}^{m_{v'}} & : \forall (v,i) \neq (v', i') \text{ with } e_v(i) = e_{v'}(i') \in \mathrm{E}, \\ c_{v,i}^{m_v} & : \forall \text{ truncated edges } e_v(i) \in \mathrm{E} \rangle_{ideal}. \end{array}$$

If we set $A_{G,t} := kQ/I_t$, then $A_G := A_{G,1}$ will be called **(Brauer) graph algebra**. This means that we have in the graph algebra A_G the following **relations**: concatenation of two arrows, not following each another directly in a cycle, is zero and different complete cycles with common starting vertex to the corresponding multiplicity are equal.

To avoid paths of length one in the relations defining the ideal I_t we delete $\alpha_{v,0}$ from the quiver (and the relations involving $\alpha_{v,0}$) where $n_v = 1 = m_v$ and add for (v', i') with $e_v(0) = e_{v'}(i')$ the relations $\alpha_{v',i'-1} c_{v',i'}^{m_{v'}}$ and $c_{v',i'}^{m_{v'}} \alpha_{v',i'}$ to I_t (which may be superfluous if $e_{v'}(i'-1)$ respectively $e_{v'}(i'+1)$ are no end edges with endvertex of multiplicity 1). This way I_t becomes an admissible ideal in kQ (cf. [ARS 95] for notation).

Example 1.2. For the locally embedded graph G from example 1.1 we give the corresponding quiver and the relations for A_G:

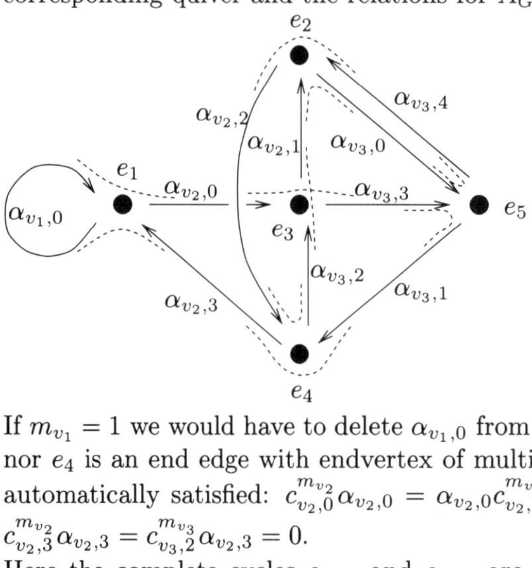

$$\begin{aligned} &\alpha_{v_1,0}\alpha_{v_2,0} = 0 = \alpha_{v_2,3}\alpha_{v_1,0}, \\ &\alpha_{v_2,1}\alpha_{v_3,0} = 0 = \alpha_{v_3,4}\alpha_{v_2,2}, \\ &\alpha_{v_2,0}\alpha_{v_3,3} = 0 = \alpha_{v_3,2}\alpha_{v_2,1}, \\ &\alpha_{v_2,2}\alpha_{v_3,2} = 0 = \alpha_{v_3,1}\alpha_{v_2,3}, \\ &\alpha_{v_3,3}\alpha_{v_3,1} = 0 = \alpha_{v_3,0}\alpha_{v_3,4}, \\ &c_{v_1,0}^{m_{v_1}} = c_{v_2,0}^{m_{v_2}}, \quad c_{v_2,2}^{m_{v_2}} = c_{v_3,0}^{m_{v_3}}, \\ &c_{v_2,1}^{m_{v_2}} = c_{v_3,3}^{m_{v_3}}, \quad c_{v_2,3}^{m_{v_2}} = c_{v_3,2}^{m_{v_3}}, \\ &c_{v_3,1}^{m_{v_3}} = c_{v_3,4}^{m_{v_3}} \end{aligned}$$

If $m_{v_1} = 1$ we would have to delete $\alpha_{v_1,0}$ from the arrowset, but because neither e_3 nor e_4 is an end edge with endvertex of multiplicity 1, the necessary relations are automatically satisfied: $c_{v_2,0}^{m_{v_2}} \alpha_{v_2,0} = \alpha_{v_2,0} c_{v_2,1}^{m_{v_2}} = \alpha_{v_2,0} c_{v_3,3}^{m_{v_3}} = 0$ and $\alpha_{v_2,3} c_{v_2,0}^{m_{v_2}} = c_{v_2,3}^{m_{v_2}} \alpha_{v_2,3} = c_{v_3,2}^{m_{v_3}} \alpha_{v_2,3} = 0$.

Here the complete cycles $c_{v_3,1}$ and $c_{v_3,4}$ are the concatenation of shorter cycles:

$c_{v_3,1} = (\alpha_{v_3,1}\alpha_{v_3,2}\alpha_{v_3,3})(\alpha_{v_3,4}\alpha_{v_3,0})$ and $c_{v_3,4} = (\alpha_{v_3,4}\alpha_{v_3,0})(\alpha_{v_3,1}\alpha_{v_3,2}\alpha_{v_3,3})$, where $\alpha_{v_3,1}\alpha_{v_3,2}\alpha_{v_3,3}$ and $\alpha_{v_3,4}\alpha_{v_3,0}$ are both paths beginning and ending at e_5, i. e. are cycles at e_5.

Remark 1.3. A graph algebra is special biserial. Graph algebras have been defined some time ago by DONOVAN [Do 79], respectively DONOVAN-FREISLICH [Do-Fr 78] in order to describe blocks of group algebras with dihedral defect groups.
The projective resolutions of the maximal uniserial modules are given by the walks on the graph G; if the graph is a tree, this is exactly GREEN's "walk around the tree".

ROGGENKAMP (cf. [Ro 96]) defined an order-analogon called graph orders to graph algebras which plays the same role as Green orders (cf. [Ro 92]) do in relation to Brauer tree algebras. These graph orders and their derived equivalences are not only useful in the context of graph algebras but also in the context of Hecke algebras.
Let R be a complete discrete valuation ring, $K = \mathrm{Quot}(R)$ its field of quotients, $(\pi) := \pi \cdot R$ its maximal ideal (generated by the parameter π) and $k := R/(\pi)$ its residue field.
For every $v \in \mathrm{V}$ take a completely ramified extension Ω_v of degree m_v with parameter π_v and let $p_v \colon \Omega_v \longrightarrow \Omega_v/(\pi_v) \cong k$ denote the canonical projection on the associated residue field concatenated with a chosen isomorphism to k. $\Omega_v - \Omega_{v'}$ is then defined by the following pullback diagram for all $v, v' \in \mathrm{V}$:

$$\begin{array}{ccc} \Omega_v - \Omega_{v'} & \longrightarrow & \Omega_{v'} \\ \downarrow & & \downarrow p_{v'} \\ \Omega_v & \xrightarrow{p_v} & k \end{array}$$

$\alpha_v \equiv \alpha_{v'}$ shall be a short hand for $(\alpha_v, \alpha_{v'}) \in \Omega_v - \Omega_{v'}$.
For every v we take the R-order

$$H_v := \begin{pmatrix} \Omega_v & \cdots & \cdots & \Omega_v \\ (\pi_v) & \ddots & & \vdots \\ \vdots & \ddots & \ddots & \vdots \\ (\pi_v) & \cdots & (\pi_v) & \Omega_v \end{pmatrix}_{n_v}$$

We denote the i-th column of H_v by $M_{v,i} := \begin{pmatrix} \Omega_v \\ \vdots \\ \Omega_v \\ (\pi_v) \\ \vdots \\ (\pi_v) \end{pmatrix}_{n_v}$. For $1 \leq i \leq n_v - 1$ we have the canonical embedding $M_{v,i} \hookrightarrow M_{v,i+1}$ denoted by 1; the canonical embedding $M_{v,n_v} \hookrightarrow M_{v,1}$ is the multiplication by π_v.

Definition 1.4. Let the local embedding at v be given by $(e_v(1), \ldots, e_v(n_v))$. In order to get $\Lambda := \Lambda_G \subseteq \prod_{v \in \mathrm{V}(G)} H_v =: H$ we replace in H for every non-truncated edge $e = e_v(i) = e_{v'}(i')$ the i-th diagonal entry Ω_v of H_v and the i'-th diagonal entry $\Omega_{v'}$ of $H_{v'}$ by the pullback $\Omega_v - \Omega_{v'}$, i. e. one identifies the diagonal entries

(i,i) of H_v and (i',i') of $H_{v'}$ "modulo π". Such an identification will be displayed as a junction. Λ will be called **graph order**.

The projective indecomposable corresponding to the edge $e_v(i) = e_{v'}(i')$ now looks as follows:

$$M_{v,i} - M_{v',i'} := i - \begin{pmatrix} \Omega_v \\ \vdots \\ \Omega_v \\ (\pi_v) \\ \vdots \\ (\pi_v) \end{pmatrix}_{n_v} \text{———} \begin{pmatrix} \Omega_{v'} \\ \vdots \\ \Omega_{v'} \\ (\pi_{v'}) \\ \vdots \\ (\pi_{v'}) \end{pmatrix}_{n_{v'}} - i',$$

Example 1.5. Here we give the graph order for the graph from example 1.1:

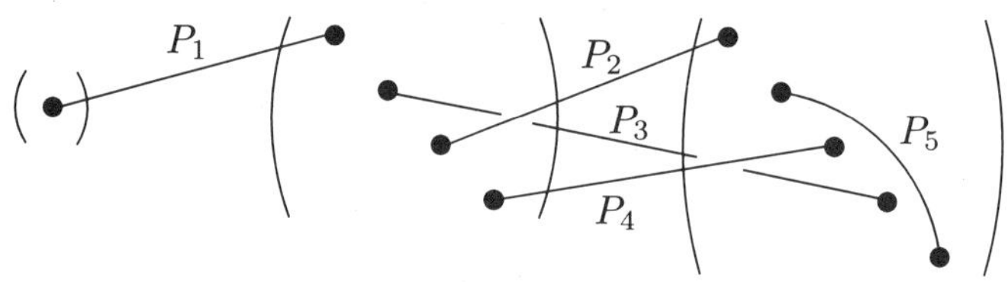

where only the diagonal entries are drawn and respective entries are joined. The above picture of the order is to be interpreted in

$$H = (\Omega_{v_1}) \times \begin{pmatrix} \Omega_{v_2} & \Omega_{v_2} & \Omega_{v_2} & \Omega_{v_2} \\ (\pi_{v_2}) & \Omega_{v_2} & \Omega_{v_2} & \Omega_{v_2} \\ (\pi_{v_2}) & (\pi_{v_2}) & \Omega_{v_2} & \Omega_{v_2} \\ (\pi_{v_2}) & (\pi_{v_2}) & (\pi_{v_2}) & \Omega_{v_2} \end{pmatrix} \times \begin{pmatrix} \Omega_{v_3} & \Omega_{v_3} & \Omega_{v_3} & \Omega_{v_3} & \Omega_{v_3} \\ (\pi_{v_3}) & \Omega_{v_3} & \Omega_{v_3} & \Omega_{v_3} & \Omega_{v_3} \\ (\pi_{v_3}) & (\pi_{v_3}) & \Omega_{v_3} & \Omega_{v_3} & \Omega_{v_3} \\ (\pi_{v_3}) & (\pi_{v_3}) & (\pi_{v_3}) & \Omega_{v_3} & \Omega_{v_3} \\ (\pi_{v_3}) & (\pi_{v_3}) & (\pi_{v_3}) & (\pi_{v_3}) & \Omega_{v_3} \end{pmatrix}$$

2. Derived equivalence of graph orders

We will call two graphs derived equivalent if the corresponding graph orders are. The next lemma will give the technical tools to manipulate a graph in its derived equivalence class:

Lemma 2.1.

(i)

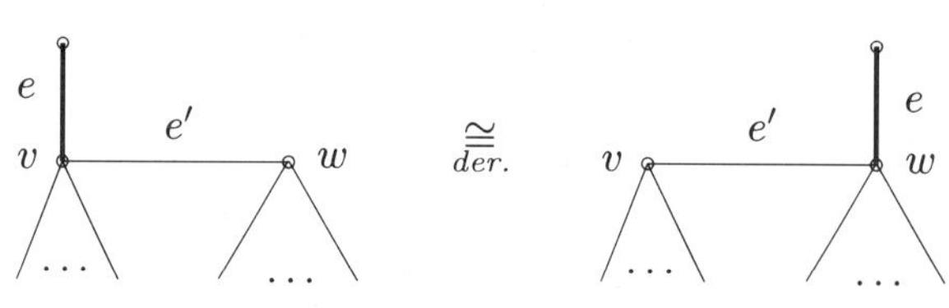

(ii)

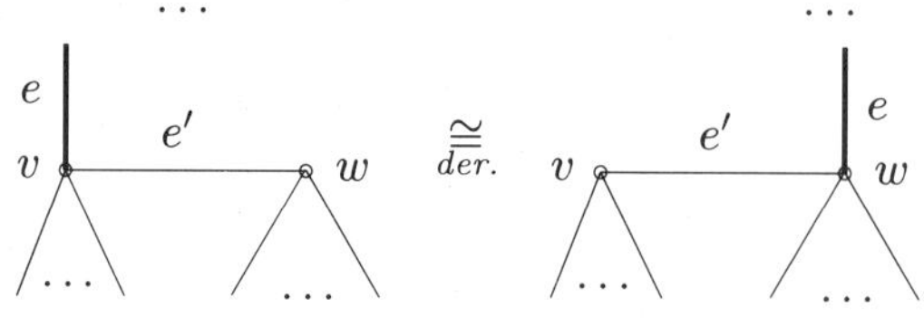

(iii) 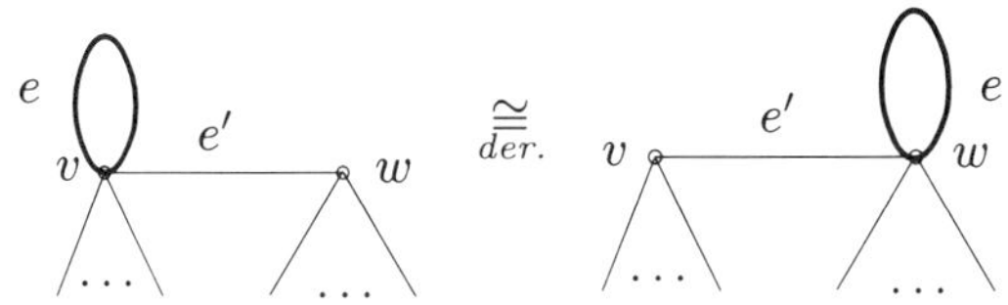

(iv) 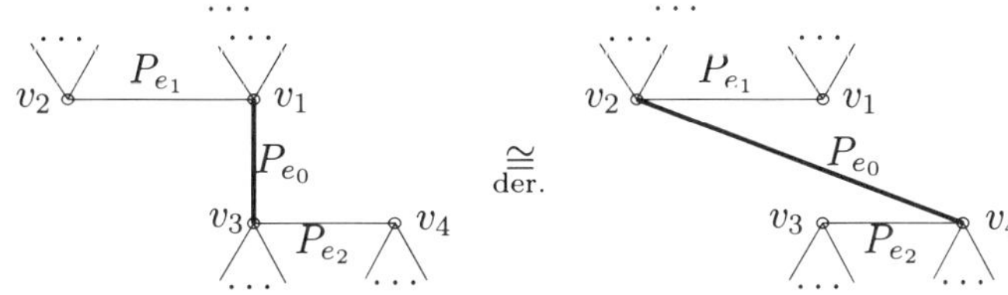

Proof: "(i)" The proof from ROGGENKAMP-ZIMMERMANN [Zi 96] of moving end edges in trees may be adapted to this situation.

"(ii)" is analog to (i).

"(iii) and (iv)": we shall give here the proof of (iv) in the case that the vertex set of $e_1, e_2 = \{v_1, v_2, v_3, v_4\}$ is a set of size 4, the other cases are done similarly. We choose a numbering at the vertices v_1 and v_3 such that P_{e_0}, P_{e_1} are first: $\varepsilon(v_1) = (P_{e_0}, P_{e_1}, \dots)$ and $\varepsilon(v_3) = (P_{e_0}, P_{e_2}, \dots)$ and therefore $P_{e_0} = M_{v_1,1} - M_{v_3,1}$, $P_{e_1} = M_{v_1,2} - M_{v_2,i'}$, $P_{e_2} = M_{v_3,2} - M_{v_4,j'}$:

$\Lambda_G =$ 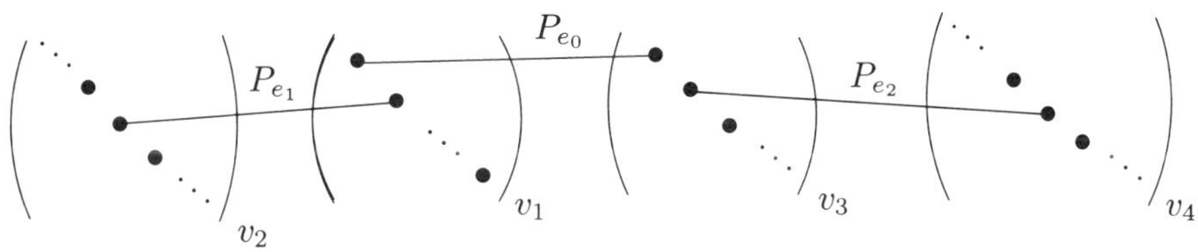

We show:

$$T^{\cdot} := T_{e_0}{}^{\cdot} \oplus \bigoplus_{e \in E \setminus \{e_0\}} P_e,$$

with $T_{e_0}{}^{\cdot} := \cdots 0 \longrightarrow P_{e_0} \xrightarrow{\left(\left(\begin{smallmatrix}1&0\\0&0\end{smallmatrix}\right)\left(\begin{smallmatrix}0&0\\1&0\end{smallmatrix}\right)\right)} P_{e_1} \oplus P_{e_2} \longrightarrow 0 \cdots$ concentrated in degrees -1 and 0, is a tilting complex, which gives the claimed derived equivalence. That $T_{e_0}{}^{\cdot}$ is a tilting complex can be checked directly, but it also fits in the frame of Prop. 2.1 in [Kö-Zi 97]. So we just have to compute $\underline{\mathrm{End}}(T^{\cdot})$:

The Pierce decomposition of $\underline{\mathrm{End}}(T^{\cdot})$ is

$$\bigoplus_{e,f \in E \setminus \{e_0\}} \underline{\mathrm{Hom}}(P_e, P_f) \oplus \bigoplus_{e \in E \setminus \{e_0\}} \underline{\mathrm{Hom}}(P_e, T_{e_0}{}^{\cdot}) \oplus \bigoplus_{e \in E \setminus \{e_0\}} \underline{\mathrm{Hom}}(T_{e_0}{}^{\cdot}, P_e) \oplus \underline{\mathrm{Hom}}(T_{e_0}{}^{\cdot}, T_{e_0}{}^{\cdot})$$

The first summand $\displaystyle\bigoplus_{e,f \in E \setminus \{e_0\}} \underline{\mathrm{Hom}}(P_e, P_f) = \bigoplus_{e,f \in E \setminus \{e_0\}} \mathrm{Hom}_{\Lambda}(P_e, P_f)$

Calculation of $\underline{\mathrm{Hom}}(T_{e_0}{}^{\cdot}, T_{e_0}{}^{\cdot})$:

$$\begin{array}{ccccccc}
0 & \longrightarrow & M_{v_1,1} - M_{v_3,1} & \xrightarrow{\left(\left(\begin{smallmatrix}1&0\\0&0\end{smallmatrix}\right)\ \left(\begin{smallmatrix}0&0\\1&0\end{smallmatrix}\right)\right)} & M_{v_1,2} - M_{v_2,i'} \oplus M_{v_3,2} - M_{v_4,j'} & \longrightarrow & 0 \\
 & & \left(\begin{smallmatrix}\alpha_1 & 0\\ 0 \ \ddots & \alpha_3\end{smallmatrix}\right)\Big\downarrow =: \alpha & \swarrow \begin{pmatrix}\sigma_1\pi_{v_1} & 0 \\ 0 & 0 \\ 0 & \sigma_3\pi_{v_3} \\ 0 & 0\end{pmatrix} & \beta := \Big\downarrow \begin{pmatrix} \left(\begin{smallmatrix}\beta_1 & 0\\ 0 \ \ddots & \beta_2\end{smallmatrix}\right) & \left(\begin{smallmatrix}0&0\\0&0\end{smallmatrix}\right) \\ \left(\begin{smallmatrix}0&0\\0&0\end{smallmatrix}\right) & \left(\begin{smallmatrix}\beta_3 & 0\\ 0 \ \ddots & \beta_4\end{smallmatrix}\right)\end{pmatrix} & & \\
0 & \longrightarrow & M_{v_1,1} - M_{v_3,1} & \xrightarrow[\left(\left(\begin{smallmatrix}1&0\\0&0\end{smallmatrix}\right)\ \left(\begin{smallmatrix}0&0\\1&0\end{smallmatrix}\right)\right)]{} & M_{v_1,2} - M_{v_2,i'} \oplus M_{v_3,2} - M_{v_4,j'} & \longrightarrow & 0
\end{array}$$

To be a complexmorphism the diagram has to be commutative, so $\left(\begin{smallmatrix}\alpha_1 & 0 & 0 & 0\\ 0 & 0 & \alpha_3 & 0\end{smallmatrix}\right) = \left(\begin{smallmatrix}\alpha_1 & 0\\ 0 & \alpha_3\end{smallmatrix}\right) \cdot \left(\begin{smallmatrix}1 & 0 & 0 & 0\\ 0 & 0 & 1 & 0\end{smallmatrix}\right) = \left(\begin{smallmatrix}1 & 0 & 0 & 0\\ 0 & 0 & 1 & 0\end{smallmatrix}\right) \cdot \begin{pmatrix}\beta_1 & 0 & 0 & 0\\ 0 & \beta_2 & 0 & 0\\ 0 & 0 & \beta_3 & 0\\ 0 & 0 & 0 & \beta_4\end{pmatrix} = \left(\begin{smallmatrix}\beta_1 & 0 & 0 & 0\\ 0 & 0 & \beta_3 & 0\end{smallmatrix}\right)$. So β determines α completely. With the given homotopy one sees that $\beta_1 \equiv \beta_3$ and that β_2, β_4 determine the homotopy class completely, therefore

$$\underline{\mathrm{Hom}}(T_{e_0}^{\cdot}, T_{e_0}^{\cdot}) \cong \Omega_{v_2} - \Omega_{v_4}$$

Calculation of $\bigoplus_{e\in \mathrm{E}\setminus\{e_0\}} \underline{\mathrm{Hom}}(P_e, T_{e_0}^{\cdot})$ and $\bigoplus_{e\in \mathrm{E}\setminus\{e_0\}} \underline{\mathrm{Hom}}(T_{e_0}^{\cdot}, P_e)$: If $v_1, v_2 \notin e$, then $\underline{\mathrm{Hom}}(P_e, T_{e_0}^{\cdot}) = \mathrm{Hom}_{\Lambda}(P_e, P_{e_1} \oplus P_{e_2})$ and $\underline{\mathrm{Hom}}(T_{e_0}^{\cdot}, P_e) = \mathrm{Hom}_{\Lambda}(P_{e_1} \oplus P_{e_2}, P_e)$. For $e \neq e_1, e_2$ we have with the Kronecker-Delta $\delta_{w,w'} = \begin{cases} 1 & \text{if } w = w' \\ 0 & \text{if } w \neq w' \end{cases}$ the following:

$$\begin{array}{ccccc}
0 & \longrightarrow & P_e = M_{w,h} - M_{w',h'} & \longrightarrow & 0 \\
\downarrow & \swarrow{\scriptstyle \begin{pmatrix} \delta_{v_1,w}\alpha_1\pi_{v_1} & \delta_{v_3,w}\alpha_3 \\ \delta_{v_1,w'}\alpha_1'\pi_{v_1} & \delta_{v_3,w'}\alpha_3' \end{pmatrix}} & \downarrow{\scriptstyle \begin{pmatrix} \delta_{v_1,w}\alpha_1\pi_{v_1} & \delta_{v_2,w}\alpha_2 & \delta_{v_3,w}\alpha_3 & \delta_{v_4,w}\alpha_4 \\ \delta_{v_1,w'}\alpha_1'\pi_{v_1} & \delta_{v_2,w'}\alpha_2' & \delta_{v_3,w'}\alpha_3' & \delta_{v_4,w'}\alpha_4' \end{pmatrix}} & & \\
M_{v_1,1} - M_{v_3,1} & \xrightarrow[\left(\left(\begin{smallmatrix}1 & 0\\ 0 & 0\end{smallmatrix}\right) \left(\begin{smallmatrix}0 & 0\\ 1 & 0\end{smallmatrix}\right)\right)]{} & M_{v_1,2} - M_{v_2,i'} \oplus M_{v_3,2} - M_{v_4,j'} & \longrightarrow & 0 \\
\downarrow & & \downarrow{\scriptstyle \begin{pmatrix} \delta_{v_1,w}\beta_1 & \delta_{v_1,w'}\beta_1' \\ \delta_{v_2,w}\beta_2 & \delta_{v_2,w'}\beta_2' \\ \delta_{v_3,w}\beta_3 & \delta_{v_3,w'}\beta_3' \\ \delta_{v_4,w}\beta_4 & \delta_{v_4,w'}\beta_4' \end{pmatrix}} & & \\
0 & \longrightarrow & P_e = M_{w,h} - M_{w',h'} & \longrightarrow & 0
\end{array}.$$

The given homotopy shows that $\underline{\mathrm{Hom}}(P_e, T^{\cdot}_{e_0})$ is given by $\delta_{v_2,w}\alpha_2$, $\delta_{v_2,w'}\alpha_2'$, $\delta_{v_4,w}\alpha_4$, $\delta_{v_4,w'}\alpha_4'$ and similarly the commutativity condition $\left(\begin{smallmatrix}0 & 0\\ 0 & 0\end{smallmatrix}\right) = \left(\begin{smallmatrix}1 & 0 & 0 & 0\\ 0 & 0 & 1 & 0\end{smallmatrix}\right) \cdot \begin{pmatrix} \delta_{v_1,w}\beta_1 & \delta_{v_1,w'}\beta_1' \\ \delta_{v_2,w}\beta_2 & \delta_{v_2,w'}\beta_2' \\ \delta_{v_3,w}\beta_3 & \delta_{v_3,w'}\beta_3' \\ \delta_{v_4,w}\beta_4 & \delta_{v_4,w'}\beta_4' \end{pmatrix} = \begin{pmatrix} \delta_{v_1,w}\beta_1 & \delta_{v_1,w'}\beta_1' \\ \delta_{v_3,w}\beta_3 & \delta_{v_3,w'}\beta_3' \end{pmatrix}$ that $\delta_{v_1,w}\beta_1 = \delta_{v_1,w'}\beta_1' = \delta_{v_3,w}\beta_3 = \delta_{v_3,w'}\beta_3' = 0$ and therefore $\underline{\mathrm{Hom}}(T^{\cdot}_{e_0}, P_e)$ is determined by $\delta_{v_2,w}\beta_2$, $\delta_{v_2,w'}\beta_2'$, $\delta_{v_4,w}\beta_4$, $\delta_{v_4,w'}\beta_4'$. So the morphisms from and to $T^{\cdot}_{e_0}$ are determined by their part to and from $M_{v_2,i'}$ and $M_{v_4,j'}$. For $e = e_1$ is

$$\begin{array}{ccccccc}
0 & \longrightarrow & 0 & \longrightarrow & P_{e_1} = M_{v_1,2} - M_{v_2,i'} & \longrightarrow & 0 \\
 & & \downarrow & \swarrow{\scriptstyle \begin{pmatrix} \sigma_1\pi_{v_1} & 0 \\ 0 & 0 \end{pmatrix}} & \downarrow{\scriptstyle \left(\begin{pmatrix} \alpha_1 & 0 \\ 0 \ddots & \alpha_2 \end{pmatrix} \left(\begin{smallmatrix}0 & 0\\ 0 & 0\end{smallmatrix}\right)\right)} & & \\
0 & \longrightarrow & M_{v_1,1} - M_{v_3,1} & \xrightarrow[\left(\left(\begin{smallmatrix}1 & 0\\ 0 & 0\end{smallmatrix}\right) \left(\begin{smallmatrix}0 & 0\\ 1 & 0\end{smallmatrix}\right)\right)]{} & M_{v_1,2} - M_{v_2,i'} \oplus M_{v_3,2} - M_{v_4,j'} & \longrightarrow & 0 \\
 & & \downarrow & & \downarrow{\scriptstyle \begin{pmatrix} \begin{pmatrix} \gamma_1 & 0 \\ 0 \ddots & \gamma_2 \end{pmatrix} \\ \left(\begin{smallmatrix}0 & 0\\ 0 & 0\end{smallmatrix}\right) \end{pmatrix}} & & \\
0 & \longrightarrow & 0 & \longrightarrow & P_{e_1} = M_{v_1,2} - M_{v_2,i'} & \longrightarrow & 0
\end{array}.$$

The given homotopy shows that $\underline{\mathrm{Hom}}(P_{e_1}, T^{\cdot}_{e_0})$ is determined by α_2.

$$\text{(1)} \qquad \underline{\mathrm{Hom}}(P_{e_1}, T^{\cdot}_{e_0}) = \Omega_{v_2}$$

Similarly the commutativity condition shows $\left(\begin{smallmatrix} 0 & 0 \\ 0 & 0 \end{smallmatrix}\right) = \left(\begin{smallmatrix} 1 & 0 & 0 & 0 \\ 0 & 0 & 1 & 0 \end{smallmatrix}\right) \cdot \left(\begin{smallmatrix} \gamma_1 & 0 \\ 0 & \gamma_2 \\ 0 & 0 \\ 0 & 0 \end{smallmatrix}\right) = \left(\begin{smallmatrix} \gamma_1 & 0 \\ 0 & 0 \end{smallmatrix}\right)$, that $\gamma_1 = 0$, so $\underline{\mathrm{Hom}}(T^{\cdot}{}_{e_0}, P_{e_1})$ is given by $\gamma_2 \in (\pi_{v_2})$:

$$\underline{\mathrm{Hom}}(T^{\cdot}{}_{e_0}, P_{e_1}) = (\pi_{v_2}) \tag{2}$$

The case $e = e_2$ is done analogously to the case $e = e_1$ and gives

$$\underline{\mathrm{Hom}}(P_{e_2}, T^{\cdot}{}_{e_0}) = \Omega_{v_4}\,, \qquad \underline{\mathrm{Hom}}(T^{\cdot}{}_{e_0}, P_{e_2}) = (\pi_{v_4}) \tag{3}$$

Now one reads of (1), (2) and (3) that

$$\underline{\mathrm{End}}(P_{e_1} \oplus P_{e_2} \oplus T_{e_0}^{\cdot}) = \begin{pmatrix} \Omega_{v_1} - \Omega_{v_2} & 0 & \Omega_{v_2} \\ 0 & \Omega_{v_3} - \Omega_{v_4} & \Omega_{v_4} \\ (\pi_{v_2}) & (\pi_{v_4}) & \Omega_{v_2} - \Omega_{v_4} \end{pmatrix} \text{ and so}$$

$\underline{\mathrm{End}}(T^{\cdot}) =$

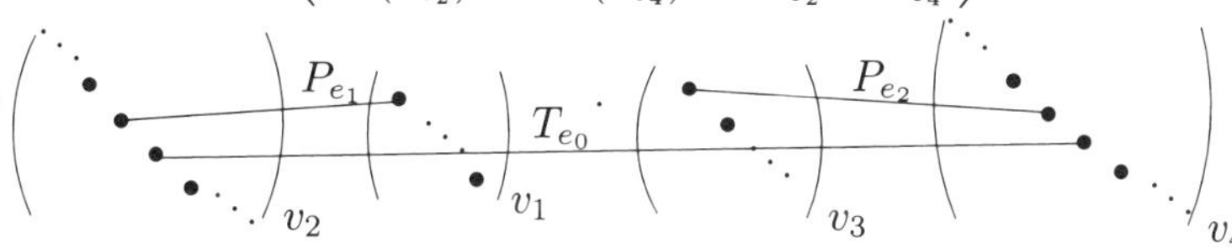

$= \Lambda_{G'}$

□

Theorem 2.2 (Rickard [Ri 89,2], Roggenkamp-Zimmermann [Zi 96]). *If T is a tree with n edges and St_n is the star with n edges (i. e. all edges incident with a certain vertex), then $T \underset{der.}{\cong} St_n$.*

Proof: Choose one vertex v_0 and move all end edges to v_0 (with the tool from Lemma 2.1, (i)). As long as T is not yet a star, there are end edges to move and after at most $n-1$ moves the star is reached. □

Theorem 2.3. *If G is a graph with n edges, then there is is a generalized star GSt_n with n edges (i. e. all edges incident with a certain vertex), such that $G \underset{der.}{\cong} GSt_n$. Moreover, except the center of this generalized star and possibly one further vertex all other vertices have valency one.*

Proof: Choose one vertex v_0.
By moving or turning an edge e (with the tools from Lemma 2.1), we can make this edge incident to every vertex of those edges which are in the same walk of the graph as e is (such a vertex will be called adjacent to this walk).
If there are walks not adjacent to v_0, then there are edges which are in two walks, one adjacent to v_0, the other not. By moving/turning this edge to v_0 both walks will become adjacent to v_0, thus reducing the number of walks not adjacent to v_0. So we may assume that all walks are adjacent to v_0. Now all edges may moved or turned to v_0 and we reach a generalized star.
If one has now a loop at v_0, then all edges which are not end edges can be turned into loops. If one doesn't have any loops at v_0, on can at least achieve that all edges which are not end edges are between two vertices v_0, v_1.
In the first case, all other vertices have valency one, in the second case one has to exclude v_1. □

Example 2.4. We visualize the mechanic of the proof by two examples. The first step in the first example makes the walk U_1 adjacent to v_0 so that any edge may be turned to v_0. In the second example it would suffice to move the end edge e_1 to v_0 to get a generalized star, but the valency of v_2 would not be as small as possible. These example illustrate also the two different possible types of resulting generalized stars.

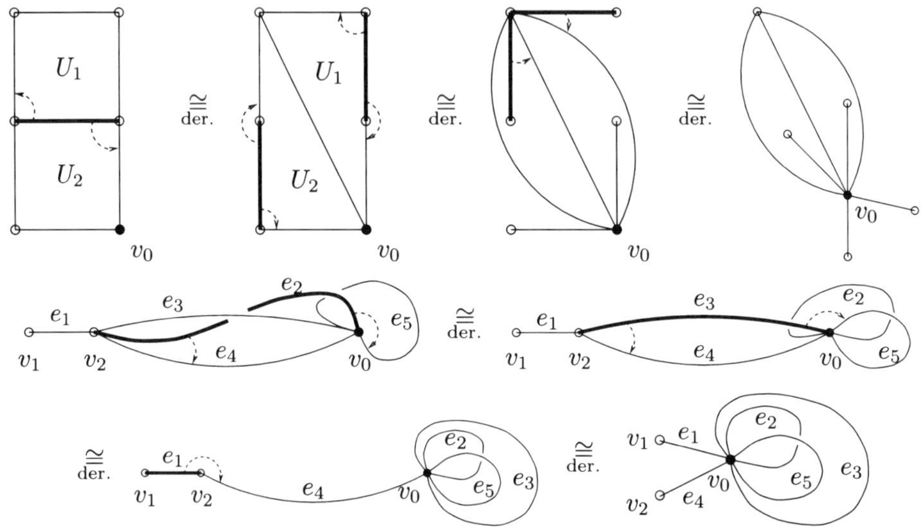

3. Derived equivalence of graph algebras†

We will show how derived equivalences of graph orders are transfered to derived equivalences on the corresponding graph algebras where we get the analogue result:

Theorem 3.1. *If G is a graph with n edges, then there is is a generalized star GSt_n with n edges, such that $A_G \underset{der.}{\cong} A_{GSt_n}$.*
Moreover, except the center of this generalized star and possibly one further vertex all other vertices have multiplicity one.

For a **Proof** we just have to show that the mechanism established by Lemma 2.1 is also valid for graph algebras.
The next lemma not only gives a considerably shortcut for the proof of the theorem in characteristic 2 but will also help to transfer the derived equivalences from the graph orders to the graph algebras.

Lemma 3.2. *If $G \underset{der.}{\cong} G'$, then $A_{G,-1} \underset{der.}{\cong} A_{G',-1}$*

Proof: By Cor. 2.2 [Ri 91,1] we have $k \otimes_R \Lambda_G \underset{\text{der.}}{\cong} k \otimes_R \Lambda_{G'}$. But $k \otimes_R \Lambda_G = A_{G,-1}$ and the lemma is proved. □

From this lemma it is immediate that

†I would like to thank the referee for pointing out an incomplete argument in a previous version of this section.

Proposition 3.3. *If* char $(k) = 2$ *and* $G \underset{der.}{\cong} G'$, *then* $A_G \underset{der.}{\cong} A_{G'}$.

Lemma 3.4. *The multiplicities of the graph may be read from the center* $\mathrm{Z}(A_{G,t})$ *of the algebra* $A_{G,t}$, $t = \pm 1$ *and is therefore a derived invariant.*

Proof: We give a basis for the center from which the claim will be apparent. We use the following abbreviations: $t_{v,i,i'} := \alpha_i \cdots \alpha_{i'-1}$, where $v \in \mathrm{V}, i, i' \in \mathbb{Z}/n_v\mathbb{Z}$, $c_{v,i} := t_{v,i,i} = \alpha_i \cdots \alpha_{i-1}$ as before a complete cycle, $p_{v,i,i',l} := t_{v,i,i'} c^l_{v,i'}$ for $i \neq i' \in \mathbb{Z}/n_v\mathbb{Z}$ and $p_{v,i,i,l} := c^l_{v,i}$ for $i \in \mathbb{Z}/n_v\mathbb{Z}$ and $1 \leq l \leq m_v$.
Thus a basis for $A_{G,t}$ is given by $\{e : e \in \mathrm{E}\} \cup \{p_{v,i,i',l} : v \in \mathrm{V}, i, i' \in \mathbb{Z}/n_v\mathbb{Z}, 0 \leq l \leq m_v - 1\} \cup \{c^{m_v}_{v,i} : e_v(i) \in \mathrm{E}\}$ and for the diagonal Pierce components $\bigoplus_{e \in \mathrm{E}} eA_{G,t}e$ by $\{e : e \in \mathrm{E}\} \cup \{c^l_{v,i} : v \in \mathrm{V}, i, i' \in \mathbb{Z}/n_v\mathbb{Z}, 0 \leq l \leq m_v - 1\} \cup \{t_{v,i,i'} c^l_{v,i'}, t_{v,i',i} c^l_{v,i} : v \in \mathrm{V}, i \neq i' \in \mathbb{Z}/n_v\mathbb{Z}$ with $e_v(i) = e_v(i'), 0 \leq l \leq m_v - 1\} \cup \{c^{m_v}_{v,i} : e_v(i) \in \mathrm{E}\}$ ($\{e_v(i) = e_v(i') : v \in \mathrm{V}, i \neq i' \in \mathbb{Z}/n_v\mathbb{Z}\}$ may be interpreted as the set of loops in G; if e is a loop in G, then the complete cycle starting at e splits in two smaller cycles).
If $m_v \geq 2$ let $z_{v,l} := \sum_{i=0}^{n_v - 1} c^l_{v,i}$ for all $1 \leq l \leq m_v - 1$. Explicit calculation shows that a basis for $\mathrm{Z}(A_{G,-1})$ in case char $(k) \neq 2$ is given by $\{1\} \cup \{z_{v,l} : v \in \mathrm{V}, 1 \leq l \leq m_v - 1\} \cup \{c^{m_v}_{v,i} : e_v(i) \in \mathrm{E}\}$ and that a basis for $\mathrm{Z}(A_{G,1})$ is given by $\{1\} \cup \{z_{v,l} : v \in \mathrm{V}, 1 \leq l \leq m_v - 1\} \cup \{t_{v,i+1,i} c^{m_v - 1}_{v,i+1} : v \in \mathrm{V}, i \in \mathbb{Z}/n_v\mathbb{Z}$ with $e_v(i) = e_v(i+1), m_v \geq 2\} \cup \{c^{m_v}_{v,i} : e_v(i) \in \mathrm{E}\}$.
To read the multiplicities $(m_v)_{v \in \mathrm{V}}$ from the center let N denote the nilpotent elements from the center and set $S_1 := \{a \in \mathrm{Z}(A_t) : a \cdot N = 0\}$ and for $i \geq 2$ set $S_i := \{a \in \mathrm{Z}(A_t) : a \cdot N \in S_{i-1}\}$. If r is the nilpotency index of N, then consider $a_i := \dim S_i / S_{i-1}$ for $1 \leq i \leq r$. Now for $2 \leq i \leq r - 1$ we have $a_i - a_{i+1} = \#\{v \in \mathrm{V} : m_v = i\}$, $a_r = \#\{v \in \mathrm{V} : m_v = r\}$ and the derived invariance of the multiplicities is clear. □

Lemma 3.5. *The algebras to the following pairs of graphs are derived equivalent:*

(i)

(ii)

(iii)

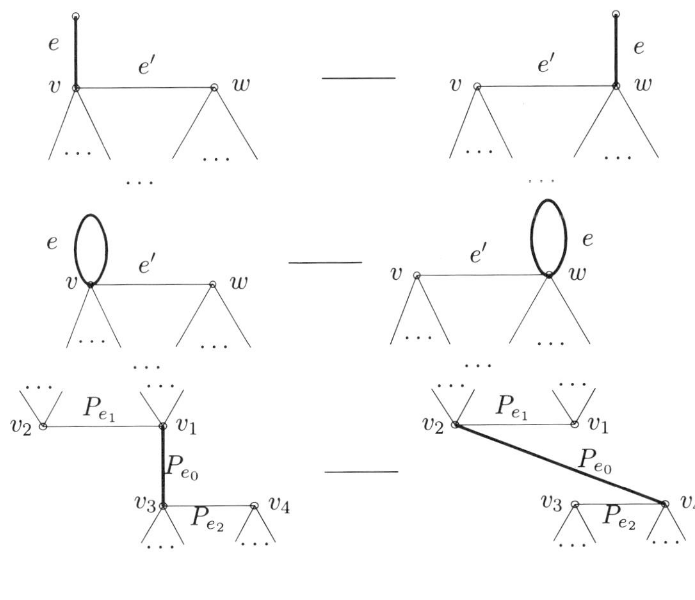

Proof: In each case call the two graphs in question G and G', then we know by Lemma 2.1 that $\Lambda_G \underset{\text{der.}}{\cong} \Lambda_{G'}$ and therefore by Lemma 3.2 that $A_{G,-1} \underset{\text{der.}}{\cong} A_{G',-1}$. Let $T^{\cdot}$ be the tilting complex from $A_{G,-1}$ to $A_{G',-1}$ (induced by the tilting complex from Lemma 2.1). Reducing the entries in $T^{\cdot}$ modulo socle we get a tilting complex $\overline{T}^{\cdot} \in K^b(\text{proj-}\, A_{G,-1}/\operatorname{soc}(A_{G,-1})) = K^b(\text{proj-}\, A_G/\operatorname{soc}(A_G))$.
By [Ri 91,2] Th. 2.1 the tilting complex $\overline{T}^{\cdot}$ can be lifted to a tilting complex $S^{\cdot} \in K^b(\text{proj-}\, A_G)$. But an algebra derived equivalent to a graph algebra is itself a graph algebra (cf. the characterization of graph algebras in [Be 91], Th. 4.18.3), so that $B := \underline{\operatorname{End}}(S^{\cdot})$ is a graph algebra $A_{G''}$. Now B is an algebra extension of $\underline{\operatorname{End}}(\overline{T}^{\cdot})$. $\underline{\operatorname{End}}(\overline{T}^{\cdot})$ may now determined explicitly as $A_{G'}/U$ where U is small enough that $G'' = G'$ is forced. As in Lemma 2.1 we will do this calculation for the situation in (iii) with vertex set of $e_1, e_2 = \{v_1, v_2, v_3, v_4\}$ is a set of size 4, the other cases are done similarly.
To streamline the notation a little bit we set $n_i := n_{v_i}$, $\alpha := \alpha_{v_1}$, $\beta := \alpha_{v_2}$, $\gamma := \alpha_{v_3}$ and $\delta := \alpha_{v_4}$ so we have

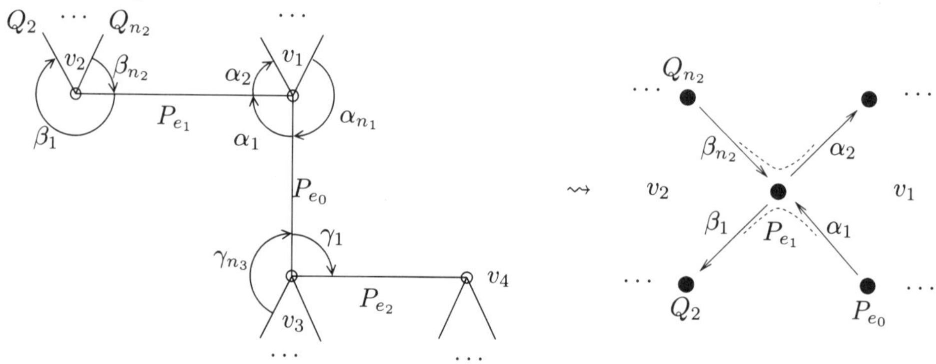

where we display the quiver in a neighborhood at the vertex P_{e_1}. With $T_{e_0}^{\cdot} = P_{e_0} \xrightarrow{(\alpha_1,\gamma_1)} P_{e_1} \oplus P_{e_2}$ displayed vertically the complete cycle corresponding to v_2 is given by

$$\begin{array}{ccccccccc}
 & & P_{e_0} & & & & & & \\
 & & \big\downarrow{\scriptstyle (\alpha_1,\gamma_1)} & & & & & & \\
P_{e_1} & \xrightarrow{(\mathrm{id},0)} & P_{e_1} \oplus P_{e_2} & \xrightarrow{\binom{\beta_1}{0}} & Q_2 & \xrightarrow{\beta_2} & \cdots & \xrightarrow{\beta_{n_2}} & P_{e_1}
\end{array}$$

Not only $(\mathrm{id} \cdot \beta_1 \cdot \ldots \cdot \beta_{n_2})^{m_{v_2}} : P_{e_1} \longrightarrow P_{e_1}$ but also $\begin{pmatrix} (\beta_1 \cdot \ldots \cdot \beta_{n_2})^{m_{v_2}} \\ 0 \end{pmatrix} = \begin{pmatrix} \beta_1 \cdot \ldots \cdot \beta_{n_2} \cdot (\mathrm{id} \cdot \beta_1 \cdot \ldots \cdot \beta_{n_2})^{m_{v_2}-1} \\ 0 \end{pmatrix} : T_{e_0}^{\cdot} \longrightarrow P_{e_1}$ is lying in the socle of $A_{G'}$ and they are therefore zero.
The same happens at the vertex v_4. We have $\underline{\operatorname{End}}(\overline{T}^{\cdot}) = A_{G'}/U$ where $U = \operatorname{soc}(A_{G'})\} \oplus k \cdot \{ \begin{pmatrix} (\beta_1 \cdot \ldots \cdot \beta_{n_2})^{m_{v_2}} \\ 0 \end{pmatrix}, \begin{pmatrix} 0 \\ (\delta_1 \cdot \ldots \cdot \delta_{n_4})^{m_{v_4}} \end{pmatrix} \}$. Because the multiplicities of the vertices are invariant (cf. 3.4), we may now conclude that the only graph algebra extending $A_{G'}/U$ is $A_{G'}$. □

4. Application – Derived equivalence of tame algebras of dihedral type

Algebras of dihedral type over an algebraically closed field k (e. g. blocks of group algebras of dihedral defect) (for notation cf. [Er 90]) with exception $\mathcal{D}(2\mathcal{A})$ and $\mathcal{D}(2\mathcal{B})$ for the parameter $c = 1$ are graph algebras.

For convenience of the reader, we include a list of these algebras and how they are interpreted as graph algebras: as usual the zero relations are given by the dotted lines in the quiver so that we give only the commutativity relations.

name	quiver	non-zero rel.	graph
$D(2\mathcal{A})^l(0)$	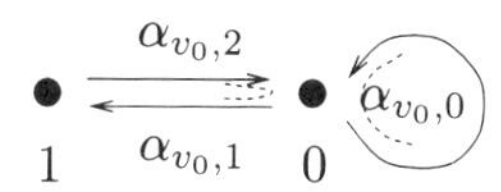	$c^l_{v_0,0} = c^l_{v_0,1}$	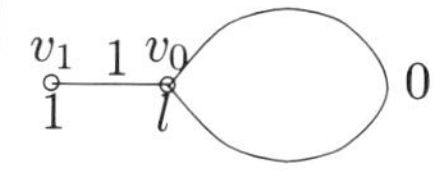
$D(2\mathcal{B})^{l,s}(0)$	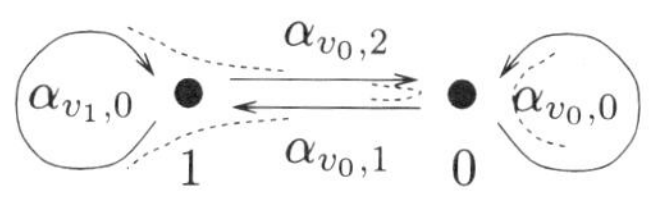	$c^s_{v_1,0} = c^l_{v_0,1}$, $c^l_{v_0,0} = c^l_{v_0,2}$	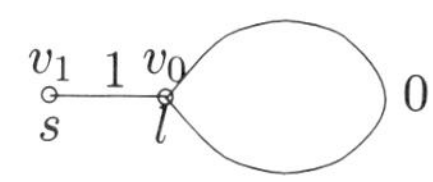
$D(3\mathcal{A})^l_1$	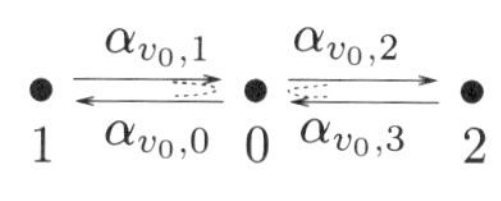	$c^l_{v_0,0} = c^l_{v_0,2}$	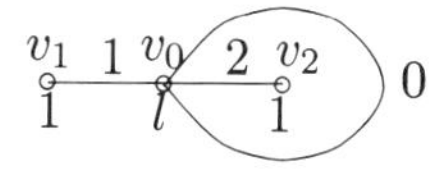
$D(3\mathcal{B})^{l,s}_1$	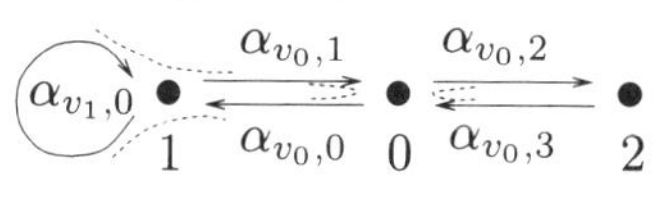	$c^s_{v_1,0} = c^l_{v_0,1}$, $c^l_{v_0,0} = c^l_{v_0,2}$	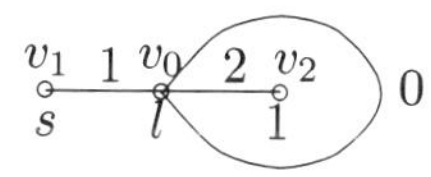
$D(3\mathcal{D})^{l,s,t}_1$	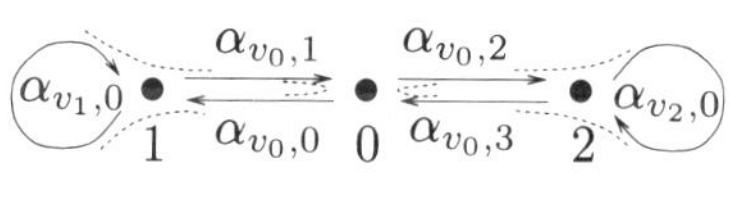	$c^s_{v_1,0} = c^l_{v_0,1}$, $c^l_{v_0,0} = c^l_{v_0,2}$, $c^l_{v_0,3} = c^t_{v_2,0}$	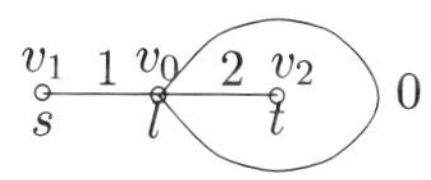
$D(3\mathcal{K})^{l,s,t}_1$	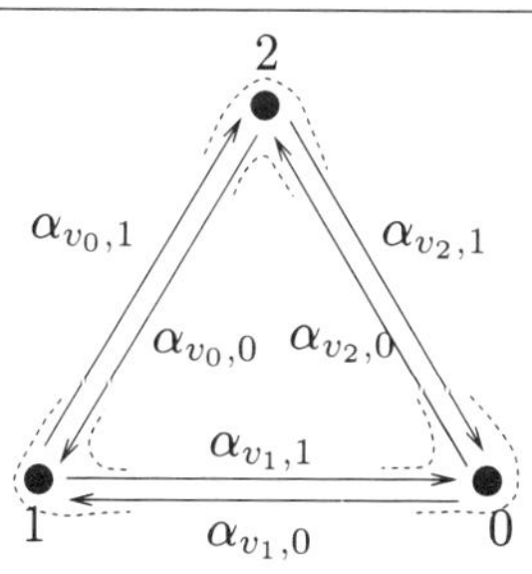	$c^l_{v_0,0} = c^t_{v_2,1}$, $c^t_{v_2,0} = c^s_{v_1,0}$, $c^s_{v_1,1} = c^l_{v_0,1}$	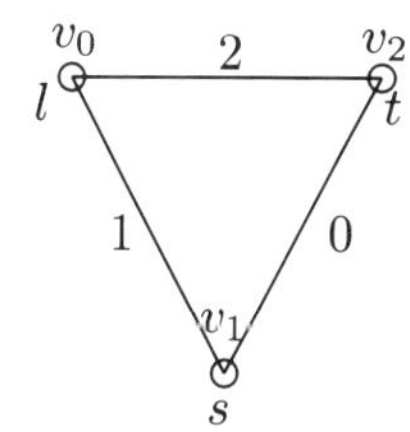
$D(3\mathcal{A})^{l,s}_2$	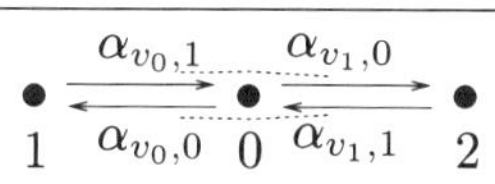	$c^l_{v_0,0} = c^s_{v_1,0}$	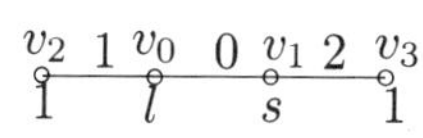
$D(3\mathcal{B})^{l,s,t}_2$	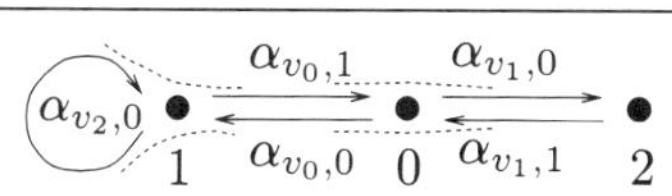	$c^t_{v_2,0} = c^l_{v_0,1}$, $c^l_{v_0,0} = c^s_{v_1,0}$	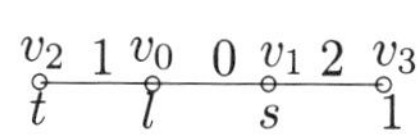

$D(3\mathcal{D})_2^{l,s,t,u}$	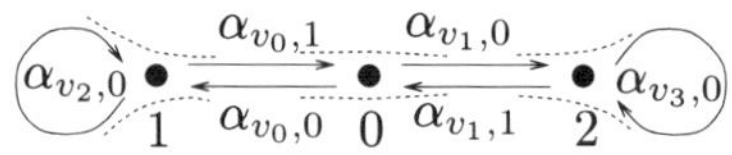	$c^t_{v_2,0} = c^l_{v_0,1}$, $c^l_{v_0,0} = c^s_{v_1,0}$, $c^s_{v_1,1} = c^u_{v_3,0}$	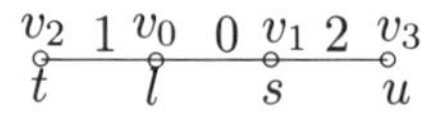
$D(3\mathcal{L})^{l,s}$	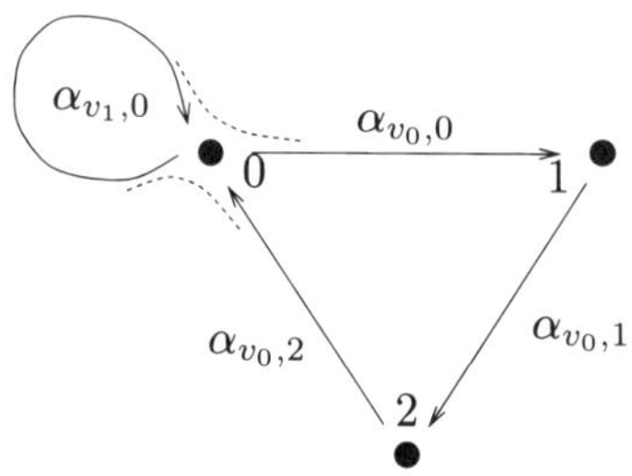	$C^s_{v_1,0} = c^l_{v_0,0}$	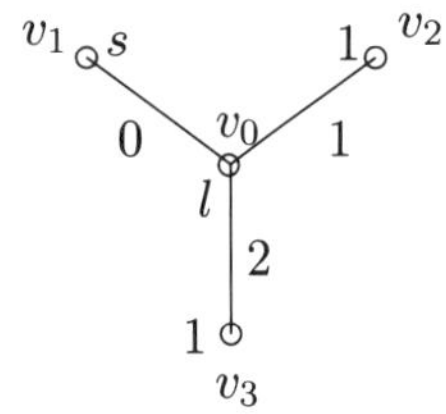
$D(3\mathcal{Q})^{l,s,t}$	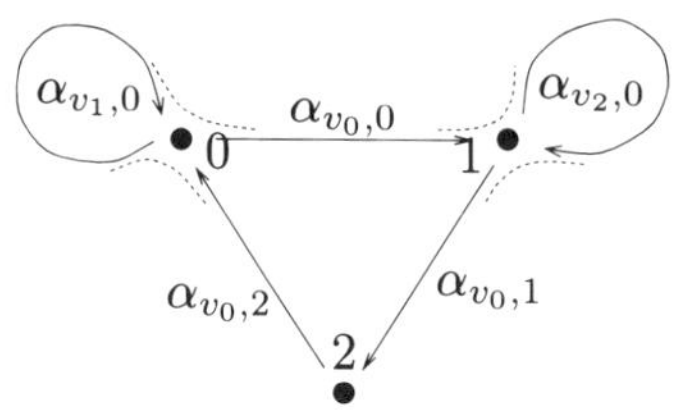	$c^s_{v_1,0} = c^l_{v_0,0}$, $c^t_{v_2,0} = c^l_{v_0,1}$	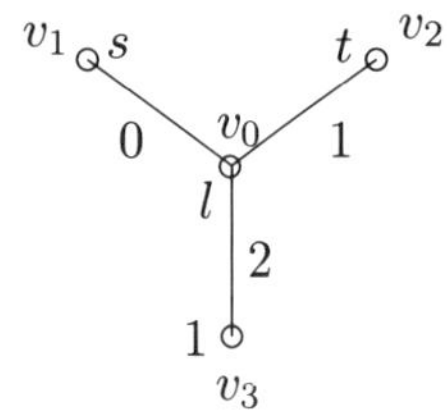
$D(3\mathcal{R})^{l,s,t,u}$	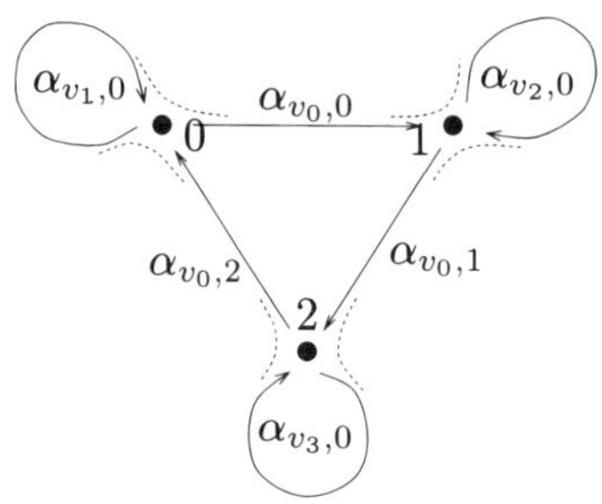	$c^s_{v_1,0} = c^l_{v_0,0}$, $c^t_{v_2,0} = c^l_{v_0,1}$, $c^u_{v_3,0} = c^l_{v_0,2}$	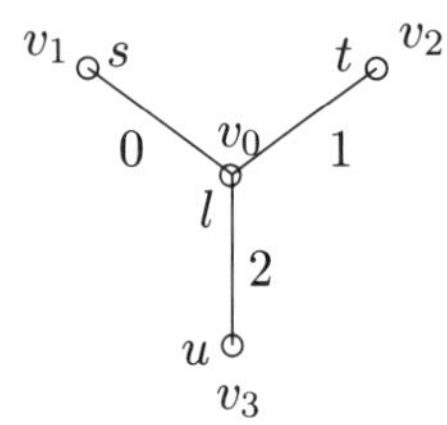

As an application we can see from this table all the (known) derived equivalences of algebras of dihedral type (cf. [Li 94, Ho 97]), for example the cases of blocks of group algebras of dihedral defect where the black vertex denotes the exceptional vertex:

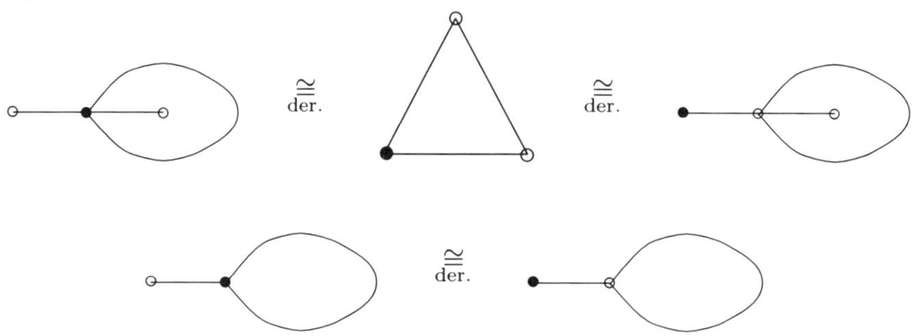

Exactly in case of char $k = 2$, there are algebras of dihedral type, namely $D(2\mathcal{A})^l(1)$ and $D(2\mathcal{B})^{l,s}(1)$, which are not graph algebras. So these can not be derived equivalent to any of the other algebras of dihedral type and we have the following

Corollary 4.1. *If* char $k = 2$, *then the algebras* $D(2\mathcal{A})^l(0)$ *and* $D(2\mathcal{A})^l(1)$, *respectively the algebras* $D(2\mathcal{B})^{l,s}(0)$ *and* $D(2\mathcal{B})^{l,s}(1)$ *are not derived equivalent.*

References

[ARS 95] AUSLANDER, M./REITEN, I./SMALØ, S., *Representation Theory of Artin Algebras*, Cambridge University Press, 1995

[Be 91] BENSON, D., *Representations and Cohomology I*, Cambridge University Press, 1991

[Do 79] DONOVAN, P. W., *Dihedral Defect Groups*, J. of Algebra **56**, (1979), 184-206

[Do-Fr 78] DONOVAN, P. W./FREISLICH, M. R., *The Indecomposable Modular Representations of Certain Groups with Dihedral Sylow Subgroup*, Math. Ann. **238** (1978), 207-216

[Er 90] ERDMANN, K., *Blocks of Tame Representation Type and Related Algebras*, Springer Lecture Notes in Mathematics 1428, Berlin 1990

[Ga-Rie 79] GABRIEL, P./RIEDTMANN, C., *Group representations without groups*, Comment. Math. Helvetici **54**(1979) 240-287

[Ho 97] HOLM, T., *Derived Equivalences Classification of Algebras of Dihedral, Semidihedral and Quaternion Type*, Preprint Magdeburg 1997, 1-42

[Kö-Zi 97] KÖNIG, S./ZIMMERMANN, A., *Tilting selfinjective algebras and Gorenstein orders*, Quat. J. of Math. **48** (1997), 351-361

[Li 94] LINCKELMANN, M., *A Derived Equivalence for Blocks with Dihedral Defect Groups*, J. of Algebra **164** (1994), 244-255

[Ri 89,1] RICKARD, J., *Morita Theory for Derived Equivalences*, J. London Math. Soc. (2) **39** (1989), 436-456

[Ri 89,2] RICKARD, J., *Derived categories and stable equivalence*, J. of Pure and Applied Algebra **61**(1989), 303-317

[Ri 91,1] RICKARD, J., *Derived equivalences as derived functors*, J. London Math. Soc. (2) **43** (1991), 37-48

[Ri 91,2] RICKARD, J., *Lifting Theorems for Tilting Complexes*, J. of Algebra **142** (1991), 383-393

[Ro 92] ROGGENKAMP, K. W., *Blocks with cyclic defect and Green orders*, Com. in Algebra **20** (6) (1992), 1715-1734

[Ro 96] ROGGENKAMP, K. W., *Generalized Brauer Tree Orders*, Coll. Math. **71** (2), (1996), 225-242

[Zi 96] ZIMMERMANN, A., *Derived Equivalences of Orders*, CMS Conference Proceedings, Vol. 18, (1996), 721-749

MATHEMATISCHES INSTITUT B, UNIVERSITÄT STUTTGART, PFAFFENWALDRING 57, D-70550 STUTTGART, GERMANY

Contemporary Mathematics
Volume **229**, 1998

Basic results on wild hereditary algebras

Otto Kerner

Abstract

The basic results on the module category H –mod of finite dimensional modules over a wild hereditary algebra are presented in this paper.

This paper is a moderately extended version of my talk, given at the conference. Let k be an algebraically closed field. A finite dimensional hereditary k-algebra H is called *wild hereditary*, provided for any finite dimensional k-algebra A, there exists a full exact embedding $F_A : A\text{-mod} \to H\text{-mod}$, where A –mod, respectively H –mod denotes the categories of finite dimensional A respectively H-modules. It is not essential for the results, with exception of 5.4, that the ground field k is algebraically closed, but the proofs become more straightforward under this assumption. Hence it is assumed here.

It is not the intention of the paper, to present new developments. All the presented results exist for more than one decade, at least in conjectural form, based on preliminary results. Those readers, who are interested in more recent developments, might for example consult the surveys [7, 11].

It is intended, to present the basic results on wild hereditary algebras. In the first three sections, proofs are omitted. If no references are given, the proofs are well documented in the standard monographs [2, 16].

The category H –reg of regular H-modules is the crucial part of the module category H –mod for a wild hereditary algebra. In the parts 4 and 5 this category is studied. In paragraph 4 the central results about homomorphism between regular modules are presented. In the final part it is shown, that all regular components of the Auslander-Reiten quiver $\Gamma(H)$ are quasi-serial of type $\mathbb{Z}A_\infty$. It is my intention, to present short and direct proofs for these results.

1991 Mathematics Subject Classification.16G20, 16G70.
This paper is in final form and no version of it will be submitted for publication elsewhere.

1 Notations and some linear algebra.

1.1 Let k be an algebraically closed field and H a finite dimensional hereditary k-algebra. The category of finitely generated left H-modules is denoted by H-mod, morphisms are written on the right side. The word module is used for finitely generated left modules.

Let $S_1, \dots, S_n$ be a complete set of representatives of simple H-modules with projective covers $P_1, \dots, P_n$ and injective hulls $I_1, \dots, I_n$.

The quiver $\mathcal{Q}(H) = (\mathcal{Q}_0, \mathcal{Q}_1)$ of H has vertices $\mathcal{Q}_0 = \{1, \dots, n\}$, and there are $\dim \mathrm{Ext}^1_H(S_i, S_j)$ arrows from i to j. If $\alpha \in \mathcal{Q}_1$ is an arrow, $s(\alpha) \in \mathcal{Q}_0$ denotes its starting vertex and $t(\alpha)$ its end. H is determined by $\mathcal{Q}(H)$ up to Morita equivalence. Indeed, the path algebra $k\mathcal{Q}(H)$ is isomorphic to the basis algebra of H. For the study of H-mod it is no restriction, to choose H basic and connected, which will be assumed throughout the paper.

The Grothendieck group $\mathrm{K}_0(H)$ has as basis the n-tuple $([S_1], \dots, [S_n])$ of iso-classes of the simple modules S_i, and we identify $\mathrm{K}_0(H)$ with $\mathbb{Z}^n$ via this basis. For any module X its image in $\mathbb{Z}^n$ is denoted by $\underline{\dim} X = (\dim \mathrm{Hom}_H(P_i, X))_{1 \leq i \leq n} = (\dim \mathrm{Hom}_H(X, I_i))_{1 \leq i \leq n}$.

A module X is called *sincere* if $\underline{\dim} X = (x_1, \dots, x_n)$ is a *strictly positive* vector, that is $x_i > 0$ for all i. It means that each S_i is a composition factor of X. Moreover a nonzero element $(x_1, \dots x_n) \in \mathbb{Z}^n$ is called *positive*, if $x_i \geq 0$ for all i. We say $x < y$ for $x, y \in \mathbb{Z}^n$, if $y - x$ is positive

1.2 Since H is hereditary, the *Cartan matrix* C of H,

$$C = (\dim \mathrm{Hom}_H(P_i, P_j))_{1 \leq i,j \leq n}$$

is in $\mathrm{GL}(n, \mathbb{Z})$. Let $p_i = \underline{\dim} P_i$ and $q_j = \underline{\dim} I_j$. The p_i as well as the q_j form a $\mathbb{Z}$-basis of $\mathbb{Z}^n$, since H is hereditary and C has columns p_i^t, respectively rows q_i.

Consider the following (non symmetric) bilinear form

$$\langle -, - \rangle : \mathbb{Z}^n \times \mathbb{Z}^n \to \mathbb{Z}$$

defined by $\langle x, y \rangle = xC^{-t}y^t$. This bilinear form has the following homological interpretation

Lemma. *For H-modules X, Y one has*

$$\langle \underline{\dim} X, \underline{\dim} Y \rangle = \dim \mathrm{Hom}_H(X, Y) - \dim \mathrm{Ext}^1_H(X, Y).$$

1.3 Let $q_H : \mathbb{Z}^n \to \mathbb{Z}$, $q_H(z) = \langle z, z \rangle$ be the quadratic form, associated to $\langle -, - \rangle$. It is simultaneously the *Tits form* of the quiver $\mathcal{Q} = \mathcal{Q}(H)$, that is $q_H(z_1, \dots, z_n) = \sum_{i \in \mathcal{Q}_0} z_i^2 - \sum_{\alpha \in \mathcal{Q}_1} z_{s(\alpha)} z_{t(\alpha)}$

By $\mathrm{rad} q_H$ we denote the *radical* of q_H that means

$$\mathrm{rad}\, q_H = \{x \in \mathbb{Z}^n \mid \langle x, - \rangle + \langle -, x \rangle = 0\}.$$

Theorem. *Let H be finite dimensional, connected and hereditary with quiver $\mathcal{Q} = \mathcal{Q}(H)$.*

(a) *q_H is positive definite if and only if the underlying graph of $\mathcal{Q}$ is a Dynkin diagram $A_n, n \geq 1$, $D_n, n \geq 4$, E_6, E_7 or E_8. In this case H is representation finite.*

(b) *q_H is positive semidefinite but not positive definite if and only if the underlying graph of $\mathcal{Q}$ is an Euclidean (or affine) graph $\tilde{A}_n, n \geq 1$, $\tilde{D}_n, n \geq 4$, $\tilde{E}_6, \tilde{E}_7$ or $\tilde{E}_8$. In this cases H is called tame hereditary, and* $\operatorname{rad} q$ *is one dimensional with a strictly positive generator h.*

(c) *In all other cases q_H is indefinite. The algebra H then is called wild hereditary and $\mathcal{Q}$ is called a wild quiver.*

1.4 Remark. Wild hereditary algebras H are characterised by the property, that for every finite dimensional k-algebra A there exists a full exact embedding $F_A : A-\text{mod} \to H-\text{mod}$. One can show this fact in the case that H has 2 simple modules, by the explicit construction of the functor F_A in [6]. If H has $n > 2$ simples, one can use for example induction on n by perpendicular categories [5, 4.2].

1.5 Lemma. *Let H be a connected hereditary algebra and assume, there exists a positive radical vector $z = (z_1, \dots z_n) \in \mathbb{Z}^n$ of q_H. Then z is strictly positive and H is tame hereditary.*

Proof. We apply the homological characterisation of $\langle -,- \rangle$. Let M be the semisimple module $M = \oplus_{i=1}^n S_i^{z_i}$. Then $\underline{\dim} M = z \in \operatorname{rad} q_H$. For all simple modules S_i therefore $0 = \langle \underline{\dim} M, \underline{\dim} S_i \rangle + \langle \underline{\dim} S_i, \underline{\dim} M \rangle = 2z_i - \sum_{\alpha \in \mathcal{Q}_1, t(\alpha)=i} z_{s(\alpha)} - \sum_{\alpha \in \mathcal{Q}_1, s(\alpha)=i} z_{t(\alpha)}$ holds. But this means, the quiver $\mathcal{Q}$ admits an additive function in the sense of [8] or [2, VII]. . The claim follows from [8], Theorem 1 or [2, VII.3].

2 Auslander-Reiten translations and Auslander-Reiten quivers.

2.1 Let D denote the duality functor $\mathrm{D} = \operatorname{Hom}_k(-,k) : H-\text{mod} \to \text{mod-}H$. Since H is hereditary, the Auslander-Reiten translation $\tau = \mathrm{DTr}$ is isomorphic to $\mathrm{D}\operatorname{Ext}_H^1(-,H)$, hence it is a left exact functor on $H-\text{mod}$ and it defines an equivalence from the full subcategory $H-\text{mod}_P$ of H-modules without nonzero projective direct summands to the category $H-\text{mod}_I$ of H-modules without nonzero injective summands. Dually, $\tau^- = Tr\mathrm{D} = \operatorname{Ext}_H^1(\mathrm{D}-,H)$ is a right exact functor. Let $\nu = \mathrm{D}\operatorname{Hom}_H(-,H)$ and $\nu^- = \operatorname{Hom}_H(\mathrm{D}-,H)$ denote the Nakayama functors. One has $\nu(P_i) = I_i$ and $\nu(X) = 0$, if X has no nonzero projective direct summand.

Lemma. *If* $0 \to X \to Y \to Z \to 0$ *is a short exact sequence in* H-mod, *then*
$0 \to \tau X \to \tau Y \to \tau Z \to \nu X \to \nu Y \to \nu Z \to 0$ *and*
$0 \to \nu^- X \to \nu^- Y \to \nu^- Z \to \tau^- X \to \tau^- Y \to \tau^- Z \to 0$ *are exact.*

2.2 A central tool is the Auslander-Reiten formula.

Theorem. *For X, Y in H-mod there are functorial isomorphisms*

$$\mathrm{Ext}^1_H(X,Y) \cong \mathrm{D}\,\mathrm{Hom}(Y, \tau X) \cong \mathrm{D}\,\mathrm{Hom}(\tau^- Y, X)$$

2.3 Definition. Let H be a connected hereditary algebra.

(a) An H-module X is called *preprojective* if it satisfies the following equivalent conditions
1) $\tau^m X = 0$ for $m \gg 0$.
2) Each indecomposable direct summand Y of X is of the form $Y = \tau^{-r} P_i$ for some r and i.
3) There exist only finitely many indecomposable modules Z with $\mathrm{Hom}_H(Z, X) \neq 0$. All these modules Z are preprojective, too.

(b) An H-module X is called *preinjective* if it satisfies the following equivalent conditions
1)$\tau^{-m} X = 0$ for $m \gg 0$.
2) Each indecomposable direct summand Y of X is of the form $Y = \tau^r I_i$ for some r and i.
3)There exist only finitely many indecomposable modules Z with $\mathrm{Hom}_H(X, Z) \neq 0$. All these modules Z are preinjective, too.

(c) A module X without nonzero preprojective or preinjective direct summands is called *regular*. In this case $X \cong \tau^m \tau^{-m} X$, for all integers m.

Theorem. *Let H be a connected hereditary algebra.*

(a) *The algebra H is representation finite, if and only if there exists an indecomposable module X, which is preprojective and preinjective, simultaneously. In this case, all modules are preprojective and preinjective, simultaneously.*

(b) *If H is representation infinite, all the modules $\tau^{-r} P_i$ and $\tau^r I_i$ for $r \geq 0$ and $1 \leq i \leq n$ are nonzero and pairwise nonisomorphic. Moreover there exist infinitely many indecomposable regular modules in this case.*

Remark. If H is tame and $h \in \mathbb{Z}^n$ is the minimal positive generator of $\mathrm{rad}\, q_H$, then there exist card$k$ pairwise nonisomorphic indecomposable regular modules $Z_\kappa, \kappa \in k$ with $\underline{\dim} Z_\kappa = h$. If H is wild hereditary, it has a tame hereditary factor-algebra $H' = H/I$ for some ideal I. Considered as H-modules, the H'-modules Z_κ are regular.

2.4 Remember that the iso-classes $[X]$ of the indecomposable H-modules X are the vertices of the Auslander-Reiten quiver $\Gamma(H)$ of H. There are r arrows from $[X]$ to $[Y]$, if the maximal number of linearly independant irreducible maps in $\mathrm{Hom}_H(X,Y)$ modulo $\mathrm{rad}^2(X,Y)$ is r.

If X is indecomposable not injective and $0 \to X \to \oplus X_i^{s_i} \to \tau^- X \to 0$ is the Auslander-Reiten sequence, starting in X, where the X_i are indecomposable and pairwise nonisomorphic, then there are exactly s_i arrows from $[X]$ to $[X_i]$, respectively from $[X_i]$ to $[\tau^- X]$. Dually, if there is an arrow $[X] \to [Y]$ in the

Auslander-Reiten quiver, then $Y \cong X_{i_0}$ for some i_0. The corresponding statement holds for an indecomposable Y, not projective and the Auslander-Reiten sequence $0 \to \tau Y \to \oplus Y_i^{r_i} \to Y \to 0$.

The irreducible maps ending in $[P_i]$, respectively starting in $[I_i]$ one gets from the inclusion $\operatorname{rad} P_i \to P_i$, respectively from the projection $I_i \to I_i/(\operatorname{soc} I_i)$.

Theorem. *Let H be a connected hereditary algebra with quiver $\mathcal{Q}(H) = \mathcal{Q}$.*

(a) *The iso-classes of indecomposable preprojective modules form one connected component in the Auslander-Reiten quiver $\Gamma(H)$ of H, the preprojective component $\mathcal{P}(H)$.*

(b) *The iso-classes of indecomposable preinjective modules form one connected component in the Auslander-Reiten quiver $\Gamma(H)$ of H, the preinjective component $\mathcal{I}(H)$.*

(c) *$\mathcal{P}(H) \neq \mathcal{I}(H)$ if and only if H is representation infinite. In this case $\mathcal{P}(H)$ has the form $\mathbb{N}\mathcal{Q}^{\mathrm{op}}$ and $\mathcal{I}(H)$ has the form $-\mathbb{N}\mathcal{Q}^{\mathrm{op}}$. All other connected components of $\Gamma(H)$ are called regular components.*

3 Coxeter transformations.

3.1 Since the dimension vectors $p_1, \dots, p_n$, respectively $q_1, \dots, q_n$ form a $\mathbb{Z}$-basis of $\mathbb{Z}^n$, the *Coxeter transformation* Φ, defined by $\Phi(p_i) = -q_i$ is in $\mathrm{GL}(n, \mathbb{Z})$.

Lemma. *Let H be a connected hereditary algebra.*
(a) *If X is indecomposable and not projective, then $\underline{\dim}\tau X = \Phi(\underline{\dim} X)$.*
(b) *If Y is indecomposable and not injective, then $\underline{\dim}\tau^{-}Y = \Phi^{-1}(\underline{\dim} Y)$.*

3.2 Since the matrix of Φ is $-C^{-t}C$, where C is the Cartan matrix, the Coxeter transformation is closely related to the homological bilinear form $\langle -, - \rangle$.

Lemma. (a) $\langle x, y \rangle = -\langle y, \Phi(x) \rangle = \langle \Phi(x), \Phi(y) \rangle$.
(b) $\Phi(x) = x$ *if and only if* $x \in \operatorname{rad} q_H$.

If X is an indecomposable module with $\underline{\dim} X = \underline{\dim}\tau^m X$ for some $m > 0$, then $\sum_{i=0}^{m-1} \underline{\dim}\tau^i X$ is a positive radical vector of q_H by 3.1(a) and 3.2(b). Hence 1.5 implies

Corollary. *Let H be connected wild hereditary, X an indecomposable H-module and $m \neq 0$ an integer. Then $\underline{\dim} X \neq \underline{\dim}\tau^m X$ holds.*

3.3 The spectral properties of Φ, considered as linear map $\Phi : \mathbb{C}^n \to \mathbb{C}^n$ are of interest. The growth number ρ_H of H is the spectral radius of Φ, that means $\rho_H = \max\{|\lambda| \mid \lambda \in \operatorname{spec}\Phi\}$. If H is not wild hereditary, then $\rho_H = 1$.

Theorem.[1, 18, 17] *Let H be a connected wild hereditary algebra.*
(a) *$1 < \rho_H$ is an eigenvalue of Φ. It is a simple root of the characteristic polynomial of Φ.*
(b) *For all eigenvalues $\lambda \in \operatorname{spec}\Phi$ with $\lambda \neq \rho_H$ one has $|\lambda| < \rho_H$.*
(c) *There exists a strictly positive eigenvector $y^+ \in \mathbb{R}^n$ for ρ_H.*

For the proofs, the Perron-Frobenius theorem is used.

The linear map Φ^{-1} has the same spectral properties as Φ has. Therefore y^- denotes the strictly positive real vector with $\Phi^{-1}(y^-) = \rho_H y^-$.

3.4 Since $\rho = \rho_H$ is a simple eigenvalue and all other eigenvalues have smaller modulus, it follows from linear algebra, that for $x \in \mathbb{R}^n$ one has $\lim_{m\to\infty} \rho^{-m}\Phi^m(x) = \alpha_x y^+$ for some $\alpha_x \in \mathbb{R}$.

Theorem.[14] *Let H be connected wild hereditary.*
(a) *Let $0 \neq X$ be an H-module without nonzero preprojective direct summand. Then $\lim_{m\to\infty} \rho^{-m}\underline{\dim}\tau^m X = \alpha_X y^+$ for some real $\alpha_X > 0$. Hence $\tau^m X$ is sincere for $m \gg 0$.*
(b) *Let $0 \neq Y$ be an H-module without nonzero preinjective direct summand. Then $\lim_{m\to\infty} \rho^{-m}\underline{\dim}\tau^{-m} Y = \beta_Y y^-$ for some real $\beta_Y > 0$. Hence $\tau^{-m} Y$ is sincere for $m \gg 0$.*

4 Morphisms between regular modules.

4.1 If H is connected wild hereditary, $R \neq 0$ is a regular and $0 \neq X$ is a regular factor module of R with minimal dimension, then X has no nontrivial regular factor modules. Dually, a minimal regular submodule Y of R has no nontrivial regular submodules.

Since the preprojective component is closed under predecessors, by 2.3(a)3, there are no nonzero maps from regulars to preprojectives. Dually preinjective modules do not map nontrivially to regular modules. Consequently the category H −reg of regular H-modules is closed under images and extensions.

Lemma. *Let H be a connected wild hereditary algebra.*

(a) *If R is indecomposable regular and $m > o$, then there are neither monos nor epis in $\mathrm{Hom}_H(R, \tau^{-m}R)$.*

(b)[3] *Let $0 \neq X$ be regular, without nontrivial regular factor modules. Then for some $m > 0$ there exists a short exact sequence $0 \to X \to \tau^m X \to Q \to 0$ with Q preinjective.*

(c)[3] *Let $0 \neq Y$ be regular, without nontrivial regular submodules. Then for some $m > 0$ there exists a short exact sequence $0 \to K \to \tau^{-m}Y \to Y \to 0$ with K preprojective.*

Proof. (a) Suppose $f : R \to \tau^{-m}R$ is surjective. Since τ^- is a right exact functor, also $\tau^{-rm}f : \tau^{-rm}R \to \tau^{-(r+1)m}R$ is surjective and none of these maps is an isomorphism, by 3.2. But there do not exist arbitrary long chains of proper epimorphisms, starting in R. If $g : R \to \tau^{-m}R$ is injective the dual argument works, since τ is left exact.

(b) Consider $X' = \oplus_{i=0}^n \tau^{2i}X$. Then X' has $n+1$ pairwise nonisomorphic direct summands, hence $\mathrm{Ext}_H^1(X', X') \neq 0$ [9]. Therefore, by the Auslander-Reiten formula, there exist some i, j with $\mathrm{Hom}_H(\tau^{2i}X, \tau^{2j+1}X) \cong \mathrm{Hom}_H(X, \tau^{2(j-i)+1}X) \neq 0$. Let $m = 2(j-i)+1$. Since X has no nontrivial regular factor module, a nonzero morphism $f : X \to \tau^m X$ has to be injective. Hence $m > 0$ by (a). Let Q be the cokernel of f. Since f is not an isomorphism, $Q \neq 0$. Since τ^- is right exact, $\tau^m X$ has no nontrivial regular factor modules, too. Therefore Q is preinjective. Part (c) is dual to (b).

Remark. If H' is tame hereditary, then the category H' –reg of regular H'-modules is an abelian category. The above lemma shows that H –reg is not closed under kernels and cokernels, if H is wild hereditary.

4.2 Lemma.[12] *Let H be connected wild hereditary and $X \neq 0$ a regular H-module. Then there exists a natural number N such that for any regular module R and any $m \geq N$ all maps $f \in \operatorname{Hom}_H(\tau^m X, R)$ have regular kernel and all maps $g \in \operatorname{Hom}_H(R, \tau^{-m}X)$ have regular cokernel.*

Proof. The first assertion will be shown, the second is dual. Since the dimensions of preprojective modules grow exponentially in τ^--direction, by 3.4, there exists N such that $\dim \tau^{-m}P > \dim X$ for all nonzero preprojective modules P and all $m \geq N$. Let R be regular and $f : \tau^m X \to R$. Let Q be the image of f and K be the kernel of f. Consider the short exact sequence $0 \to K \to \tau^m X \to Q \to 0$. Since Q is regular, application of τ^{-m} gives the short exact sequence $0 \to \tau^{-m}K \to X \to \tau^{-m}Q \to 0$. By the choice of N the kernel K of f therefore is regular.

4.3 Theorem. *Let H be a connected wild hereditary algebra and X, Y be nonzero regular H-modules.*
(a)[11] $\operatorname{Hom}_H(\tau^m X, Y) = 0$ *for* $m \gg 0$.
(b)[4] $\operatorname{Hom}_H(X, \tau^m Y) \neq 0$ *for* $m \gg 0$.

Proof. (a)[13] There exists only a finite set $\{x^{(1)}, \dots x^{(t)}\}$ of dimension vectors $x^{(i)}$ of regular modules such that $x^{(i)} \leq \underline{\dim} Y$. By 3.4, the vectors $\Phi^{-m}x^{(i)}$ grow exponentially with m. Take N with $\Phi^{-m}x^{(i)} > \underline{\dim} X$ for all $m \geq N$ and all $i = 1, \dots t$. Suppose, there is a nonzero map $f : \tau^m X \to Y$ for some $m \geq N$. Consider the short exact sequence $0 \to \ker f \to \tau^m X \to \operatorname{Im} f \to 0$. The module $\operatorname{Im} f$ is a regular submodule of Y. hence $\underline{\dim} \operatorname{Im} f = x^{(i_0)}$ for some i_o. Application of τ^{-m} gives the short exact sequence

$$0 \to \tau^{-m} \ker f \to X \to \tau^{-m} \operatorname{Im} f \to 0$$

But $\underline{\dim} \tau^{-m} \operatorname{Im} f = \Phi^{-m}(x^{(i_0)}) > \underline{\dim} X$, a contradiction.

(b) Consider first the case, when X has no nontrivial regular factors. By 4.1(b) there exists a short exact sequence

$$(\eta) \quad 0 \to X \to \tau^m X \to Q \to 0$$

with $m > 0$ and Q preinjective. If $Y \neq 0$ is a regular module, then $\tau^t Y$ is sincere for $t \gg 0$, by 3.4(b). Take N such that
(1)$\operatorname{Ext}^1_H(Q, \tau^s Y) \cong \operatorname{D}\operatorname{Hom}_H(\tau^o Y, \tau Q) \neq 0$, for all $s \geq N$, by 3.4(b), and
(2) $\operatorname{Ext}^1_H(\tau^m X, \tau^s Y) \cong \operatorname{D}\operatorname{Hom}_H(\tau^s Y, \tau^{m+1} X) = 0$, for all $s \geq N$, by (a).

The isomorphisms in (1) and (2) are given by the Auslander-Reiten formula 2.2. Application of $\operatorname{Hom}_H(-, \tau^s Y)$ with $s \geq N$ to (η) gives $\cdots \to \operatorname{Hom}_H(X, \tau^s Y) \to \operatorname{Ext}^1_H(Q, \tau^s Y) \to \operatorname{Ext}^1_H(\tau^m X, \tau^s Y) = 0$, Hence $\operatorname{Hom}_H(X, \tau^s Y) \neq 0$.

If X has nontrivial regular factor modules, take a minimal regular factor X' of X. Then $\operatorname{Hom}_H(X', \tau^s Y) \neq 0$ for $s \gg 0$ and consequently $\operatorname{Hom}_H(X, \tau^s Y) \neq 0$ for $s \gg 0$.

Remark. The assertion 4.3(b) also implies, that there are nonzero maps between any two regular components of $\Gamma(H)$, provided H is wild hereditary. For tame hereditary algebras all regular components are pairwise orthogonal.

4.4 If X, Y are nonzero regular, for $s \gg 0$ one has $\text{Ext}^1_H(X, \tau^s Y) = 0$, by 4.3(a) and 2.2. Therefore 1.2 implies $\dim \text{Hom}_H(X, \tau^s Y) = \langle \underline{\dim} X, \underline{\dim} \tau^s Y \rangle$. On the other hand, one has $\lim_{s\to\infty} \rho_H^{-s} \langle \underline{\dim} X, \underline{\dim} \tau^s Y \rangle = \alpha$ for some $0 < \alpha \in \mathbb{R}$, by [14]. Hence the dimensions of the Hom-spaces $\text{Hom}_H(X, \tau^s Y)$ grow exponentially in s.

5 The regular components.

5.1 If A is any finite dimensional k-algebra, a connected component $\mathcal{C}$ of the Auslander-Reiten quiver $\Gamma(A)$ of A is called *regular*, if $\mathcal{C}$ contains neither projective nor injective vertices.

Proposition. *Let A be a finite dimensional algebra, $\mathcal{C}$ a regular component in the Auslander-Reiten quiver $\Gamma(A)$ of A. Assume, for any indecomposable $X \in \mathcal{C}$ there exist A-modules U_X and V_X, with $\text{Hom}_A(U_X, X) \neq 0$, $\text{Hom}_A(U_X, \tau^- X) = 0$ and with $\text{Hom}_A(X, V_X) = 0$, $\text{Hom}_A(\tau^- X, V_X) \neq 0$. Let $X \in \mathcal{C}$ be indecomposable and*

$$0 \to X \xrightarrow{f} \oplus_{i=1}^r E_i \xrightarrow{g} \tau^- X \to 0$$

the Auslander-Reiten sequence, starting in X, with E_i indecomposable. Then $r \leq 2$ holds. For $r = 2$ one has

$$\dim E_2 < \min\{\dim X, \dim \tau^- X\} \text{ and } \max\{\dim X, \dim \tau^- X\} < \dim E_1.$$

Proof. The proof is done in 3 steps. Let $U = U_X$ and $V = V_X$.

(1) Let $E = \oplus_{i=1}^r E_i = M \oplus N$ with $M \neq 0$, and $f = (f_M, f_N) : X \to M \oplus N$, respectively $g = (g_M, g_N)^t : M \oplus N \to \tau^- X$. All the maps f_M, f_N, g_M, g_N are irreducible, hence any of them is injective or surjective. The claim is:

(a) f_M is mono, if and only if g_M is epi, f_M is epi, if and only if g_M is mono.

(b) f_M or f_N is mono, but not both.

Indeed, if f_M and g_M are monos, simultaneously, then $f_M g_M : X \to \tau^- X$ is a mono, too and $(U, f_M g_M) : \text{Hom}_A(U, X) \to \text{Hom}_A(U, \tau^- X)$ is nonzero. Dually if both f_M and g_M are surjective, then $\text{Hom}_A(\tau^- X, V) \neq 0$ implies $\text{Hom}_A(X, V) \neq 0$, since $f_M g_M$ is surjective, a contradiction. This proves (a).

Suppose f_M and f_N both are monos, respectively epis. Then g_M and g_N both are epis, respectively monos, by (a). The equation $\dim X + \dim \tau^- X = \dim M + \dim N$ gives a contradiction, hence (b) holds.

(2) Let $f = (f_1, \ldots, f_r) : X \to \oplus_{i=1}^r E_i$ and $g = (g_1, \ldots, g_r)^t : \oplus_{i=1}^r E_i \to \tau^- X$. The claim is, that exactly one of the f_i is injective.

If $f_i : X \to E_i$ is injective, then $\dim \text{Hom}_A(U, X) \leq \dim \text{Hom}_A(U, E_i)$. On the other hand, $\text{Hom}_A(U, \tau^- X) = 0$ implies $\dim \text{Hom}_A(U, X) = \dim \text{Hom}_A(U, \oplus E_i) = \sum \dim \text{Hom}_A(U, E_i)$. Hence at most one f_i is injective.

In any case there exists an i_0 with $\text{Hom}_A(U, E_{i_0}) \neq 0$. If all f_i are surjective, then all g_i are injective, by (1). But then $(U, g_{i_0}) : \text{Hom}_A(U, E_{i_0}) \to \text{Hom}_A(U, \tau^- X)$ is nonzero, a contradiction.

Choose f_1 to be injective. Consequently $\dim \text{Hom}_A(U, X) = \dim \text{Hom}_A(U, E_1)$ and $\text{Hom}_A(U, E_i) = 0$ for $i > 1$.

(3) $r \leq 2$. Consider the Auslander-Reiten sequences

$$0 \to \tau E_i \stackrel{(g_i',p_i)}{\longrightarrow} X \oplus X' \stackrel{(f_i,q_i)^t}{\longrightarrow} E_i \to 0$$

For $i \geq 2$, application of $\mathrm{Hom}_A(U,-)$ to this short exact sequence gives

$$\dim \mathrm{Hom}_A(U,\tau E_i) = \dim \mathrm{Hom}_A(U, X \oplus X') \geq \dim \mathrm{Hom}_A(U,X)$$

since $\mathrm{Hom}_A(U,E_i) = 0$ by (2).

For $i = 1$ the irreducible map f_1 is injective, consequently, by (1), $g_1' : \tau E_1 \to X$ is surjective. Therefore, again by (1), now applied to the Auslander-Reiten sequence $0 \to \tau X \stackrel{f'}{\to} \oplus \tau E_i \stackrel{g'}{\to} X \to 0$, the map $(g_2', \dots g_r')^t : \oplus_{i=2}^r \tau E_i \to X$ is injective, which means $\dim \mathrm{Hom}_A(U,X) \geq \dim \mathrm{Hom}(U, \oplus_{i=2}^r \tau E_i) = \sum_{i=2}^r \dim \mathrm{Hom}_A(U,\tau E_i) \geq (r-1) \dim \mathrm{Hom}_A(U,X)$, which implies $r \leq 2$

In case $r = 2$, it follows from (1) that f_2, g_1 are surjective and f_1, g_2 are injective, which gives the final claim.

Let $\mathcal{C}$ be as in the proposition. Following [15], an indecomposable $X \in \mathcal{C}$ is called *quasi-simple*, if the Auslander-Reiten sequence $0 \to \tau X \to E \to X \to 0$ has indecomposable middle term E. If X has minimal k-dimension among all indecomposables in $\mathcal{C}$, then X and hence all $\tau^m X$ are quasi-simple.

For X quasi-simple in $\mathcal{C}$, there is an infinite chain of irreducible monos respectively epis

$$X = X(1) \to X(2) \to X(3) \to \cdots \to X(n) \to \cdots$$
$$\cdots \to [m]X \to [m-1]X \to \cdots \to [2]X \to [1]X = X$$

For $Y = X(i)$ (respectively $Z = [i]X$), we call i the *quasi-length* of Y (respectively Z) and X the *quasi-socle* (respectively *quasi-top*) of Y(respectively Z). If follows from the definition and the proposition, that $\tau^m X(i) = (\tau^m X)(i)$ and $\tau^m [i]X = [i](\tau^m X)$ for all integers m.

5.2 Let now H be a connected wild hereditary algebra and $\mathcal{C}$ be a regular component in $\Gamma(H)$. If X is indecomposable in $\mathcal{C}$, there exists by 4.3(a) some $m \geq 0$ with $\mathrm{Hom}_H(\tau^m X, X) \neq 0$ but $\mathrm{Hom}_H(\tau^m X, \tau^- X) = 0$. Simultaneously one has $\mathrm{Hom}_H(X, \tau^{-m-1}X) = 0$ but $\mathrm{Hom}_H(\tau^- X, \tau^{-m-1}X) \neq 0$. Therefore 5.1 applies for all regular components of wild hereditary algebras.

Let now X be quasi-simple in $\mathcal{C}$ and let $X = X(1) \to X(2) \to \cdots \to X(m) \to \cdots$ be a chain of irreducible monos. After having fixed these irreducible maps, we may consider them as inclusions. All the modules $\tau^i X$ are pairwise nonisomorphic, for example by 3.2, hence all the modules $\tau^i X(m)$ are pairwise nonisomorphic.

Theorem.[15] (a) *If Z is indecomposable in $\mathcal{C}$ then $Z \cong \tau^m X(i)$ for some m, i, that is $\mathcal{C}$ is of type $\mathbb{Z}A_\infty$.*

(b) *For $i > j$ we have $X(i)/X(j) \cong \tau^{-j} X(i-j)$.*

Proof (a) Each indecomposable module Z in $\mathcal{C}$ has exactly two immediate predecessors and successors or it is quasi-simple.

(b) We use induction on j. The case $j = 1$ and $i = 2$ is clear, assume $i > 2$ and consider the Auslander-Reiten sequence

$$0 \to X(i) \stackrel{(\epsilon,\pi)}{\to} X(i+1) \oplus \tau^- X(i-1) \stackrel{(\pi',\epsilon')^T}{\to} \tau^- X(i) \to 0$$

Since $X = \ker\pi$ by induction, the condition $\epsilon\pi' = -\pi\epsilon'$ implies $X = \ker\pi'$.

For $j > 1$ we have $X(i)/X(j) \cong (X(i)/X(j-1))/(X(j)/X(j-1)) \cong (\tau^{-j+1}X(i-j+1))/(\tau^{-j+1}X) \cong \tau^{-j}X(i-j)$, the last two isomorphisms by induction.

Property (b) is the reason that these components are called *quasi-serial.*

5.3 Theorem.[19] *Let H be a connected wild hereditary algebra and $\mathcal{C}$ be a regular component of $\Gamma(H)$. If X, Y are indecomposable in $\mathcal{C}$ with $X \not\cong Y$, then $\underline{\dim}X \neq \underline{\dim}Y$.*

Proof. Suppose, the assertion is false. Let U be the quasi-socle of X, say $X = U(r)$ and $Y = \tau^m U(r+i)$ with $-r < i$. Clearly $m \neq 0$ and we may assume $m > 0$. By 3.2 we have $i \neq 0$.

Consider first the case $i > 0$. From $\underline{\dim}U(r) = \underline{\dim}\tau^m U(r+i)$, one gets for all $s \geq 0$ the equation $\underline{\dim}\tau^{sm}U(r) = \Phi^{sm}(\underline{\dim}U(r)) = \Phi^{sm}(\underline{\dim}\tau^m U(r+i)) = \underline{\dim}\tau^{(s+1)m}U(r+i)$, where Φ is the Coxeter transformation. Since $\underline{\dim}\tau^{sm}U(r) < \underline{\dim}\tau^{sm}U(r+i)$, we get an infinite, strictly decreasing chain

$$\underline{\dim}U(r) > \underline{\dim}\tau^m U(r) > \underline{\dim}\tau^{2m}U(r) > \underline{\dim}\tau^{3m}U(r) > \cdots$$

a contradiction.

For $i < 0$, we apply the Coxeter transformations Φ^{-sm} and get dually an infinite, strictly decreasing chain $\underline{\dim}\tau^m U(r+i) > \underline{\dim}U(r+i) > \underline{\dim}\tau^{-m}U(r+i) > \cdots$, again a contradiction.

5.4 Since k is algebraically closed, there exist at most card$k$ iso-classes of indecomposable H-modules, hence there exist at most card$k$ regular components in $\Gamma(H)$, if H is wild hereditary. As already mentioned in 2.3, H has a tame hereditary factor algebra H' and there exist card$k$ pairwise nonisomorphic indecomposable regular H'-modules $Z_\kappa, \kappa \in k$, all of them with the same dimension vector h. Hence all these modules have the same dimension vector, considered as H-modules. Consequently, since they are indecomposable regular H-modules, they are in different Auslander-Reiten components of $\Gamma(H)$, by 5.3.

Corollary. *If H is a connected wild hereditary algebra over some algebraically closed field, then there exist exactly* card k *regular components in* $\Gamma(H)$.

References

[1] N. A'Campo. Sur les valeurs propres de la transformation de Coxeter. Invent. math. **33** (1976), 61–67.

[2] M. Auslander, I. Reiten and S. Smalø. Representation theory of artin algebras. Cambridge Studies in Advanced Mathematics, 1994.

[3] D. Baer. Homological properties of wild hereditary Artin algebras. Proc. ICRA IV, Ottawa 1984, Springer Lecture Notes in Math. **1177** (1986), 1–12.

[4] D. Baer. Wild hereditary artin algebras and linear methods. manuscripta math. **55** (1986), 69–82.

[5] D. BAER. A note on wild quiver algebras and tilting modules. Comm. Algebra **17** (1989), 751–757.

[6] S. BRENNER. Decomposition properties of small diagrams of modules. Symp. Math. Ist. Naz. Alta Mat. **13** (1974), 127–141.

[7] W. CRAWLEY-BOEVEY. Lectures on representations of quivers. More lectures on representations of quivers. Preprints, Oxford 1992.

[8] D. HAPPEL, U. PREISER AND C.M. RINGEL. Vinberg's characterization of Dynkin diagrams using subadditive functions with applications to DTr-periodic modules. Proc. ICRA II, Ottawa 1979, Springer Lect. Notes **832**(1980), 280–294.

[9] D. HAPPEL AND C.M. RINGEL. Tilted algebras. Trans. Amer. Math. Soc. **274** (1982), 399–443.

[10] O. KERNER. Tilting wild algebras. J. London Math. Soc. **39** (1989), 29–47.

[11] O. KERNER, Representations of wild quivers. CMS Conf. Proc. **19** (1996), 65–107.

[12] O. KERNER AND F. LUKAS. Elementary modules, Math. Z. **223** (1996), 421–434.

[13] F. LUKAS. Infinite dimensional modules over wild hereditary algebras. J. London Math. Soc. **44** (1991) 401–419.

[14] J.A. DE LA PEÑA AND M. TAKANE. Spectral properties of Coxeter transformations and applications. Arch. Math. **55** (1990), 120–134.

[15] C.M. RINGEL. Finite dimensional hereditary algebras of wild representation type. Math.Z. **161**(1978), 235–255.

[16] C.M. RINGEL. Tame algebras and integral quadratic forms. Springer Lecture Notes in Math. **1099**, (1984).

[17] C.M. RINGEL. The spectral radius of the Coxeter transformations for a generalized Cartan matrix. Math. Ann. **300** (1994), 331–339.

[18] V.F. SUBBOTIN AND R.B. STEKOL'SHCHIK. Jordan form of Coxeter transformations and applications to representations of finite graphs. Funct. Anal. Appl. **12** (1978), 67–68.

[19] Y. ZHANG. The modules in any component of the AR-quiver of a wild hereditary artin algebra are uniquely determined by their composition factors. Archiv Math. **53** (1989), 250–251.

Mathematisches Institut, Heinrich-Heine-Universität, D-40225 Düsseldorf
email: kerner@mx.cs.uni-duesseldorf.de

Contemporary Mathematics
Volume **229**, 1998

ON MINIMAL APPROXIMATIONS OF MODULES

HENNING KRAUSE AND MANUEL SAORÍN

Let R be a ring and consider the category $\operatorname{Mod} R$ of (right) R-modules. Given a class $\mathcal{C}$ of R-modules, a morphism $M \to N$ in $\operatorname{Mod} R$ is called a left $\mathcal{C}$-approximation of M provided that N belongs to $\mathcal{C}$ and the induced morphism $\operatorname{Hom}_R(N, C) \to \operatorname{Hom}_R(M, C)$ is surjective for every C in $\mathcal{C}$. This concept has been introduced by Auslander and Smalø [2], and independently by Enochs [5]. Approximations arise naturally in various situations, for instance in representation theory of finite dimensional algebras, in particular in tilting theory, or in commutative algebra. In this note we address the problem of finding minimal approximations. In general, a left $\mathcal{C}$-approximation $M \to N$ of M is not uniquely determined by M, but it happens quite often that there exists an approximation $\phi\colon M \to N$ which is left minimal in the sense that any endomorphism ψ of N satisfying $\psi \circ \phi = \phi$ is an isomorphism. Note that such a minimal approximation of M is unique up to isomorphism.

This paper is divided into four sections. In Section 1 we characterize the morphisms $\phi\colon M \to N$ which admit a decomposition $\phi = (\phi', \phi'')\colon M \to N' \coprod N'' = N$ such that ϕ' is left minimal and $\phi'' = 0$. For example, such a decomposition always exists if N is pure-injective. In Section 2 we discuss criteria for a class $\mathcal{C}$ of R-modules such that every R-module admits a left $\mathcal{C}$-approximation. In Section 3 we introduce a new class of modules M which we call product-complete since the class $\operatorname{Add} M$ of direct summands of coproducts of copies of M is closed under taking products. We prove that an R-module M is product-complete if and only if every R-module admits a (minimal) left $\operatorname{Add} M$-approximation. In fact, we provide a host of characterizations of such modules which indicate that product-complete modules are particulary well-behaved. For example, every endofinite module is product-complete. More precisely, we prove in Section 4 that a module M is endofinite if and only if every direct summand of M is product-complete.

This paper originates from discussions during a visit of the first named author at the University of Murcia. He would like to thank for financial support and hospitality which made this stay highly pleasant.

1. Left minimal morphisms

Let R be a an associative ring with identity. We consider the category $\operatorname{Mod} R$ of (right) R-modules and denote by $\operatorname{mod} R$ the full subcategory of finitely presented R-modules. A morphism $\phi\colon M \to N$ in any category is said to be *left minimal* if any endomorphism ψ of N satisfying $\psi \circ \phi = \phi$ is an isomorphism. A concept which

1991 *Mathematics Subject Classification.* 16D70, 16D90, 16P70, 18A40.

The second named author has been partially supported by the D.G.E.S. of Spain (proyecto PB96-0961-C02-02).

is closely related is that of an injective envelope. Recall that a monomorphism $\phi\colon M \to N$ in any abelian category is an *injective envelope* of M provided that N is an injective object and every morphism $N \to N'$ is a monomorphism if the composition with ϕ is a monomorphism.

Lemma 1.1. *A monomorphism $\phi\colon M \to N$ into an injective object N is an injective envelope if and only if ϕ is left minimal.*

Proof. Straightforward. □

The following result gives the precise relation between left minimal morphisms and injective envelopes.

Proposition 1.2. *Let $\phi\colon M \to N$ be a morphism in some additive category $\mathcal{A}$ and suppose that idempotents in* End(N) *split. Then the following are equivalent:*

(1) *There exists a decomposition $\phi = (\phi', \phi'')\colon M \to N' \coprod N'' = N$ such that ϕ' is left minimal and $\phi'' = 0$.*
(2) *There exists an abelian category $\mathcal{B}$ and an additive functor $F\colon \mathcal{A} \to \mathcal{B}$ such that*
   (a) *$F(N)$ is injective;*
   (b) *F induces an isomorphism* $\operatorname{Hom}(M \coprod N, N) \to \operatorname{Hom}(F(M \coprod N), F(N))$;
   (c) *the image* $\operatorname{Im} F(\phi)$ *has an injective envelope.*

A decomposition $\phi = (\phi', \phi'')\colon M \to N' \coprod N'' = N$ of a morphism ϕ such that ϕ' is left minimal and $\phi'' = 0$ is unique up to isomorphism. The morphism ϕ' is sometimes called the *minimal version* of ϕ.

Proof. We apply the characterization of injective envelopes from Lemma 1.1.

(1) $\Rightarrow$ (2) Let $S = \operatorname{End}(N)^{\mathrm{op}}$ and consider the functor

$$F\colon \mathcal{A} \longrightarrow \mathcal{B} = (\operatorname{Mod} S)^{\mathrm{op}}, \quad X \mapsto \operatorname{Hom}(X, N).$$

It is clear that (a) and (b) are satisfied. A decomposition $\phi = (\phi', \phi'')\colon M \to N' \coprod N'' = N$ of ϕ such that ϕ' is left minimal and $\phi'' = 0$ induces a morphism $\operatorname{Im} F(\phi) \to F(N')$ in $\mathcal{B}$ which is an injective envelope by Lemma 1.1.

(2) $\Rightarrow$ (1) Let $F\colon \mathcal{A} \to \mathcal{B}$ be an additive functor and suppose that $\varepsilon\colon \operatorname{Im} F(\phi) \to Q$ is an injective envelope. Consider the induced morphism $\psi\colon \operatorname{Im} F(\phi) \to F(N)$. There are morphisms $\alpha\colon Q \to F(N)$ and $\beta\colon F(N) \to Q$ satisfying $\alpha \circ \varepsilon = \psi$ and $\beta \circ \psi = \varepsilon$ since Q and $F(N)$ are injective. We may assume that $\beta \circ \alpha = \mathrm{id}_Q$ since ε is an injective envelope. Thus $\alpha \circ \beta$ is an idempotent in $\operatorname{End}(F(N))$ and we obtain therefore a decomposition $N = N' \coprod N''$ in $\mathcal{A}$ with $F(N') = \operatorname{Im} \alpha$. It is now easily checked that in the induced decomposition $\phi = (\phi', \phi'')\colon M \to N' \coprod N'' = N$ of ϕ the morphism ϕ' is left minimal and $\phi'' = 0$. □

Our main application of the preceding result depends on the well-known fact that the fully faithful functor

$$\operatorname{Mod} R \longrightarrow D(R) = (\operatorname{mod} R^{\mathrm{op}}, \mathrm{Ab}), \quad M \mapsto M \otimes_R -$$

from $\operatorname{Mod} R$ into the category of additive functors $\operatorname{mod} R^{\mathrm{op}} \to \mathrm{Ab}$ identifies the pure-injective R-modules with the injective objects in $D(R)$. Recall that a map

$M \to N$ in $\operatorname{Mod} R$ is a *pure monomorphism* if $M \otimes_R X \to N \otimes_R X$ a monomorphism for every X in $\operatorname{mod} R^{\mathrm{op}}$, and M is *pure-injective* if every pure monomorphism $M \to N$ splits.

Corollary 1.3. *Let $\phi\colon M \to N$ be a morphism in $\operatorname{Mod} R$ and suppose that N is pure-injective. Then there exists a decomposition*

$$\phi = (\phi', \phi'')\colon M \longrightarrow N' \coprod N'' = N$$

such that ϕ' is left minimal and $\phi'' = 0$.

Proof. The Grothendieck category $D(R)$ has injective envelopes, and the assertion is therefore a consequence of Proposition 1.2. □

Another application of Proposition 1.2 shows that any morphism in a Krull-Schmidt category has a minimal version.

Corollary 1.4. *Let $\phi\colon M \to N$ be a morphism in some additive category and suppose that idempotents in $\operatorname{End}(N)$ split. If N is a finite coproduct of indecomposable objects with local endomorphism ring, then there exists a decomposition*

$$\phi = (\phi', \phi'')\colon M \longrightarrow N' \coprod N'' = N$$

such that ϕ' is left minimal and $\phi'' = 0$.

Proof. We consider as in the proof of Proposition 1.2 the functor

$$F\colon \mathcal{A} \longrightarrow \mathcal{B} = (\operatorname{Mod} S)^{\mathrm{op}}, \quad X \mapsto \operatorname{Hom}(X, N)$$

where $S = \operatorname{End}(N)^{\mathrm{op}}$. The assumption on N implies that S is a semi-perfect ring, and therefore every finitely generated S-module has a projective cover in $\operatorname{Mod} S$. By duality, it follows that $\operatorname{Im} F(\phi)$ has an injective envelope in $\mathcal{B}$, and the assertion is now a consequence of Proposition 1.2. □

Remark 1.5. We leave it to the reader to formulate the dual statements of Proposition 1.2 and Corollary 1.4. For instance, the dual of Corollary 1.4 generalizes the well-known fact that any finitely generated module over a semi-perfect ring has a projective cover.

2. Left approximations

Let $\mathcal{A}$ be any category and $\mathcal{C}$ be any class of objects in $\mathcal{A}$. Following [2], a morphism $M \to N$ in $\mathcal{A}$ is called a *left $\mathcal{C}$-approximation* of M provided that N belongs to $\mathcal{C}$ and the induced morphism $\operatorname{Hom}(N, C) \to \operatorname{Hom}(M, C)$ is surjective for every C in $\mathcal{C}$. If every object in $\mathcal{A}$ has a left $\mathcal{C}$-approximation, then $\mathcal{C}$ is said to be *covariantly finite* in $\mathcal{A}$. Note that this concept has been introduced independently in [5].

We begin our discussion with a criterion for a class of modules to be covariantly finite. Suppose that $\mathcal{C}$ is a class of R-modules which is closed under direct summands. It is easily seen that $\mathcal{C}$ needs to be closed under taking products if $\mathcal{C}$ is covariantly finite in $\operatorname{Mod} R$.

Lemma 2.1. *Let $\mathcal{C}$ be any class of R-modules which is closed under direct summands. If $\mathcal{C}$ is covariantly finite in* $\operatorname{Mod} R$, *then $\mathcal{C}$ is closed under taking products.*

Proof. Let $M = \prod_{i \in I} M_i$ be a product of modules in $\mathcal{C}$ and let $\phi\colon M \to N$ be a left $\mathcal{C}$-approxiamation. Every projection $M \to M_i$ factors through ϕ and therefore ϕ is a split monomorphism. It follows that M belongs to $\mathcal{C}$ since $\mathcal{C}$ is closed under direct summands by assumption. □

In [13], it has been shown that also the converse is true if $\mathcal{C}$ is closed under pure submodules.

Lemma 2.2. *Let $\mathcal{C}$ be any class of R-modules which is closed under pure submodules. If $\mathcal{C}$ is closed under taking products, then $\mathcal{C}$ is covariantly finite in* $\operatorname{Mod} R$.

Proof. See Corollary 3.5 (c) in [13]. □

The next result is essentially a reformulation of Corollary 1.3.

Theorem 2.3. *Let $\mathcal{C}$ be any class of modules which is closed under direct summands. Let M be an R-module and suppose that M has a left $\mathcal{C}$-approximation which is pure-injective. Then M has a minimal left $\mathcal{C}$-approximation.*

Proof. Apply Corollary 1.3. □

We present an interesting consequence.

Corollary 2.4. *Let $\mathcal{C}$ be any class of pure-injective R-modules which is closed under direct summands. Then the following are equivalent:*

(1) *Every R-module has a left $\mathcal{C}$-approximation.*
(2) *Every R-module has a minimal left $\mathcal{C}$-approximation.*
(3) *$\mathcal{C}$ is closed under taking products.*

Proof. (1) $\Rightarrow$ (2) Use Theorem 2.3.

(2) $\Rightarrow$ (3) Use Lemma 2.1.

(3) $\Rightarrow$ (1) Let $\mathcal{D}$ be the class of pure submodules of modules in $\mathcal{C}$. Then every R-module M has a left $\mathcal{D}$-approximation $M \to N$ by Lemma 2.2. Suppose that N is the pure submodule of $N' \in \mathcal{C}$. It is clear that the composition $M \to N'$ is a left $\mathcal{C}$-approximation. □

Example 2.5. Let $\mathcal{C}$ be the class of all pure-injective R-modules. Then a morphism $\phi\colon M \to N$ is a minimal left $\mathcal{C}$-approximation if and only if ϕ is a pure-injective envelope of M (i.e. N is pure-injective and any map $\psi\colon N \to N'$ is a pure monomorphism if and only if $\psi \circ \phi$ is a pure monomorphism). Indeed, the canonical map $\phi_M\colon M \to D^2M$ (where $D = \operatorname{Hom}_{\mathbb{Z}}(-, \mathbb{Q}/\mathbb{Z})$) is a left $\mathcal{C}$-approximation since D^2M is pure-injective and ϕ_M is a pure monomorphism. Therefore a minimal left $\mathcal{C}$-approximation $M \to N$ is a pure monomorphism, hence a pure-injective envelope. Conversely, if $\phi\colon M \to N$ is a pure-injective envelope, then any endomorphism ψ of N satisfying $\phi = \psi \circ \phi$ needs to be a pure-monomorphism, hence a split monomorphism, and therefore an isomorphism. The fact that every module has a pure-injective envelope is a classical result of Kiełpiński [10].

3. Product-complete modules

Given any R-module M we denote by Add M the full subcategory of R-modules which are direct summands of coproducts of copies of M. Note that Add M is the smallest full subcategory of Mod R which contains M and is closed under forming arbitrary coproducts and direct summands. Our main result is a characterization of the fact that Add M is closed under taking products.

Theorem 3.1. *The following are equivalent for an R-module M:*

(1) *Add M is closed under taking products.*
(2) *Every product of copies of M is a direct summand of a coproduct of copies of M.*
(3) *Every product of copies of M is a coproduct of (indecomposable) direct summands of M.*
(4) *Every R-module has a (minimal) left* Add *M-approximation.*

We call a module M *product-complete* if M satisfies the equivalent conditions of the preceding theorem.

Recall that a module M is *Σ-pure-injective* if every coproduct of copies of M is pure-injective. We shall work with the following characterization of a Σ-pure-injective module.

Lemma 3.2. *A module M is Σ-pure-injective if and only if there exists a cardinal κ such that every product of copies of M is a pure submodule of a coproduct of modules having cardinality at most κ.*

Proof. See [9]. □

We shall also need the following property of a Σ-pure-injective module.

Lemma 3.3. *Any Σ-pure-injective module is a coproduct of indecomposable modules with local endomorphism ring.*

Proof. See [9, 16, 17]. □

Proof of Theorem 3.1. (1) $\Rightarrow$ (2) Clear.

(2) $\Rightarrow$ (3) M is Σ-pure-injective by Lemma 3.2. It follows from Lemma 3.3 in combination with the Krull-Remak-Schmidt-Azumaya Theorem that every direct summand of a coproduct of copies of M decomposes into indecomposable modules which are direct summands of M. Thus (2) implies (3).

(3) $\Rightarrow$ (4) M is Σ-pure-injective by Lemma 3.2. Thus Add M consists of pure-injective modules and is closed under taking products. It follows from Corollary 2.4 that every R-module has a minimal left Add M-approximation.

(4) $\Rightarrow$ (1) Use Lemma 2.1. □

We give an example of a Σ-pure-injective module which is not product-complete.

Example 3.4. Let $R = \mathbb{Z}$ and let $M = \mathbb{Z}_{p^\infty}$ be a Prüfer group. This module is Σ-pure-injective but not product-complete. However $M \coprod \mathbb{Q}$ is product-complete.

We list some basic properties of product-complete modules.

(1) The class of product-complete modules is closed under finite coproducts.

(2) Let I be any non-empty set. Then a module M is product-complete if and only if $M^{(I)}$ is product-complete.

(3) If M is product-complete, then M^I is product-complete for every set I.

We have the following categorical characterization of product-complete modules.

Proposition 3.5. *An R-module M is product-complete if and only if there exists a (locally noetherian) Grothendieck category $\mathcal{A}$ and a functor $F\colon \mathcal{A} \to \operatorname{Mod} R$ which commutes with products and coproducts, such that F induces an equivalence between the full subcategory* $\operatorname{Inj}\mathcal{A}$ *of injective objects in $\mathcal{A}$ and* $\operatorname{Add} M$.

Proof. We use the functor category $D(R) = (\operatorname{mod} R^{\mathrm{op}}, \mathrm{Ab})$. Suppose that M is product-complete and consider the full subcategory $\mathcal{T}$ of all objects X in $D(R)$ with $\operatorname{Hom}(X, M \otimes_R -) = 0$. This is a localizing subcategory and we can therefore form the quotient category $\mathcal{A} = D(R)/\mathcal{T}$, e.g. see [6]. The composition of the embedding $\operatorname{Mod} R \to D(R)$, $N \mapsto N \otimes_R -$, with the quotient functor $D(R) \to \mathcal{A}$ identifies $\operatorname{Add} M$ with $\operatorname{Inj}\mathcal{A}$ since $M \otimes_R -$ is an injective cogenerator for $\mathcal{A}$, and therefore $\mathcal{A}$ is locally noetherian since $\operatorname{Inj}\mathcal{A}$ is closed under forming coproducts [14]. The composition of the right adjoint of the quotient functor with the functor $D(R) \to \operatorname{Mod} R$, $X \mapsto X(R)$, gives the functor $F\colon \mathcal{A} \to \operatorname{Mod} R$ which has the desired properties. Conversely, suppose there exists a functor $F\colon \mathcal{A} \to \operatorname{Mod} R$ which commutes with products and coproducts, such that F induces an equivalence $\operatorname{Inj}\mathcal{A} \to \operatorname{Add} M$. It follows that $\operatorname{Add} M$ is closed under taking products since $\operatorname{Inj}\mathcal{A}$ is closed under taking products, and therefore $\operatorname{Add} M$ is product-complete. □

Corollary 3.6. *If M is product-complete, then* $\operatorname{Add} M$ *is closed under taking direct limits.*

Proof. If $\mathcal{A}$ is a locally noetherian category, then $\operatorname{Inj}\mathcal{A}$ is closed under taking direct limits [6, p.358]. The assertion is therefore a consequence of the construction given in the preceding proof. □

In [18], Ziegler introduced a topology on the set $\operatorname{Ind} R$ of isomorphism classes of indecomposable pure-injective R-modules, and it is also possible to characterize the product-complete modules in terms of this topology.

Proposition 3.7. *A module M is product-complete if and only if M is Σ-pure-injective and the indecomposable direct summands of M form a Ziegler-closed subset of* $\operatorname{Ind} R$.

Proof. In [11, Proposition 8.7] it is shown that for every Σ-pure-injective module M the indecomposable direct summands of products of copies of M form a Ziegler-closed set. Using the characterization (2) in Theorem 3.1 the assertion follows. □

The argument in the preceding proof leads to a new characterization of Σ-pure-injective modules.

Corollary 3.8. *A module M is Σ-pure-injective if and only if there exists a non-empty set I such that M^I is product-complete.*

Proof. One direction is clear. Suppose therefore that M is Σ-pure-injective. The argument given in the proof of Proposition 3.7 shows that we can choose a non-empty set I such that the indecomposable direct sumands of M^I form a Ziegler-closed set. Thus M^I is product-complete by Proposition 3.7. $\square$

We now characterize the finitely generated modules which are product-complete.

Proposition 3.9. *A finitely generated module M is product-complete if and only if $S = \operatorname{End}_R(M)$ is left coherent and right perfect, and M is finitely presented as S^{op}-module.*

Proof. The functors $\operatorname{Hom}_R(M,-)$ and $-\otimes_S M$ induce mutually inverse equivalences between $\operatorname{Add} M$ and the full subcategory $\operatorname{Proj} S$ of projective S-modules since M is finitely generated. In particular, the canonical map $\varepsilon_N\colon \operatorname{Hom}_R(M,N)\otimes_S M \to N$ is an isomorphism for all N in $\operatorname{Add} M$. Now suppose that M is product-complete. Since $M^I \in \operatorname{Add} M$, we get that $S^I \in \operatorname{Proj} S$, for every set I. A classical result of Chase says that S is right perfect and left coherent [3, Theorem 3.3]. Moreover, since ε_{M^I} is an isomorphism for every set I, due to our above observation, the canonical map $S^I \otimes_S M \to M^I$ is an isomorphism, for every set I. This is well-known to be equivalent to the fact that M is finitely presented as S^{op}-module; see, e.g., [15, Lemma I.13.2]. To prove the converse, we need to check that $M^I \in \operatorname{Add} M$ for every set I. We have $S^I \in \operatorname{Proj} S$ since S is left coherent and right perfect, and $S^I \otimes_S M \simeq M^I$ since M is finitely presented over S^{op}. Thus $M^I \in \operatorname{Add} M$, and therefore M is product-complete. $\square$

Remark 3.10. (1) When $M = R$ in the above proposition, we get that every R-module has a minimal projective left approximation if and only if R is right perfect and left coherent [1, Proposition 3.5].

(2) The hypothesis that M is finitely generated cannot be omitted in the above proposition. Indeed, if R is a right noetherian ring and M is a coproduct of a representative set of indecomposable injective R-modules, then $\operatorname{Add} M = \operatorname{Inj} R$, so that M is product-complete. However, in general, M has an infinite number of direct summands and so $S = \operatorname{End}_R(M)$ is not right perfect.

4. Endofinite modules

Recall that an R-module M is *endofinite* if it has finite length as a module over its endomorphism ring $\operatorname{End}_R(M)^{\mathrm{op}}$. This class of modules has been studied by various authors. Crawley-Boevey noticed their importance in representation theory of finite dimensional algebras, and he introduced their name [4]. In this section we discuss the relation between endofinite and product-complete modules. The basic result is the following.

Theorem 4.1. *A module M is endofinite if and only if every direct summand of M is product-complete.*

Proof. Suppose first that M is indecomposable. It is a well-known result due to Gruson [8] and Garavaglia [7] that M is endofinite if and only if $\operatorname{Add} M$ is closed under taking products. We refer to [12, Corollary 10.5] for a proof since

[8] contains no proof and [7] has never been published. Now suppose that M is an arbitrary module. We use the fact that M is endofinite if and only if there are indecomposable endofinite modules $M_1, \dots, M_n$ and sets $I_1, \dots, I_n$ such that $M \simeq \coprod_{i=1}^{n} M_i^{(I_i)}$; see [4, Proposition 4.5]. The fact that every endofinite module is product-complete is now a consequence of the basic properties of product-complete modules listed after Theorem 3.1, and the fact that every indecomposable endofinite module is product-complete. Conversely, let M be a module such that every direct summand is product-complete. It follows from Proposition 3.7 that the indecomposable direct summands of M form a Ziegler-closed subset $\mathbf{U}$ of $\operatorname{Ind} R$ having the property that every subset $\mathbf{V} \subseteq \mathbf{U}$ is also Ziegler-closed. However, $\operatorname{Ind} R$ is a quasi-compact space [18, Theorem 4.9] and therefore $\mathbf{U}$ needs to be finite. Thus M is endofinite since each $N \in \mathbf{U}$ is endofinite and M is a coproduct of modules in $\mathbf{U}$ by Theorem 3.1. This completes the proof. □

We fix an R-module M with $S = \operatorname{End}_R(M)^{\mathrm{op}}$. Recall that an additve subgroup of M is of *finite definition* if it arises as the kernel of a morphism $M \to M \otimes_R X$, $m \mapsto m \otimes x$, for some finitely presented R^{op}-module X, and some element $x \in X$. Note that a subgroup of finite definition is a S-submodule of M. If M is Σ-pure-injective, then any finitely generated S-submodule of M is a subgroup of finite definition. We shall now prove that every subgroup of finite definition is a finitely generated S-module provided that M is product-complete.

Proposition 4.2. *A module M is product-complete if and only if the following conditions hold:*

1. *(1) M has the descending chain condition on subgroups of finite definition.*
2. *(2) Every subgroup of finite definition is a finitely generated $\operatorname{End}_R(M)^{\mathrm{op}}$-module.*

Proof. We use again the embedding $\operatorname{Mod} R \to D(R)$, $N \mapsto T_N = N \otimes_R -$. Identifying $M = \operatorname{Hom}(T_R, T_M)$, any additive subgroup U is of finite definition if and only if $U = \operatorname{Hom}(T_R/F, T_M)$ for some finitely generated subfunctor $F \subseteq T_R$. Note that U is finitely generated as S-module if and only if there exists a map $\phi\colon T_R/F \to T_{M^n}$ for some $n \in \mathbb{N}$ such that every map $T_R/F \to T_M$ factors through ϕ. Now suppose that M is product-complete. Then M is Σ-pure-injective and it is well-known that (1) holds [9, 16]. To check (2) let $U = \operatorname{Hom}(T_R/F, T_M)$ be a subgroup of finite definition. There exists a map $\phi\colon T_R/F \to T_{M^I}$ for some set I such that every map $T_R/F \to T_M$ factors through ϕ. We find a set J such that M^I is a direct summand of $M^{(J)}$, and the map $T_R/F \to T_{M^I} \to T_{M^{(J)}}$ factors through $T_{M^{(J')}}$ for some finite subset $J' \subseteq J$ since F/T_R is finitely generated. Denoting by $\phi'\colon T_R/F \to T_{M^{(J')}}$ the corresponding map it is clear that every map $T_R/F \to T_M$ factors through ϕ'. Thus U is a finitely generated S-module and (2) holds. Using similar arguments it is not hard to check that M is product-complete if (1) and (2) hold. This is left to the reader. □

Lemma 4.3. *M is endofinite if and only if M is Σ-pure-injective and a noetherian module over S.*

Proof. It is well-known that M is Σ-pure-injective if and only if M has the dcc on subgroups of finite definition [9, 16]. Thus every endofinite module is Σ-pure-injective since every subgroup of finite definition of M is a S-submodule. Conversely, if M is Σ-pure-injective, then every finitely generated S-submodule is a subgroup of finite definition; e.g. see [1, Lemma 4.1]. It follows that M is artinian over S since every S-submodule of M is finitely generated if M is noetherian over S. Thus M has finite endolength. □

Corollary 4.4. *Let M be an R-module and suppose that $\operatorname{End}_R(M)$ is left noetherian. Then M is endofinite if and only if M is product-complete.*

Proof. If M is product-complete, then M is finitely generated over $S = \operatorname{End}_R(M)^{\mathrm{op}}$ by Proposition 4.2. Thus M is noetherian over S and the assertion follows from Lemma 4.3. □

The semi-simple modules form another class of modules where endofinite and product-complete modules coincide.

Proposition 4.5. *For a semi-simple R-module M the following are equivalent:*

(1) *M is endofinite.*
(2) *M is product-complete.*
(3) *Every product of copies of M is semi-simple.*
(4) *The ring $R/\operatorname{ann} M$ is semi-simple.*

Proof. (1) $\Rightarrow$ (2) Use Theorem 4.1.

(2) $\Rightarrow$ (3) Clear.

(3) $\Rightarrow$ (4) Consider the canonical map $R \to M^M$ which induces a monomorphism $R/\operatorname{ann} M \to M^M$. Any submodule of a semi-simple module is semi-simple, and therefore $R/\operatorname{ann} M$ is a semi-simple ring.

(4) $\Rightarrow$ (1) Any module over a semi-simple ring is endofinite. □

References

[1] J. Asenio Mayor and J. Martinez Hernandez, On flat and projective envelopes, J. Algebra **160** (1993) 434–440.

[2] M. Auslander and S.O. Smalø, Preprojective modules over artin algebras, J. Algebra **66** (1980) 61–122.

[3] S.U. Chase, Direct products of modules, Trans. Amer. Math. Soc. **97** (1960) 457–473.

[4] W.W. Crawley-Boevey, Modules of finite length over their endomorphism ring, in: Representations of algebras and related topics, eds. S. Brenner and H. Tachikawa, London Math. Soc. Lec. Note Series **168** (1992) 127–184.

[5] E.E. Enochs, Injective and flat covers, envelopes and resolvents, Israel J. Math. **39** (1981) 189–209.

[6] P. Gabriel, Des catégories abéliennes, Bull. Soc. Math. France **90** (1962) 323–448.

[7] S. Garavaglia, Dimension and rank in the theory of modules, East Lansing, preprint (1979).

[8] L. Gruson, Simple coherent functors, in: Representations of algebras, Springer Lecture Notes in Math. **488** (1975) 156–159.

[9] L. Gruson and C.U. Jensen, Deux applications de la notion de L-dimension, C. R. Acad. Sci. Paris Ser. A **282** (1976) 23–24.

[10] R. KIEŁPIŃSKI, On Γ-pure injective modules, Bull. Acad. Polon. Sci. Sér. Math. Astr. Phys. **15** (1967) 127–131.
[11] H. KRAUSE, Generic modules over artin algebras, Proc. London Math. Soc. **76** (1998) 276–306.
[12] H. KRAUSE, Exactly definable categories, J. Algebra **201** (1998) 456–492.
[13] J. RADA AND M. SAORÍN, Rings characterized by (pre)envelopes and (pre)covers of their modules, Comm. Algebra **26** (1998) 899–912.
[14] J.-E. ROOS, Sur la condition AB6 et ses variantes dans les catégories abéliennes, C.R. Acad. Sc. Paris **264** (1967) 991–994.
[15] B. STENSTRØM, Rings of quotients, Springer-Verlag (1975).
[16] W. ZIMMERMANN, Rein injektive direkte Summen von Moduln, Comm. Algebra **5** (1977) 1083–1117.
[17] W. ZIMMERMANN AND B. ZIMMERMANN-HUISGEN, Algebraically compact rings and modules, Math. Z. **161** (1978) 81–93.
[18] M. ZIEGLER, Model theory of modules, Ann. of Pure and Appl. Logic **26** (1984) 149–213.

FAKULTÄT FÜR MATHEMATIK, UNIVERSITÄT BIELEFELD, 33501 BIELEFELD, GERMANY
E-mail address: henning@mathematik.uni-bielefeld.de

DEPARTAMENTO DE MATEMÁTICAS, UNIVERSIDAD DE MURCIA, 30100 ESPINARDO - MU, SPAIN
E-mail address: msaorinc@fcu.um.es

Contemporary Mathematics
Volume **229**, 1998

Serre Duality for Generalized Auslander Regular Algebras

Roberto Martínez-Villa

ABSTRACT. It is well known, by a classical paper by J. P. Serre, [36] that it is possible to characterize the category of coherent sheaves on projective varieties in a purely algebraic way: the category of coherent sheaves is equivalent to the category of graded modules over a graded algebra Λ, modulo the finite dimensional ones. P. Gabriel in [19] developed the module approach as a non commutative version of algebraic geometry; it has been recently used and further investigated by several authors [2], [14], [15], [29], [35].

The motivation to study non commutative geometry comes from different areas: differential geometry, Lie theory, quantum groups and so on... [29], [32]. H. Lenzing and his coauthors, [8], [9], [10], [20] have related the indecomposable modules with certain algebraic objects like: weighted projective spaces, following an approach similar to the infinite dimensional one.

We feel the study of Koszul algebras and its application to concrete examples of finite dimensional algebras, is related with both approaches [12], [14], [15],[16], [17], [23], [28], [34]. In [31] we study the infinite dimensional algebras that appear as Yoneda algebras of a finite dimensional selfinjective Koszul algebra, they generalize both the preprojective and the symmetric algebras and constitute non commutative generalizations of the Auslander regular algebras [1], [4], [18]. The aim of this paper is to prove that for generalized Auslander regular algebras, even in the non noetherian case, the category of torsion free sheaves satisfies Serre duality. For finite dimensional algebras the Auslander Reiten theory of almost split sequences is related with Serre duality [3], here we prove that for generalized Auslander regular algebras of global dimension two Serre duality implies the existence of almost split sequences in the category of torsion free sheaves. Our results closely relate with [37], [38].

1. Basic notation and known results.

We will start by recalling some basic notation and known results.

Let K be a field and Q a finite quiver. The quiver algebra KQ is a positively graded algebra with grading given by the lengths of the paths. We denote by J the ideal generated by the arrows of KQ.

DEFINITION *Let KQ be a quiver algebra with I an homogeneous ideal, $I \subset J^2$. Then the algebra KQ/I is called a graded quiver algebra.*

DEFINITION *Let $\Lambda = KQ/I$ be a graded quiver algebra. A graded module $M = \{M_i\}_{i\in Z}$ is locally finite if for each $i \in Z$, the K-vector space M_i is finite dimensional.*

1991 Mathematics Subject Classification. 16E99, 18G60
Part of this paper was written during my visit to the University of Washington on July, 1997 as part of the collaboration project CONACyT-NSF in Representation Theory, I thank the funding agencies for supporting the program.

If $M = \{M_i\}_{i\in Z}$ is a graded Λ-module and k is an integer, then the k-shift of M is the module $M[k]$ defined by $M[k]_i = M_{i+k}$.

If M and N are graded modules and n is an integer, then a degree n map $f: M \to N$ is a Λ-map such that for all integers i there is an incluison $f(M_i) \subseteq M_{i+n}$. We denote by $\mathrm{Hom}^n_\Lambda(M,N)$ the K-vector space of all degree n maps. It is clear $\mathrm{Hom}^n_\Lambda(M,N) \cong \mathrm{Hom}^0_\Lambda(M,N[n])$.

It is known [31] that if M is finitely generated, then there exists an isomorphism $\mathrm{Hom}_\Lambda(M,N) \cong \bigoplus_{n\in Z} \mathrm{Hom}^n_\Lambda(M,N)$ and $\mathrm{Hom}_\Lambda(M,N)$ is a graded vector space.

Let $\mathrm{Gr}\,\Lambda$ be the category of graded modules and degree zero maps and let $\mathrm{l.f.Gr}\,\Lambda$ be the full subcategory of locally finite graded modules, denote by Λ^{op} the opposite algebra. Then there exists a duality $D\colon \mathrm{l.f.Gr}\,\Lambda \to \mathrm{l.f.Gr}\,\Lambda^{\mathrm{op}}$ given by $D(M)_j = \mathrm{Hom}_K(M_{-j},K)$.

DEFINITION *Let $\Lambda = \bigoplus_{i\geq 0} \Lambda_i$ be a positively graded algebra. We say that a graded module $M = \{M_i\}_{i\in Z}$ is generated in highest degree k, with $k \in Z$, if $M_i = 0$ if $i < k$ and $\Lambda_i M_k = M_{i+k}$ for all $i \geq 0$.*

Let M be a graded module over a graded quiver algebra Λ. We will denote by $M_{\geq k}$ the graded submodule of M defined by: $(M_{\geq k})_i = M_i$ if $i \geq k$ and 0 otherwise.

If M is generated in highest degree k, then for any positive integer i, we have an equality $J^i M = M_{\geq k+i}$.

EXAMPLES Let Q be the quiver with one vertex and n loops $x_1, x_2 \ldots x_n$ and K a field, KQ is the free algebra $K\langle x_1, x_2, \ldots x_n\rangle$. Let I be the ideal of KQ generated by the relations $x_i x_j + c_{ij} x_j x_i$, with $c_{ij} \in K - \{0\}$ for $i \neq j$. Then KQ/I consists of the quantum polynomials. If $c_{ij} = -1$ for all i and j, with $i \neq j$, then KQ/I is the usual polynomial algebra $K[x_1, x_2, \ldots x_n]$.

If I is the ideal of $K\langle x_1, x_2, \ldots x_n\rangle$ generated by $x_i x_j + x_j x_i$, with $i \neq j$ and x_i^2 then KQ/I is the exterior algebra.

From these examples it is clear that quiver algebras generalize polynomial algebras and their quotients and non commutative finite dimensional basic algebras over algebraically closed fields.

Serre proved the following: [36]

THEOREM 1.1. *Let K be a field and $\Lambda = K[x_0, x_1, x_2 \ldots x_n]/I$ be the quotient of the polynomial algebra $K[x_0, x_1, x_2, \ldots x_n]$ by the homogeneous ideal I, let X be the subscheme of P^n defined by the ideal and $\mathcal{O}_X$ the sheaf of regular functions on X. Then the category $\mathrm{Coh}\,(X)$ of coherent $\mathcal{O}_X$-module sheaves is equivalent to the category of finitely generated graded Λ-modules, modulo the full subcategory of the finite dimensional modules.* □

In [19], Gabriel studied quotient categories as non commutative generalizations of algebraic geometry. This approach has recently received a lot of attention because of its possible applications to field theory in physics and other branches of mathematics and has been systematical studied by many authors [2], [15], [29], [35].

The graded quiver algebras given above are examples of Koszul algebras, we recall some of the basic properties of these algebras:

DEFINITION *A graded quiver algebra $\Lambda = KQ/I$ is called quadratic if the ideal I is generated by linear combinations of paths of length 2.*

For a graded quiver algebra $\Lambda = KQ/I$ we will consider the Yoneda algebra $\Gamma = \bigoplus_{i\geq 0} \mathrm{Ext}^i_\Lambda(\Lambda/\overline{J}, \Lambda/\overline{J})$, with $\overline{J} = J/I$ the graded Jacobson radical of Λ.

Denote by Γ^{op} the opposite algebra of Γ. Let $\mathrm{Gr}\,\Lambda$ and $\mathrm{Gr}\,\Gamma^{\mathrm{op}}$ be the categories of graded Λ and Γ^{op}-modules, respectively, and degree zero maps. Then there exists a functor: $F : \mathrm{Gr}\,\Lambda \to \mathrm{Gr}\,\Gamma^{\mathrm{op}}$ defined in objects as $F(M) = \bigoplus_{i\geq 0} \mathrm{Ext}^i_\Lambda(M, \Lambda/\overline{J})$.

DEFINITION [22], [23], [24] *Let $\Lambda = KQ/I$ be a graded quiver algebra and let M be a finitely generated Λ-module generated in highest degree zero, we say M is a Koszul module if $F(M)$ is generated in highest degree zero. We say Λ is Koszul if all graded simples generated in degree zero are Koszul.*

It was proved in [22], [23] that a graded module M generated in highest degree zero is Koszul if and only if there exists a graded projective resolution of M : $\longrightarrow P^{(n)} \xrightarrow{f_n} P^{(n-1)} \longrightarrow \cdots P^{(1)} \xrightarrow{f_1} P^{(0)} \xrightarrow{f_0} M \longrightarrow 0$ with $P^{(j)}$ finitely generated, graded projective modules generated in highest degree j, and f_j degree zero maps.

DEFINITION *We say that a graded module M is Koszul up to shifting k if there exists an integer k such that $M[k]$ is Koszul.*

Koszul algebras are quadratic [11], [23], [34].

We will use the following results from [22], [23], [24]:

THEOREM 1.2. *The following statements hold:*

i) *If a quiver algebra $\Lambda = KQ/I$ is Koszul, then its Yoneda algebra Γ is Koszul.*

ii) *If $\Lambda = KQ/I$ is Koszul and Q^{op} denotes the opposite quiver of Q, then the Yoneda algebra Γ is isomorphic to KQ^{op}/L, where L is the ideal defined as follows:*

   Let V_2 be the vector subspace of KQ generated by the paths of length 2 and $I_2 = I \cap V_2$. Denote by V_2^0 the corresponding subspace of KQ^{op} and let $\langle , \rangle : V_2 \times V_2^0 \to K$ be the bilinear function defined as follows:

$$\langle \alpha\beta, \beta'\alpha' \rangle = \begin{cases} 1 & \text{if } \alpha^{\mathrm{op}} = \alpha',\ \beta^{\mathrm{op}} = \beta' \\ 0 & \text{otherwise} \end{cases}$$

   Then the orthogonal of I_2 with respect to $\langle , \rangle$, denoted by $I_2^{\perp} = L_2$, generates L.

iii) *If $\Lambda = KQ/I$ is Koszul and Γ is the Yoneda algebra of Λ, then the Yoneda algebra of Γ is isomorphic to Λ.*

iv) *If K_Λ and $K_{\Gamma^{\mathrm{op}}}$ denote the full subcategories of $\mathrm{Gr}\,\Lambda$ and $\mathrm{Gr}\,\Gamma^{\mathrm{op}}$ of Koszul Λ and Γ^{op}-modules respectively, then the functor $F : \mathrm{Gr}\,\Lambda \to \mathrm{Gr}\,\Gamma^{\mathrm{op}}$ induces a duality $F : K_\Lambda \to K_{\Gamma^{\mathrm{op}}}$.*

F has the following properties:

If P is an indecomposable graded projective generated in degree zero, then $F(P) = S$ is a graded simple generated in degree zero.

If S is simple generated in degree zero, then $F(S)$ is an indecomposable projective generated in degree zero.

If M is a Koszul module and J the graded radical of Λ, then $F(J^k M[k]) \cong \Omega^k F(M)[k]$, where $\Omega^k(M)$ denotes the k th syzygy of M.

A Koszul module M has graded projective dimension n if and only if $F(M)$ has Lowey length n. □

It was proved in [13] that the algebras $K[x_1, x_2, \ldots x_n]$ and $K\langle x_1, x_2, \ldots x_n\rangle / \langle x_i x_j + x_j x_i,\ x_i^2\rangle$, with $i \neq j$, are derived equivalent. More generally, it is known by [11] that if Λ is a Koszul algebra and Γ its Yoneda algebra, then Λ and $\Gamma^{\rm op}$ are derived equivalent, from this fact, it was proved in [37] that the functor $F : K_\Lambda \to K_{\Gamma^{\rm op}}$ is a duality. The proof given in [23] is more straightforward.

In any case, when Λ is an infinite dimensional Koszul algebra of finite global dimension its Yoneda algebra is finite dimensional and by means of the duality F we can apply the methods and concepts of representation theory like: almost split sequences, Auslander Reiten sequences, contravariantly finite subcategories, etc. and the classification of finite dimensional algebras to study the module structure of Λ, this has been the approach of [30], [31].

2. Koszul algebras: approximation by Koszul modules.

Since the algebras we are considering are generally non noetherian, we will use the graded small global dimension of a graded quiver algebra, [31] this is: the maximum of the projective dimensions of the graded simples, and consider instead of the category of finitely generated modules, the full subcategory, $\operatorname{Finres}\Lambda$, of the category of all graded modules and graded degree zero maps, $\operatorname{Gr}\Lambda$, consisting of all modules with graded finitely generated projective resolutions.

In this section we will give some basic properties of $\operatorname{Finres}\Lambda$ and prove that, for Koszul algebras, all modules in $\operatorname{Finres}\Lambda$ can be approximated by Koszul modules, this is: given a module M in $\operatorname{Finres}\Lambda$, there exists a Koszul submodule L of M such that M/L is of finite length.

We have the following:

LEMMA 2.1. *Let Λ be a graded quiver algebra. Then* $\operatorname{Finres}\Lambda$ *contains all finitely generated projectives and it is closed under extensions, cokernels of monomorphisms (it is resolving) and kernels of epimorphisms. Assume* $\operatorname{Finres}\Lambda$ *contains all graded simple modules, then it contains all finite dimensional modules and if $L \in \operatorname{Finres}\Lambda$, then for all $k \geq 0$ the module $J^k L \in \operatorname{Finres}\Lambda$ and the module $L_{\geq k} \in \operatorname{Finres}\Lambda$.*

PROOF: It is clear $\operatorname{Finres}\Lambda$ is closed under extensions. Let $0 \to L \to M \to N \to 0$ be exact with L and $M \in \operatorname{Finres}\Lambda$ and let P_0 be the projective cover of M. We have a commutative exact diagram:

$$\begin{array}{ccccccccc} & & 0 & & 0 & & & & \\ & & \downarrow & & \downarrow & & & & \\ 0 & \to & K & \to & K & \to & 0 & & \\ & & \downarrow & & \downarrow & & \downarrow & & \\ 0 & \to & H & \to & P_0 & \to & N & \to & 0 \\ & & \downarrow & & \downarrow & & \downarrow & & \\ 0 & \to & L & \to & M & \to & N & \to & 0 \\ & & \downarrow & & \downarrow & & \downarrow & & \\ & & 0 & & 0 & & 0 & & \end{array}$$

The module M in $\operatorname{Finres}\Lambda$ implies the module K is in $\operatorname{Finres}\Lambda$. Since the subcategory $\operatorname{Finres}\Lambda$ is closed under extensions H is in $\operatorname{Finres}\Lambda$, therefore: $N \in \operatorname{Finres}\Lambda$.

Now if $\operatorname{Finres}\Lambda$ contains the graded simples, then it is clear, by what we already proved, $\operatorname{Finres}\Lambda$ contains all finite dimensional graded modules.

Let $0 \to L \to M \to N \to 0$ be exact with M and $N \in \operatorname{Finres}\Lambda$. Let P be the projective cover of M. Then we have an exact commutative diagram:

$$\begin{array}{ccccccccccc} & & & & 0 & & 0 & & & & \\ & & & & \downarrow & & \downarrow & & & & \\ & & 0 & \to & \Omega(M) & \to & \Omega(N) \oplus Q & \to & L & \to & 0 \\ & & \downarrow & & \downarrow & & \downarrow & & & & \\ & & 0 & \to & P & \to & P & \to & 0 & & \\ & & \downarrow & & \downarrow & & \downarrow & & & & \\ 0 & \to & L & \to & M & \to & N & \to & 0 & & \\ & & & & \downarrow & & \downarrow & & & & \\ & & & & 0 & & 0 & & & & \end{array}$$

where Q is a finitely generated graded projective.

Since M and $N \in \operatorname{Finres}\Lambda$, the modules $\Omega(M)$ and $\Omega(N) \oplus Q$ are in $\operatorname{Finres}\Lambda$. But we already proved $\operatorname{Finres}\Lambda$ is closed under cokernel of monomorphisms, it follows $L \in \operatorname{Finres}\Lambda$.

If we assume $\operatorname{Finres}\Lambda$ contains the graded simples and $L \in \operatorname{Finres}\Lambda$, then it follows from the previous statement, $JL \in \operatorname{Finres}\Lambda$.

For any integer k the modules L and $L/L_{\geq k} \in \operatorname{Finres}\Lambda$, as above, the module $L_{\geq k} \in \operatorname{Finres}\Lambda$. ∎

DEFINITION *Let $\Lambda = KQ/I$ be a graded quiver algebra. We say a graded module M is bounded below if there exists an integer b such that $M_j = 0$ for $j < b$. In this case we call to the largest such b the bound. Similarly, M is bounded above if there exists c such that $M_j = 0$ for all $j > c$. The module M is bounded if and only if it is bounded above and below.*

LEMMA 2.2. *Let $\Lambda = KQ/I$ be a graded quiver algebra. Then any graded module bounded below, has a minimal projective cover.*

PROOF: Let M be bounded below with bound b, define $N_b = M_b$.

The module $M/\Lambda M_b$ is bounded below, let k_1 be the bound.

We have an exact sequence of vector spaces: $0 \to (\Lambda M_b)_{k_1} \to M_{k_1} \to (M/\Lambda M_b)_{k_1} \to 0$.

Let N_{k_1} be a vector space complement of $(\Lambda M_b)_{k_1}$ in M_{k_1}, we have a decomposition $M_{k_1} = (\Lambda M_b)_{k_1} \oplus N_{k_1}$.

Define by induction an increasing sequence: $k_0 < k_1 < k_2 \cdots < k_t$ by $k_0 = b$ and for $t > 0$ the integer k_t is such that $\left(M/\sum_{i=0}^{t-1} \Lambda N_{k_i}\right)_{k_t} \neq 0$ and $\left(M/\sum_{i=0}^{t-1} \Lambda N_{k_i}\right)_j = 0$ if $j < k_t$.

It is clear $\sum_{i=0}^{\infty} \Lambda N_{k_i} = M$.

We have the following isomorphisms: $\Lambda/J \bigotimes_{\Lambda} M \cong M/JM \cong \Lambda/J \bigotimes_{\Lambda} \sum_{i=0}^{\infty} \Lambda N_{k_i} \cong \sum_{i=0}^{\infty} \Lambda/J \bigotimes_{\Lambda} \Lambda N_{k_i} \cong \sum_{i=0}^{\infty} N_{k_i}$, where the last sum is direct.

Each $N_{k_i} = \bigoplus_{i=0}^{n} S_i^{(A_i)}[k_i]$, with S_i a graded simple generated in degree zero and $S_i^{(A_i)}$ a direct sum of A_i copies of S_i. Let P_i be the projective cover of S_i and $P_i^{(A_i)}$ a direct sum of A_i copies of P_i. Then $N_{k_i} = \bigoplus_{i=0}^{n} P_i^{(A_i)}[k_i]/JP_i^{(A_i)}[k_i] \cong \left(\bigoplus_{i=0}^{n} P_i^{(A_i)}[k_i]\right)/J\left(\bigoplus_{i=0}^{n} P_i^{(A_i)}[k_i]\right)$.

We have an epimorphism: $\phi\colon \oplus_{k_i} \oplus_{i=0}^{n} P_i^{(A_i)}[k_i] \to M \to 0$,

with $\ker \phi \subseteq J\left(\oplus_{k_i} \oplus_{i=0}^{n} P_i^{(A_i)}[k_i]\right)$. □

COROLLARY *If $\Lambda = KQ/I$ is a graded quiver algebra, then a graded module M bounded below has a minimal projective resolution.*

PROOF: It is clear from the fact that M is bounded below and if $0 \to \Omega(M) \to P \to M \to 0$ is exact with P the graded projective cover of M, then the syzygy $\Omega(M)$ is bounded below. □

We have the following description of $\operatorname{Finres}\Lambda$.

PROPOSITION 2.3. *Let $\Lambda = KQ/I$ be a graded quiver algebra. Then a graded Λ-module M bounded below is in* $\operatorname{Finres}\Lambda$ *if and only if* $\dim_K Ext^n_\Lambda(M, \Lambda/J) < \infty$ *for all $n \geq 0$.*

PROOF: The module M bounded below implies it has a minimal projective resolution: $\to P^{(n)} \to P^{(n-1)} \to \cdots P^{(1)} \to P^{(0)} \to M \to 0$ and $\operatorname{Ext}^n_\Lambda(M, \Lambda/J) = \operatorname{Ext}^1_\Lambda(\Omega^{n-1}(M), \Lambda/J)$.

We have an exact sequence:

$0 \to \Omega^n(M) \to P^{(n-1)} \to \Omega^{n-1}(M) \to 0$ which induces an exact sequence:

$0 \to \operatorname{Hom}_\Lambda(\Omega^{n-1}(M), \Lambda/J) \to \operatorname{Hom}_\Lambda(P^{(n-1)}, \Lambda/J) \to \operatorname{Hom}_\Lambda(\Omega^n(M), \Lambda/J) \to \operatorname{Ext}^1_\Lambda(\Omega^{n-1}(M), \Lambda/J) \to 0$

The isomorphism $P^{(n)}/JP^{(n)} \cong \Omega^n(M)/J\Omega^n(M)$ implies $\mathrm{Hom}_\Lambda(\Omega^n(M),\Lambda/J) \cong \mathrm{Ext}^n_\Lambda(M,\Lambda/J) \cong \mathrm{Hom}_\Lambda(P^{(n)}/JP^{(n)},\Lambda/J)$.

It follows $\dim_K \mathrm{Ext}^n_\Lambda(M,\Lambda/J) < \infty$ if and only if $P^{(n)}$ is finitely generated. □

The following proposition was used in [26]:

PROPOSITION 2.4. *Let $\Lambda = KQ/I$ be a Koszul algebra, M a graded module with $pdM = n < \infty$ and $M \in \mathrm{Finres}\,\Lambda$. Then there is an integer m such that $M_{\geq m}$ is Koszul up to shifting.*

PROOF: Let $0 \to P^{(n)} \to P^{(n-1)} \to \cdots P^{(1)} \to P^{(0)} \to M \to 0$ be the minimal graded projective resolution of M, each $P^{(j)}$ has a decomposition: $P^{(j)} = Q_1^{(j)} \oplus Q_2^{(j)} \oplus \cdots Q_{t_j}^{(j)}$, with $Q_k^{(j)}$ finitely generated projective, generated in highest degree $m_k^{(j)}$.

Let $m_j = \max\left\{m_k^{(j)} \mid 1 \leq k \leq t\right\}$. Then $P^{(j)}_{\geq m_j}$ is generated in highest degree m_j. Let $m = \max\{m_j\}$. Then for each j the module $P^{(j)}_{\geq m}$ is generated in highest degree m.

It is known [23] that if P is a finitely generated projective module generated in highest degree, then for any integer k the module $P_{\geq k}$ is Koszul up to k shifting and that a direct sum of Koszul modules up to shifting k is Koszul up to shifting k.

Hence; we have an exact sequence of Koszul modules and degree zero maps:

$0 \to P^{(n)}_{\geq m}[m] \to P^{(n-1)}_{\geq m}[m] \to \cdots \to P^{(0)}_{\geq m}[m]$

Each $P^{(j)}_{\geq m}[m]$ is a Koszul module. It was proved in [24] that given a degree zero monomorphism of Koszul modules, the cokernel is Koszul. It follows by induction $M_{\geq m}$ is Koszul up to shifting. □

We obtain the following.

COROLLARY *Let $\Lambda = KQ/I$ be a Koszul algebra of finite graded small global dimension n. Then a finitely generated graded module M is in* $\mathrm{Finres}\,\Lambda$ *if and only if there exists some integer k such that $M_{\geq k}\,[k]$ is Koszul.*

PROOF: Let M be a finitely generated module in $\mathrm{Finres}\,\Lambda$, since Λ has finite graded small global dimension, M has a finite projective resolution. By the theorem, there exists an integer k such that $M_{>k}$ is Koszul.

Conversely, if M is finitely generated and for some integer k the module $M_{\geq k}$ is Koszul up to shifting, then $M/M_{\geq k}$ is finite dimensional and hence, both $M_{\geq k}$ and $M/M_{\geq k}$ are in $\mathrm{Finres}\,\Lambda$, therefore M is in $\mathrm{Finres}\,\Lambda$. □

3. Generalized Auslander regular algebras.

In [26], [30], [31] we studied the graded algebras that appear as Yoneda algebras of selfinjective Koszul algebras and prove that they have similar properties to the Auslander regular algebras. In this section we recall the main results and prove some lemmas that will be needed later on.

DEFINITION *Let $\Lambda = KQ/I$ be a graded quiver algebra, we say Λ is generalized Auslander regular if the following statements are true:*

i) *The algebra Λ has finite graded small global dimension n.*

ii) *All graded simples are in* $\operatorname{Finres}\Lambda$ *and have projective dimension n.*

iii) *For any graded simple S the equality $Ext^k_\Lambda(S,\Lambda) = 0$ holds for $0 \le k < n$.*

iv) *The functor $Ext^n_\Lambda(-,\Lambda)$ induces a bijection between the Λ and the Λ^{op}-graded simples.*

In [16], [31], [37] the following theorem was proved:

THEOREM 3.1. *Let $\Lambda = KQ/I$ be an indecomposable finite dimensional Koszul algebra. Then Λ is selfinjective if and only if its Yoneda algebra Γ is a generalized Auslander regular Koszul algebra.* □

We know the following facts about selfinjective Koszul algebras: [26], [30], [31].

THEOREM 3.2. *Let $\Lambda = KQ/I$ be an indecomposable graded quiver algebra with $r^2 \neq 0$. Then:*

i) *If Λ is Koszul, then it is of infinite representation type.*

ii) *If r^3 =0 and Λ is of infinite representation type, then it is Koszul.* □

DEFINITION *If R is a ring and M a $R-R$ bimodule, then the trivial extension is the abelian group $R \oplus M$ with multiplication $(r,m).(r',m') = (rr', rm' + mr')$.*

Examples of selfinjective Koszul algebras are the following:

a) The exterior algebra $K\langle x_1, x_2 \dots x_n\rangle / \langle x_i x_j + x_j x_i,\, x_i^2\rangle$ with $i \neq j$.

b) Let Q be a connected quiver with only sinks and sources, K a field. Consider the quiver algebra KQ. Since the trivial extension $KQ \rhd D(KQ)$ is a selfinjective algebra with $r^2 \neq 0$ and $r^3 = 0$, it is Koszul if and only if Q is not Dynkin.

The Yoneda algebra of the algebra given in a) is the polynomial algebra $K[x_1, x_2, \dots x_n]$ and the Yoneda algebra of the algebra given in b) is the preprojective algebra, whose construction we recall now:

Given a finite quiver $Q = (Q_0, Q_1)$, with Q_0 the set of vertices and Q_1 the set of arrows, we obtain another quiver $\hat{Q} = (\hat{Q}_0, \hat{Q}_1)$, with set of vertices $\hat{Q}_0 = Q_0$ and set of arrows $\hat{Q}_1 = Q_1 \cup Q_1^{\mathrm{op}}$, this is, $\hat{Q}$ contains the arrows of Q and for each arrow $e \xrightarrow{\alpha} f$ of Q we introduce an arrow $f \xrightarrow{\alpha^{-1}} e$. The preprojective algebra is the quiver algebra $K\hat{Q}/I$, where I is the ideal generated by the following relations: if

$$e \begin{array}{c} \overset{\alpha_1}{\nearrow} \\ \overset{\alpha i}{\to} \\ \underset{\alpha_n}{\searrow} \end{array} \begin{array}{c} f_1 \\ \vdots \\ f_i \\ \vdots \\ f_n \end{array}$$

are all arrows of Q starting at e, then $\sum \alpha_i^{-1}\alpha_i \in I$, similarly for all arrows α ending at e' we have $\sum \alpha_i \alpha_i^{-1} \in I$.

Preprojective algebras have been considered by many authors [6], [7], [8], [9], [10], [21]. The following theorem was proved in [8], [9], [26].

THEOREM 3.3. *Let Q be a finite quiver, K a field and $K\hat{Q}/I$ the preprojective algebra corresponding to Q, then the following statements hold:*

i) *The algebra $K\hat{Q}/I$ is finite dimensional if and only if Q is Dynkin.*

ii) *The algebra $K\hat{Q}/I$ is noetherian if and only if Q is Euclidean, or Dynkin.* □

By theorem 3.2, part ii) of this theorem provide us with many examples of non noetherian generalized Auslander regular algebras.

LEMMA 3.4. *Let $\Lambda = KQ/I$ be a generalized Auslander algebra. Then Λ^{op} is generalized Auslander regular.*

PROOF: Since the graded simples correspond to the vertices of Q the algebras Λ and Λ^{op} have the same number of graded simples.

Given a minimal projective resolution of the graded Λ-module S:

$0 \to P_n \xrightarrow{f_n} P_{n-1} \to \cdots P_1 \xrightarrow{f_1} P_0 \xrightarrow{f_0} S \to 0$ we obtain by iii) and iv) a minimal projective resolution of the simple $S' = Ext^n_\Lambda(S, \Lambda)$:

$0 \to P_0^* \xrightarrow{f_1^*} P_1^* \to \cdots P_{n-1}^* \xrightarrow{f_n^*} P_n^* \to S' \to 0$

Then, by iv) all graded Λ^{op}-simples have projective dimension n.

Applying the functor $\mathrm{Hom}_{\Lambda^{\mathrm{op}}}(-, \Lambda^{\mathrm{op}}) = (-)^*$ to the sequence we obtain a commutative diagram:

$$\begin{array}{ccccccccccccc} 0 \to & P_n & \xrightarrow{f_n} & P_{n-1} & \to \cdots & P_1 & \xrightarrow{f_1} & P_0 & \xrightarrow{f_0} & S & \to 0 \\ & \downarrow\cong & & \downarrow\cong & & \downarrow\cong & & \downarrow\cong & & \downarrow & \\ 0 \to & P_n^{**} & \xrightarrow{f_n^{**}} & P_{n-1}^{**} & \to \cdots & P_1^{**} & \xrightarrow{f_1^{**}} & P_0^{**} & \to & C & \to 0 \end{array}$$

Where C is the cokernel of f_1^{**}.

It follows the bottom row is exact and $C = \mathrm{Ext}^n_{\Lambda^{\mathrm{op}}}(S', \Lambda^{\mathrm{op}})$ is isomorphic to S.

Then $\mathrm{Ext}^i_{\Lambda^{\mathrm{op}}}(S', \Lambda^{\mathrm{op}}) = 0$ for all $i \neq n$. □

Let Λ be a graded algebra, let M be a Λ-Λ bimodule and let $\sigma : \Lambda \to \Lambda$ be an automorphism of graded algebras. Then we will denote by $M\sigma$ the bimodule with the same left structure and right structure given by: $m * \lambda = m.\sigma(\lambda)$, where the dot denotes the original right operation.

We have the following theorem from [31].

THEOREM 3.5. *Let $\Lambda = KQ/I$ be a generalized Auslander regular algebra of graded small global dimension n. Then the following statement is true: There exists a graded ring isomorphism $\sigma : \Lambda \to \Lambda$ and a natural isomorphism of graded vector spaces:*

$Hom_\Lambda(M, D(\Lambda)[-n]\sigma) \cong Ext^n_\Lambda(M, \Lambda)$, for all graded modules M of finite length.

□

DEFINITION [19], [33] *Let M be a graded Λ-module, where $\Lambda = KQ/I$ is a graded quiver algebra. We define the torsion part of M, $t(M)$, as: $t(M) = \sum_{\alpha} M_a$, where the sum runs through all graded submodules M_a of M of finite length. A graded module M is torsion if $t(M) = M$ and torsion free if $t(M) = 0$.*

For noetherian graded algebras the next proposition is well known. We will prove it for generalized Auslander regular Koszul algebras.

PROPOSITION 3.6. *Let $\Lambda = KQ/I$ be a generalized Auslander regular Koszul algebra and let M be a graded module, $M \in \operatorname{Finres}\Lambda$. Then $t(M)$ is of finite length.*

PROOF: Since Λ has graded small global dimension n, the module M in $\operatorname{Finres}\Lambda$ implies pd$M$ is finite and there exists an integer s such that $M_{\geq s}[s]$ is Koszul.

It is easy to see that graded summands of Koszul modules are Koszul, $M_{\geq s}$ is generated in highest degree and $\dim_K(M_{\geq s})_s$ finite implies $M_{\geq s} = N_1 \oplus N_2 \cdots \bigoplus N_t$, where for $1 \leq i \leq t$ each N_i is an indecomposable Koszul module up to shifting by s.

Let Γ be the Yoneda algebra of Λ and $F = \bigoplus_{n \geq 0} \operatorname{Ext}^n_\Gamma(-, \Gamma/J)$ the Koszul duality. Then there exists indecomposable Koszul Γ^{op}-modules L_i such that $F(L_i) \cong N_i$.

Since Γ is selfinjective, either L_i is projective or $\Omega^k(L_i)$ is an indecomposable module for all k. Hence; $M_{\geq s} = N_1 \oplus N_2 \oplus \cdots N_l \oplus T$, where T is semisimple and $J^m N_i = F(\Omega^m(L_i))$ is indecomposable for all $m \geq 0$ and $1 \leq i \leq l$. Therefore: $M_{\geq s+1} = JN_1 \oplus JN_2 \oplus \cdots JN_l$ has no semisimple summands.

Assume $t(M) \cap M_{\geq s+1} \neq 0$. Let X be a graded finitely generated submodule of $t(M) \cap M_{>>s+1}$. Then $X \subseteq \sum_a M_a$, with M_a of finite length. The generators of X are in a finite sum therefore X is finite dimensional and soc$X \neq 0$. This is: there exists a graded simple $S \subseteq M_{\geq s+1}$. It follows that for some i and some m, there is an incluison $S \subseteq J^m N_i$ and the simple S is not contained in $J^{m+1}N_i$, it follows S is a summand of $J^m N_i$.

The submodule $J^m N_i$ indecomposable implies $J^m N_i = S_i \cong F(\Omega^m(L_i))$. A contradiction, because $\Omega^m(L_i)$ is non projective unless $m = 0$.

We proved $t(M) \cap M_{>>s+1} = 0$.

It follows there exists a monomorphism: $t(M) \to M/M_{\geq s+1}$ and $t(M)$ is of finite length. □

We also have the following

PROPOSITION 3.7. *Let $\Lambda = KQ/I$ be an indecomposable generalized Auslander regular Koszul algebra of finite graded global dimension n, let X be an indecomposable non simple Koszul module. Then the graded radical JX of X has projective dimension $n - 1$.*

PROOF: Let Γ be the Yoneda algebra of Λ and $F = \bigoplus_{n \geq 0} \operatorname{Ext}^n_\Gamma(-, \Gamma/J)$ the Koszul duality.

We have an isomorphism $X \cong F(Y)$ with Y a Koszul module.

The module X non simple implies by [23], Y is non projective.

We have an exact sequence of Γ-modules: $0 \to \Omega(Y) \to P \to Y \to 0$, with P a finitely generated projective generated in highest degree zero.

We proved in [31], for all indecomposable selfinjective graded quiver algebras, the indecomposable projective modules have the same Lowey length. The algebra Γ has Lowey length n, then $socP = soc\Omega(Y)$ is generated in degree n. The syzygy $\Omega(Y)$ is generated in degree 1. It follows the Lowey length of $\Omega(Y)$ is $n-1$. Therefore: $F(\Omega(Y)) \cong JF(Y) = JX$ has projective dimension $n-1$ by [25].

We need the following:

LEMMA 3.8. *Let Λ be any ring. Given a module M of projective dimension n and $M \in \operatorname{Finres}\Lambda$, for any Λ-module L there exists a natural isomorphism:*
$Ext^n_\Lambda(M,L) \cong Ext^n_\Lambda(M,\Lambda) \bigotimes_\Lambda L$.

If Λ is a graded ring and M and L are graded modules, then the isomorphism is a graded isomorphism.

PROOF: By assumption M has a minimal projective resolution:
$0 \to P^{(n)} \to P^{(n-1)} \to \cdots P^{(1)} \to P^{(0)} \to M \to 0$, with each $P^{(j)}$ finitely generated.

We have a natural isomorphism: $\operatorname{Ext}^n_\Lambda(M,L) \cong \operatorname{Ext}^1_\Lambda(\Omega^{n-1}(M),L)$.

The exact sequence $*)$ $0 \to P^{(n)} \to P^{(n-1)} \to \Omega^{n-1}(M) \to 0$ induces exact sequences:

$**)$ $0 \to \operatorname{Hom}_\Lambda(\Omega^{n-1}(M),\Lambda) \to \operatorname{Hom}_\Lambda(P^{(n-1)},\Lambda) \to \operatorname{Hom}_\Lambda(P^{(n)},\Lambda) \to \operatorname{Ext}^n_\Lambda(M,\Lambda) \to 0$

and: $0 \to \operatorname{Hom}_\Lambda(\Omega^{n-1}(M),L) \to \operatorname{Hom}_\Lambda(P^{(n-1)},L) \to \operatorname{Hom}_\Lambda(P^{(n)},L) \to \operatorname{Ext}^n_\Lambda(M,L) \to 0$

Tensoring $**)$ with L, the natural isomorphisms: $\operatorname{Hom}_\Lambda(P^{(i)},\Lambda) \otimes L \to \operatorname{Hom}_\Lambda(P^{(i)},L)$, $i = n-1, n$, induce a natural isomorphism: $\operatorname{Ext}^n_\Lambda(M,L) \cong \operatorname{Ext}^n_\Lambda(M,\Lambda) \bigotimes_\Lambda L$.

It is clear that in the graded situation the isomorphisms are graded isomorphisms. □

4. Non commutative projective schemes and generalized Auslander regular algebras.

The notion of non commutative projective schemes has been defined and used by several authors [2], [15], [19], [20], [28], [32], [33] they constitute an algebraic generalization of classical commutative projective schemes. We will use the notation and concepts introduced in [2] to prove for generalized Auslander regular algebras [1], [4], [18], a non commutative version of Serre duality [15], [27].

Let $\Lambda = KQ/I$ be a graded quiver algebra, $\operatorname{Gr}\Lambda$ will denote the category of graded modules and degree zero maps, $\operatorname{gr}\Lambda$ is the full subcategory of $\operatorname{Gr}\Lambda$ of all finitely generated modules. By $\operatorname{Finres}\Lambda$ denote the full subcategory of $\operatorname{gr}\Lambda$ of all modules with finitely generated graded projective resolutions.

Observe $\operatorname{Finres}\Lambda = \operatorname{gr}\Lambda$ if and only if Λ is noetherian.

We say that $x \in M$ is torsion if there exists some integer s such that $\Lambda_{>>s}x = 0$. It is easy to see that the submodule $t(M)$ of M defined in section 2 is the set of torsion elements and that $t(M)$ is the smallest submodule of M such that $M/t(M)$ has no torsion elements different from zero.

Denote by Tors the full subcategory of $\operatorname{Gr}\Lambda$ of all torsion submodules and define:

$\operatorname{QGr}\Lambda = \operatorname{Gr}\Lambda/\text{Tors}$, $\operatorname{QGr}\Lambda = \operatorname{gr}\Lambda/\text{Tors}$ and $\operatorname{Q}\operatorname{Finres}\Lambda = \operatorname{Finres}\Lambda/\text{tors}$, the quotient categories in the sense of [19], [33].

We have contentions: $\operatorname{Q}\operatorname{Finres}\Lambda \subseteq \operatorname{QGr}\Lambda \subseteq \operatorname{QGr}\Lambda$.

We recall the definition of the quotient category.

Given two graded Λ-modules X and Y we denoted by $\operatorname{Hom}^n_\Lambda(X,Y)$ the set of graded degree n maps from X to Y and $\operatorname{Hom}_\Lambda(X,Y) = \bigoplus\limits_n \operatorname{Hom}^n_\Lambda(X,Y)$.

The category $\operatorname{QGr}\Lambda$ has the same objects as $\operatorname{Gr}\Lambda$. Denote by $\pi : \operatorname{Gr}\Lambda \to \operatorname{QGr}\Lambda$ the quotient functor, the degree n maps in $\operatorname{QGr}\Lambda$ are defined by:

$\operatorname{Hom}^n_{\operatorname{QGr}\Lambda}(\pi N, \pi M) = \varinjlim \operatorname{Hom}^n_\Lambda(N', M/M')$ where the direct limit is taken over the set I of all pairs (N', M') of submodules of N and M, respectively for which N/N' and M' are torsion.

We define $\operatorname{Hom}_{\operatorname{QGr}\Lambda}(\pi N, \pi M) = \bigoplus\limits_n \varinjlim \operatorname{Hom}^n_\Lambda(N', M/M')$.

Since the set $(N', t(M))$ is cofinal in I, it is known by [19], [33]

$\varinjlim \operatorname{Hom}^n_\Lambda(N', M/M') = \varinjlim \operatorname{Hom}^n_\Lambda(N', M/t(M))$

hence; $\operatorname{Hom}_{\operatorname{QGr}\Lambda}(\pi N, \pi M) = \varinjlim \operatorname{Hom}_\Lambda(N', M/t(M))$. □

We will need the following:

LEMMA 4.1. *Let $\Lambda = KQ/I$ be a graded quiver algebra and let M be a graded torsion module. Then the following statements hold:*

i) *Any graded submodule L of M and any factor module M/L are torsion.*

ii) *The socle of M, $socM$, is essential.*

iii) *For any graded simple S its injective envelope is torsion.*

iv) *A graded injective I is torsion free if and only if $socI = 0$.*

v) *Let $\{N^{(i)}\}_{i\in N}$ be a direct system of finitely presented graded modules such that $N^{(k)}_j = 0$ for all $j < k$. Then $\varinjlim_i Hom_\Lambda(N^{(i)}, M) = 0$. Assume all graded simples are in* $\operatorname{Finres}\Lambda$, *if N is a graded module, $N \in \operatorname{Finres}\Lambda$, then $\varinjlim Hom_\Lambda(N_{\geq n}, M) = 0$.*

PROOF: Since M can be described, both as a set of torsion elements and as a sum of modules of finite length, i) follows.

ii) Let X be a submodule of M and L a finitely generated submodule of X. The module L is finite dimensional, hence, $\operatorname{soc}L \neq 0$ and $\operatorname{soc}L \subseteq \operatorname{soc}X \subseteq \operatorname{soc}M \cap X$.

iii) Any graded simple Λ^{op}-module has finitely generated projective cover, hence any graded simple Λ-module S has finitely cogenerated injective envelope $I(S)$. If S is concentrated in degree k, then $I(S)_j = 0$ for $j > k$. If $x \in I(S)_j$, then $J^{k-j+1}x = 0$ and $I(S)$ is torsion.

iv) Let I be graded injective. Then $\operatorname{soc}I \subseteq t(I)$. Hence I torsion free implies $\operatorname{soc}I = 0$.

Conversely, since $\mathrm{soc}(t(I)) \subseteq \mathrm{soc} I$, the equality $\mathrm{soc} I = 0$ implies $t(I) = 0$.

v) The module M is a direct limit $M = \varinjlim M_\alpha$ of finitely generated torsion modules, hence of finite dimensional modules M_α. Since each $N^{(j)}$ is finitely presented, we have the following chain of isomorphisms: $\varinjlim_i \mathrm{Hom}_\Lambda(N^{(i)}, M) \cong \varinjlim_i \mathrm{Hom}_\Lambda(N^{(i)}, \varinjlim_\alpha M) \cong \varinjlim_i \varinjlim_\alpha \mathrm{Hom}_\Lambda(N^{(i)}, M_\alpha) \cong \varinjlim_\alpha \varinjlim_i \mathrm{Hom}_\Lambda(N^{(i)}, M_\alpha)$.

It is clear that for sufficiently large k, the equality $\mathrm{Hom}^0_\Lambda(N^{(k)}, M_\alpha) = 0$ holds. Then $\varinjlim_i \mathrm{Hom}^0_\Lambda(N^{(i)}, M_\alpha) = 0$.

Hence; $\varinjlim_\alpha \varinjlim_i \mathrm{Hom}^0_\Lambda(N^{(i)}, M_\alpha) = 0$.

If $\mathrm{Hom}^p_\Lambda(N^{(i)}, M_\alpha)$ denotes the graded maps of degree p, then we have: $\varinjlim \mathrm{Hom}^p_\Lambda(N^{(i)}, M_\alpha) \cong \varinjlim \mathrm{Hom}^0_\Lambda(N^{(i)}, M_\alpha[p]) = 0$ for all p.

Therefore: $\varinjlim_i \mathrm{Hom}_\Lambda(N^{(i)}, M_\alpha) = \bigoplus_p \varinjlim_i \mathrm{Hom}^p_\Lambda(N^{(i)}, M_\alpha) = 0$.

It follows $\varinjlim_i \mathrm{Hom}_\Lambda(N^{(i)}, \varinjlim_i M) \cong \varinjlim_i \varinjlim_\alpha \mathrm{Hom}_\Lambda(N^{(i)}, M_\alpha) \cong \varinjlim_\alpha \varinjlim_i \mathrm{Hom}_\Lambda(N^{(i)}, M_\alpha) = 0$

If $N \in \mathrm{Finres}\,\Lambda$, then by lemma 2.1, $N_{\geq k} \in \mathrm{Finres}\,\Lambda$ and the second statement follows. □

If Λ is a noetherian graded algebra, N is a finitely generated graded module and M is an arbitrary graded module, then it is known [33],

$\mathrm{Hom}_{\mathrm{QGr}\,\Lambda}(\pi N, \pi M) = \varinjlim \mathrm{Hom}_\Lambda(N_{\geq n}, M)$.

We will prove the following version of this statement:

PROPOSITION 4.2. *Let $\Lambda = KQ/I$ be a graded quiver algebra. Then if N is a graded finitely generated module and M is an arbitrary graded module, then we have:*

$\mathrm{Hom}_{\mathrm{QGr}\,\Lambda}(\pi N, \pi M) = \varinjlim \mathit{Hom}_\Lambda(N_{\geq n}, M/t(M))$.

Furthermore, if we assume all graded Λ simples in $\mathrm{Finres}\,\Lambda$ *and* $N \in \mathrm{Finres}\,\Lambda$, *then we have natural isomorphisms:* $\varinjlim \mathit{Ext}^t_\Lambda(N_{\geq k}, M) \cong \varinjlim \mathit{Ext}^t_\Lambda(N_{\geq k}, M/t(M))$, *for all* $t \geq 0$.

PROOF: Let N' be a submodule of N such that N/N' is torsion. The module N/N' is a sum of submodules of finite length and it is finitely generated, hence of finite length. It follows there exists some integer n such that $N_{\geq n} \subseteq N'$. Then the "tails" $N_{\geq n}$ are cofinal in the set I and we have:

$\mathrm{Hom}_{\mathrm{QGr}\,\Lambda}(\pi N, \pi M) = \varinjlim \mathrm{Hom}_\Lambda(N_{\geq n}, M/t(M))$

If all graded simples and N are in $\mathrm{Finres}\,\Lambda$, then we have by lemma 4.1: $\varinjlim \mathrm{Hom}_\Lambda(N_{\geq k}, t(M)) = 0$.

By Proposition 2.1, $N_{\geq k} \in \mathrm{Finres}\,\Lambda$ and for each k we have part of a minimal projective resolution:

*) $0 \to \Omega^t(N_{\geq k}) \to P^{(k)}_{t-1} \to \cdots \to P^{(k)}_1 \to P^{(k)}_0 \to N_{\geq k} \to 0$.

Each $P_s^{(k)}$ satisfies $(P_s^{(k)})_j = 0$ if $j < k$, hence, $\Omega^s(N_{\geq k})_j = 0$ if $j < k$.

The inclusion maps: $N_{\geq k} \to N_{\geq k-1}$ induce maps: $P_s^{(k)} \to P_{s-1}^{(k)}$ and $\Omega^s(N_{\geq k}) \to \Omega^{s-1}(N_{\geq k})$.

The short exact sequence: $0 \to \Omega^t(N_{\geq k}) \to P_{t-1}^{(k)} \to \Omega^{t-1}(N_{\geq k}) \to 0$

Induces an exact sequence of direct systems:

$0 \to \operatorname{Hom}_\Lambda(\Omega^{t-1}(N_{\geq k}), t(M)) \to \operatorname{Hom}_\Lambda(P_{t-1}^{(k)}, t(M)) \to \operatorname{Hom}_\Lambda(\Omega^t(N_{\geq k}), t(M)) \to \operatorname{Ext}^t_\Lambda(N_{\geq k}, t(M)) \to 0.$

Taking direct limits we have an epimorphism:

$\varinjlim_k \operatorname{Hom}_\Lambda(\Omega^t(N_{\geq k}), t(M)) \to \varinjlim_k \operatorname{Ext}^t_\Lambda(N_{\geq k}, t(M)) \to 0$

By lemma 4.1, $\varinjlim_k \operatorname{Hom}_\Lambda(\Omega^t(N_{\geq k}), t(M)) = 0$.

Therefore: $\varinjlim \operatorname{Ext}^t_\Lambda(N_{\geq k}, t(M)) = 0$.

From the exact sequence: $0 \to t(M) \to M \to M/t(M) \to 0$ we obtain exact sequences:

$\varinjlim \operatorname{Ext}^t_\Lambda(N_{\geq k}, t(M)) \to \varinjlim \operatorname{Ext}^t_\Lambda(N_{\geq k}, M) \to \varinjlim \operatorname{Ext}^t_\Lambda(N_{\geq k}, M/t(M)) \to \varinjlim \operatorname{Ext}^{t+1}_\Lambda(N_{\geq k}, t(M))$.

It follows $\varinjlim \operatorname{Ext}^t_\Lambda(N_{\geq k}, M) \cong \operatorname{Ext}^t_\Lambda(N_{\geq k}, M/t(M))$. □

The next proposition is well known in the noetherian case [33].

PROPOSITION 4.3. *Let $\Lambda = KQ/I$ be a generalized Auslander regular Koszul algebra and let M and N be in* $\operatorname{Finres}\Lambda$. *Then $\pi M \cong \pi N$ if and only if there exists an integer n such that $M_{\geq n} \cong N_{\geq n}$.*

PROOF: Suppose $f : N_{\geq n} \to M_{\geq n}$ is an isomorphism in $\operatorname{Gr}\Lambda$. Then πf is an isomorphism in $\operatorname{QGr}\Lambda$, but $M/M_{\geq n}$ and $N/N_{\geq n}$ are torsion so $\pi N \cong \pi(N_{\geq n}) \cong \pi(M_{\geq n}) \cong \pi M$.

Suppose $\pi N \cong \pi M$. Then the morphism is of the form πf with $f : N' \to M/M'$, where N/N' and M' are torsion.

It is known that if πf is isomorphism [33], then $\ker f$ and $\operatorname{coker} f$ are torsion.

But $\ker f \subseteq N' \subseteq N$ implies $\ker f \subseteq t(N)$. Since $N \in \operatorname{Finres}\Lambda$, the submodule $t(N)$ is finite dimensional, hence; $\ker f$ is finite dimensional.

The module M is finitely generated, then $\operatorname{coker} f$ is finitely generated and torsion; therefore finite dimensional.

It follows, there exists some integer n such that the restriction $f : N'_{\geq n} \to (M/M')_{\geq n}$ is an isomorphism.

We have exact sequences: $0 \to N' \to N \to N/N' \to 0$ and $0 \to M' \to M \to M/M' \to 0$ with N/N' and M' torsion. The module N/N' is finitely generated and $M' \subseteq t(M)$. Because $M \in \operatorname{Finres}\Lambda$, the submodule $t(M)$ is finite dimensional, hence for large enough n we have isomorphisms: $N'_{\geq n} \cong N_{\geq n}$ and $M_{\geq n} \cong (M/M')_{\geq n}$.

It follows $N_{\geq n} \cong M_{\geq n}$. □

COROLLARY *Let $\Lambda = KQ/I$ be a generalized Auslander regular Koszul algebra. Then for any module $M \in$* $\operatorname{Finres}\Lambda$, *with M of infinite length, there exists some integer n such that $\pi M \cong \pi M_{\geq n}$ and $M_{\geq n}$ Koszul up to shifting.* □

We have the following non Koszul version of the previous proposition:

PROPOSITION 4.4. [33] *Let $\Lambda = KQ/I$ be a generalized Auslander regular algebra and let M and $N \in \operatorname{Finres}\Lambda$ be two torsion free modules. Then $\pi M \cong \pi N$ if and only if there exists an integer n such that $M_{\geq n} \simeq N_{\geq n}$.* □

We will use the following:

PROPOSITION 4.5. [2], [19], [33] *Let $\Lambda = KQ/I$ be a graded quiver algebra. The quotient category, QGr_Λ, has injective objets. Then for any pair of graded modules, L and M and any integer $k \geq 0$, the extension groups $Ext^k_{\mathrm{Gr}\,\Lambda}(\pi L, \pi M)$ can be defined.*

Assume all graded simples are in $\operatorname{Finres}\Lambda$. If $L \in \operatorname{Finres}\Lambda$ and M in is any graded module, then $Ext^k_{\mathrm{Gr}\,\Lambda}(\pi L, \pi M) \cong \varinjlim_l Ext^k_\Lambda(L_{\geq l}, M)$.

PROOF: It is known by [19], [33] QGr_Λ has injective envelopes and the extension groups $\mathrm{Ext}^k\mathrm{Gr}\,\Lambda(\pi L, \pi M)$ are obtained as derived functors of the homomorphism functor.

We will prove the claim by induction on k, for $k = 0$ it has been already proved. Assume $\mathrm{Ext}^{k-1}_{\mathrm{Gr}\,\Lambda}(\pi L, \pi M) \cong \varinjlim_l \mathrm{Ext}^{k-1}_\Lambda(L_{\geq l}, M)$ for all $k \geq 1$.

The injective envelope of a torsion free module is torsion free.

Consider the exact sequence:

*) $0 \to M/t(M) \to I \to \Omega^{-1}(M/t(M)) \to 0$ with I the injective envelope of $M/t(M)$.

Applying the exact functor $\pi. \mathrm{Gr}_\Lambda \to \mathrm{QGr}_\Lambda$ we obtain an exact sequence in QGr_Λ:

**) $0 \to \pi(M/t(M)) \to \pi(I) \to \pi(\Omega^{-1}(M/t(M))) \to 0$.

It follows from [33] $\pi(I)$ is the injective envelope of $\pi(M/t(M))$.

By the long homology sequence we obtain from **) an exact sequence:

$\mathrm{Ext}^{k-1}_{\mathrm{Gr}\,\Lambda}(\pi L, \pi I) \to \mathrm{Ext}^{k-1}_{\mathrm{Gr}\,\Lambda}(\pi L, \pi(\Omega^{-1}(M/t(M)))) \to \mathrm{Ext}^{k}_{\mathrm{Gr}\,\Lambda}(\pi L, \pi M/t(M)) \to 0$.

By the long homology sequence we obtain from *) an exact sequence:

$\mathrm{Ext}^{k-1}_\Lambda(L_{\geq l}, I) \to \mathrm{Ext}^{k-1}_\Lambda(L_{\geq l}, \Omega^{-1}(M/t(M))) \to \mathrm{Ext}^{k}_\Lambda(L_{\geq l}, M/t(M)) \to 0$.

Taking limits, we obtain an exact sequence:

$\varinjlim \mathrm{Ext}^{k-1}_\Lambda(L_{\geq l}, I) \to \varinjlim \mathrm{Ext}^{k-1}_\Lambda(L_{\geq l}, \Omega^{-1}(M/t(M))) \to \varinjlim \mathrm{Ext}^{k}_\Lambda(L_{\geq l}, M/t(M)) \to 0$.

By induction hypothesis there are natural isomorphisms:

$$\varinjlim \mathrm{Ext}^{k-1}_\Lambda(L_{\geq l}, I) \cong \mathrm{Ext}^{k-1}_{\mathrm{Gr}\,\Lambda}(\pi L, \pi I), \text{ and}$$

$$\varinjlim \mathrm{Ext}^{k-1}_\Lambda(L_{\geq l}, \Omega^{-1}(M/t(M))) \cong \mathrm{Ext}^{k-1}_{\mathrm{Gr}\,\Lambda}(\pi L, \pi(\Omega^{-1}(M/t(M)))).$$

It follows $\varinjlim \mathrm{Ext}^{k}_\Lambda(L_{\geq l}, M/t(M)) \cong \mathrm{Ext}^{k}_{\mathrm{Gr}\,\Lambda}(\pi L, \pi(M/t(M)))$.

By Proposition 4.2, $\varinjlim \mathrm{Ext}^{k}_\Lambda(L_{\geq l}, M/t(M)) \cong \varinjlim \mathrm{Ext}^{k}_\Lambda(L_{\geq l}, M)$, hence using the isomorphism $\pi(M/t(M)) \cong \pi(M)$ the result follows. □

COROLLARY *Let $\Lambda = KQ/I$ be a graded quiver algebra, such that all graded simples are in $\operatorname{Finres}\Lambda$. Let M be a torsion module and $L \in \operatorname{Finres}\Lambda$. Then $\varinjlim_l Ext^k_\Lambda(L_{\geq l}, M) = 0$, for any integer $k \geq 0$.* □

In order to motivate a few ideas in non commutative geometry we recall first some of the main results from Serre's paper [36].

Let K be a field and $\Lambda = K\left[x_0, x_1, x_2, \ldots x_n\right]/I$ be the quotient of the polynomial algebra modulo an homogeneous ideal I, let X be the subscheme of P^n defined by I and $\mathcal{O}_X$ the sheaf of regular functions. Let $\operatorname{Coh}(X)$ be the category of coherent sheaves, $\mathcal{O}_X(n)$ the nth-power of the twisting sheaf. Define a functor: $\Gamma_* : \operatorname{Coh}(X) \to \operatorname{QGr}\Lambda$ by $\Gamma_*(\mathcal{F}) = \bigoplus_{d=-\infty}^{d=\infty} H^0(X, \mathcal{F} \otimes O_X(d))$, where $H^0(X, \mathcal{F})$ denotes the global sections of the sheaf $\mathcal{F}$.

The functor Γ_* is an equivalence of categories with inverse functor $\sim$: $\operatorname{QGr}\Lambda \to \operatorname{Coh}(X)$ defined as follows:

The set $\{X_f\}_{f\in\Lambda}$ with $X_f = \{x \in X \mid f(x) \neq 0\}$ form a basis of open sets for the topology of X.

The structure sheaf $\mathcal{O}$ is defined by $\mathcal{O}(X_f) = \Lambda\left[f^{-1}\right]_0$, the degree zero part of $\Lambda\left[f^{-1}\right]$, which consists of the rational functions g/h in $K(X)$ having no pole on X_f, the polynomials g and h are homogeneous with $\deg g = \deg h$. If M is a graded module, define the sheaf $\widetilde{M}$ by $\widetilde{M}(X_f) = (\Lambda\left[f^{-1}\right] \bigotimes_\Lambda M)_0 = M[f^{-1}]_0$.

For a coherent sheaf $\mathcal{F}$ cohomology groups $H^q(\mathcal{F})$ were defined by Čech, we refer the reader to Serre's paper for the definition and for the proof of the isomorphisms $H^q(\mathcal{F}) \cong H^q(X, \mathcal{F})$, where $H^q(X, -)$ denotes the derived functors of the global section functor $H^0(X, -)$.

The groups $H^q(X, \mathcal{F})$ can be defined in terms of the extension groups as follows:

Let M be a graded module such that $\widetilde{M} = \mathcal{F}$.

The category $\operatorname{QGr}\Lambda$ has enough injectives, hence derived functors of the graded homorphism functor exist. It is proved in [36] $H^q(X, \mathcal{F}) \cong \operatorname{Ext}^q_{\operatorname{QGr}\Lambda}(\mathcal{O}_X, \mathcal{F})_0$, the degree zero part of the graded extension group.

Let $J = (x_0, x_1, x_2, \ldots x_n)/I$. Then $\operatorname{Ext}^q_{\operatorname{QGr}\Lambda}(\mathcal{O}_X, \mathcal{F}) \cong \varinjlim \operatorname{Ext}^q_\Lambda(J^k, M)$.

Define $\underline{H}^q(X, \mathcal{F}) = \bigoplus_n H^q(X, \mathcal{F}(n))$, where $\mathcal{F}(n) = \mathcal{F} \bigotimes \mathcal{O}_X(n)$ and $\widetilde{\mathcal{F}(n)} = \widetilde{M[n]}$, $M[n]$ the n-th shift of M.

With this definition $\underline{H}^q(X, \mathcal{F}) = \bigoplus_n \varinjlim \operatorname{Ext}^q_\Lambda(J^k, M[n])_0 \cong \varinjlim \bigoplus_n \operatorname{Ext}^q_\Lambda(J^k, M)_n \cong \varinjlim \operatorname{Ext}^q_\Lambda(J^k, M)$.

In particular
$\Gamma_*(\mathcal{F}) = \underline{H}^0(X, \mathcal{F}) = \bigoplus_n \operatorname{Hom}^n_{\operatorname{QGr}\Lambda}(\pi\Lambda, \pi M) = \operatorname{Hom}_{QGr\Lambda}(\pi\Lambda, \pi M)$.

From these results one can see that for a noetherian graded quiver algebra $\Lambda = KQ/I$ the category $\operatorname{QGr}\Lambda$ is the generalization of the category of coherent sheaves, and the extension groups $\varinjlim \operatorname{Ext}^q_\Lambda(J^k, M) \cong \operatorname{Ext}^q_{QGr\Lambda}(\pi\Lambda, \pi M)$, where $\pi : \operatorname{gr}\Lambda \to \operatorname{QGr}\Lambda$ is the quotient functor, generalize Čech cohomology.

In case $\Lambda = KQ/I$ is a non noetherian generalized Auslander regular algebra we have the quotient functor $\pi : \operatorname{Finres}\Lambda \to \operatorname{Q}\operatorname{Finres}\Lambda$ and $\varinjlim \operatorname{Ext}^q_\Lambda(J^k, M) \cong \operatorname{Ext}^q_{\operatorname{QGr}\Lambda}(\pi\Lambda, \pi M)$ for $M \in \operatorname{Finres}\Lambda$.

After these remarks we will devote the rest of the paper to prove Serre duality for generalized Auslander regular algebras.

As an application of theorem 3.5 we obtain the following, non commutative, non noetherian version of a classical theorem:

PROPOSITION 4.6. *(Cyclic homology) Let $\Lambda = KQ/I$ be an indecomposable generalized Auslander regular algebra of graded small global dimension n. Then for any graded bounded below locally finite module M and any graded module L there exists a natural isomorphism:*
$Hom_K(\varinjlim Ext^{n-i}_\Lambda(M/M_{\geq k}, L), K) \cong Ext^i_\Lambda(L, \sigma^{-1}M\,[n])$.

PROOF: By induction on i.

$i = 0$

By lemma 3.8 $\mathrm{Ext}^n_\Lambda(M/M_{\geq k}, L) \cong \mathrm{Ext}^n_\Lambda(M/M_{\geq k}, \Lambda) \bigotimes_\Lambda L$.

By proposition 3.5 we have a natural isomorphism:

$$\begin{aligned}\mathrm{Ext}^n_\Lambda(M/M_{\geq k}, \Lambda) &\cong \mathrm{Hom}_\Lambda(M/M_{\geq k}, D(\Lambda)[-n])\sigma \cong\\ \mathrm{Hom}_\Lambda(M/M_{\geq k}[n], D(\Lambda))\sigma &\cong D(M/M_{\geq k})[-n]\sigma\end{aligned}$$

Hence we have isomorphisms:
$\mathrm{Ext}^n_\Lambda(M/M_{\geq k}, L) \cong D(M/M_{\geq k}\,[-n])\sigma \bigotimes_\Lambda L \cong D(M/M_{\geq k}) \bigotimes_\Lambda \sigma L\,[-n]$.

Then $\varinjlim \mathrm{Ext}^n_\Lambda(M/M_{\geq k}, L) \cong \varinjlim D(M/M_{\geq k}) \bigotimes_\Lambda \sigma L\,[-n]$.

It follows:
$\mathrm{Hom}_K(\varinjlim \mathrm{Ext}^n_\Lambda(M/M_{\geq k}, L), K) \cong \mathrm{Hom}_K(\varinjlim D(M/M_{\geq k}) \otimes_\Lambda \sigma L\,[-n]\,, K) \cong$
$\mathrm{Hom}_\Lambda(\sigma L[-n], \mathrm{Hom}_K(\varinjlim D(M/M_{\geq k}), K) \cong \mathrm{Hom}_\Lambda(\sigma L[-n], \varprojlim D^2(M/M_{\geq k})) \cong$
$\mathrm{Hom}_\Lambda(\sigma L[-n], \varprojlim M/M_{\geq k})$.

But by [31], $\varprojlim M/M_{\geq k} \cong M$.

We have isomorphisms: $\mathrm{Hom}_\Lambda(\sigma L\,[-n]\,, M) \cong \mathrm{Hom}_\Lambda(L, \sigma^{-1}M\,[n]) \cong \mathrm{Hom}_K(\varinjlim \mathrm{Ext}^n_\Lambda(M/M_{\geq k}, L), K)$

The claim is true for $n = 0$.

Assume the result for i, with $n > i \geq 0$, then
$\mathrm{Hom}_K(\varinjlim \mathrm{Ext}^{n-i}_\Lambda(M/M_{>>k}, L), K) \cong \mathrm{Ext}^i_\Lambda(L, \sigma^{-1}M\,[n])$.

Let $0 \to \Omega(L) \to P \to L \to 0$ be exact with P the projective cover of L.

We have an exact sequence:

$$\begin{aligned}\mathrm{Ext}^{n-i-1}_\Lambda(M/M_{\geq k}, P) &\to \mathrm{Ext}^{n-i-1}_\Lambda(M/M_{\geq k}, L) \to \mathrm{Ext}^{n-i}_\Lambda(M/M_{\geq k}, \Omega(L))\\ \to \mathrm{Ext}^{n-i}_\Lambda(M/M_{\geq k}, P)\end{aligned}$$

By proposition 3.5, $\mathrm{Ext}^{n-i-1}_\Lambda(M/M_{\geq k}, P) = 0$.

The exact sequence:
$0 \to \varinjlim \mathrm{Ext}^{n-i-1}_\Lambda(M/M_{\geq k}, L) \to \varinjlim \mathrm{Ext}^{n-i}_\Lambda(M/M_{\geq k}, \Omega(L)) \to$
$\varinjlim \mathrm{Ext}^{n-i}_\Lambda(M/M_{\geq k}, P)$

induces an exact sequence:

$$D(\varinjlim \mathrm{Ext}_\Lambda^{n-i}(M/M_{\geq k}, P)) \;\to\; D(\varinjlim \mathrm{Ext}_\Lambda^{n-i}(M/M_{\geq k}, \Omega(L)))$$
$$\to D(\varinjlim \mathrm{Ext}_\Lambda^{n-i-1}(M/M_{\geq k}, L)) \to \;0$$

For each $i \geq 0$ we have an exact sequence:

$$\mathrm{Ext}_\Lambda^i(P, \sigma^{-1}M[n]) \to \mathrm{Ext}_\Lambda^i(\Omega(L), \sigma^{-1}M[n]) \to \mathrm{Ext}_\Lambda^{i+1}(L, \sigma^{-1}M[n]) \to 0$$

By induction hypothesis we have natural isomorphisms:

$$\mathrm{Ext}_\Lambda^i(P, \sigma^{-1}M\,[n]) \;\cong\; D(\varinjlim \mathrm{Ext}_\Lambda^{n-i}(M/M_{\geq k}, P))$$
$$\mathrm{Ext}_\Lambda^i(\Omega(L), \sigma^{-1}M\,[n]) \;\cong\; D(\varinjlim \mathrm{Ext}_\Lambda^{n-i}(M/M_{\geq k}, \Omega(L))).$$

It follows there exists a natural isomorphism:
$\mathrm{Ext}_\Lambda^{i+1}(L, \sigma^{-1}M\,[n]) \cong D(\varinjlim \mathrm{Ext}_\Lambda^{n-i-1}(M/M_{\geq k}, L)).$ □

COROLLARY 1. *Let Λ be a generalized Auslander regular algebra of graded small global dimension n. Then for any pair of graded Λ-modules L and $M \in$ Finres Λ and any non negative integer i the graded module $\varinjlim_k Ext_\Lambda^i(L_{\geq k}, M)$ is locally finite.*

PROOF: The exact sequence: $0 \to L_{\geq k} \to L \to L/L_{\geq k} \to 0$ induces exact sequences: $\mathrm{Ext}_\Lambda^i(L, M) \to \mathrm{Ext}_\Lambda^i(L_{\geq k}, M) \to \mathrm{Ext}_\Lambda^{i+1}(L/L_{\geq k}, M)$.

Taking direct limits and dualizing we obtain an exact sequence:
$D(\varinjlim \mathrm{Ext}_\Lambda^{i+1}(L/L_{\geq k}, M)) \to D(\varinjlim \mathrm{Ext}_\Lambda^i(L_{\geq k}, M)) \to D(\mathrm{Ext}_\Lambda^i(L, M))$

By the proposition, $D(\varinjlim \mathrm{Ext}_\Lambda^{i+1}(L/L_{\geq k}, M)) \cong \mathrm{Ext}_\Lambda^{n-i-1}(M, \sigma^{-1}L\,[n])$.

Hence; $D(\varinjlim \mathrm{Ext}_\Lambda^{i+1}(L/L_{\geq k}, M))$ and $D(\mathrm{Ext}_\Lambda^i(L, M))$ are locally finite.

It follows $D(\varinjlim \mathrm{Ext}_\Lambda^i(L_{\geq k}, M))$ is locally finite, for any $0 \leq i \leq n$. Then $\varinjlim \mathrm{Ext}_\Lambda^i(L_{\geq k}, M)$ is locally finite.

COROLLARY 2. *Let Λ be a generalized Auslander regular algebra of graded small global dimension n and let L be a graded bounded below locally bounded torsion free Λ-module. Then for all integer $k > 0$ the module $L_{\geq k}$ has projective dimension $n - 1$.*

PROOF: The exact sequence $0 \to L_{\geq k} \to L \to L/L_{\geq k} \to$ induces an exact sequence:

$$\mathrm{Ext}_\Lambda^{n-1}(L/L_{\geq k}, \Lambda) \;\to\; \mathrm{Ext}_\Lambda^{n-1}(L, \Lambda) \to \mathrm{Ext}_\Lambda^{n-1}(L_{\geq k}, \Lambda)$$
$$\to \mathrm{Ext}_\Lambda^{n}(L/L_{\geq k}, \Lambda) \;\to\; \mathrm{Ext}_\Lambda^{n}(L, \Lambda)$$

By the proposition, we have a natural isomorphism:

$$\mathrm{Ext}_\Lambda^n(L, \Lambda) \cong \mathrm{Ext}_\Lambda^n(\sigma^{-1}L\,[n]\,, \sigma^{-1}\Lambda\,[n]) \cong D(\varinjlim \mathrm{Hom}_\Lambda(\Lambda/\Lambda_{\geq k}, \sigma^{-1}L\,[n]))$$

The module L torsion free implies $\mathrm{Hom}_\Lambda(\Lambda/\Lambda_{\geq k}, \sigma^{-1}L[n]) = 0$. Then $\mathrm{Ext}^n_\Lambda(L, \Lambda) = 0$.

By a similar argument, we have $\mathrm{Ext}^n_\Lambda(L_{\geq k}, \Lambda) = 0$.

Since $\mathrm{Ext}^n_\Lambda(L/L_{\geq k}, \Lambda) \neq 0$ it follows $\mathrm{Ext}^{n-1}_\Lambda(L_{\geq k}, \Lambda) \neq 0$.

Therefore: $\mathrm{pd}L_{\geq k} = n - 1$. □

We have the following:

LEMMA 4.7. *Let $\Lambda = KQ/I$ be a graded quiver algebra, and let $\{L_\alpha\}_{\in I}, \{M_\alpha\}_{\in I}$, and $\{N_\alpha\}_{\in I}$ be inverse systems of locally finite graded Λ-modules, and let $0 \to L_\alpha \to M_\alpha \to N_\alpha \to 0$ be an exact sequence of inverse systems. Then taking inverse limits, we have an exact sequence:*

$0 \to \varprojlim L_\alpha \to \varprojlim M_\alpha \to \varprojlim N_\alpha \to 0.$

PROOF: Applying the duality D we obtain an exact sequence of direct systems:

$0 \to D(N_\alpha) \to D(M_\alpha) \to D(L_\alpha) \to 0$

Taking direct limits we obtain an exact sequence:

$0 \to \varinjlim D(N_\alpha) \to \varinjlim D(M_\alpha) \to \varinjlim D(L_\alpha) \to 0$ applying the functor $\mathrm{Hom}_K(-, K)$ again we obtain an exact sequence:

$$0 \to \mathrm{Hom}_K(\varinjlim D(L_\alpha), K) \to \mathrm{Hom}_K(\varinjlim D(M_\alpha), K) \to \mathrm{Hom}_K(\varinjlim D(N_\alpha), K) \to 0$$

Then the sequence :

*) $0 \to \varprojlim \mathrm{Hom}_K(D(L_\alpha), K) \to \varprojlim \mathrm{Hom}_K(D(M_\alpha), K) \to \varprojlim \mathrm{Hom}_K(D(N_\alpha), K) \to 0$

is exact.

But the sequence *) is isomorphic to the sequence:

$0 \to \varprojlim L_\alpha \to \varprojlim M_\alpha \to \varprojlim N_\alpha \to 0$ □

PROPOSITION 4.8. *Let Λ be an indecomposable generalized Auslander regular algebra of graded small global dimension n. Let M and L be torsion free Λ modules, with M and $L \in$ Finres Λ. Then for any integer i with $0 \leq i \leq n-1$ there exist natural homomorphisms:*

$$\mu_{M,L} : \varinjlim \mathrm{Ext}^i_\Lambda(M_{\geq k}, L) \to \mathrm{Hom}_K(\varprojlim_l \mathrm{Ext}^{n-1-i}_\Lambda(J^l L, \sigma^{-1}M[n]), K).$$

If $M = \Lambda$, then $\mu_{\Lambda,L}$ is an isomorphism.

PROOF: By the long homology sequence applied to the exact sequence: $0 \to M_{\geq k} \to M \to M/M_{\geq k} \to 0$ we have the connecting map:

$$\mathrm{Ext}^i_\Lambda(M_{\geq k}, L) \to \mathrm{Ext}^{i+1}_\Lambda(M/M_{\geq k}, L).$$

Taking direct limits, we have a natural map:

$$\varinjlim \mathrm{Ext}^i_\Lambda(M_{\geq k}, L) \to \varinjlim \mathrm{Ext}^{i+1}_\Lambda(M/M_{\geq k}, L).$$

By cyclic homology, we have an isomorphism:

$$\varinjlim \mathrm{Ext}^{i+1}_\Lambda(M/M_{\geq k}, L) \cong D(\mathrm{Ext}^{n-i-1}(L, \sigma^{-1}M[n]).$$

Composing with the above homomorphism, we obtain a natural homomorphism: $\varinjlim \operatorname{Ext}^i_\Lambda(M_{\geq k}, L) \to D(\operatorname{Ext}^{n-i-1}(L, \sigma^{-1}M[n])$.
Substituting L by $L_{\geq l}$ we obtain natural homomorphisms:
$\gamma_{M,L} : \varinjlim \operatorname{Ext}^i_\Lambda(M_{\geq k}, L_{\geq l}) \to D(\operatorname{Ext}^{n-i-1}(L_{\geq l}, \sigma^{-1}M[n])$.
Consider now the exact sequence: $0 \to L_{\geq l} \to L \to L/L_{\geq l} \to 0$.
By the long homology sequence again we obtain an exact sequence:
$\operatorname{Ext}^{i-1}_\Lambda(M_{\geq k}, L/L_{\geq l}) \to \operatorname{Ext}^i_\Lambda(M_{\geq k}, L_{\geq l}) \to \operatorname{Ext}^i_\Lambda(M_{\geq k}, L) \to \operatorname{Ext}^i_\Lambda(M_{\geq k}, L/L_{\geq l})$
Taking limits, it induces an exact sequence:
$\to \varinjlim_k \operatorname{Ext}^{i-1}_\Lambda(M_{\geq k}, L/L_{\geq l}) \to \varinjlim_k \operatorname{Ext}^i_\Lambda(M_{\geq k}, L_{\geq l}) \to \varinjlim_k \operatorname{Ext}^i_\Lambda(M_{\geq k}, L)$
$\to \varinjlim_k \operatorname{Ext}^i_\Lambda(M_{\geq k}, L/L_{\geq l}) \to$
By Corollary 4.5, $\varinjlim_k \operatorname{Ext}^{i-1}_\Lambda(M_{\geq k}, L/L_{\geq l}) = 0 = \varinjlim_k \operatorname{Ext}^i_\Lambda(M_{\geq k}, L/L_{\geq l})$.
Therefore: $\varinjlim_k \operatorname{Ext}^i_\Lambda(M_{\geq k}, L) \cong \varinjlim_k \operatorname{Ext}^i_\Lambda(M_{\geq k}, L_{\geq l})$

Composing this isomorphism with $\gamma_{M,L}$ we obtain a natural homomorphism:
$\varinjlim_k \operatorname{Ext}^i_\Lambda(M_{\geq k}, L) \to D(\operatorname{Ext}^{n-i-1}(L_{\geq l}, \sigma^{-1}M[n]))$.
Taking inverse limits, and composing with the natural isomorphism:
$\varprojlim_l D(\operatorname{Ext}^{n-i-1}(L_{\geq l}, \sigma^{-1}M[n]) \cong D(\varinjlim_l \operatorname{Ext}^{n-i-1}(L_{\geq l}, \sigma^{-1}M[n])$.
we obtain a natural homomorphism:

$\eta_{M,L}: \varinjlim_k \operatorname{Ext}^i_\Lambda(M_{\geq k}, L) \to D(\varinjlim_l \operatorname{Ext}^{n-i-1}(L_{\geq l}, \sigma^{-1}M[n])$.
We prove now that for $\Lambda = M$ the above homomorphism is an isomorphism:
We consider first the case $i = 0$.
The exact sequence:
$0 \to J^k \to \Lambda \to \Lambda/J^k \to 0$ induces an exact sequence:
$0 \to \operatorname{Hom}_\Lambda(\Lambda/J^k, L) \to \operatorname{Hom}_\Lambda(\Lambda, L) \to \operatorname{Hom}_\Lambda(J^k, L) \to \operatorname{Ext}^1_\Lambda(\Lambda/J^k, L) \to 0$.
Since L is torsion free $\operatorname{Hom}_\Lambda(\Lambda/J^k, L) = 0$.
Changing L by $L_{\geq l}$ we obtain exact sequences:
$0 \to L_{\geq l} \to \operatorname{Hom}_\Lambda(J^k, L_{\geq l}) \to \operatorname{Ext}^1_\Lambda(\Lambda/J^k, L_{\geq l}) \to 0$.
Taking direct limits and composing with the natural isomorphism:
$\varinjlim_k \operatorname{Ext}^1_\Lambda(\Lambda/J^k, L_{\geq l}) \cong D(\operatorname{Ext}^{n-1}_\Lambda(L_{\geq l}, \sigma^{-1}\Lambda[n])$
we have an exact sequence:
$0 \to L_{\geq l} \to \varinjlim_k \operatorname{Hom}_\Lambda(J^k, L_{\geq l}) \to D(\operatorname{Ext}^{n-1}_\Lambda(L_{\geq l}, \sigma^{-1}\Lambda[n]) \to 0$.

As above, we have natural isomorphism:
$\varinjlim_k \operatorname{Hom}_\Lambda(J^k, L_{\geq l}) \cong \varinjlim_k \operatorname{Hom}_\Lambda(J^k, L)$.
Hence an exact sequence:
$0 \to L_{\geq l} \to \varinjlim_k \operatorname{Hom}_\Lambda(J^k, L) \to D(\operatorname{Ext}^{n-1}_\Lambda(L_{\geq l}, \sigma^{-1}\Lambda[n]) \to 0$.

Taking inverse limits we have, by lemma 4.7, an exact sequence:

$$0 \to \varprojlim_{l} L_{\geq l} \to \varinjlim_{k} \operatorname{Hom}_\Lambda(J^k, L) \to \varprojlim_{l} D(\operatorname{Ext}_\Lambda^{n-1}(L_{\geq l}, \sigma^{-1}\Lambda[n]) \to 0\,.$$

But $\varprojlim_{l} L_{\geq l} - \bigcap_{l} L_{\geq l} - 0$ and $\varprojlim_{l} D(\operatorname{Ext}_\Lambda^{n-1}(L_{\geq l}, \sigma^{-1}\Lambda[n])$ is isomorphic to $D(\varinjlim_{l} \operatorname{Ext}_\Lambda^{n-1}(L_{\geq l}, \sigma^{-1}\Lambda[n]))$.

Then the natural homomorphism $\eta_{\Lambda,L}$ is an isomorphism.

For $i > 0$ we have an exact sequence:

$$\operatorname{Ext}_\Lambda^i(\Lambda, L) \to \operatorname{Ext}_\Lambda^i(J^k, L) \to \operatorname{Ext}_\Lambda^{i+1}(\Lambda/J^k, L) \to \operatorname{Ext}_\Lambda^{i+1}(\Lambda, L) \to,$$

with $\operatorname{Ext}_\Lambda^i(\Lambda, L) = 0 = \operatorname{Ext}_\Lambda^{i+1}(\Lambda, L)$.

Taking limits we have an isomorphism:
$\varinjlim \operatorname{Ext}_\Lambda^i(J^k, L) \to \varinjlim \operatorname{Ext}_\Lambda^{i+1}(\Lambda/J^k, L)$.

Using the cyclic homology we have an isomorphism:
$\varinjlim \operatorname{Ext}_\Lambda^i(J^k, L) \to D(\varinjlim \operatorname{Ext}_\Lambda^{n-1-i}(L, \sigma^{-1}\Lambda[n]))$.

Changing L by $L_{\geq l}$ and taking limits it follows, as above, the map:
$\eta_{\Lambda,L}: \varinjlim_{k} \operatorname{Ext}_\Lambda^i(J^k, L) \to D(\varinjlim_{l} \operatorname{Ext}_\Lambda^{n-1-i}(L_{\geq l}, \sigma^{-1}L[n]))$ is an isomorphism.

We have proved the proposition. □

Given the quotient functor $\pi : \operatorname{gr}\Lambda \to \operatorname{QGr}\Lambda$ and an object X in $\operatorname{gr}\Lambda$ we will denote $\pi(X)$ by $\underline{X}$.

We can prove now our main theorem:

THEOREM 4.9. *Let Λ be an indecomposable generalized Auslander regular algebra of graded small global dimension n, let M and L be torsion free graded modules with M and $L \in \operatorname{Finres}\Lambda$. Then for all integer i with $0 \leq i \leq n-1$ there exists a natural isomorphism:*
$\operatorname{Ext}^i_{\operatorname{QGr}\Lambda}(\underline{M}, \underline{L}) \to D(\operatorname{Ext}^{n-1-i}_{\operatorname{QGr}\Lambda}(\underline{L}, \underline{\sigma^{-1}M[n]}))$.

PROOF: By induction on projective dimension of M.

If M is projective, then the natural homomorphism $\eta_{M,L}$ is an isomorphism.

Assume it is true for all modules of projective dimension $< k$ and $\operatorname{pd}M = k$.

Consider the exact sequence: $0 \to \Omega(M) \to P \to M \to 0$, with P the projective cover of M. For each integer s induces exact sequences: $0 \to \Omega(M)_{\geq s} \to P_{\geq s} \to M_{\geq s} \to 0$ which in turn induces exact sequences:

$\operatorname{Ext}_\Lambda^i(P_{\geq s}, L) \to \operatorname{Ext}_\Lambda^i(\Omega(M)_{\geq s}, L) \to \operatorname{Ext}_\Lambda^{i+1}(M_{\geq s}, L) \to \operatorname{Ext}_\Lambda^{i+1}(P_{\geq s}, L) \to \operatorname{Ext}_\Lambda^{i+1}(\Omega(M)_{\geq s}, L)$. Taking limits we have an exact sequence:

$\varinjlim \operatorname{Ext}_\Lambda^i(P_{\geq s}, L) \to \varinjlim \operatorname{Ext}_\Lambda^i(\Omega(M)_{\geq s}, L) \to \varinjlim \operatorname{Ext}_\Lambda^{i+1}(M_{\geq s}, L) \to$
$\varinjlim \operatorname{Ext}_\Lambda^{i+1}(P_{\geq s}, L) \to \varinjlim \operatorname{Ext}_\Lambda^{i+1}(\Omega(M)_{\geq s}, L)$

The exact sequence $0 \to \sigma^{-1}\Omega(M)[n] \to \sigma^{-1}P[n] \to \sigma^{-1}M[n] \to 0$ induces exact sequences:

$$\operatorname{Ext}_\Lambda^{n-i-2}(L_{\geq k}, \sigma^{-1}\Omega(M)[n]) \to \operatorname{Ext}_\Lambda^{n-i-2}(L_{\geq k}, \sigma^{-1}P[n]) \to$$

$\mathrm{Ext}_\Lambda^{n-i-2}(L_{\geq k},\sigma^{-1}M[n]) \to \mathrm{Ext}_\Lambda^{n-i-1}(L_{\geq k},\sigma^{-1}\Omega(M)[n]) \to$
$\mathrm{Ext}_\Lambda^{n-i-1}(L_{\geq k},\sigma^{-1}P[n])$.

Taking direct limits and dualizing we obtain an exact sequence:

$D(\varinjlim \mathrm{Ext}^{n-i-1}(L_{\geq k},\sigma^{-1}P[n])) \to D(\varinjlim \mathrm{Ext}_\Lambda^{n-i-1}(L_{\geq k},\sigma^{-1}\Omega(M)[n])) \to$

$D(\varinjlim \mathrm{Ext}_\Lambda^{n-i-2}(L_{\geq k},\sigma^{-1}M[n])) \to D(\varinjlim \mathrm{Ext}_\Lambda^{n-i-2}(L_{\geq k},\sigma^{-1}P[n])) \to$

$D(\varinjlim \mathrm{Ext}_\Lambda^{n-i-2}(L_{\geq k},\sigma^{-1}\Omega(M)[n]))$.

By induction hypothesis, we have the following chain of natural isomorphisms:

$$\varinjlim \mathrm{Ext}_\Lambda^{i}(P_{\geq s},L) \cong D(\varinjlim \mathrm{Ext}^{n-i-1}(L_{\geq k},\sigma^{-1}P[n])$$

$$\varinjlim \mathrm{Ext}_\Lambda^{i}(\Omega(M)_{\geq s},L) \cong D(\varinjlim \mathrm{Ext}_\Lambda^{n-i-1}(L_{\geq k},\sigma^{-1}\Omega(M)[n]))$$

$$\varinjlim \mathrm{Ext}_\Lambda^{i+1}(P_{\geq s},L) \cong D(\varinjlim \mathrm{Ext}_\Lambda^{n-i-2}(L_{\geq k},\sigma^{-1}P[n]))$$

$$\varinjlim \mathrm{Ext}_\Lambda^{i+1}(\Omega(M)_{\geq s},L) \cong D(\varinjlim \mathrm{Ext}_\Lambda^{n-i-2}(L_{\geq k},\sigma^{-1}\Omega(M)[n]))$$

It follows from five's lemma, there is a natural isomorphism:

$$\varinjlim \mathrm{Ext}_\Lambda^{i+1}(M_{\geq s},L) \cong D(\varinjlim \mathrm{Ext}_\Lambda^{n-i-2}(L_{\geq k},\sigma^{-1}M[n])).$$

We consider now the special case of generalized Auslander regular Koszul algebras of global dimension 2.

M. Auslander and I. Reiten proved in [5], that for a finite dimensional algebra Λ over a field K there exists almost split sequences, by using the following formula:

Given indecomposable, finitely generated modules M and L, there exists a natural isomorphism:

$\underline{\mathrm{Hom}}_\Lambda(M,L) \cong \mathrm{Hom}_k(\mathrm{Ext}_\Lambda^1(L,\tau M),K)$, where $\underline{\mathrm{Hom}}_\Lambda(M,L)$ denotes $\mathrm{Hom}_\Lambda(M,L)$ modulo the maps from M to L that factors through projectives and τM is the following module:

Given a finitely generated module X denote by $X^* = \mathrm{Hom}_\Lambda(X,\Lambda)$ and let $P_1 \xrightarrow{f} P_0 \to M \to 0$ be a minimal projective presentation of M then $\mathrm{tr}M$ is the cokernel of the natural map $P_0^* \xrightarrow{f^*} P_1^*$. Then $\tau M = \mathrm{Hom}_K(\mathrm{tr}M,K)$. In case Λ is selfinjective then $\tau M \cong \Omega^2\nu(M)$, where ν is the Nakayama equivalence obtained by composing the two dualities, $\nu(M) = \mathrm{Hom}_K(\mathrm{Hom}_\Lambda(M,\Lambda),K)$.

Serre duality resembles the formula given above, specially in the case $n = 2$. In [20] the existence of almost split sequences for the category of sheaves was established in the noetherian case. In [25] the existence of almost split sequences for the category of Koszul modules over Koszul selfinjective algebras was studied, we intent to explore the connection of this with Serre duality as established here.

We will consider only the case $n = 2$.

THEOREM 4.10. *Let $\Lambda = KQ/\bar{I}$ be a indecomposable Auslander regular Koszul algebra of graded small global dimension* 2 *and* $\mathrm{Q}\,\mathrm{Finres}\,\Lambda$ *the quotient category defined in section 3. Then* $\mathrm{Q}\,\mathrm{Finres}\,\Lambda$ *has almost split sequences. Moreover, if* $\underline{M} \in$

$\mathrm{Q\,Finres}\,\Lambda$ is an indecomposable object and $\pi : \mathrm{Gr}\,\Lambda \to \mathrm{QGr}\,\Lambda$ the quotient functor, then there exists a torsion free Koszul module M and an almost split sequence of Koszul Λ-modules: $0 \to M \to E \to N \to 0$, such that $\pi M \cong \underline{M}$ and $0 \to \pi M \to \pi E \to \pi N \to 0$, is an almost split sequence in $\mathrm{Q\,Finres}\,\Lambda$, with $\pi N \cong \sigma\underline{M}\,[-2]$, and $\sigma : \Lambda \to \Lambda$ is an automorphism of Λ.

PROOF: Let Λ be an algebra satisfying the conditions of the theorem and Γ its corresponding Yoneda algebra, which we know, is selfinjective with $r^2 \neq 0$ and $r^3 = 0$.

Let $\underline{M} \in \mathrm{Q\,Finres}\,\Lambda$. By propositions 2.4 and 3.7, $\underline{M} = \pi M$, with M a Koszul Λ-module of projective dimension 1.

Let $F : K_\Lambda \to K_{\Gamma^{op}}$ and $G : K_{\Gamma^{op}} \to K_\Lambda$ be the Koszul dualities, $F(M) = A$ is a Koszul Γ-module of Lowey length 2. Then the almost split sequence in $\mathrm{mod}_{\Gamma^{op}}$, $0 \to \tau A \to B \to A \to 0$, consists of Koszul modules, $\tau A, B$ and A. There is an isomorphism $\tau A \cong \Omega^2 \nu A\,[-2]$, with ν the Nakayama equivalence.

Since G is an equivalence of categories, applying G, we have an almost split sequence in K_Λ : $0 \to G(A) \to G(B) \to J^2\sigma G(A)\,[-2] \to 0$, and $G(A) \cong M$. Applying π we obtain an exact sequence: $0 \to \pi M \to \pi L \to \pi\sigma J^2 M\,[-2] \to 0$, with $L = G(B)$.

Set $\pi M = \underline{M}$ and $\pi L = \underline{E'}$, we have an isomorphism $\pi\sigma J^2 M\,[-2] \cong \sigma\underline{M}\,[-2]$. With this notation we must prove the exact sequence: $0 \to \underline{M} \to \underline{E'} \to \sigma\underline{M}\,[-2] \to 0$ is almost split, or equivalently, that the sequence: $0 \to \sigma^{-1}\underline{M}\,[2] \to \underline{E} \to \underline{M} \to 0$, with $\sigma^{-1}\underline{E'}\,[2] = \underline{E}$ is almost split.

Set $N = \sigma J^2 M\,[-2]$. The sequence $0 \to M \to L \to N \to 0$ is almost in K_Λ. For any integer k we have exact sequences of Koszul modules: *) $0 \to J^k M\,[-k] \overset{j'_k}{\to} J^k L\,[-k] \overset{p'_k}{\to} J^k N\,[-k] \to 0$, with $j'_k = \mathrm{res} j\,|_{J^k M[-k]}, p'_k = \mathrm{res} p\,|_{J^k L[-k]}$.

Applying the functor F we obtain an exact sequence of Koszul Λ-modules: **) $0 \to \Omega^k F(N)\,[-k] \overset{\Omega^k p[-k]}{\to} \Omega^k F(L)\,[-k] \overset{\Omega^k j[-k]}{\to} \Omega^k F(M)\,[-k] \to 0$.

The algebra Γ selfinjective and the fact that syzygies take almost split sequences to almost split sequences, implies that the exact sequences **) are almost split in mod_Γ, applying G to the sequences **) it follows the sequences *) are almost split in K_Λ.

If X is an indecomposable non simple Koszul Λ-module with $t(X) \neq 0$, then $\mathrm{soc}X \neq 0$ and there is an integer k such that $J^k X$ has a simple summand, therefore: $F(J^k X) \cong \Omega^k F(X)$ has a projective summand, contradicting Γ is selfinjective.

It follows the modules: M, L and N are torsion free.

If X, Y are torsion free modules in $\mathrm{Finres}\,\Lambda$, then we have $\mathrm{Hom}_{\mathrm{QGr}_\Lambda}(X, Y) \cong \bigcup_{k\geq 0} \mathrm{Hom}_\Lambda(J^k X, Y)$.

Since the functor $\pi : \mathrm{Gr}\,\Lambda \to \mathrm{QGr}\,\Lambda$ is exact, we have an exact sequence: $0 \to \pi M \overset{\pi j}{\to} \pi L \overset{\pi p}{\to} \pi N \to 0$, if the sequence splits, then there exists $q \in \bigcup_{k\geq 0} \mathrm{Hom}_\Lambda(J^k L, M)$ with $q\pi j = 1$. This implies that for some integer k the exact sequence: $0 \to J^k M \to J^k L \to J^k N \to 0$ splits. A contradiction.

Let $\underline{X}$ be an indecomposable object in $\mathrm{Q\,Finres}\,\Lambda$ and $\phi : \underline{X} \to \underline{N}$ be a non isomorphism. Then $\underline{X} \cong \pi X$ with X an indecomposable, torsion free, Koszul module and $\phi \in \mathrm{Hom}_{\mathrm{QGr}\,\Lambda}(X, N) = \bigcup_{k\geq 0} \mathrm{Hom}_\Lambda(J^k X, N)$. Assume $\phi : J^k X \to N$ is a degree zero map. The module N generated in degree zero implies $\mathrm{Im}\,\phi \subseteq J^k N$ and $\mathrm{res}\phi = f$, where $f : J^k X \to J^k N$ is a non isomorphism. Then there exists a lifting of f to $J^k L$, this is there exists a map $g : J^k X \to J^k L$ such that $p'_k g = f$. Hence a lifting of $\phi : \underline{X} \to \underline{N}$ to πL.

We have proved the sequence $0 \to \pi M \overset{\pi j}{\to} \pi L \overset{\pi p}{\to} \pi N \to 0$ is almost split in $\mathrm{Q\,Finres}\,\Lambda$. Therefore the sequence: $0 \to \sigma^{-1}\underline{M}[2] \to \underline{E} \to \underline{M} \to 0$ is almost split. □

LEMMA 4.11. *Let $\Lambda = KQ/I$ be an indecomposable generalized Auslander regular Koszul algebra of graded small global dimension 2, let M be an indecomposable torsion free Koszul module. Then $\mathrm{End}_{\mathrm{QGr}\,\Lambda}(M) = \sum_{i\in Z} R_i$ is a graded, locally finite, local ring.*

PROOF: The vector space R_i is defined as:
$R_i = \varinjlim_k \mathrm{Hom}^0_\Lambda(J^k M, M[i]) = \bigcup_k \mathrm{Hom}^0_\Lambda(J^k M, M[i])$

By Corollary 1 of Proposition 4.6, each R_i is finite dimensional over K.

Let $f \in R_0$, be a degree zero map $f : J^k M \to M$. Then $\mathrm{Im}\, f \subseteq J^k M$. Call f' the restriction of f to its image. The Λ-module $J^k M$ is indecomposable, hence; $\mathrm{End}^0_\Lambda(J^k M)$ is local and the map $f' \in \mathrm{End}^0_\Lambda(J^k M)$ is either nilpotent or invertible. Therefore: f is either nilpotent or invertible.

Let m be the maximal ideal of R_0.

Assume $R_j R_{-j}$ is not contained in m. Then there exist x, y with $x \in R_j$, $y \in R_{-j}$ and xy invertible. It follows y is invertible and there exists an isomorphism $h : J^s M \to M[-j]$, hence there exists an isomorphism $h' : J^s M \to J^s M[-j]$. Then, shifting by $-j$ we have an isomorphism $J^s M[-j] \to J^s M[-2j]$ and composing the two isomorphisms, we have an isomorphism: $J^s M \to J^s M[-2j]$. For any positive integer t, we have by induction, a non zero map $J^s M \to J^s M[-tj]$. But since $J^s M$ is finetely generated, it is easy to see [31], there exists an integer k such that the graded ring $\mathrm{End}_\Lambda(J^s M) = \sum_{i\in Z} S_i$ has $S_{-n} = 0$ for all $n \geq k$.

So we have obtained a contradiction. Therefore: $R_j R_{-j} \subseteq m$.

Similarly, $R_{-j} R_j \subseteq m$. □

Using the standard Auslander-Reiten argument [5], Serre duality:
$\mathrm{Hom}_{\mathrm{QGr}_\Lambda}(\underline{M}, \underline{N}) \cong D(\mathrm{Ext}^1_{\mathrm{QGr}\,\Lambda}(\underline{N}, \underline{\sigma^{-1}M[2]}))$, proves the existence of almost split sequences in $\mathrm{Q\,Finres}\,\Lambda$. By uniqueness of almost split sequences, the sequence we get from the duality is isomorphic to the sequence: $0 \to \sigma^{-1}\underline{M}[2] \to \underline{E} \to \underline{M} \to 0$.

References

[1] M. Artin and W. Schelter, *Graded algebras of dimension 3*, Advances in Math. 66 (1987), 172–216.

[2] M. Artin and J. Zhang, *Noncommutative Projective Schemes*, Advances in Math. Vol 109, Dec. 1994, 229–287.

[3] M. Auslander, *Coherent functors*, Proc. La Jolla Conference on Categorical Algebra (1965) 189–231.

[4] M. Auslander and M. Bridger, *Stable Module Theory*, Mem. of AMS 94, Providencc 1969.

[5] M. Auslander and I. Reiten, *Representation theory of artin algebras III.* Comm. in Algebra, 3 (1975) 239–294.

[6] D. Baer, *Noetherian Categories and Representation Theory of Hereditary Artin Algebras*, Comm. in Algebra, 13(1), (1985), 247–258.

[7] D. Baer, *Homological properties of wild hereditary artin algebras*, Lecture Notes in Mathematics, vol. 1177, Springer Verlag 1986.

[8] D. Baer and H. Lenzing, *A homological approach to representation of algebras I: the wild case*, Jour. of Pure and Applied Algebra 24 (1982), 227–233

[9] D. Baer, H. Brune and H. Lenzing, *A homological approach to representation of algebras II: tame hereditary algebras II*, Jour. of Pure and Applied Algebra 26 (1982), 141–153.

[10] D. Baer, W. Geigle, H. Lenzing, *The preprojective algebra of a tame hereditary algebra*, Comm. in Algebra, 15 (1&2) (1987), 425–457.

[11] A. Beilinson, V. Ginsburg, and W. Soergel, *Koszul duality patterns in representation theory.* J. Amer. Math. Soc. 9 (1996), 473–527.

[12] J. Bernstein, I. Gelfand and S. Gelfand, *A category of G-modules.* Functional Anal. Appl. 10(1976), 87–92.

[13] I. N. Bernstein, I. M. Gelfand and S. I. Gelfand, *Algebraic bundles over* $\mathcal{P}^n$ *and problems of linear algebra.* Funkts. Anal. Prilozh. 12, No. 3 (1978), 66–67. English translation: Funct. Anal. Appl. 12 (1979), 212–214.

[14] A.I. Bondal, *Representation of Associative Algebras and Coherent Sheaves.* Izv. Akad. Nauk SSSR, Ser. Mat. 53, No. 1 (1989), 25–44. English translation: Math USSR, Izv. 34 (1990), 23–42.

[15] A. I. Bondal, M. M. Kapranov, *Representable functors, Serre functors and mutations.* Izv. Akad. Nauk SSSR, Ser. Mat. 53, No. 6 (1989), 1183–1205. English translation: Math. USSR, Izv. 35 (1990), 519–541.

[16] A.I. Bondal, E. Politshchuk, *Homological properties of associative algebras: the method of helices.* Izv. Akad. Nauk. Russia, Ser. Mat. 57, No. 2 (1993), 3–50. English translation: Russ. Acad. Sci. Izv. Math. 42 , No. 2, (1994), 219–260.

[17] P. Dowbor, H. Meltzer, *On equivalences of Bernstein-Gelfand-Gelfand, Beilinson and Happel.* Comm. in Algebra 20 (9), (1992), 2513–2532.

[18] R. Fossum, P. Griffith and I. Reiten, *Trivial extensions of abelian categories*, SLN 456 (1975).

[19] P. Gabriel, *Des catégories abéliennes*, Bull. Soc. Math. France 90 (1962), 323–448.

[20] W. Geigle and H. Lenzing, *A class of weighted projective curves arising in representation theory of finite dimensional algebras.* Singularities, Representations of Algebras and Vector Bundles, Lecture Notes in Math. Vol. 1273, Springer Verlag, (1987), 265–297.

[21] I. M. Gelfand, and V. A. Ponomarev, *Model algebras and representation of graphs*, Funct. Anal. Appl. 13 (1979), 157–166.

[22] E. L. Green, *Introduction to Koszul Algebras*, preprint (1996).

[23] E. L. Green, R. Martínez-Villa, *Koszul and Yoneda algebras I*, Rep. Theory of Algebras, CMS Conference Proceedings, Vol. 18 (1996), 247–306.

[24] E. L. Green, R. Martínez-Villa, *Koszul and Yoneda algebras II*, to appear in ICRA VIII, CMS Conference Proceedings.

[25] E. L. Green, R. Martínez-Villa, I. Reiten, Ø. Solberg and D. Zacharia, *On Modules with Linear Presentations*, to appear in J. Algebra.

[26] J. Y. Guo, R. Martínez-Villa and M. Takane, *Preprojective algebras and generalized Auslander regular algebras*, to appear in ICRA VIII, CMS Conference Proceedings.

[27] R. Hartshorne, *Residues and Duality*, LNM 20, Springer-Verlag, (1966).

[28] L. Hille, *Consistent algebras and special tilting sequences*, Math. Z. 220, (1995), 189–205.

[29] Y. Manin, *Quantum groups and non-commutative geometry*, University of Montreal 1989.

[30] R. Martínez-Villa, *Applications of Koszul algebras: the preprojective algebra*, Rep. Theory of Algebras, CMS Conference Proceedings, Vol. 18 (1996), 487–504.

[31] R. Martínez-Villa, *Graded, Selfinjective and Koszul Algebras*, preprint (1996).

[32] H. Nakajima, *Varieties Associated with quivers*, Rep. Theory of Algebras and Related Topics, CMS Conference Proceedings, Vol. 19 (1996), 139–157.

[33] N. Popescu, *Abelian Categories with Applications to Rings and Modules*, Academic Press (1973).

[34] S. Priddy, *Koszul resolutions.* Trans. Amer. Math. Soc. 152 (1970), 39–60.

[35] A. Rudakov, *Helices and vector bundles.* LMS Lect. Note Series 148, Cambridge, 1990.

[36] J.P. Serre, *Faisceaux algébriques cohérents*, Ann. Math. 61 (1955), 197–278.

[37] S.P. Smith, *Some finite dimensional algebras related to elliptic curves*, Rep. Theory of Algebras and Related Topics, CMS Conference Proceedings, Vol. 19 (1996), 315–348.

[38] A. Yekutieli and J. Zhang, *Serre duality for Noncommutative Projective Schemes*, preprint (1994).

Contemporary Mathematics
Volume **229**, 1998

CLASSIFYING FINITE-DIMENSIONAL SEMISIMPLE HOPF ALGEBRAS

SUSAN MONTGOMERY[1]

0. Introduction

This paper is primarily a survey of what is currently known about the classification of finite-dimensional semi-simple Hopf algebras H over an algebraically closed field k of characteristic 0, although we briefly discuss the non-semisimple case at the end.

Aside from its intrinsic algebraic interest, being able to classify Hopf algebras would have applications to a number of other areas of mathematics. In mathematical physics, Drinfeld's and Jimbo's original work on quantum groups was to provide solutions to the quantum Yang-Baxter equation. More recently, in conformal field theory, I. Frenkel and Y. Zhu have shown how to assign a Hopf algebra as an invariant to any conformal field theory model; this assignment goes by way of a correspondence between conformal field theory models and vertex operator algebras [FZ]. For the so-called "rational" models, the assigned Hopf algebra is finite-dimensional.

In topology, quasi-triangular and ribbon Hopf algebras provide many invariants of knots, links, tangles, and 3-manifolds; this is work of Hennings, Kauffman, and Radford and of Reshetikhin and Turaev; see for example [He], [Kf], [R2], [ReT].

Moreover in operator algebras, a finite-dimensional (semisimple) Hopf algebra can be assigned as an invariant for certain extensions $N \subset M$ of von Neumann algebras; M must be a factor and N a subfactor of finite Jones index such that various centralizer conditions are satisfied in the Jones tower constructed from the extension. This assignment was proposed by Ocneanu and proved by Szymanski [Sz]. Thus

[1]The author was supported by NSF Grant DMS 95-00649.

1991 *Mathematics Subject Classification*: 16W30, 16K20

these subfactors can be classified by Hopf algebras, and conversely new Hopf algebras can be constructed from known subfactors. The correspondence of subfactors and Hopf algebras has also been studied by David, Longo, Izumi and Kosaki[IK], and Yamagouchi. We note that the Hopf algebras arising here are Kac algebras; that is, Hopf algebras which are also C^*-algebras.

Finally in algebra itself, the Hopf Galois theory of fields can provide a finer invariant than classical Galois theory, as there exist field extensions which have trivial Galois group in the usual sense but are Hopf Galois with respect to a non-trivial Hopf algebra [GrP].

We first give some preliminaries and general results, then discuss some specific dimensions, and then describe some of the tools used.

Throughout k is algebraically closed of characteristic 0, and H is a finite-dimensional Hopf algebra with multiplication $m : H \otimes H \to H$, unit $u : k \to H$, comultiplication $\triangle : H \to H \otimes H$, counit $\varepsilon : H \to k$, and antipode $S : H \to H$. The counit ε is also called the *augmentation map* of H, and $H^+ := \ker \varepsilon$ the *augmentation ideal.* The linear dual H^* of H is also a Hopf algebra, with multiplication $\triangle^*$, unit ε^*, comultiplication m^*, counit u^*, and antipode S^*, where $*$ denotes the transpose map. For example, this means that if $f \in H^*$, then $\triangle f(h \otimes l) = m^* f(h \otimes l) = f(hl)$, for all $h, l \in H$. H is called *cocommutative* if $\triangle = \tau \circ \triangle$, where τ is the usual twist map $\tau(a \otimes b) = b \otimes a$; equivalently H^* is a commutative algebra.

The most basic example is that of $H = kG$, the group algebra of the group G. In this case $\triangle g = g \otimes g$, $\varepsilon(g) = 1$, and $Sg = g^{-1}$ for all $g \in G$. Note that kG is cocommutative.

We may also consider $H = (kG)^* = k^G$, its dual; H is a commutative algebra under pointwise multiplication. Letting $\{p_g, g \in G\}$ denote the dual basis to G, the Hopf structure is given by

$$\triangle p_g = \sum_{hl=g} p_h \otimes p_l,\ \varepsilon(p_g) = \delta_{g,1},\ S(p_g) = p_{g^{-1}}.$$

We say that H is *trivial* if either $H = kG$ or $H = (kG)^*$. Note that when G is abelian, $(kG)^* \cong kG$.

For any H, $G(H) = \{0 \neq h \in H \mid \triangle h = h \otimes h\}$ is the set of *group-like* elements in H. $G(H)$ is a group and distinct group-like elements are linearly independent. Thus $kG(H)$ is a Hopf subalgebra of H isomorphic to the group algebra of $G(H)$. Moreover $f \in G(H^*)$ iff $f(hl) = f(h)f(l)$, using the above expression for $\triangle f$; that is, f is an algebra map. A fundamental approach to many classification results is to try to determine $G(H)$ and $G(H^*)$. Not only are they invariants of

H, but it is also useful to know the orders of $G(H)$ and $G(H^*)$ even if the group structure cannot be determined, as this can help in various counting arguments; see §3. Moreover, some things are known about the Hopf extensions of $(kG)^*$ by kF, where F is some other group; see §4.

We give a few useful facts that are true for any finite-dimensional Hopf algebra over k:

(0.1) "*Lagrange's Theorem*" [NZ]: If K is a Hopf subalgebra of H, then H is free over K.

We note that this is false if H is not finite dimensional. An immediate consequence of (0.1) is that $|G(H)|$ divides dim H when H is finite-dimensional; in fact this special case was known earlier.

(0.2) [ZY] If $\dim H = p$, a prime, then $H \cong k\mathbb{Z}_p$.
This was proved for Kac algebras in [K2].

(0.3) [LR1, LR2] H is semi-simple $\Leftrightarrow$ H^* is semi-simple $\Leftrightarrow$ $S^2 = Id$.
A short proof of this result is given in [S2].

1. Results for special dimensions

From now on we assume H (and so H^*) is semisimple. We first state a few general results, then proceed to list what is known about various dimensions.

(1.1) H is commutative $\Leftrightarrow$ $H \cong (kG)^*$, for some G (see [Mo, 2.3.1]). Consequently H is cocommutative $\Leftrightarrow$ $H = kG$.

(1.2) [St] For each $n > 0$, there are only finitely many isomorphism classes of n-dimensional (semisimple) Hopf algebras.
This is false in the non-semisimple case; see §5.

(1.3) [Ma4] If $\dim H = p^2$, then $H \cong kG$.
More generally, if $\dim H = p^n$, then H contains a central group-like element. Consequently [MoW], H is solvable, in the sense that it has a normal series of Hopf subalgebras such that each Hopf quotient is commutative.

(1.4) [EG2] If $\dim H = pq$, where p, q are distinct primes, then $H = kG$ or $(kG)^*$.
A number of special cases of this were known earlier. [Ma3] did the case of dim $2p$. For dim $3p$, the case of Kac algebras was done in

[IK]. The general result for $3p$ was shown independently by [GW] and by Masuoka (private communication, using the methods of the dim 21 case [Ma6]).

The proof in [EG2] uses both [GW], in which the problem is reduced to showing that both H and H^* are of Frobenius type (see §2), and the work of [EG1] on when quasitriangular Hopf algebras are of Frobenius type. [GW] uses work on extensions, as well as [NR] and [MoW]. To show how knowing H is of Frobenius type is useful, we will give a short proof of the cases of dim 15, 35, and 55 in §2.

(1.5) [Ma2] If $\dim H = p^3, p$ odd, then there are exactly $p+8$ isomorphism classes. 7 of these are trivial, and the other $p+1$ are all non-commutative, non-cocommutative, and self-dual.

(1.6) [Ma5] There is an infinite family of non-trivial examples of dim $2p^2$, for p odd.

(1.7) [G1] If $\dim H = pq^2$, for p, q primes with $p \equiv 1 \pmod{q}$, then there exist two non-trivial, non-isomorphic, self-dual examples, A_{qp} and $\mathcal{A}_{qp}$. In A_{qp} one has $G(H) \cong \mathbb{Z}_{q^2}$ and in $\mathcal{A}_{qp}$ one has $G(H) \cong \mathbb{Z}_q \times \mathbb{Z}_q$.

Using the above results, it follows that all the (semisimple) Hopf algebras of odd dimension < 60 are now known except for dimension 45.

Much less is known about the even-dimensional cases. In addition to the more general results above, only the classification for 8, 12, and 18 is complete; we discuss these cases below. However even for dim < 30, the cases 16, 20, 24, and 28 remain open.

(1.8) $\dim H = 8$

In 1966, [KP] constructed an example of dim 8 which was not kG or $(kG)^*$; it was the first example of a "non-trivial" semi-simple Hopf algebra. It was shown in [W] by combinatorial methods that this was the only exceptional case; in fact [W] classifies all H of dim ≤ 11. [Ma1] proved independently that the dim 8 exception was unique, using extension theory. [EV] show that it may be constructed as a cocycle deformation of $k(\mathbb{Z}_2^2 \rtimes \mathbb{Z}_2)$.

(1.9) $\dim H = 12$

[F] shows that there are exactly two non-trivial cases. They are, by necessity, the two self-dual examples of [G1] mentioned in (1.7); in fact one of these dim 12 examples appears in [R1]. The proof in [F] uses extensions. It was shown in [EV] that one of the exceptions is a cocycle deformation of $k(Z_3 \rtimes \mathbb{Z}_2^2)$ and in [V] that the other one is a deformation of $k(Z_3 \rtimes Z_4)$.

(1.10) $\dim H = 18$

Independently, [Ma5] and [Se] show that there are only two exceptions, with one being the dual of the other. In one case $|G(H)| = 6$ and in the other case $|G(H)| = 9$. [V] shows that the first exception is a deformation of $k(\mathbb{Z}_3^2 \rtimes \mathbb{Z}_2)$.

All of the above examples are *semi-solvable* in the sense of [MoW]; that is, H has a normal series of Hopf subalgebras such that each Hopf quotient is commutative or cocommutative. We ask, analogously to the situation for finite groups:

QUESTION. If $\dim H < 60$, must H be semi-solvable?

If $\dim H = 60$, then it is possible for H to be a *simple* Hopf algebra; that is, H has no proper normal Hopf subalgebras. For, if G is any simple group, then any cocycle deformation H of kG is also a simple Hopf algebra; see §4. [Ni1] shows there exist non-commutative, non-cocommutative cocycle deformations of $kA_n, n \geq 5$, which are simple. This has recently been extended to any non-abelian simple group in [H].

In [Ni2] a stronger definition of simple is considered; in this definition, H has no non-trivial quotients as well as no non-trivial normal Hopf subalgebras (this is more restrictive since it is possible to have a Hopf quotient which does not come from a normal Hopf subalgebra). [Ni2] shows that no H of dim < 60 is simple in this stronger sense.

2. Counting and the "Class Equation"

The first method we discuss concerns dimensions of irreducible modules. Let $R(H)$ denote the character ring of H; that is, the k-subalgebra of H^* generated by the characters χ_V of H-modules V. It can be shown that $R(H)$ is also semisimple.

2.1 THEOREM [K2][ZY](the "Class Equation"). *For every primitive idempotent* $e = e^2$ *of* $R(H)$, $\dim(eH^*)$ *divides* $\dim(H)$. *If* $e_1, \cdots, e_k$ *are the primitive idempotents of* $R(H)$, *where* e_1 *is an integral in* H^*, *then*

$$\dim(H) = 1 + \sum_{i>1} \dim(e_i H^*).$$

A more elementary proof of the theorem appears in [Lo]. Here an *integral* $t \in H$ is an element such that $th = \varepsilon(h)t$, all $h \in H$. By old work of Larson-Sweedler, the space of integrals is always one-dimensional; when H is semisimple, we may choose an integral e with $\varepsilon(e) = 1$. It follows that $e^2 = e$. When e is an idempotent integral in H^*, then $(\dim H)e = Tr_H$, where Tr_H is the trace of the (left) regular representation of H; in particular $e \in R(H)$ (this is a result of Larson).

We show what the theorem means in two special cases:

2.2(i) When $H = kG$, $R(H)$ consists of those $f \in H^*$ which are constant on the conjugacy classes $C_1, \cdots, C_k$ of G. Thus for each i, $e_i = \sum_{g \in C_i} p_g$, and so $\dim(e_i H^*) = |C_i|$. Now $e_1 = p_1$ and so $\dim(p_1 H^*) = 1$. Thus the theorem is just the usual class equation

2.2(ii) When $H = (kG)^*$, then $R(H) = kG$ since every $g \in G$ acts as a character on H. Thus the $\{e_i\}$ are the usual primitive idempotents in kG, and $\dim(e_i H^*) = \dim(e_i(kG))$ is the degree of the irreducible representation associated to e_i. Thus the theorem reduces to Frobenius' classical theorem for finite groups. Observe that in this case, $e_1 = (1/|G|) \sum_{g \in G} g$.

Note that the semisimple case of (0.2) folows immediately from the theorem. For, since $\dim H = p$, $\dim\, e_i H^* = 1$ for all i; thus $k = p$ and $R(H) = H^*$ is spanned by the degree one characters of H. Thus $H^* = kG = k\mathbb{Z}_p \cong H$.

By analogy with Frobenius' theorem for kG, we say that H is of *Frobenius type* if for any simple H-module V, $\dim V$ divides $\dim H$. It is an old conjecture of Kaplansky [Ka] that any (semisimple) H is of this type. Recently it has been shown in [NR] that the conjecture is true whenever $\dim V = 2$, and in [MoW] that it is true whenever H is semisolvable. In [EG1] it is shown that it is true if H is quasitriangular (and semisimple); in fact, for the Drinfel'd double $D(H)$ of H, they show that the dimension of every simple $D(H)$-module divides $\dim H$. This was shown in [ZY2] for $D(H)$-submodules of H.

We next summarize some known results which will help in counting arguments. Write $H = k^{(n_0)} \oplus M_{d_1}(k)^{(n_1)} \oplus \cdots \oplus M_{d_s}(k)^{(n_s)}$, so that the $\{d_i\}$ are the degrees of the irreducible characters of H.

2.3 PROPOSITION. *With the decomposition of H as above:*
(1) $n_0 = |G(H^*)|$ *and divides* $\dim H$.
(2) [NR] *If dim H is odd, then no $n_i = 2$.*
(3) [ZS] *If* $\dim H > 1$ *and the set $\{d_i\}$ contains less than four distinct*

integers, then $n_0 > 1$.
(4) [Ma2, Lemma 3.4] *For all* i, $n_0 | n_i d_i^2$.

To see (1), note that a one-dimensional representation ρ of H is just an algebra map from H to k, and thus is a group-like element of H^* as noted in the introduction. Thus $n_0 = |G(H^*)|$. As mentioned in (0.1), this number must divide dim H.

We now apply the proposition to give short proofs of some special cases of (1.4). These arguments for dim 15 and 35 were done jointly with E. Kashina, and for dim 55 were shown to us by Martin Lorenz.

2.4 COROLLARY *If dim* $H = 15$, 35, *or* 55, *then* H *is trivial.*

Proof. In all three cases, dim H is odd, and so no $d_i = 2$.

For the case of dim 15, the only simple summands are of dim 1 or 9. If there is a summand of dim 9, then $n_0 = 6$, which contradicts 2.3(1). Thus 1 is the only possibility, so $n_0 = 15$. But then H^* is a group algebra, and we are done. In fact in this case G must be abelian and so H is a group algebra.

When dim $H = 35$, any simple summands have dimension 1, 9, 16, or 25. The only possibility using 25 is $35 = 25 + 9 + 1$, but then $\{d_i\} \subseteq \{1, 3, 5\}$, so 2.4(3)[ZS] applies to get that $n_0 = |G(H^*)| > 1$. Not using 25, we are always in a case to which [ZS] applies, and thus $|G(H^*)| > 1$ in either case. Thus $|G(H^*)| = 5, 7$ or 35; this is also the number of one-dimensional summands. Now it is straightforward to check that no possible combination works, except for $|G(H^*)| = 35$. That is, H^* is a group algebra. By group theory, G is abelian, and so also $H = kG$.

For the case of dim 55, any simple summands have dimension 1, 9, 16, 25, or 36. Also $n_0 = 1, 5, 11$, or 55, since $n_0|$ dim H. If $n_0 = 55$ then H^* is a group algebra and we are done. It is easy to eliminate $n_0 = 11$ as a possiblity. If $n_0 = 1$, then one can check that the only possibility is $55 = 1 + 9 + 9 + 36$; however, this contradicts 2.3(3). Thus $n_0 = 5$ is the only possibility. In this case there are two possibilities: $55 = 5 + 9 + 9 + 16 + 16$ and $55 = 5 + 25 + 25$. The first case is impossible by 2.3(4)[Ma2]. The second case is possible, but now we are done by [GW] since H (and similarly H^*) is of Frobenius type; see (1.4).

It is clear from the above computations that a positive answer to Kaplansky's conjecture would be very helpful in the classification problem.

3. Extensions

The most general type of Hopf extension comes from a Hopf surjection; that is,

$$A \hookrightarrow H \xrightarrow{\pi} \overline{H}$$

where π is a surjection of Hopf algebras and

$$A = H^{co\overline{H}} = \{h \in H | (id \otimes \pi)\triangle(h) = h \otimes 1\}$$

is the subalgebra of (right)$\overline{H}$-coinvariants (also called the (right) Hopf kernel by some authors). Note that in the case of a group algebra $H = kG$ with $\overline{G}$ a homomorphic image of G and $\overline{H} = k\overline{G}$, A will be precisely kN, where N is the kernel of the group homomorphism $G \to \overline{G}$. But in general A is not a Hopf subalgebra of H, although it is a subalgebra, and very little is known about such arbitrary extensions. In two cases, however, something can be said.

(3.1) Normal Hopf subalgebras.

One may define the left (and right) adjoint action of H on itself, generalizing the notion of conjugation in a group, as follows: $(ad_l h)(g) = \sum h_1 g(Sh_2)$ and $(ad_r h)(g) = \sum (Sh_1)gh_2$, for $g, h \in H$, where $\triangle h = \sum h_1 \otimes h_2$. Then a Hopf subalgebra K of H is *normal* if it is closed under the left and right adjoint actions of H on itself. When H is finite-dimensional, it can be shown that K is normal if and only if $HK^+ = K^+H$. In this case HK^+ is a Hopf ideal and we may form the *Hopf quotient* $\overline{H} := H/HK^+$. Now if we begin with such a normal Hopf subalgebra K of H, and consider the quotient map $\pi : H \to \overline{H}$, then in fact one recovers $A = K$. Thus the above sequence becomes

$$K \hookrightarrow H \xrightarrow{\pi} \overline{H}.$$

By [S1], $H \cong K\#_\sigma \overline{H}$, a crossed product of $\overline{H}$ over K, where $\sigma : \overline{H} \otimes \overline{H} \to K$ is a Hopf 2-cocycle.

Given K and $\overline{H}$, one needs to determine which extensions H can occur. The case in which some progress has been made is that in which G, F are groups, $K = (kG)^* = k^G$, and $\overline{H} = kF$. [T1] considers "matched pairs" of groups G, F; that is, the groups are given with actions of each on the other satisfying certain compatibility conditions, such that an extension H of the above form exists. This is the method used for dimensions $2p, p^3, 12$ and 18. More generally, work on classifying these extensions via cohomology was done in [K1], but it is hard to apply directly. This work on extensions has recently been extended in [Ma7]. Some other results on extensions may be found in [AD].

(3.2) Biproducts

In this case we have an extension of the form

$$A \hookrightarrow H \underset{\mu}{\overset{\pi}{\rightleftarrows}} \overline{H}$$

where π is a Hopf projection of H onto the Hopf algebra $\overline{H}$. That is, μ is also a Hopf map and $\pi\mu = id$. [R1] gives necessary and sufficient conditions for such a set-up; in current terminology, A must be a Hopf algebra in the Yetter-Drinfeld category $\mathcal{YD}_{\overline{H}}^{\overline{H}}$. In particular $H = A \star \overline{H}$, the "biproduct" of A with $\overline{H}$.

In most of the examples considered in previous sections, $\overline{H} = kF$, where the group F has a given bicharacter $\chi : F \times F \to k^*$, and we may work in $\mathcal{M}^{kF}$, the category of F-graded modules. For $M = \oplus_{g\in F} M_g$ in $\mathcal{M}^{kF}$, the F-action on M is given by $g \cdot m_l = \chi(g,l)m_l$, for $g, l \in F, m_l \in M_l$, and the braiding in $\mathcal{M}^{kF}$ is given by

$$\tau : M \otimes N \to N \otimes M, \; m_g \otimes n_l \mapsto \chi(g,l) n_l \otimes m_g.$$

Then A is simply an F-graded Hopf algebra; that is, A is a graded algebra, a graded coalgebra, and $\triangle_A$ is a graded algebra map using the braiding.

4. Cocycle deformations

These are related to ideas of [Dr] in his work on quasi-Hopf algebras. We first deform the comultiplicaton, as in [EV] and [V]. Consider a Hopf algebra H and an invertible element $\Omega \in H \otimes H$. H_Ω is called a *cocycle deformation* of H with respect to Ω if:

(i) $H_\Omega = H$ as an algebra
(ii) H_Ω has comultiplication $\triangle_\Omega(h) := \Omega(\triangle h)\Omega^{-1}$

It is straightforward to verify that $\triangle_\Omega$ is coassociative if and only if the element

$$\partial_2\Omega := [(id \otimes \triangle)(\Omega^{-1})](1 \otimes \Omega^{-1})(\Omega \otimes 1)[(\triangle \otimes id)(\Omega)]$$

belongs to the centralizer of $(\triangle \otimes id)\triangle(H)$ in $H^{\otimes 3}$. Such an element Ω is called a *pseudo-cocycle*. If $\partial_2\Omega = 1 \otimes 1 \otimes 1$, then Ω is an ordinary Hopf 2-cocycle on H^*.

The methods of [EV], [V], and [Ni1], following [La], involve first finding a pseudo-cocycle Ω on an abelian subgroup L of a group G. Since Ω is also a pseudo-cocycle on G, it may be used to obtain a deformation of $H = kG$; this deformation will be non-trivial provided that Ω does not centralize the image of $\triangle(H)$ in $H \otimes H$. [Ni1] shows that a sufficient condition for this to hold is that L contains no proper normal subgroups of G and that $\tau \circ \Omega \neq \Omega$, where τ is the twist mentioned in the introduction. In particular he shows this is true for the alternating groups $G = A_n$, $n \geq 5$.

Now for any H and Ω, it is easy to see that there is a one-to-one correspondence between Hopf ideals of H and those of H_Ω. As a consequence, when $H = kG$ and G is simple, then H_Ω is a simple Hopf algebra. Thus for $H = kA_n$, $n \geq 5$, it follows that H_Ω is simple, non-commutative, and non-cocommutative.

More generally, [Ni1] connects these deformations to $K_0(H)$; note that for any Hopf algebra H, the group $K_0(H)$ becomes an algebra in a natural way, as H acts on tensor products of modules via $\triangle$. [Ni1] shows that for two finite-dimensional semisimple Hopf algebras H_1 and H_2, then H_1 and H_2 are twistings of each other (that is $H_1 = H_2$ as algebras and $H_2 = (H_1)_\Omega$ for some Ω) $\Leftrightarrow$ $K_0(H_1) \cong K_0(H_2)$ as ordered rings with involution with a "marked element".

The dual situation, at least for ordinary cocycles, has been studied in [Do], [T2]. In this case the coalgebra structure is unchanged but the algebra structure is twisted by a Hopf 2-cocycle $\sigma : H \otimes H \to k$. That is, H_σ has new multiplication

$$h \cdot l := \sum \sigma(h_1, l_1) h_2 l_2 \sigma^{-1}(h_3, l_3), \tag{$*$}$$

where here we have used the summation notation $\triangle h = \Sigma h_1 \otimes h_2$ and $(id \otimes \triangle)\triangle(h) = \Sigma h_1 \otimes h_2 \otimes h_3$.

More generally, one could consider pseudo-cocycles in this situation as well: here a pseudo-cocycle is a convolution invertible map $\sigma : H \otimes H \to k$ such that the product in $(*)$ is associative. One may verify that σ is a pseudo-cocycle $\Leftrightarrow$

$$\sum \omega(h_1, l_1, m_1) h_2 l_2 m_2 = \sum h_1 l_1 m_1 \omega(h_2, l_2, m_2),$$

where

$$\omega(h, l, m) := \sum \sigma^{-1}(h_1, l_1, m_1) \sigma^{-1}(l_2, m_2) \sigma(h_2, l_3) \sigma(h_3 l_4, m_3).$$

When $\omega(h, l, m) = \varepsilon(h, l, m), \sigma$ is an ordinary Hopf 2-cocyle.

5. The non-semisimple case

In fact a few things are known about the non-semisimple case. First, the result of [ZY] in (0.2) says that if H has dim p, it is semisimple.

When dim $H = p^2$, the only known non-semisimple example is the Taft algebra [Tf]. This can be described as follows. Let ω denote a primitive pth root of 1; then as an algebra,

$$H_{p^2} = k < x, g | x^p = 0, g^p = 1, xg = \omega gx >.$$

The coalgebra structure of H is given by letting g be group-like and x be g-skew-primitive; that is, $\triangle x = x \otimes 1 + g \otimes x$, $\varepsilon(x) = 0$, and $S(x) = -xg^{-1}$. Note that such skew-primitive elements appear in the quantum enveloping algebras.

This example was known to Sweedler in the case $p = 2$, and in this case [Ka] showed that H_4 is the only non-semisimple example. A similar result for $p = 3$ appears in [W]. Recently progress has been made on the $p > 3$ case: [AS1] have shown that if both H and H^* have non-trivial group-like elements, then H is the Taft algebra.

In the dim p^3 case, something is known. Independently, both [AS2] and [CD] have classified the pointed Hopf algebras of dim p^3; there are only a finite number of them.

In dim p^4, however, there exist infinitely many isomorphism classes. This was shown independently in [AS2], [BDG], and [G2]. All of these give counterexamples to a conjecture of Kaplansky that there were only finitely many types in each dimension [Ka]. Contrast these results with the semisimple case (1.2).

REFERENCES

[AD] N. Andruskiewitsch and J. Devoto, Extensions of Hopf algebras, *St. Petersburg Math J.* 7 (1996), 17-52.

[AS1] N. Andruskiewitsch and H.-J. Schneider, Hopf algebras of order p^2 and braided Hopf algebras of order p, *J. Algebra* 199 (1998), 430-454.

[AS2] N. Andruskiewitsch and H.-J. Schneider, Lifting of quantum linear spaces and pointed Hopf algebras of order p^3, *J. Algebra*, to appear.

[BDG] M. Beattie, S. Dascalescu, and L. Grünenfelder, On the number of types of finite-dimensional Hopf algebras, Inventiones Math, to appear.

[CD] S. Caenepeel and S. Dascalescu, Pointed Hopf algebras of dimension p^3, *J. Algebra*, to appear.

[Do] Y. Doi, Braided bialgebras and quadratic bialgebras, *Comm. Alg.* 12 (1993), 1731-1749.

[Dr] V.G. Drinfeld, Quasi-Hopf algebras, *Leningrad Math J.* 1 (1990), 1419-1457.

[EV] M. Enock and L. Vainerman, Deformation of a Kac algebra by an abelian subgroup, *Comm. Math. Phys.* 178 (1996), 571-596.

[EG1] P. Etingof and S. Gelaki, Some properties of finite-dimensional semisimple Hopf algebras, *Math Research Letters*, to appear.

[EG2] P. Etingof and S. Gelaki, Semisimple Hopf algebras of dimension pq are trivial, *J. of Algebra*, to appear.

[FZ] I. Frenkel and Y. Zhu, Vertex operator algebras associated to representations of affine and Virasoro algebras, *Duke Math J.* 66 (1992), 123-168.

[F] N. Fukuda, Semisimple Hopf algebras of dimension 12, *Tsukuba J. Math* 21 (1997), 43-54.

[G1] S. Gelaki, Quantum groups of dimension pq^2, *Israel J* 102 (1997), 227-267.

[G2] S. Gelaki, On pointed Hopf algebras and Kaplansky's tenth conjecture, J. Algebra, to appear.

[GW] S. Gelaki and S. Westreich, On semisimple Hopf algebras of dimension pq, *Proc. AMS*, to appear.

[GrP] C. Greither and B. Pareigis, Hopf Galois theory for separable field extensions, *J. Algebra* 106 (1987), 239-258.

[He] M. A. Hennings, Hopf algebras and regular isotopy invariants for link diagrams, *Proc. Cambridge Phil. Soc* 109 (1991), 59-77.

[H] C. Hoffman, On some examples of simple quantum groups, preprint.

[IK] M. Izumi and H. Kosaki, Finite-dimensional Kac algebras arising from certain group actions on a factor, *International Math. Research Notes* 8 (1996), 357-370.

[K1] G.I. Kac, Extensions of groups to ring groups, *Mat. USSR Sbornik* 5(1968), 451-474 (English translation).

[K2] G.I. Kac, Certain arithmetic properties of ring groups, *Functional Anal. Appl.* 6 (1972), 158-160.

[KP] G.I. Kac and V.G. Paljutkin, Finite ring-groups, *Trans. Moscow Math Soc.* (1966), 251-294. (English translation).

[Ka] I. Kaplansky, *Bialgebras*, University of Chicago Lecture Notes in Math, 1975.

[Kf] L. Kauffman, Invariants of links and 3-manifolds via Hopf algebras, in *Geometry and Physics*, Marcel Dekker Lecture Notes in Pure and Appl. Math vol 184, 1997, 471-479.

[La] M.B. Landstad, Quantization arising from abelian subgroups, *International J. of Math* 5 (1994), 879-936.

[LR1] R. Larson and D. Radford, Semisimple cosemisimple Hopf algebras, *Amer. J. Math* 109 (1987), 287-295.

[LR2] R. Larson and D. Radford, Finite-dimensional cosemisimple Hopf algebras in characteristic 0 are semisimple, *J. Algebra* 117 (1988), 267-289.

[LR3] R. Larson and D. Radford, Semisimple Hopf algebras, *J. Algebra* 171 (1995), 5-35.

[Lo] M. Lorenz, On the class equation for Hopf algebras, *Proc. AMS*, to appear.

[Ma1] A. Masuoka, Semisimple Hopf algebras of dimension 6, 8, *Israel J. Math* 92 (1995), 361-373.

[Ma2] A. Masuoka, Self-dual Hopf algebras of dimension p^3 obtained by extension, *J. Algebra* 178 (1995), 791-806.

[Ma3] A. Masuoka, Semisimple Hopf algebras of dimension $2p$, *Comm. Alg.* 23 (1995), 1931-1940.

[Ma4] A. Masuoka, The p^n theorem for semi-simple Hopf algebras, *Proc. AMS* 124 (1996), 735-737.

[Ma5] A. Masuoka, Some further classification results on semisimple Hopf algebras, *Comm. Alg.* 24 (1996), 307-329.

[Ma6] A. Masuoka, Some further classification results on semisimple Hopf algebras II, unpublished notes.

[Ma7] A. Masuoka, Calculations of some groups of Hopf algebra extensions, *J. Algebra* 191 (1997), 568-588.

[MaD] A. Masuoka and Y. Doi, Generalization of cleft comodule algebras, *Comm. Alg.* 20 (1992), 3703-3721.

[Mo] S. Montgomery, *Hopf algebras and their actions on rings*, CBMS Lecture Notes 82, American Math Society, 1993.

[MoW] S. Montgomery and S. Witherspoon, Irreducible representations of crossed products, *J. Pure Appl. Alg.* 129 (1998).

[NZ] W. Nichols and B. Zoeller, A Hopf algebra freeness theorem, *Amer. J. Math* 111 (1989), 381-385.

[NR] W. Nichols and B. (Zoeller) Richmond, The Grothendieck group of a Hopf algebra, *J. Pure Appl. Algebra* 106 (1996), 297-306.

[Ni1] D. Nikshych, K_0-rings and twisting of finite-dimensional semisimple Hopf algebras, *Comm. Alg.* 26 (1998), 321-342; Corrigendum 26 (1998), 2019.

[Ni2] D. Nikshych, On finite-dimensional simple Hopf algebras, preprint.

[R1] D. Radford, The structure of Hopf algebras with a projection, *J. Algebra* 92 (1985), 322-347.

[R2] D. Radford, On Kauffman's knot invariants arising from finite-dimensional Hopf algebras, in *Advances in Hopf Algebras*, Marcel Dekker Lecture Notes in Pure and Appl. Math vol 158 (1994), 205-266.

[ReT] N. Reshetikhin and V. G. Turaev, Ribbon graphs and their invariants derived from quantum groups, *Comm. Math. Physics* 127 (1990), 1-26.

[S1] H.-J. Schneider, Normal basis and transitivity of crossed products for Hopf algebras, *J. Algebra* 151 (1992), 289-312.

[S2] H.-J. Schneider, *Lectures on Hopf algebras*, Universidad de Cordoba Trabajos de Matematica, No. 31/95, Cordoba (Argentina), 1995.

[Se] Y. Sekine, An example of finite dimensional Kac algebras of Kac-Paljutkin type, *Proc. AMS* 124 (1996), 1139-1147.

[St] D. Stefan, The set of types of n-dimensional semisimple and cosemisimple Hopf algebras is finite, *J. Algebra* 193 (1997), 571-580.

[Sz] W. Szymanski, Finite index subfactors and Hopf algebra crossed products, *Proc. AMS* 120 (1994), 519-528.

[Tf] E. Taft, The order of the antipode of a finite-dimensional Hopf algebra, *Proc. Nat. Acad. Sci. USA* 68 (1971), 2631-2633.

[T1] M. Takeuchi, Matched pairs of groups and bismash products of Hopf algebras, *Comm. Alg.* 9 (1981), 841-882.

[T2] M. Takeuchi, Cocycle deformations of coordinate rings of quantum matrices, *J. Algebra* 189 (1997), 23-33.

[V] L. Vainerman, 2-cocycles and twisting of Kac algebras, *Comm. Math Physics* 191 (1998), 697-721.

[W] R. Williams, Finite-dimensional Hopf algebras, Ph.D. Thesis, Florida State University, 1988.

[ZS] S. Zhu, On finite-dimensional Hopf algebras, *Comm. Algebra* 21 (1993), 3871-3885.

[ZY1] Y. Zhu, Hopf algebras of prime dimension, *International Math. Res. Notes* 1 (1994), 53-59.

[ZY2] Y. Zhu, A commuting pair in Hopf algebras, Proc. AMS 125 (1997), 2847 - 2851.

University of Southern California, Los Angeles, CA 90089-1113
email: smontgom@math.usc.edu

Contemporary Mathematics
Volume **229**, 1998

Geometry of Modules: Degenerations

Christine Riedtmann
University of Berne, Sidlerstrasse 5, CH-3012 Bern
e-mail: riedtm@math-stat.unibe.ch

ABSTRACT. The aim of this expository note is to describe and compare some of the algebraic conditions related to the geometric notion of degeneration for finite dimensional modules over an associative algebra. [1]

1. Degenerations

Let k be an algebraically closed field and Λ a finite dimensional associative k-algebra with a unit element. We view Λ as a quotient of a free associative algebra $k\langle\lambda_1, \dots, \lambda_r\rangle$ by a two-sided ideal of relations, generated by $\rho_1, \dots, \rho_s$. A d-dimensional Λ-module is given by r k-linear endomorphisms $m_1, \dots, m_r$ of some d-dimensional vector space satisfying the relations $\rho_1, \dots, \rho_s$. By fixing a basis, a Λ-module becomes an r-tupel of $d \times d$-matrices satisfying the polynomial relations $\rho_1, \dots, \rho_s$.

Let $\mathrm{mod}_d\Lambda$ be the affine variety of d-dimensional Λ-modules; i.e., the set of all r-tupels of $d \times d$-matrices satisfying $\rho_1, \dots, \rho_s$. The group $GL(d)$ acts on $\mathrm{mod}_d\Lambda$ by conjugation. Orbits are isomorphism classes of modules. The variety $\mathrm{mod}_d\Lambda$ depends on the choice of generators and relations for Λ only up to $GL(d)$-equivariant isomorphism.

A module N is called a degeneration of a module M – in symbols $M \leq_{deg} N$ – if N belongs to the Zariski closure of the $GL(d)$-orbit of M in $\mathrm{mod}_d\Lambda$. This defines a partial order on the set of isomorphism classes, since orbits are irreducible and open in their closure. Our goal is to find algebraic conditions on M and N related to the existence of a degeneration $M \leq_{deg} N$.

---

[1] 1991 *Mathematics Subject Classification.* Primary 16G10; Secondary 14L30.

2. Examples of degenerations

If

$$0 \to N' \to M \to N'' \to 0$$

is a short exact sequence of Λ-modules, there is a degeneration from M to $N' \oplus N''$. Indeed, choosing a basis for N' and completing it to one for M, we use conjugation by the 1-parameter subgroup

$$\lambda(t) = \begin{pmatrix} t \cdot \mathbf{1} & 0 \\ 0 & \mathbf{1} \end{pmatrix}, \; t \in k^*$$

to conclude that $\lambda(t) \cdot M$ is isomorphic to M for $t \neq 0$ and tends to $N' \oplus N''$ for $t \to 0$.

Another source for examples of degenerations $M \leq_{deg} N$ are short exact sequences

$$0 \to S \to S \oplus M \to N \to 0.$$

The reason is that "generically" the quotient of a monomorphism $S \to S \oplus M$ is M [**R**]. An exact sequence

$$0 \to N \to S \oplus M \to S \to 0$$

yields a degeneration $M \leq_{deg} N$ as well.

3. Partial orders stronger than $\leq_{deg}$

The transitive closure of the relation $M \leq N' \oplus N''$ if there is an exact sequence

$$0 \to N' \to M \to N'' \to 0$$

is obviously a partial order. We denote it by $\leq_{ext}$.

Surprisingly, the relation $M \leq_R N$ if there is an exact sequence

$$0 \to N \to M \oplus S \to S \to 0$$

is a partial order as well. In addition, it coincides with the "dual" one obtained from exact sequences

$$0 \to S \to M \oplus S \to N \to 0.$$

These facts can be found in Zwara's recent preprint [**ZW3**].

Obviously $M \leq_{ext} N$ implies $M \leq_R N$. Indeed, it is enough to add up

$$0 \to N' \to M \to N'' \to 0$$

and

$$0 \to N'' \xrightarrow{\text{id}} N'' \to 0 \to 0$$

and to put $S = N''$.

In general $M \leq_R N$ is not equivalent to $M \leq_{ext} N$: if $M \leq_{ext} N$ and N is not isomorphic to M, N is necessarily decomposable, whereas $M \leq_R N$ can happen nontrivially for an indecomposable N (see [**R**] for an example).

4. The Hom order

For any Λ-module X, consider the variety of pairs (M, f), where M lies in $\text{mod}_d\Lambda$ and $f : X \to M$ is Λ-linear. The fiber over M of the projection onto $\text{mod}_d\Lambda$ is $\text{Hom}_\Lambda(X, M)$. From the semicontinuity of the dimensions of the fibers of an algebraic map we conclude that $M \leq_{deg} N$ implies

$$[X, M] = \dim_k \text{Hom}_\Lambda(X, M) \leq [X, N] = \dim_k \text{Hom}_\Lambda(X, N).$$

We define $M \leq N$ if $[X, M] \leq [X, N]$ holds for all modules X – or equivalently for all indecomposables. An elementary proof of the implication

$$M \leq_{deg} N \Longrightarrow M \leq N$$

can be found in [**R**]. It shows in addition that $M \leq N$ is equivalent to $[M, X] \leq [N, X]$ for all X.

In order to prove that $\leq$ actually is a partial order we use almost split sequences: an exact sequence

$$0 \to Y \xrightarrow{\alpha} E \xrightarrow{\beta} Y' \to 0$$

is an almost split sequence starting in Y if it does not split, Y and Y' are indecomposable and any non-isomorphism $f : Y \to Z$ to an indecomposable Z factors through α. Almost split sequences or "Auslander–Reiten sequences" play a fundamental role in representation theory. They have been invented and studied by Auslander and Reiten in the seventies [**AR1**]. For a condensed approach see [**Ga**].

If Y is non-injective there exists an almost split sequence starting in Y. By definition, the sequence

$$0 \to \text{Hom}_\Lambda(Y', M) \to \text{Hom}_\Lambda(E, M) \to \text{Hom}_\Lambda(Y, M) \to k^{\mu(Y,M)} \to 0$$

of k-vector spaces is exact for any M if $\mu(Y, M)$ denotes the multiplicity of Y as a direct summand of M (Krull–Schmidt decomposition). If I is indecomposable injective, the sequence of k-vector spaces

$$0 \to \text{Hom}_\Lambda(I/\text{soc}I, M) \to \text{Hom}_\Lambda(I, M) \to k^{\mu(I,M)} \to 0$$

is exact as well.

Now assume that $[X, M] = [X, N]$ for M, N in $\text{mod}_d\Lambda$ and all X. Then the multiplicities of any indecomposable Y in M and in N must be the same, and therefore M and N are isomorphic. Compare [**AR2**], [**Bo1**].

Let $\Lambda = \oplus P_i$ be a decomposition of Λ into a sum of indecomposable projectives. If $M \leq N$ we have in particular that $[P_i, M] \leq [P_i, N]$ for all i. On the other hand, we know that $d = [\Lambda, M] = [\Lambda, N]$. Therefore $M \leq N$ implies $[P, M] = [P, N]$ for all projectives P and dually $[M, I] = [N, I]$ for all injectives I. So M and N have the same image in the Grothendieck group of Λ.

5. Virtual degenerations

We say that M degenerates virtually to N, in symbols $M \leq_{virt} N$, if $M \oplus R$ degenerates to $N \oplus R$ for a suitable module R. Clearly $\leq_{virt}$ is a partial order, and there is a chain of implications:

$$M \leq_{deg} N \Longrightarrow M \leq_{virt} N \Longrightarrow M \leq N.$$

In case Λ is of finite representation type; i.e., admits only finitely many indecomposables up to isomorphism, the last implication is an equivalence [**R**]. The proof is characteristic of the type of arguments used in this subject. Here it is.

Suppose $M \leq N$. We know that then $[M, I] = [N, I]$ for all injectives I. Let

$$\sum : 0 \to A \to B \to C \to 0$$

be the direct sum over all (representatives of the isomorphism classes of) non-injective indecomposables Y of $[N, Y] - [M, Y]$ copies of an almost split sequence starting in Y. The sequence

$$0 \to \mathrm{Hom}(C, Y) \to \mathrm{Hom}(B, Y) \to \mathrm{Hom}(A, Y) \to k^{[N,Y]-[M,Y]} \to 0$$

of k-vector spaces is exact by construction of $\sum$. From

$$[C, Y] - [B, Y] + [A, Y] = [N, Y] - [M, Y]$$

for any Y, we conclude that $N \oplus B$ is isomorphic to $M \oplus A \oplus C$. Since $\sum$ is exact, $N \oplus B$ degenerates to $N \oplus A \oplus C$, and therefore $M \leq_{virt} N$.

6. Cancellation

When does $M \oplus R \leq_{deg} N \oplus R$ imply $M \leq_{deg} N$; i.e., when can the common direct summand R be cancelled? Certainly not always, as the following modified version of an example of J. Carlson's [**R**] shows: Let Q be the quiver

$$\underset{\bullet}{1} \;\underset{\gamma}{\overset{\alpha}{\rightrightarrows}}\; \underset{\bullet}{2} \;\underset{\delta}{\overset{\beta}{\rightrightarrows}}\; \underset{\bullet}{3},$$

I the ideal generated by $\delta\alpha, \beta\gamma$ and $\beta\alpha - \delta\gamma$ in the quiver algebra kQ, and set $\Lambda = kQ/I$. There is an exact sequence

i) $$0 \to \mathrm{rad}P \to P \oplus \mathrm{rad}P/\mathrm{soc}P \to P/\mathrm{soc}P \to 0,$$

where P denotes the unique projective-injective Λ-module

$$P = k \;\underset{\binom{0}{1}}{\overset{\binom{1}{0}}{\rightrightarrows}}\; k^2 \;\underset{(0\ 1)}{\overset{(1\ 0)}{\rightrightarrows}}\; k.$$

For $t \in k$, we set

$$M_t = k \;\underset{t}{\overset{1}{\rightrightarrows}}\; k \qquad 0$$

and

$$N_t = 0 \qquad k \underset{1}{\overset{t}{\rightrightarrows}} k.$$

It is easy to construct exact sequences

ii) $$0 \to 0\;k\;0 \to P/\mathrm{soc}P \to M_t \to 0$$

and

iii) $$0 \to N_s \to \mathrm{rad}P \to 0\;k\;0 \to 0$$

for any $s,\ t \in k$. Then by i), ii), and iii), resp., we have

$$P \oplus \mathrm{rad}P/\mathrm{soc}P \leq_{deg} \mathrm{rad}P \oplus P/\mathrm{soc}P \leq_{deg}$$

$$\mathrm{rad}P \oplus M_t \oplus 0\;k\;0 \leq_{deg} M_t \oplus N_s \oplus (0\;k\;0)^2,$$

for any $s,\ t \in k$.

We claim that the common direct summand $\mathrm{rad}P/\mathrm{soc}P \xrightarrow{\sim} (0\;k\;0)^2$ cannot always be cancelled. There are several ways to do this. If we work with the group $GL(4, k)$, the easiest is to count dimensions: the dimension of the orbit of P is 15, $1 = \dim_k \mathrm{End}_\Lambda P$ less than the dimension of the group. The dimension of the orbit of $M_s \oplus N_t$ is 13, and these orbits form a 2-parameter family.
But the orbit of P is irreducible and open in its closure, and the dimension of the complement is ≤ 14.

If we work with the smaller group

$$\begin{pmatrix} k^* & 0 & 0 \\ 0 & GL(2) & 0 \\ 0 & 0 & k^* \end{pmatrix} \subset GL(4)$$

of base changes for representations of Q with "dimension vector" $(1,2,1)$ instead, which leads to equivalent theories by [**Bo2**], we can show more explicitly that if P degenerates to $M_s \oplus N_t$ we must have $s + t = 0$. It is easy to construct a degeneration from P to $M_s \oplus N_{-s}$.

For a module

$$k \underset{f'}{\overset{f}{\rightrightarrows}} k^2 \underset{g'}{\overset{g}{\rightrightarrows}} k$$

in the orbit of P under the small group, the identity

$$g\begin{pmatrix} 0 & 1 \\ 0 & 0 \end{pmatrix} f + g'\begin{pmatrix} 0 & 1 \\ 0 & 0 \end{pmatrix} f' = 0$$

must hold by the invariance of the trace of a matrix under conjugation. Since this identity is polynomial, it is still satisfied in the closure of the orbit of P. But it holds for $M_s \oplus N_t$ in case $s + t = 0$ only.

Notice that this example shows that $M \leq N$ does not necessarily imply $M \leq_{deg} N$.

There are conditions, however, under which cancellation is possible: if $M \oplus R \leq_{deg} N \oplus R$ and $[R, M] = [R, N]$, then $M \leq_{deg} N$. This is Bongartz' cancellation theorem [**Bo4**], [**Bo5**]. So projective – and dually injective – common direct summands can always be cancelled.

7. Nilpotent matrices

The example $\Lambda = k[T]/(T^N)$ of nilpotent matrices has been treated in [**Ge**] and [**He**]. An indecomposable Λ-module is isomorphic to $k[T]/(T^n)$, $1 \leq n \leq N$, and

$$[k[T]/(T^n),\ k[T]/(T^m)] = \min\{m, n\}.$$

A d-dimensional Λ-module M is isomorphic to a direct sum

$$M \xrightarrow{\sim} \bigoplus_{i=1}^{r} k[T]/(T^{p_i})$$

of indecomposables; up to isomorphism M is determined by the partition $\pi = \pi_M = (p_1, \ldots, p_r)$ of d:

$$N \geq p_1 \geq p_2 \geq \ldots \geq p_r > 0, \quad p_1 + \ldots + p_r = d.$$

We represent π by the Young diagram with p_i boxes in the i-th row.

There is a well-known partial order on the set of partitions of d:

$$\pi = (p_1, \ldots, p_r) \geq \pi' = (p'_1, \ldots, p'_s)$$

if $p_1 + \ldots + p_n \geq p'_1 + \ldots + p'_n$ for $n \geq 1$. We agree that $p_m = p'_{m'} = 0$ for $m > r$, $m' > s$. This order is the transitive closure of "moving down a box":

$$\pi = (p_1, \ldots, p_r) \mapsto \pi' = (p_1, \ldots, p_{i-1}, p_i - 1, p_{i+1}, \ldots, p_{j-1}, p_j + 1, p_{j+1}, \ldots, p_r),$$

which is possible if $p_i - 1 \geq p_{i+1}$ and $p_{j-1} \geq p_j + 1$, $i < j$.

The dual partition $\pi^\vee = (q_1, \ldots, q_s)$ of $\pi = (p_1, \ldots, p_r)$ is defined by

$$q_i = \#\{p_j \geq i\}.$$

Its Young diagram is obtained from the one of π by "transposing"; i.e., exchanging rows and columns. Clearly $\pi \geq \pi'$ if and only if $\pi^\vee \leq \pi'^\vee$.

If M is the module

$$M = \bigoplus_{i=1}^{r} k[T]/(T^{p_i})$$

corresponding to $\pi_M = (p_1, \ldots p_r)$, we have

$$\begin{aligned}
[k[T]/(T^n), M] &= \sum_{i=1}^{r} [k[T]/(T^n),\ k[T]/(T^{p_i})] \\
&= \sum_{p_i \le n} p_i + \#\{p_i > n\} \cdot n \\
&= \sum_{j=1}^{n} \#\{p_k \ge j\},
\end{aligned}$$

which is the sum of the first n elements in $\pi_M^\vee$. So $M \le N$ is equivalent to $\pi_M^\vee \le \pi_N^\vee$ and hence to $\pi_M \ge \pi_N$.

For $N \ge p > q+1$ there is an exact sequence

$$0 \to k[T]/(T^{p-1}) \to k[T]/(T^p) \oplus k[T]/(T^q) \to k[T]/(T^{q+1}) \to 0$$

in which maps in $\mathrm{Hom}_\Lambda(k[T]/(T^r),\ k[T]/(T^s))$ are given by truncating if $r > s$ and by sending T^i to T^{i+s-r} if $s > r$, up to a sign. So "moving down a box" can be realized as an exact sequence. This means that $\le_{ext}$, which is the transitive closure of "moving down boxes", coincides with $\le$.

Therefore all partial orders introduced before coincide for $k[T]/(T^N)$.

8. First algebraic descriptions

There is a chain of implications

$$M \le_{ext} N \Longrightarrow M \le_R N \Longrightarrow M \le_{deg} N \Longrightarrow M \le N.$$

The first attempts to characterize $M \le_{deg} N$ algebraically were proofs that $M \le N$ implies $M \le_{ext} N$. The strategy is the following: Suppose that Λ is of finite representation type and use induction on

$$\delta(M,N) = \sum([X,N] - [X,M])$$

over all indecomposables X. If $\delta(M,N) = 0$, N is isomorphic to M. Otherwise, look for a non-split exact sequence

$$0 \to A \to B \to C \to 0$$

such that

i) $N = A \oplus C \oplus N'$

and

ii) $[X, A \oplus C] - [X,B] \le [X,N] - [X,M]$ for all indecomposables X.

Then $M \le N' \oplus B$, $\delta_{M,N'\oplus B} = \delta_{M,N} - \delta_{B,A\oplus C} < \delta_{M,N}$, and $N' \oplus B \le_{ext} N = N' \oplus A \oplus C$.

In this way it was shown that all four partial orders are the same for representations of

– a quiver with underlying non-oriented graph A_n in [**AD1**],

– $\begin{smallmatrix}1\\2\end{smallmatrix} \rightrightarrows 3 \to 4 \ldots \to n$ in [**AD2**],

– quivers with underlying graph A_n or D_n in [**R**].

The problem is to find an exact sequence with i) and ii). All proofs mentioned above are based on a complete knowledge of all indecomposables and all homomorphisms between them. So in order to answer the relatively "coarse" question about degenerations we needed to know much "finer" facts about all modules, encoded in the "Auslander–Reiten quiver" [**Ga**].

In [**R**] the starting point for finding a suitable exact sequence is a "left most" indecomposable direct summand A of N; i.e., one for which $\operatorname{Hom}(A', A) \neq 0$ implies $A' \xrightarrow{\sim} A$ for all indecomposable direct summands A' of N. Such a "starting point" always exists if Λ is simply connected [**Ga**], but nobody has succeeded in applying covering techniques to geometric questions. (There is one exception in [**Bo3**], §3).

9. Bongartz' approach

The first step away from constructing suitable exact sequences as above was taken by Bongartz [**Bo4**]. He still needed a "left most" direct summand of N though.

We reproduce his argument in a special case:

Proposition. *If Λ is representation directed, the partial orders $\leq$ and $\leq_{deg}$ coincide.*

This condition means that there are no chains of non-isomorphisms

$$X_0 \xrightarrow{f_0} X_1 \xrightarrow{f_1} \ldots \longrightarrow X_r \xrightarrow{f_r} X_0$$

between indecomposables. It implies that Λ is of finite representation type. All hereditary algebras of finite representation type are representation directed.

Proof. Suppose that $M \leq N$ and use induction on $\delta'(M,N) = \sum([N,X] - [M,X])$ over all indecomposables. We may assume that M and N have no direct summand in common. Since Λ is representation directed, N contains an indecomposable direct summand U with no "predecessors": $\operatorname{Hom}(U', U) \neq 0$ for an indecomposable direct summand U' of N implies $U' \xrightarrow{\sim} U$. Then $[N,U] > [M,U]$, and U cannot be injective. Consider an almost split sequence starting in U

$$0 \to U \to X \to U' \to 0.$$

Clearly

$$M \oplus X \leq_{ext} M \oplus U \oplus U' \leq N \oplus X.$$

Moreover $\delta'(M \oplus U \oplus U', N \oplus X) = \delta'(M,N) - 1$, which implies $M \oplus X \leq_{deg} N \oplus X$ by induction. Finally, one can show that $[X,M] = [X,N]$ and apply Bongartz' cancellation theorem (see §6) in order to obtain $M \leq_{deg} N$.

In the same paper [**Bo4**] Bongartz shows that $\leq$ and $\leq_{ext}$ coincide if N is "preprojective". His construction of a "suitable" exact sequence is based on the existence of a special morphism from M to N. The information he needs on the indecomposables and the morphisms between them is much less precise than in

[**R**]; it is essentially the fact that $\delta'(M, N)$ is finite and that the set of direct summands of M and N carries a partial order given by $X \leq Y$ if $\mathrm{Hom}(X, Y) \neq 0$.

This was the first positive result for which the algebra may be representation infinite. The paper [**Bo3**] contains a much more transparent proof of the same result (in §2) as well as the equivalence of $\leq$ and $\leq_{deg}$ for tame hereditary algebras. Again the proof is based on the classification of all indecomposables and maps between them. As Bongartz used his cancellation theorem in the proof, he could not show that $\leq$ is equivalent to $\leq_{ext}$. But they are by Zwara's paper [**ZW1**].

10. Zwara's theorem

We conclude with Zwara's spectacular theorem [**ZW2**]:

Theorem. *If Λ is of finite representation type, $M \leq N$ implies $M \leq_R N$.*

So $M \leq_{deg} N$ is equivalent to $M \leq_R N$ over algebras of finite representation type. Apparently there is no counterexample known in infinite representation type.[2] We explain the idea of Zwara's proof.

For $M \leq N$ consider the non-negative function

$$\delta_{M,N} = [N, \ ] - [M, \ ]$$

on the set of isomorphism classes of indecomposables. If

$$\Sigma : 0 \to A \to B \to C \to 0$$

is exact, set

$$\delta_\Sigma = \delta_{B, A\oplus C}.$$

Start from the following question: Given an exact sequence

$$\Phi : 0 \to N \to B \to C \to 0,$$

when is B isomorphic to $M \oplus C$? This is certainly the case if $[B, X] = [M \oplus C, X]$ for all indecomposables X; i.e., if $\delta_\Phi = \delta_{M,N}$. Suppose that $M \oplus C \leq_{deg} B$ and $\delta_\Phi(B) = \delta_{M,N}(B)$. Now the identity of B is a surjection of B to B, and by the dual of theorem 2.4 in [**Bo4**], there is a surjection of B to $M \oplus C$. Since $\dim B = \dim(M \oplus C)$ this means that B is isomorphic to $M \oplus C$.

First ingredient. (Lemma 3.2 of [**ZW2**]). If $M \oplus C \leq B$ and $\delta_\Phi(B) = \delta_{M,N}(B)$, B is isomorphic to $M \oplus C$.

Thus we need to find an exact sequence

$$\Phi : 0 \to N \to B \to C \to 0$$

with $\delta_\Phi \leq \delta_{M,N}$ and $\delta_\Phi(B) = \delta_{M,N}(B)$.

Second ingredient. (Lemma 3.1 of [**ZW2**]). Let

$$\Sigma : 0 \to A \xrightarrow{f} B \xrightarrow{g} C \to 0$$

[2] Actually, Zwara proved in December 1997 that $M \leq_{\mathrm{deg}} N$ implies $M \leq_R N$ over any Λ [**ZW4**].

be an exact sequence such that f and g lie in the radical of $\operatorname{mod}\Lambda$. Suppose $\delta_\Sigma \leq \delta_{M,N}$ and $\delta_\Sigma(B) < \delta_{M,N}(B)$. Then there exists an exact sequence

$$\Sigma^* : 0 \to A \xrightarrow{f'} B' \xrightarrow{g'} C' \to 0$$

with f', g' still in the radical such that $\delta_\Sigma \leq \delta_{\Sigma^*} \leq \delta_{M,N}$ and $\delta_{\Sigma^*}(B) = \delta_{M,N}(B)$.

Admitting these two ingredients, the proof goes as follows: Decompose N into indecomposables

$$N = \oplus N_i^{\mu_i}.$$

Set

$$\begin{aligned} \nu_i &= \min\{\mu_i, \delta_{M,N}(N_i)\}, \\ N' &= \oplus N_i^{\nu_i}, \\ N'' &= \oplus N_i^{\mu_i - \nu_i}. \end{aligned}$$

Obviously then $N = N' \oplus N''$. If $N' = 0$, $\delta_{M,N}(N) = 0$, which is impossible (Lemma 3.2 of [**ZW2**]).

If N_i is a direct summand of N', $\delta_{M,N}(N_i) > 0$, and N_i is non-injective. Let Σ be the direct sum over i of ν_i copies of an almost split sequence starting in N_i. By definition

$$\delta_\Sigma(X) = \begin{cases} \nu_i & X \xrightarrow{\sim} N_i \\ 0 & \text{otherwise} \end{cases}$$

for X indecomposable and therefore $\delta_\Sigma \leq \delta_{M,N}$. Take

$$\Phi : 0 \to N' \to B \to C \to 0$$

with $\delta_\Sigma \leq \delta_\Phi \leq \delta_{M,N}$ and such that $\sum_X \delta_\Phi(X)$ is as big as possible. By the second ingredient, we must have $\delta_\Phi(B) = \delta_{M,N}(B)$. Finally, take

$$\Phi^* : 0 \to N \to B \oplus N'' \to C \to 0$$

to be the direct sum of Φ with

$$0 \to N'' \xrightarrow{\text{id}} N'' \to 0 \to 0.$$

Clearly $\delta_{\Phi^*} = \delta_\Phi \leq \delta_{M,N}$ and thus $\delta_{\Phi^*}(B) = \delta_{M,N}(B)$. In order to finish by applying the first ingredient, we need to show

$$M \oplus C \leq N'' \oplus B \text{ and } \delta_{\Phi^*}(N'' \oplus B) = \delta_{M,N}(N'' \oplus B).$$

The former is equivalent to $\delta_{\Phi^*} \leq \delta_{M,N}$, and the latter follows from $\delta_{\Phi^*}(N'') = \delta_{M,N}(N'')$: If N_i is a direct summand of N'', we must have

$$\mu_i > \nu_i = \delta_{M,N}(N_i).$$

Then

$$\delta_\Sigma(N_i) = \nu_i = \delta_{M,N}(N_i)$$

and thus

$$\delta_{\Phi^*}(N_i) = \delta_\Phi(N_i) = \delta_{M,N}(N_i).$$

References

[AD1] S. Abeasis and A. Del Fra: *Degenerations for the representations of a quiver of type A_m*, J. of Algebra 93 (1985), p. 376-412.

[AD2] S. Abeasis and A. Del Fra: *Degenerations for representations of an equioriented quiver of type D_m*, Adv. Math. 52 (1984), p. 81-172.

[AR1] M. Auslander and I. Reiten: *Representation theory of Artin algebras IV*, Comm. Algebra 5 (1977), p. 443-518.

[AR2] M. Auslander and I. Reiten: *Modules determined by their composition factors*, Illinois J. Math. 29 (1985), p. 289-301.

[Bo1] K. Bongartz: *A generalization of a theorem by M. Auslander*, Bull. London Math. Soc. 21 (1989), p. 255-256.

[Bo2] K. Bongartz: *A geometric version of Morita theory*, J. Algebra 139 (1991), p. 159-179.

[Bo3] K. Bongartz: *Degenerations for representations of tame quivers*, Ann. Scient. Ecole Norm. Sup. (1995), p. 647-668.

[Bo4] K. Bongartz: *On degenerations and extensions of finite dimensional modules*, Adv. Math. 121 (1996), p. 245-287.

[Bo5] K. Bongartz: *Some geometric aspects of representation theory*, preprint 1996.

[Ga] P. Gabriel: *Auslander–Reiten sequences and representation–finite algebras*, in "Representation Theory I", Springer Lecture Notes 831 (1980), p. 1-71.

[Ge] M. Gerstenhaber: *On dominance and varieties of commuting matrices*, Ann. of Math. 73 (1961), p. 324-348.

[He] W. Hesselink: *Singularities in the nilpotent scheme of a classical group*, TAMS 222 (1976), p. 1-32.

[R] Ch. Riedtmann: *Degenerations for representations of quivers with relations*, Ann. Sci. Ecole Norm. Sup. 4 (1986), p. 275-301.

[ZW1] G. Zwara: *Degenerations for representations of extended Dynkin quivers*, Comment. Math. Helv., to appear.

[ZW2] G. Zwara: *Degenerations for modules over representation–finite algebras*, Proceedings of the AMS, to appear.

[ZW3] G. Zwara: *New partial order in the varieties of finite dimensional modules*, Preprint, 1997.

[ZW4] G. Zwara: *Degenerations of finite dimensional modules are given by extensions*, preprint, 1998.

Contemporary Mathematics
Volume **229**, 1998

The Preprojective Algebra of a Tame Quiver: The Irreducible Components of the Module Varieties

CLAUS MICHAEL RINGEL

ABSTRACT. We consider a tame quiver Q and its preprojective algebra $\mathcal{P}_k(Q)$. In his study of the affine quantum groups, G. Lusztig has determined the irreducible components of the variety of all nilpotent $\mathcal{P}_k(Q)$-modules with fixed dimension vector. The aim of our paper is to present a proof of this result using the Auslander-Reiten translation and the structure of the Auslander-Reiten quiver of the category of all kQ-modules.

1. Introduction

Let k be an algebraically closed field. Let $Q = (Q_0, Q_1, s, t)$ be a finite quiver (here, Q_0, Q_1 are finite sets and $s, t\colon Q_1 \to Q_0$ are maps). We will assume that Q is connected and that there are no cyclic paths. We denote by kQ the path algebra of Q. Given a finite dimensional Q_0-graded vector space V, let

$$\mathcal{R}(Q;V) = \bigoplus_{\alpha\in Q_1} \operatorname{Hom}_k(V_{s\alpha}, V_{t\alpha});$$

this is the set of all representations of Q realized on V; its elements will be written in the form (V, x) where $x = (x_\alpha)_\alpha$ is a family of k-linear maps $x_\alpha\colon V_{s\alpha} \to V_{t\alpha}$; the family $(\dim V_i)_{i\in Q_0}$ is called the *dimension vector* of (V, x). The group

$$G(V) = \prod_{i\in Q_0} \mathrm{GL}(V_i)$$

operates on $\mathcal{R}(Q;V)$ via

$$(g * x)_\alpha = g_{t\alpha} x_\alpha g_{s\alpha}^{-1},$$

1991 *Mathematics Subject Classification.* Primary 16G20; Secondary 14D20, 14L30, 15A30, 16G70, 17B37.

for $g = (g_i)_i \in G(V)$ and $x = (x_\alpha)_\alpha \in \mathcal{R}(Q; V)$. Elements $x, y \in \mathcal{R}(Q; V)$ belong to the same $G(V)$-orbit if and only if the representations (V, x) and (V, y) are isomorphic.

To Q there also is attached the preprojective algebra $\mathcal{P}_k(Q)$. We recall the definition: Let $\overline{Q}$ be obtained from Q by adding for every arrow $\alpha\colon x \to y$ a formal inverse $\alpha^*\colon y \to x$, the set of new arrows will be denoted by Q_1^*. We consider the following element

$$\rho = \sum_{\alpha \in Q_1} [\alpha^*, \alpha]$$

in $k\overline{Q}$ (the summands of ρ are just the usual commutators $[\alpha^*, \alpha] = \alpha^*\alpha - \alpha\,\alpha^*$). Let (ρ) be the ideal of $k\overline{Q}$ generated by ρ; the corresponding factor algebra $\mathcal{P}_k(Q) = k\overline{Q}/(\rho)$ is called the *preprojective algebra of the quiver Q*. We write any representation of $\overline{Q}$ in the form $(V, x, \xi) = (V, x_\alpha, \xi_\alpha)_{\alpha \in Q_1}$; here, (V, x) is a representation of Q and (V, ξ) is a representation of Q^*. We denote by $\mathcal{R}_0(\overline{Q}, \rho\,; V)$ the subset of $\mathcal{R}(\overline{Q}, V)$ of all elements (V, x, ξ) which satisfy the relation ρ and which are nilpotent (a representation of $k\overline{Q}$ is said to be nilpotent provided it is annihilated by some power of the ideal of $k\overline{Q}$ generated by the arrows of $\overline{Q}$).

The path algebra kQ can be considered as a subalgebra of $\mathcal{P}_k(Q)$, thus the restriction yields a functor from the category $\operatorname{mod}\mathcal{P}_k(Q)$ of $\mathcal{P}_k(Q)$-modules to the category $\operatorname{mod} kQ$ of kQ-modules; it sends the representation (V, x, ξ) of $\overline{Q}$ to the representation $(V, x) = \pi(V, x, \xi)$ of Q. This map

$$\pi\colon \mathcal{R}_0(\overline{Q}, \rho\,; V) \longrightarrow \mathcal{R}(Q; V),$$

will be of interest to us.

Recall that the quiver Q is *tame*, provided the underlying graph of Q is one of the following:

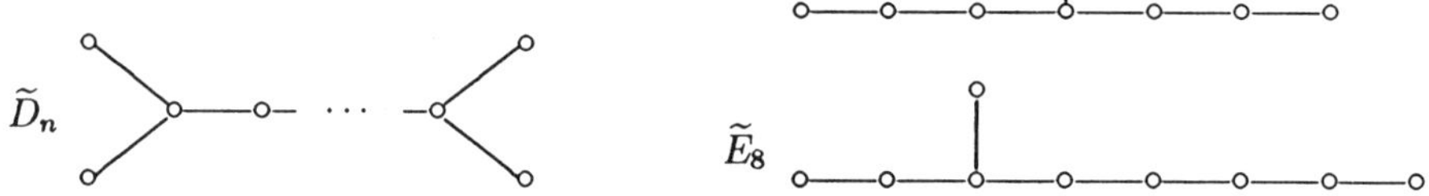

For these tame quivers we are going to describe a stratification of the affine spaces $\mathcal{R}(Q; V)$ by suitable subsets $\Xi(M, \mu)$ so that all the irreducible components of $\mathcal{R}_0(\overline{Q}, \rho\,; V)$ are obtained as closures of certain inverse images $\pi^{-1}(\Xi(M, \mu))$.

This classification of the irreducible components of $\mathcal{R}_0(\overline{Q},\rho;V)$ (with some minor restrictions on Q) is due to Lusztig, see [L2], Proposition 4.14; it has been an essential prerequisite for his investigation of the affine quantum groups attached to a symmetric generalized Cartan matrix. The version presented here is streamlined in order to allow a parallel discussion of arbitrary symmetrizable generalized Cartan matrices as indicated in [R3]. Our proof relies on a description of the fibers of π in terms of the Auslander-Reiten translation in the category $\operatorname{mod} kQ$: for any representation (V,x) of kQ, the fiber $\pi^{-1}(V,x)$ may be identified with the set of nilpotent elements of degree 1 in the orbit ring $\mathcal{O}^{\tau^-}(V,x)$, see [R4]. This description of the fibers of π and the consequences for the tame quivers as outlined below have been presented first at ICRA VI, Ottawa 1992 and in lectures at Brandeis University in the same year.

2. The deficiency of a module.

We consider finite dimensional kQ-modules. Given such a module M, let $\mathcal{O}(M) = \mathcal{O}^{\tau^-}(M)$ denote the orbit ring with respect to the Auslander-Reiten translation τ^-, see [R4]. The multiplication in $\mathcal{O}(M)$ will be denoted by $*$; correspondingly, the t-fold power of an element $f \in \mathcal{O}(M)$ will be denoted by f^{*t}. Recall that $\mathcal{O}(M)$ is a graded ring and we are mainly interested in the set $\mathcal{O}(M)_1 = \operatorname{Hom}(\tau^- M, M)$ of elements of degree 1. We denote by $\mathcal{N}(M) = \mathcal{N}^{\tau^-}(M)$ the set of elements of $\mathcal{O}(M)_1$ which are nilpotent in $\mathcal{O}(M)$. We denote by $\gamma(M)$ the codimension of $\mathcal{N}(M)$ in $\operatorname{Hom}(\tau^- M, M)$ and call it the *deficiency* of M. We are going to derive some recipes for calculating $\gamma(M)$. First, we recall three properties which have been mentioned already in [R4]:

2.1. *Let M and N be kQ-modules with $\operatorname{Hom}(\tau^- N, M) = 0$. Write $\mathcal{O}(M\oplus N)_1$ in matrix form as follows:*

$$\mathcal{O}(M \oplus N)_1 = \begin{bmatrix} \operatorname{Hom}(\tau^- M, M) & 0 \\ \operatorname{Hom}(\tau^- M, N) & \operatorname{Hom}(\tau^- N, N) \end{bmatrix},$$

then $\mathcal{N}(M \oplus N)$ is just the subset

$$\mathcal{N}(M \oplus N) = \begin{bmatrix} \mathcal{N}(M) & 0 \\ \operatorname{Hom}(\tau^- M, N) & \mathcal{N}(N) \end{bmatrix}.$$

In particular, $\gamma(M \oplus N) = \gamma(M) + \gamma(N)$.

PROOF. The triangular form of the matrix presentation of $\mathcal{O}(M\oplus N)_1$ implies that the nilpotency in $\mathcal{O}(M \oplus N)$ of an element of $\mathcal{O}(M \oplus N)_1$ has to be checked only for M and N separately.

Let M be a kQ-module. Recall that M is said to be *preprojective* provided there exists some $n \geq 0$ such that $\tau^n(M) = 0$. Similarly, M is said to be *preinjective* provided there exists some $n \geq 0$ such that $\tau^{-n}(M) = 0$. Finally, M is said to be *regular* provided no indecomposable direct summand of M is preprojective or preinjective. Of course, by definition, any module M is isomorphic to a module of the form $P \oplus R \oplus I$, where P is preprojective, R regular and I preinjective.

2.2. *If M is a preprojective or a preinjective kQ-module, then*

$$\mathcal{N}(M) = \mathrm{Hom}(\tau^- M, M),$$

thus $\gamma(M) = 0$.

PROOF. If M is preinjective, then $\tau^{-n}M = 0$ for some $n \geq 0$. Thus, for any $f \in \mathrm{Hom}(\tau^- M, M)$, the t-fold power f^{*t} in the orbit ring $\mathcal{O}(M)$ is zero.

Similarly, in case M is preprojective, then $\tau^n M = 0$ for some $n \geq 0$. Since τ^{-n} is left adjoint to τ^n, it follows that $\mathrm{Hom}(\tau^{-n}M, M) \simeq \mathrm{Hom}(M, \tau^n M) = 0$. Thus for any element in $f \in \mathrm{Hom}(\tau^- M, M)$, we also have $f^{*n} = 0$ in $\mathcal{O}(M)$.

2.3. *For any kQ-module M, we have $\gamma(M) = \gamma(R)$, where R is a maximal regular direct summand of M.*

PROOF. We write $M = P \oplus R \oplus I$ where P is preprojective, R regular and I preinjective and we use well-known assertions: First of all, there are no non-zero homomorphisms from a preinjective module to a preprojective or a regular module, and with I also $\tau^- I$ is preinjective. Second, there are no non-zero homomorphisms from a regular module to a preprojective module, and with R also $\tau^- R$ is regular. Thus we can apply first property (2.1) twice in order to see that $\gamma(P \oplus R \oplus I) = \gamma(P) + \gamma(R) + \gamma(I)$, and then we use property (2.2).

This shows that it is sufficient to consider the case of a regular module R. In general, the problem of determining $\mathcal{N}(R)$ inside $\mathcal{O}(R)_1$ seems to be difficult. *Let us assume from now on that Q is a tame quiver.*

We recall part of the structure theory for the category of all regular kQ-modules. A kQ-module M is said to be *homogeneous* provided any indecomposable direct summand N of M is isomorphic to τN. Of course, all the homogeneous modules are regular (and non-zero homogeneous modules do exist only in case Q is tame). The kQ-module M will be called *completely inhomogeneous* provided no indecomposable direct summand N of M is homogeneous. Clearly, any kQ-module is a direct sum of a homogeneous module and a completely inhomogeneous module.

2.4. *Let P be preprojective and I preinjective. Let R be a completely inhomogeneous regular module and H homogeneous. Then*

$$\mathcal{N}(P \oplus R \oplus H \oplus I) = \begin{bmatrix} \mathrm{Hom}(\tau^- P, P) & 0 & 0 & 0 \\ \mathrm{Hom}(\tau^- P, R) & \mathcal{N}(R) & 0 & 0 \\ \mathrm{Hom}(\tau^- P, H) & 0 & \mathcal{N}(H) & 0 \\ \mathrm{Hom}(\tau^- P, I) & \mathrm{Hom}(\tau^- R, I) & \mathrm{Hom}(\tau^- H, I) & \mathrm{Hom}(\tau^- I, I) \end{bmatrix}.$$

PROOF. If a module M is preprojective, or preinjective, or completely inhomogeneous regular, or homogeneous, the same is true for $\tau^- M$. Note that $\mathrm{Hom}(M, N) = 0$ in the following cases: M is preinjective and N is regular or preprojective; M is regular and N is preprojective; M is completely inhomogeneous regular and N is homogeneous; M is homogeneous and N is completely inhomogeneous regular. According to (2.1), this yields the triangular structure of the matrix as well as the additional zero entry. For the first and the last diagonal entry, we use (2.2).

2.4′. *Let M and H be kQ-modules and assume that M is completely inhomogeneous, whereas H is homogeneous. Then $\gamma(M \oplus H) = \gamma(M) + \gamma(H)$.*

PROOF. We decompose $M = P \oplus R \oplus I$ with P preprojective, R regular and I preinjective and apply the previous result to $P \oplus R \oplus H \oplus I$.

Recall that in the tame case, all regular kQ-modules are τ-periodic. A module M is said to be *aperiodic* provided there does not exist any indecomposable module N of τ-period p such that the direct sum $\bigoplus_{i=0}^{p-1} \tau^i N$ is isomorphic to a direct summand of M. This terminology is due to Lusztig [L2] (in [R2] we used the word 'separated'). A module $M = P \oplus R \oplus I$ with P preprojective, R regular and I preinjective is aperiodic, if and only if R is aperiodic. This means that the problem of aperiodicity only concerns the regular modules. Of course, all the aperiodic modules are completely inhomogeneous.

2.5. *Let M be a kQ-module, where Q is tame. Then M is aperiodic if and only if $\gamma(M) = 0$.*

PROOF. First, let us assume that M is not aperiodic. Let N be an indecomposable module of τ-period p such that the direct sum $M' = \bigoplus_{i=0}^{p-1} \tau^i N$ is a direct summand of M. It is sufficient to construct an element f of $\mathcal{O}(M')_1$ which does not belong to $\mathcal{N}(M')$. Since N is τ-periodic with τ-period p, it follows that $\tau^- M'$ is isomorphic to M' (the direct summands of $\tau^- M'$ are the same as those of M', but cyclically shifted). Let $f\colon \tau^- M' \to M'$ be an isomorphism. Then $\tau^{-t} f$ is an isomorphism, for all natural numbers t. The t-fold power f^{*t} of f is just the composition of these isomorphisms and therefore non-zero, for all t. Thus, we have shown that $\gamma(M) > 0$.

Conversely, assume that $\gamma(M) > 0$. Write $M = P \oplus R \oplus I$ with P preprojective, R regular and I preinjective. According to (2.3), we have $\gamma(R) = \gamma(M)$, thus there exists an element $f\colon \tau^- R \to R$ which does not belong to $\mathcal{N}(R)$. Write $R = \bigoplus_{i=1}^m R_i$ with indecomposable modules R_i. Any regular module is τ-periodic, thus there is some integer $q \geq m$ such that $\tau^q R$ is isomorphic to R. Consider the q-fold power f^{*q} of f in $\mathcal{O}(R)$. We may consider f^{*q} as a homomorphism $R \to R$. If this endomorphism f^{*q} of R belongs to the radical of $\operatorname{End}(R)$, then it is nilpotent, and then also the corresponding power of f^{*q} vanishes in $\mathcal{O}(R)$. Since we assume that f does not belong to $\mathcal{N}(R)$, it follows that f^{*q} does not belong to the radical of $\operatorname{End}(R)$. Recall that f^{*q} has a factorization of the following kind:

$$R = \tau^{-q} R \to \tau^{-q+1} R \to \cdots \to \tau^- R \to R,$$

It follows that there is a sequence of numbers $i(j)$ with $1 \leq i(j) \leq m$ and $0 \leq j \leq q$, such that there are isomorphisms

$$\tau^{-q}R_{i(q)} \to \tau^{-q+1}R_{i(q-1)} \to \cdots \to \tau^{-}R_{i(1)} \to R_{i(0)}.$$

Since $q \geq m$, we deal with at least $m+1$ modules $R_{i(j)}$, thus at least two of these modules have to be equal, say $R_{i(j)}$ and $R_{i(j+p)}$, where $0 \leq j < j+p \leq q$. Let $N = R_{i(j)}$. Then, for $1 \leq t \leq p$, we apply τ^{j+t} to the isomorphism

$$\tau^{-j-t}R_{i(j+t)} \to \tau^{-j}R_{i(j)} = \tau^{-j}N$$

and obtain an isomorphism

$$R_{i(j+t)} \to \tau^{t}N.$$

Since $R_{i(j+p)} = R_{i(j)} = N$, we see that N is of τ-period p. Of course, we may assume that p is chosen minimal, thus the modules $R_{i(j)}, R_{i(j+1)}, \ldots R_{i(j+p-1)}$ are pairwise different, and their direct sum is a direct summand of R. This shows that $\bigoplus_{t=0}^{p-1} \tau^{t}N$ is isomorphic to a direct summand of R, and thus of M. Altogether, we have shown that M is not aperiodic.

2.6. *If H is a homogeneous kQ-module, then $\mathcal{O}(H)_1 = \mathrm{Hom}(\tau^{-}H, H)$ can be identified with the set* $\mathrm{End}(H)$ *of all endomorphisms of H such that, under this identification, the set $\mathcal{N}(H)$ corresponds to the set of nilpotent endomorphisms. In particular, the variety $\mathcal{N}(H)$ is irreducible.*

REMARK. If one writes $\mathcal{O}(H)_1 = \mathrm{End}(H)$, one has to be aware that this is an equality only of additive groups (or vector spaces), but that the multiplicative structures are different: $\mathrm{End}(H)$ is a ring, whereas the $*$-multiplication of elements from $\mathcal{O}(H)_1$ yields elements in $\mathcal{O}(H)_2$.

PROOF OF 2.6. Consider the full subcategory of all homogeneous kQ-modules. On this subcategory, the Auslander-Reiten translation is the identity functor. It follows that for any homogeneous module H, the orbit ring $\mathcal{O}(H)$ is the polynomial ring $\mathrm{End}(H)[t]$ in one commuting indeterminate t with coefficients in $\mathrm{End}(H)$, and $\mathcal{O}(H)_1 = \mathrm{End}(H) \cdot t$. Of course, given any ring Λ and its polynomial ring $\Lambda[t]$, an element $c \in \Lambda$ is nilpotent if and only if $c \cdot t$ is a nilpotent element of $\Lambda[t]$.

Recall that the set $\mathcal{N}$ of nilpotent endomorphisms of a vector space V is irreducible: let V be n-dimensional; then $\mathcal{N}$ is just the closure of the orbit of any nilpotent endomorphism of V of nilpotency index n under conjugation by $\mathrm{GL}(V)$. This completes the proof.

2.6′. *Let H be a homogeneous kQ-module. Then $\gamma(H)$ is the number of indecomposable direct summands in a direct decomposition of H.*

PROOF. Write $H = \bigoplus_{i=1}^{t} H_i^{m_i}$, with pairwise non-isomorphic indecomposable modules H_i and natural numbers m_i. Let E be the endomorphism ring of H and J its radical. Note that E/J is the product of t rings $\overline{E}_i$, where $\overline{E}_i$ is the ring of all $(m_i \times m_i)$-matrices over k. Of course, an element of E is nilpotent if and only if its residue class in all the factor rings $\overline{E}_i$ is nilpotent. However, the set of nilpotent matrices in the space of all $(m_i \times m_i)$-matrices has codimension m_i, thus the set of nilpotent elements of E has codimension $\sum_{i=1}^{t} m_i$ in E.

2.6″. *If H is a homogeneous kQ-module, and the direct sum of pairwise non-isomorphic modules, then $\mathcal{N}(H)$ is an affine space.*

PROOF. According to (6), the set $\mathcal{N}(H)$ can be identified with the set of nilpotent endomorphisms of H, thus with the radical of $\operatorname{End}(H)$.

2.7. *Let M be aperiodic and let H be homogeneous. Then $\mathcal{N}(M \oplus H)$ is irreducible.*

PROOF. As in the proof of (2.3), we decompose $M = P \oplus R \oplus I$ with P preprojective, R regular and I preinjective. The fact that M, thus R, is aperiodic means that $\mathcal{N}(R) = \operatorname{Hom}(\tau^{-}R, R)$. Thus, we see that $(M \oplus H)$ is the product of an affine space and $\mathcal{N}(H)$. However, $\mathcal{N}(H)$ is irreducible, according to (6).

We need the precise structure of the homogeneous kQ-modules. Recall that a *partition* is a sequence $\lambda = (\lambda_1, \lambda_2, \ldots, \lambda_s)$ of natural numbers $\lambda_1 \geq \lambda_2 \geq \cdots \geq \lambda_s \geq 1$, and we write $|\lambda| = \sum_i \lambda_i$; the number $s = l(\lambda)$ of parts is called the *length* of λ. We denote by $\mathcal{P}$ the set of all partitions. Given a function $\mu \colon \mathcal{P} \to \mathbb{N}_0$ with finite support, let $|\mu| = \sum \mu(\lambda)$ and let $||\mu|| = \sum \mu(\lambda)|\lambda|$. The function μ is called *regular* provided $\mu(\lambda) \neq 0$ implies that $l(\lambda) = 1$.

The full subcategory of all homogeneous modules is an abelian category, and the simple objects in this category are said to be the *simple homogeneous* modules (the simple homogeneous modules are those non-zero homogeneous modules which have no non-zero proper regular submodule; note that they are not necessarily simple kQ-modules). For any simple homogeneous module S and any $n \in \mathbb{N}_1$, there exists (up to isomorphism) a unique indecomposable module $S[n]$ which has a filtration of length n such that all the factors are isomorphic to S. In this way, one obtains all the indecomposable homogeneous modules. Given a partition $\lambda = (\lambda_1, \lambda_2, \ldots, \lambda_s)$, and a simple homogeneous module S, let

$$S[\lambda] = \bigoplus\nolimits_i S[\lambda_i],$$

we obtain in this way all the homogeneous kQ-modules which have a filtration where all factors are isomorphic to S.

An arbitrary homogeneous module is of the form $H = \bigoplus_{j=1}^{t} S_j[\lambda^{(j)}]$, where $S_1, \ldots, S_t$ are pairwise non-isomorphic simple homogeneous modules and $\lambda^{(j)}$ is a partition, for $1 \leq j \leq t$. For any partition λ, let $\mu(\lambda)$ denote the number of indices j with $\lambda^{(j)} = \lambda$. We obtain in this way a function $\mu \colon \mathcal{P} \to \mathbb{N}_0$ with finite support which we call the *type* of H.

2.8. *Let M be completely inhomogeneous, let H be homogeneous of type μ. Then*

$$\gamma(M \oplus H) = \gamma(M) + \sum_{\lambda \in \mathcal{P}} \mu(\lambda) \cdot l(\lambda).$$

PROOF. Since $l(\lambda)$ denotes the number of parts of λ, this is the number of summands in the decomposition $S[\lambda] = \bigoplus_i S[\lambda_i]$. Also, $\mu(\lambda)$ denotes the number of indices j with $\lambda^{(j)} = \lambda$. Thus $\sum_{\lambda\in\mathcal{P}} \mu(\lambda)\cdot l(\lambda)$ denotes the number of indecomposable direct summands in the decomposition $H = \bigoplus_j \bigoplus_i S_j[\lambda_i^{(j)}]$ and therefore it is equal to $\gamma(H)$, according to (2.6′). The assertion now follows from (2.4′).

3. The dimension of endomorphism rings and Ext-groups.

3.1. Proposition. *Let M be completely inhomogeneous, let $\mu\colon \mathcal{P} \to \mathbb{N}_0$ be a function with finite support. There exist numbers $e^0(M,\mu)$ and $e^1(M,\mu)$ with the following property: For H homogeneous of type μ, the dimension of* $\operatorname{End}(M\oplus H)$ *is equal to $e^0(M,\mu)$ and the dimension of* $\operatorname{Ext}^1(M\oplus H, M\oplus H)$ *is equal to $e^1(M,\mu)$.*

REMARK. We are going to present explicit formulae for $e^0(M,\mu)$ and $e^1(M,\mu)$. Recall that the simple homogeneous modules all have the same dimension vector, say $\mathbf{h}$. Let $\mathbf{dim}(M,\mu) = \mathbf{dim}\, M + ||\mu||\cdot\mathbf{h}$: thus $\mathbf{dim}(M\oplus H) = \mathbf{dim}(M,\mu)$ for all homogeneous kQ-modules H of type μ. The dimension vectors of kQ-modules may be considered as elements of the Grothendieck group $K_0(kQ)$ of all finite length modules modulo exact sequences. We use the bilinear form $\langle -,-\rangle$ on $K_0(kQ)$ defined by

$$\langle \mathbf{dim}\, N_1, \mathbf{dim}\, N_2, \rangle = \dim \operatorname{Hom}(N_1,N_2) - \dim \operatorname{Ext}^1(N_1,N_2)$$

for kQ-modules N_1, N_2, and we write $q(\mathbf{d}) = \langle \mathbf{d},\mathbf{d}\rangle$; of course, we obtain in this way a quadratic form q on $K_0(kQ)$.

THE FORMULA FOR $e^0(M,\mu)$.. Let $M = P\oplus R\oplus I$, where P is preprojective, R regular and I preinjective:

$$\begin{aligned} e^0(M,\mu) = \dim\operatorname{End}(M\oplus H) = &\dim\operatorname{End}(M) \\ &+ ||\mu||\langle \mathbf{dim}\, P, \mathbf{h}\rangle + ||\mu||\langle \mathbf{h}, \mathbf{dim}\, I\rangle \\ &+ \sum_{\lambda\in\mathcal{P}} \Big(\mu(\lambda)\sum_i (2i-1)\lambda_i\Big). \end{aligned}$$

The formula shows that the dimension of $\operatorname{End}(M\oplus H)$ only depends on the isomorphism class of M, P, I and on μ. But, of course, M determines uniquely the isomorphism classes of P and I.

THE FORMULA FOR $e^1(M,\mu)$. We use the definition of the bilinear form $\langle -,-\rangle$:

$$e^1(M,\mu) = \dim\operatorname{Ext}^1(M\oplus H, M\oplus H) = e^0(M,\mu) - q(\mathbf{dim}(M,\mu)).$$

PROOF OF THE FORMULA FOR $e^0(M,\mu)$. Since H is of type μ, the dimension vector of H is $||\mu||\cdot\mathbf{h}$. The dimension e of $\operatorname{End}(M\oplus H)$ is the sum of the dimensions of $\operatorname{End}(M)$, $\operatorname{Hom}(M,H)$, $\operatorname{Hom}(H,M)$ and $\operatorname{End}(H)$ and we will show that these numbers are given by the four summands on the right side of the formula.

First, let us consider $\mathrm{Hom}(M,H)$ and $\mathrm{Hom}(H,M)$. Note that $\mathrm{Hom}(R\oplus I,H)=0=\mathrm{Ext}^1(P,H)$, thus

$$\dim \mathrm{Hom}(M,H)=\dim \mathrm{Hom}(P,H)=\langle \mathbf{dim}\,P,\mathbf{dim}\,H\rangle=\langle \mathbf{dim}\,P,||\mu||\cdot \mathbf{h}\rangle.$$

Similarly, $\mathrm{Hom}(H,P\oplus R)=0=\mathrm{Ext}^1(H,I)$ yield

$$\dim \mathrm{Hom}(H,M)=\langle \mathbf{dim}\,H,\mathbf{dim}\,I\rangle=\langle ||\mu||\cdot \mathbf{h},\mathbf{dim}\,I\rangle.$$

Finally, we have to calculate the dimension of $\mathrm{End}(H)$. Let $H=\bigoplus_{j=1}^t H_j$ and $H_j=S_j[\lambda^{(j)}]$, where $S_1,\dots,S_t$ are pairwise non-isomorphic simple homogeneous modules and $\lambda^{(j)}$ are partitions. The endomorphism ring of H is the product of the endomorphism rings of the modules H_j, and the k-dimension of $\mathrm{End}(H_j)$ is $\sum_i(2i-1)\lambda_i^{(j)}$. Since the type of H is μ, the number of indices j with $\lambda^{(j)}=\lambda$ is just $\mu(\lambda)$, for any partition λ, thus

$$\dim \mathrm{End}(H)=\sum_{\lambda\in\mathcal{P}}\Big(\mu(\lambda)\sum_i(2i-1)\lambda_i\Big).$$

This completes the proof of the formula, and therefore of the Proposition.

4. A stratification of $\operatorname{mod} kQ$.

We want to present a stratification of the category of all kQ-modules, or better stratifications of the affine spaces $\mathcal{R}(Q;V)$, where V is a finite dimensional Q_0-graded vector space V. The strata $\Xi(M,\mu)$ will be indexed by pairs (M,μ) where M is a completely inhomogeneous kQ-module and $\mu\colon \mathcal{P}\to\mathbb{N}_0$ has finite support. Given such a pair (M,μ), let $\mathbf{dim}(M,\mu)=\mathbf{dim}\,M+||\mu||\cdot\mathbf{h}$, of course this is the dimension vector of all kQ-modules of the form $M\oplus H$ where H is homogeneous and of type μ.

Let M be a completely inhomogeneous kQ-module and assume that $\mu\colon\mathcal{P}\to\mathbb{N}_0$ has finite support. Let V be a Q_0-graded vector space with dimension vector $\mathbf{dim}(M,\mu)$. Let $\Xi(M,\mu)$ be the set of all elements (V,x) in $\mathcal{R}(Q;V)$ which are isomorphic to a module of the form $M\oplus H$ where H is homogeneous and of type μ.

4.1. Let V be a Q_0-graded vector space with dimension vector $\mathbf{d}$. Then $\mathcal{R}(Q;V)$ *is the disjoint union of the sets* $\Xi(M,\mu)$ *with* $\mathbf{dim}(M,\mu)=\mathbf{d}$. This decomposition is a refinement of the sheet decomposition: Recall that the *sheets* $\mathcal{R}(Q;V)_e$ of $\mathcal{R}(Q;V)$ are the subsets of all elements (V,x) such that the dimension of $\mathrm{End}(V,x)$ is just e (here, e is a natural number). Namely, according to (3.1), $\Xi(M,\mu)$ is a subset of $\mathcal{R}(Q;V)_{e^0(M,\mu)}$.

4.2. *For any dimension vector* $\mathbf{d}$, *there are only finitely many pairs* (M,μ) *with* $\mathbf{dim}(M,\mu)=\mathbf{d}$.

PROOF. Let Q be a quiver with $n+1$ vertices, thus dimension vectors are $(n+1)$-tuples. Let us assume that the pair (M,μ) satisfies $\mathbf{dim}(M,\mu) = \mathbf{d} = (d_0,\ldots,d_n)$, and let $d = \sum_i d_i$. If λ is a partition such that $\mu(\lambda) \neq 0$, then $|\lambda| \le d$ and $\mu(\lambda) \le d$. Thus, there are only finitely many possibilities for μ. Also, we must have $\mathbf{dim}\, M \le \mathbf{d}$. By assumption, M is completely inhomogeneous. Note that the number of isomorphism classes of indecomposable modules N with fixed dimension vector and such that τN is not isomorphic to N is bounded by $n+1$, see [DR]. This shows that there are only finitely many possibilities for the indecomposable direct summands of M.

REMARK. The completely inhomogeneous modules M are given by combinatorial invariants as follows: It is sufficient to describe the set Γ'_0 of isomorphism classes of indecomposable modules N which are not homogeneous, since the completely inhomogeneous modules M are given by the functions $\Gamma'_0 \to \mathbb{N}_0$ with finite support. The set Γ'_0 consists of complete components of the Auslander-Reiten quiver of kQ and the number c of such components satisfies $2 \le c \le 5$. To be more precise: Always, there are the preprojective component and the preinjective component. The case $\widetilde{A}_1$ is the only case with $c=2$; for the remaining cases $\widetilde{A}_n$, one has $c=3$ or $c=4$. Finally, for $\widetilde{D}_n, \widetilde{E}_6, \widetilde{E}_7$ and $\widetilde{E}_8$, we have $c=5$, see [DR].

4.3. Theorem. *Let M be a completely inhomogeneous kQ-module, and assume that $\mu\colon \mathcal{P} \to \mathbb{N}_0$ has finite support. Let V be a Q_0-graded vector space with dimension vector $\mathbf{dim}(M,\mu)$. Then $\Xi(M,\mu)$ is a constructible subset of $\mathcal{R}(Q;V)$. It is irreducible, $G(V)$-invariant and its dimension is*

$$\dim G(V) - e^0(M,\mu) + |\mu|.$$

PROOF. By definition, $\Xi(M,\mu)$ is a union of isomorphism classes, thus of $G(V)$-orbits. All these orbits have the same dimension, namely

$$w = \dim G(V) - e^0(M,\mu)$$

The isomorphism classes of the homogeneous, simple regular modules can be parametrized by a cofinite subset $\mathbb{X}$ of the projective line $\mathbb{P}_1(k)$, say S_t with $t \in \mathbb{X}$. Let $n = |\mu|$. We take a sequence $\lambda^{(1)},\ldots,\lambda^{(n)}$ of partitions such that the partition λ occurs precisely $\mu(\lambda)$ times. Take pairwise different elements $t(1),\ldots,t(n)$ in $\mathbb{X}$ and consider the module

$$H(t(1),\ldots,t(n)) = \bigoplus_{j=1}^{n} S_{t(j)}[\lambda^{(j)}]$$

We obtain in this way all the isomorphism classes of modules H of type μ, and any isomorphism class contains only finitely many modules of the form $H(t(1),\ldots,t(n))$. More precisely: Let $\mathbb{X}^{[n]}$ be the set of n-tuples of pairwise different elements in $\mathbb{X}$. The construction of the modules $M \oplus H(t(1),\ldots,t(n))$

yields a (linear) embedding $\iota\colon \mathbb{X}^{[n]} \to \mathcal{R}(Q;V)$, the image is a subset of $\Xi(M,\mu)$ and is locally closed in $\mathcal{R}(Q;V)$. Since all the isomorphism classes of $\Xi(M,\mu)$ are obtained in this way, we see that $\Xi(M,\mu)$ is the image of the multiplication map $G(V)\times\mathbb{X}^{[n]} \to \mathcal{R}(Q;V)$. Since $G(V)\times\mathbb{X}^{[n]}$ is irreducible, also the image of this map is irreducible.

We denote by Σ_m the symmetric group on m letters. The product group $\Sigma = \prod \Sigma_{\mu(\lambda)}$ operates on $\mathbb{X}^{[n]}$ freely in such a way that the orbits are just the fibers of ι (the elements of the group $\Sigma_{\mu(\lambda)}$ permute the indices j with $\lambda^{(j)} = \lambda$). It follows that the dimension of $\Xi(M,\mu)$ is equal to the sum of the orbit dimension and the number $n = |\mu|$ of parameters.

4.4. We see that we may identify the sets $\mathbb{X}^{[n]}/\Sigma$ and $\Xi(M,\mu)/G(V)$.

5. The module category $\operatorname{mod}\mathcal{P}_k(Q)$.

5.1. Theorem. *Let M be a completely inhomogeneous kQ-module, and assume that $\mu\colon \mathcal{P}\to\mathbb{N}_0$ has finite support. Let V be a Q_0-graded vector space with dimension vector $\mathbf{dim}(M,\mu)$. Then $\pi^{-1}(\Xi(M,\mu))$ is a constructible subset of $\mathcal{R}_0(\overline{Q},\rho\,;V)$ of dimension*

$$\dim\mathcal{R}(Q;V) - \gamma(M) - \sum_{\lambda\in\mathcal{P}} \mu(\lambda)\cdot(l(\lambda)-1).$$

In case M is aperiodic, $\pi^{-1}(\Xi(M,\mu))$ is irreducible.

PROOF. According to [R4] one may identify the fiber $\pi^{-1}(V,x)$ of π at the element $(V,x)\in\mathcal{R}(Q;V)$ with $\mathcal{N}(V,x)$. Assume now (V,x) belongs to $\Xi(M,\mu)$. In this case, the dimension of $\mathcal{N}(V,x)$ is

$$\dim\mathcal{N}(V,x) = \mathrm{e}^1(M,\mu) - \gamma(M,\mu)$$

with

$$\gamma(M,\mu) = \gamma(M) + \sum_{\lambda\in\mathcal{P}} \mu(\lambda)\cdot l(\lambda),$$

according to (2.8). In particular, this fiber dimension is constant on $\Xi(M,\mu)$. The dimension of $\pi^{-1}(\Xi(M,\mu))$ therefore is

$$\dim G(V) - \mathrm{e}^0(M,\mu) + |\mu| + \mathrm{e}^1(M,\mu) - \gamma(M,\mu).$$

On the one hand, the definition of the bilinear form $\langle -,-\rangle$ yields

$$q(\mathbf{dim}(M,\mu)) = \mathrm{e}^0(M,\mu) - \mathrm{e}^1(M,\mu),$$

on the other hand, q has also the following geometrical interpretation:

$$q(\mathbf{dim}(M,\mu)) = \dim G(V) - \dim\mathcal{R}(Q;V),$$

and therefore

$$\dim\mathcal{R}(Q;V) = \dim G(V) - \mathrm{e}^0(M,\mu) + \mathrm{e}^1(M,\mu).$$

Thus, the dimension of $\pi^{-1}(\Xi(M,\mu)$ can be written in the form

$$\begin{aligned}
&\dim G(V) - e^0(M,\mu) + |\mu| + e^1(M,\mu) - \gamma(M,\mu)\\
&\quad = \dim \mathcal{R}(Q;V) + |\mu| - \gamma(M,\mu)\\
&\quad = \dim \mathcal{R}(Q;V) + |\mu| - \gamma(M) - \sum_{\lambda\in\mathcal{P}} \mu(\lambda)\cdot l(\lambda)\\
&\quad = \dim \mathcal{R}(Q;V) - \gamma(M) - \sum_{\lambda\in\mathcal{P}} \mu(\lambda)\cdot (l(\lambda)-1).
\end{aligned}$$

Thus we have obtained the dimension formula.

Finally, let us assume that M is aperiodic. In this case all the fibers are irreducible, according to (2.7). The fact that all the fibers are irreducible and have the same dimension implies that $\pi^{-1}(\Xi(M,\mu))$ is irreducible.

5.2. Corollary. *Let M be a completely inhomogeneous kQ-module, and assume that $\mu\colon \mathcal{P}\to\mathbb{N}_0$ has finite support. Let V be a Q_0-graded vector space with dimension vector* $\mathbf{dim}(M,\mu)$. *Then $\pi^{-1}(\Xi(M,\mu))$ is of dimension* $\dim\mathcal{R}(V,Q)$ *if and only if M is aperiodic and μ is regular. All other sets $\pi^{-1}(\Xi(M,\mu))$ have strictly smaller dimension.*

PROOF. Note that $\gamma(M)\geq 0$ and also $\mu(\lambda)\cdot(l(\lambda)-1)\geq 0$ for any partition λ. Thus the dimension of $\pi^{-1}(\Xi(M,\mu))$ is equal to the dimension of $\mathcal{R}(Q;V)$ if and only if first of all $\gamma(M)=0$ and second $\mu(\lambda)\cdot(l(\lambda)-1)=0$ for any partition λ. According to (2.5), we know that $\gamma(M)=0$ just means that M is aperiodic. On the other hand, the partitions λ with $l(\lambda)=1$ are just those with one part. This shows both assertions.

REMARK 1. If M is aperiodic and μ is regular, then $\mathcal{N}(M\oplus H)$ is even an affine space, namely a subspace of $\mathcal{O}(M\oplus H)_1$ of codimension r, according to (2.6″).

REMARK 2. If M is not aperiodic, then $\Xi(M,\mu)$ may be reducible. As a typical example, consider the case of the affine quiver Q of type $\widetilde{A}_{12}$

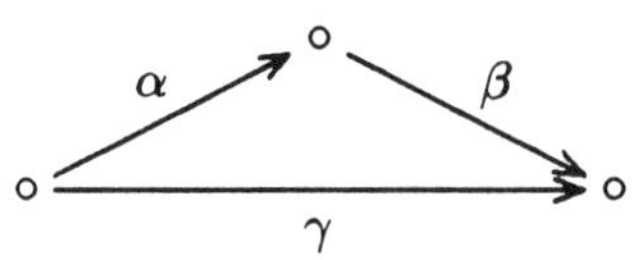

and the module $M=M_1\oplus M_2$, where M_1 is the simple module with dimension vector $[0\ 1\ 0]$ and $M_2=\tau M_1$, thus M_2 is indecomposable and has dimension vector $[1\ 0\ 1]$. Let V be the Q_0-graded vector space $V=k\times k\times k$, thus $\mathcal{R}(Q;V)=k^3$. We have

$$\Xi(M,0)=\{(V,x)\mid x_\alpha=x_\beta=0, x_\gamma\neq 0\},$$

and

$$\begin{aligned}\pi^{-1}(\Xi(M,0)) &= \{(V,x,\xi) \mid x_\alpha = x_\beta = \xi_\gamma = 0, x_\gamma \neq 0, x_\gamma\xi_\beta\xi_\alpha = 0\} \\ &= \{(V,x,\xi) \mid x_\alpha = x_\beta = \xi_\alpha = \xi_\gamma = 0,\ x_\gamma \neq 0\} \\ &\quad \cup \{(V,x,\xi) \mid x_\alpha = x_\beta = \xi_\beta = \xi_\gamma = 0,\ x_\gamma \neq 0\},\end{aligned}$$

thus reducible.

As an immediate consequence of Theorem (5.1) and Corollary (5.2) we obtain:

5.3. Corollary (Lusztig). *Let V be a Q_0-graded vector space. The irreducible components of $\mathcal{R}_0(\overline{Q},\rho;V)$ are just the closures of the sets $\pi^{-1}(\Xi(M,\mu))$, where M is an aperiodic kQ-module and $\mu\colon \mathcal{P} \to \mathbb{N}_0$ is a regular function with* $\mathbf{dim}(M,\mu) = \mathbf{dim}\,V$.

PROOF. Lusztig has shown in [L1] that all irreducible components of $\overline{\mathcal{R}} = \mathcal{R}_0(\overline{Q},\rho;V)$ have dimension equal to $r = \dim \mathcal{R}(Q;V)$. If M is aperiodic and μ is regular, then we know that $\pi^{-1}(\Xi(M,\mu))$ is irreducible and has dimension r, thus its closure is an irreducible component of $\overline{\mathcal{R}}$. Conversely, let Θ be an irreducible component of $\overline{\mathcal{R}}$. Since $\overline{\mathcal{R}}$ is the disjoint union of a finite number of sets of the form $\pi^{-1}(\Xi(M',\mu'))$, it follows that Θ is the disjoint union of a finite number of sets $\Theta \cap \pi^{-1}(\Xi(M',\mu'))$. All these sets are constructible, thus one of them must be of dimension r; and since Θ is irreducible, there is precisely one pair (M',μ') such that $\Theta \cap \pi^{-1}(\Xi(M',\mu'))$ has dimension r. In particular, $\pi^{-1}(\Xi(M',\mu'))$ itself has to have dimension r, thus M' is aperiodic and μ' regular. Also, Θ has to be the closure of $\pi^{-1}(\Xi(M',\mu'))$. This completes the proof.

REMARK. A regular function $\mu\colon \mathcal{P} \to \mathbb{N}_0$ with finite support can be codified by a single partition, say $\overline{\mu}$; namely, let $\overline{\mu}$ be the partition which has $\mu((t))$ parts equal to t, for $t \in \mathbb{N}_1$; it is a partition of $||\mu||$. This shows that the irreducible components of $\mathcal{R}_0(\overline{Q},\rho;V)$ can be indexed by pairs (M,λ), where M is an aperiodic kQ-module and λ is a partition.

References

[DR] V. Dlab, C. M. Ringel: Indecomposable representations of graphs and algebras. Mem. Amer. Math. Soc. 173 (1976).

[L1] G. Lusztig: Quivers, perverse sheaves, and quantized enveloping algebras. J. Amer. Math. Soc. 4 (1991), 365-421.

[L2] G. Lusztig: Affine quivers and canonical bases. Publ. Math. IHES 76 (1992), 111-163.

[R1] C. M. Ringel: Tame algebras and integral quadratic forms. Springer LNM 1099 (1984).

[R2] C. M. Ringel: The composition algebra of a cyclic quiver. Towards an explicit description of the quantum group of type $\tilde{A}_n$. Proceedings London Math. Soc. (3) 66 (1993), 507–537.

[R3] C. M. Ringel: From representations of quivers via Hall and Loewy algebras to quantum groups. Proceedings Novosibirsk Conference 1989. Contemporary Mathematics 131 (Part 2) (1992), 381-401.

[R4] C. M. Ringel: The preprojective algebra of a quiver. Proceedings ICRA VIII. Amer. Math. Soc. (To appear)

FAKULTÄT FÜR MATHEMATIK, UNIVERSITÄT BIELEFELD, POBOX 100 131, D-33 501 BIELEFELD, GERMANY

E-mail address: ringel@mathematik.uni-bielefeld.de

Contemporary Mathematics
Volume **229**, 1998

Representation types, Tits reduced quadratic forms and orbit problems for lattices over orders

Daniel Simson*
Faculty of Mathematics and Informatics
Nicholas Copernicus University
87-100 Toruń, ul. Chopina 12/18, Poland

Abstract

Let D be a complete discrete valuation domain with the residue field K. We study in the paper a class of subamalgams $\Lambda^\bullet$ (1.4) of tiled D-orders Λ (1.1) by means of an integral quadratic Tits form $q_{\Lambda^\bullet} : \mathbb{Z}^m \to \mathbb{Z}$ (1.5) and a family (1.6) of parabolic algebraic groups $G_{\Lambda^\bullet}(v)$ acting on irreducible algebraic K-varieties $\mathcal{M}_{\Lambda^\bullet}(v)$. Simple criteria for the finite lattice type and tame lattice type of $\Lambda^\bullet$ are given in terms of the Tits form $q_{\Lambda^\bullet}$, in terms of a two-peak poset $(I^{*+}_{\Lambda^\bullet}, \mathfrak{Z}_{\Lambda^\bullet})$ with zero-relations associated to $\Lambda^\bullet$ in (3.3), and in terms of $G_{\Lambda^\bullet}(v)$-orbits of $\mathcal{M}_{\Lambda^\bullet}(v)$ (see Theorems 1.7 and 1.8).

1. Introduction

Throughout K is an algebraically closed field. We denote by $\mathbb{Z}$ the ring of integers, and by $\mathbb{N}$ the subset of $\mathbb{Z}$ consisting of non-negative integers. Given a finite dimensional K-algebra R we denote by $\text{mod}(R)$ the category of finitely generated right R-modules. Let D be a complete discrete

---

*1991 Mathematics Subject Classification. Primary 16G30, 16G50, 15A21. Secondary 15A63, 14L30. E-mail address: simson@mat.uni.torun.pl
Partially supported by Polish KBN Grant 2 P0 3A 007 12.
This paper is in final form and no version of it will be submitted for publication elsewhere.

valuation domain which is a K-algebra such that $D/\mathfrak{p} \cong K$, where $\mathfrak{p}$ is the unique maximal ideal of D. We denote by $F = D_0$ the field of fractions of D.

We recall that a **D-order** Λ in a finite dimensional semisimple F-algebra C is a subring Λ of C which is a finitely generated free D-submodule of C and Λ contains an F-basis of C [7]. We denote by $\mathrm{latt}(\Lambda)$ the category of right Λ-lattices, that is, finitely generated right Λ-modules which are free as D-modules. It is well-known that any D-order is a noetherian semiperfect ring and the category $\mathrm{latt}(\Lambda)$ has the Krull-Schmidt property [7].

A D-order Λ is said to be of **finite lattice type** if the category $\mathrm{latt}(\Lambda)$ has finitely many isomorphism classes of indecomposable modules. Assume that K is an algebraically closed field and D is a complete discrete valuation domain which is a K-algebra such that $D/\mathfrak{p} \cong K$. A D-order Λ is said to be of **tame lattice type** if the indecomposable Λ-lattices of any fixed D-rank form a finite set of at most one-parameter families (see Section 4).

The orders of tame lattice type are divided into two classes: the orders of polynomial growth and the orders of non-polynomial growth (see [9], [31, Section 3], [35, Section 7] and Section 4). A D-order Λ is said to be of **wild lattice type** if there exists a representation embedding functor $\mathrm{mod}(R) \to \mathrm{latt}(\Lambda)$ for every finite dimensional K-algebra R (see [9], [31], [35]).

Our aim is to define for a class of D-orders $\Lambda^\bullet$, a positive integer $m_{\Lambda^\bullet}$, a Tits reduced quadratic form

$$q_{\Lambda^\bullet} : \mathbb{Z}^{m_{\Lambda^\bullet}} \longrightarrow \mathbb{Z}$$

and a system

$$* : G_{\Lambda^\bullet}(v) \times \mathcal{M}_{\Lambda^\bullet}(v) \longrightarrow \mathcal{M}_{\Lambda^\bullet}(v), \quad v \in \mathbb{N}^{m_{\Lambda^\bullet}}$$

of algebraic groups $G_{\Lambda^\bullet}(v)$ acting on irreducible algebraic K-varieties $\mathcal{M}_{\Lambda^\bullet}(v)$ in such a way that $\Lambda^\bullet$ is of tame lattice type if and only if $\dim G_{\Lambda^\bullet}(v) \geq \dim \mathcal{M}_{\Lambda^\bullet}(v)$, or equivalently, if and only if the quadratic form $q_{\Lambda^\bullet}$ is weakly non-negative, that is, $q_{\Lambda^\bullet}(v) \geq 0$ for all vectors $v \in \mathbb{N}^{m_{\Lambda^\bullet}}$.

In particular we do it for three-partite subamalgams of tiled D-orders $\Lambda^\bullet$ defined in [36] as follows. Denote by $\mathbb{M}_n(D)$ the full $n \times n$-matrix ring with coefficients in D. Suppose that $n, n_1, n_2 > 0$ and $n_3 \geq 0$ are natural numbers, and Λ is a tiled D-suborder of $\mathbb{M}_n(D)$ of the form

$$\Lambda = \left.\begin{pmatrix} D & {}_1D_2 & \dots & {}_1D_{n-1} & {}_1D_n \\ \mathfrak{p} & D & \dots & {}_2D_{n-1} & {}_2D_n \\ \vdots & \vdots & \ddots & \vdots & \vdots \\ \mathfrak{p} & \mathfrak{p} & \dots & D & {}_{n-1}D_n \\ \mathfrak{p} & \mathfrak{p} & \dots & \mathfrak{p} & D \end{pmatrix}\right\} n \tag{1.1}$$

such that

(a) ${}_iD_j$ is either D or $\mathfrak{p}$,

(b) Λ admits a three-partition

$$\Lambda = \left(\begin{array}{c|c|c} \Lambda_1 & \mathcal{X} & \mathbb{M}_{n_1}(D) \\ \hline \mathbb{M}_{n_3\times n_1}(\mathfrak{p}) & \Lambda_3 & \mathcal{Y} \\ \hline \mathbb{M}_{n_1}(\mathfrak{p}) & \mathbb{M}_{n_1\times n_3}(\mathfrak{p}) & \Lambda_2 \end{array}\right) \begin{array}{l} \}n_1 \\ \}n_3 \\ \}n_2 \end{array} \tag{1.2}$$

where $\Lambda_2 = \Lambda_1$, $n_1 = n_2$, $n_1 + n_2 + n_3 = n$ and Λ_3 is a hereditary $n_3 \times n_3$-matrix D-order

$$\Lambda_3 = \left.\begin{pmatrix} D & D & \dots & D & D \\ \mathfrak{p} & D & \dots & D & D \\ \vdots & \ddots & \ddots & \vdots & \vdots \\ \mathfrak{p} & \mathfrak{p} & \dots & D & D \\ \mathfrak{p} & \mathfrak{p} & \dots & \mathfrak{p} & D \end{pmatrix}\right\} n_3 \tag{1.3}$$

In particular ${}_iD_j = D$ holds in Λ for $1 \leq i \leq n_1$ and $n_1 + n_3 < j \leq n$.

Note that $1 = \varepsilon_1 + \varepsilon_3 + \varepsilon_2$, where ε_1, ε_3 and ε_2 are the matrix idempotents of Λ corresponding to the identity elements of Λ_1, Λ_3 and Λ_2, respectively. Following [36] by a **three-partite subamalgam** of Λ we shall mean the D-suborder

$$\Lambda^\bullet = \{\lambda = [\lambda_{ij}] \in \Lambda; \quad \varepsilon_1\lambda\varepsilon_1 - \varepsilon_2\lambda\varepsilon_2 \in \mathbb{M}_{n_1}(\mathfrak{p})\} \tag{1.4}$$

of Λ consisting of all matrices $\lambda = [\lambda_{ij}]$ of Λ such that the left upper corner $n_1 \times n_1$ submatrix $\varepsilon_1\lambda\varepsilon_1$ of λ is congruent modulo $\mathbb{M}_{n_1}(\mathfrak{p})$ to the right lower corner $n_1 \times n_1$ submatrix $\varepsilon_2\lambda\varepsilon_2$ of λ.

To any such a D-order $\Lambda^\bullet$ we have associated in [36] the Tits reduced quadratic form

$$q_{\Lambda^\bullet} : \mathbb{Z}^{n_1+2n_3+2} \longrightarrow \mathbb{Z} \tag{1.5}$$

in the indeterminates $x_*, x_+, x_1, \ldots, x_{n_1+n_3}, \overline{x}_{n_1+1}, \ldots, \overline{x}_{n_1+n_3}$ defined by the formula

$$\begin{aligned} q_{\Lambda^\bullet}(x_1, \ldots, x_{n_1+n_3}, \overline{x}_{n_1+1}, \ldots, \overline{x}_{n_1+n_3}, x_*, x_+) &= \\ = x_*^2 + x_+^2 + \sum_{j=1}^{n_1+n_3} x_j^2 + \sum_{j=n_1+1}^{n_1+n_3} \overline{x}_j^2 & \\ + \sum_{\substack{{}_iD_j=D \\ 1\le i<j\le n_1+n_3}} x_i x_j + \sum_{s<t} \overline{x}_s \overline{x}_t + \sum_{\substack{{}_tD_s=D \\ n_1<t\le n_1+n_3<s}} x_{s-n_1-n_3}\overline{x}_t & \\ - x_+(\sum_{j=1}^{n_1+n_3} x_j) - x_*(\sum_{j=1}^{n_1} x_j + \sum_{j=n_1+1}^{n_1+n_3} \overline{x}_j). & \end{aligned}$$

In [36] we have also associated with $\Lambda^\bullet$ a finite poset (i.e. a partially ordered set) $(I^{*+}_{\Lambda^\bullet}, \preceq)$ consisting of n_1+2n_3+2 elements and having precisely two maximal elements p_1, p_2, and a set $\mathfrak{Z}_{\Lambda^\bullet} \subseteq \{(i,j) \in I^{*+}_{\Lambda^\bullet} \times I^{*+}_{\Lambda^\bullet}; i \preceq j\}$ of zero-relations in $I^{*+}_{\Lambda^\bullet}$ (see Section 3 for precise definitions).

Further, we associate with $\Lambda^\bullet$ a family

$$(1.6) \qquad * : G_{\Lambda^\bullet}(v) \times \mathcal{M}_{\Lambda^\bullet}(v) \longrightarrow \mathcal{M}_{\Lambda^\bullet}(v)$$

$v \in \mathbb{N}^{I^{*+}_{\Lambda^\bullet}} = \mathbb{N}^{n_1+2n_3+2}$, of parabolic algebraic groups $G_{\Lambda^\bullet}(v)$ acting on irreducible algebraic K-varieties $\mathcal{M}_{\Lambda^\bullet}(v)$ defined as follows. The set $\mathcal{M}_{\Lambda^\bullet}(v)$ is the K-vector space of all partitioned K-matrices of the form (compare with [30])

$$A = \underbrace{\begin{array}{|c|c|c|c|} \hline A_{1p_1} & A_{2p_1} & \cdots & A_{mp_1} \\ \hline A_{1p_2} & A_{2p_2} & \cdots & A_{mp_2} \\ \hline \end{array}}_{v(1)\quad v(2)\quad \cdots \quad v(m)} \begin{array}{l} \}v(p_1) \\ \}v(p_2) \end{array}$$

with coefficients in K, where $m = n_1 + 2n_3$, $A_{ip} = 0$ if either $i \not\preceq p$, or $(i,p) \in \mathfrak{Z}_{\Lambda^\bullet}$, $p \in \{p_1, p_2\}$. We define $G_{\Lambda^\bullet}(v)$ to be the parabolic group

$$G_{\Lambda^\bullet}(v) = H_{\Lambda^\bullet}(v) \times \mathrm{Gl}(v(p_1), K) \times \mathrm{Gl}(v(p_2), K)$$

where $H_{\Lambda^\bullet}(v)$ is a group consisting of all K-matrices of the form

$$h = \begin{bmatrix} h_{11} & h_{12} & \ldots & h_{1m} \\ 0 & h_{22} & \ldots & h_{2m} \\ \vdots & \vdots & \ddots & \vdots \\ 0 & 0 & \ldots & h_{mm} \end{bmatrix} \in \mathrm{Gl}(v(1) + \cdots + v(m),\ K)$$

where $h_{ii} \in \mathrm{Gl}(v(i), K)$ and $h_{ij} = 0$ if either i and j are not comparable in $I^{*+}_{\Lambda^\bullet}$, or $(i,j) \in \mathfrak{Z}_{\Lambda^\bullet}$. We suppose here that the partial order $\preceq$ in $I^{*+}_{\Lambda^\bullet}$ is such that $i \prec j$ implies that $i < j$ in the natural order. We can always achieve this by a suitable permutation of the elements in I^{*+}.

The multiplication of two matrices $h = [h_{ij}]$ and $h' = [h'_{ij}]$ in $H_{\Lambda^\bullet}(v)$ is the matrix $h'' = [h''_{ij}]$, where

$$(h \cdot h')_{ij} = h''_{ij} = \begin{cases} \sum\limits_{i \preceq s \preceq j} h_{is} h'_{sj} & \text{if } i \preceq j \text{ and } (i,j) \notin \mathfrak{Z}_{\Lambda^\bullet} \\ 0 & \text{if } i \not\preceq j \text{ or } (i,j) \in \mathfrak{Z}_{\Lambda^\bullet}. \end{cases}$$

Note that the group $G_{\Lambda^\bullet}(v)$ is parabolic. In case the set $\mathfrak{Z}_{\Lambda^\bullet}$ is not empty the group $H_{\Lambda^\bullet}(v)$ is not a subgroup of the group $\mathrm{Gl}(v(1)+\cdots+v(m), K)$. The action (1.6) is defined by the formula

$$(h, (g_1, g_2)) * A = \begin{pmatrix} g_1 & 0 \\ 0 & g_2 \end{pmatrix} \cdot A \circ h^{-1}$$

where $h \in H_v^{(I,\mathfrak{Z})}$, $g_1 \in \mathrm{Gl}(v(p_1), K)$, $g_2 \in \mathrm{Gl}(v(p_2), K)$, $A \circ h^{-1} = A'$ is a partitioned matrix of the form shown above with $A'_{sp} = (Ah^{-1})_{sp}$, if $s \preceq p$ and $(s,p) \notin \mathfrak{Z}$, and $A'_{sp} = 0$, if $s \not\preceq p$ or $(s,p) \in \mathfrak{Z}$.

One of our main results is the following characterisation of three-partite subamalgam D-orders $\Lambda^\bullet$ of finite lattice type. It essentially depends on the main result of [36].

THEOREM 1.7. *Assume that D is a complete discrete valuation domain with the residue field K. Let Λ be a three-partite D-order* (1.2) *and let $\Lambda^\bullet$ be the subamalgam* (1.3) *of Λ, where $\Lambda_1 = \Lambda_2$, Λ_3 and n, $n_1 = n_2$, n_3 are as in introduction. Let $(I^{*+}_{\Lambda^\bullet}, \mathfrak{Z}_{\Lambda^\bullet})$ be the two-peak poset with zero-relations associated with $\Lambda^\bullet$ in* (3.3). *The following conditions are equivalent.*

(a) *The D-order $\Lambda^\bullet$ is of finite lattice type.*

(a′) *For any positive integer the set* $\mathrm{ind}_s(\mathrm{latt}(\Lambda^\bullet))$ *of the isomorphism classes of indecomposable $\Lambda^\bullet$-lattices of D-rank s is finite.*

(b) *The number of $G_{\Lambda^\bullet}(v)$-orbits in $\mathcal{M}_{\Lambda^\bullet}(v)$ with respect to the algebraic group action* (1.6) *is finite for all vectors $v \in \mathbb{N}^{n_1+2n_3+2}$.*

(c) $\dim G_{\Lambda^\bullet}(v) > \dim \mathcal{M}_{\Lambda^\bullet}(v)$ *for all $v \in \mathbb{N}^{n_1+2n_3+2}$, where* dim *is the variety dimension.*

(d) *The Tits reduced quadratic form $q_{\Lambda^\bullet} : \mathbb{Z}^{n_1+2n_3+2} \longrightarrow \mathbb{Z}$* (1.5) *is weakly positive, that is, $q_{\Lambda^\bullet}(v) > 0$ for all non-zero vectors $v \in \mathbb{N}^{n_1+2n_3+2}$.*

(e) *If* $R = KI^{*+}_{\Lambda^\bullet}/\mathfrak{z}_{\Lambda^\bullet}$ *is the factor algebra of the incidence* K*-algebra* $KI^{*+}_{\Lambda^\bullet}$ *of the poset* $I^{*+}_{\Lambda^\bullet}$ *(see* (3.3)*) modulo the ideal* $\mathfrak{z}_{\Lambda^\bullet}$ *of zero-relations (see* (2.1)*) then*

$$\dim_K \operatorname{Ext}^1_R(U_1, U_2) \leq 1$$

for any pair of modules U_1, U_2 *in* $\operatorname{mod}(R)$ *satisfying the following conditions:*

(i) $\operatorname{End}(U_1) \cong \operatorname{End}(U_2) \cong K$,

(ii) $\operatorname{Hom}_R(U_1, U_2) = 0$ *and* $\operatorname{Hom}_R(U_2, U_1) = 0$,

(iii) *the kernel of the projective cover* $P(U_j) \to U_j$ *of* U_j *in* $\operatorname{mod}(R)$ *is a semisimple projective* R*-module for* $j = 1, 2$.

(f) *The set* $\mathcal{R}^+_{\Lambda^\bullet} = \{v \in \mathbb{N}^{n_1+2n_3+2};\ q_{\Lambda^\bullet}(v) = 1\}$ *of positive roots of the quadratic form* $q_{\Lambda^\bullet} : \mathbb{Z}^{n_1+2n_3+2} \longrightarrow \mathbb{Z}$ (1.5) *is finite.*

(g) *The* D*-order* Λ_1 *is hereditary of the form* (1.3) *and the two-peak poset* $(I^{*+}_{\Lambda^\bullet}, \mathfrak{z}_{\Lambda^\bullet})$ *with zero-relations associated to* $\Lambda^\bullet$ *in* (3.3) *does not contain as a two-peak subposet with zero-relations any of the following eight forms:*

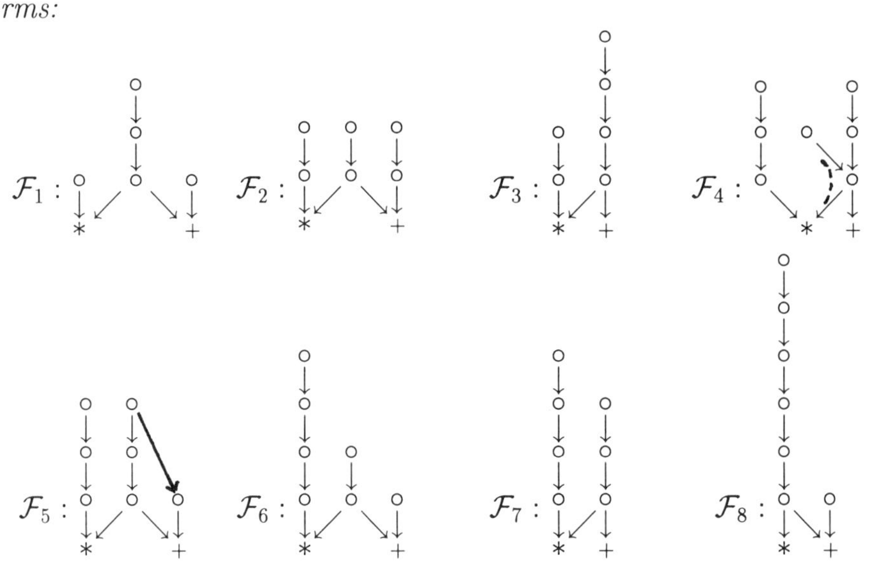

where the dotted line in $\mathcal{F}_4$ *means a zero-relation.*

We note that the equivalence (a)$\Leftrightarrow$(a′) is a version of the second Brauer-Thrall conjecture for lattices over orders (see [29, p.4], [39] and [40]).

A characterisation of three-partite subamalgam D-orders $\Lambda^\bullet$ of tame lattice type analogous to that one in Theorem 1.7 is the following result proved in Section 4 (see Theorem 4.1).

THEOREM 1.8. *Assume that K is an algebraically closed field and D is a complete discrete valuation domain which is a K-algebra such that $D/\mathfrak{p} \cong K$, where $\mathfrak{p}$ is the unique maximal ideal of D. Let Λ, $\Lambda^\bullet$, $q_{\Lambda^\bullet}$ and $* : G_{\Lambda^\bullet}(v) \times \mathcal{M}_{\Lambda^\bullet}(v) \to \mathcal{M}_{\Lambda^\bullet}(v)$, $v \in \mathbb{N}^{I^{*+}_{\Lambda^\bullet}} = \mathbb{N}^{n_1+2n_3+2}$ be as above. If the part $\mathcal{X}$ or the part $\mathcal{Y}$ of the D-order Λ in (1.2) consists of matrices with coefficients in $\mathfrak{p}$ then the following conditions are equivalent.*

(a) *The D-order $\Lambda^\bullet$ is of tame lattice type.*

(b) $\dim G_{\Lambda^\bullet}(v) \geq \dim \mathcal{M}_{\Lambda^\bullet}(v)$ *for every $v \in \mathbb{N}^{n_1+2n_3+2}$, where* dim *is the variety dimension.*

(c) *The Tits reduced quadratic form $q_{\Lambda^\bullet} : \mathbb{Z}^{n_1+2n_3+2} \to \mathbb{Z}$ is weakly non-negative, that is, $q_{\Lambda^\bullet}(v) \geq 0$ for all vectors $v \in \mathbb{N}^{n_1+2n_3+2}$.*

(d) *If $R = KI^{*+}_{\Lambda^\bullet}/\mathfrak{z}_{\Lambda^\bullet}$ (2.1) is the factor of the incidence K-algebra $KI^{*+}_{\Lambda^\bullet}$ of the poset $I^{*+}_{\Lambda^\bullet}$ (3.3) modulo the ideal $\mathfrak{z}_{\Lambda^\bullet}$ of zero-relations then*

$$\dim_K \operatorname{Ext}^1_R(U_1, U_2) \leq 2$$

for any pair of modules U_1, U_2 in mod(R) *satisfying the following conditions:*

(i) $\operatorname{End}(U_1) \cong \operatorname{End}(U_2) \cong K$,

(ii) $\operatorname{Hom}_R(U_1, U_2) = 0$ *and* $\operatorname{Hom}_R(U_2, U_1) = 0$,

(iii) *the kernel of the projective cover $P(U_j) \to U_j$ of U_j in* mod(R) *is a semisimple projective R-module for $j = 1, 2$.*

(e) *Either $n_3 = 0$ and the D-order Λ_1 in (1.2) does not contain minor D-suborders of one of the forms $\Delta_0, \Delta_1, \Delta_2, \Delta_3$ shown in Theorem 4.1, or else $n_3 \geq 1$, Λ_1 is hereditary of the form (1.3) and the three-partite subamalgam D-orders $\Lambda^\bullet$ and $rt(\Lambda)^\bullet$ do not contain three-partite minor D-suborders being dominated by any of the 17 three-partite subamalgam D-orders $\Omega^\bullet_1, \ldots, \Omega^\bullet_{17}$ of size at most 15×15 listed in* [37, Table 1].

In this case the structure of the Auslander-Reiten quiver of the category latt$(\Lambda^\bullet)$ can be described like in [36, Theorem 6.1] and the shape of it is presented in Section 6 of [36, Figure 1].

2. Preliminaries

In this section we recall from [36] basic definitions and facts we need throughout this paper. We shall use here the terminology and notation introduced in [36].

The study of the category lattices over D-orders reduces to a matrix problem or a class of modules over artin algebras (see [9], [12], [26], [2], [13] and [35]). For some classes of orders the study of lattices reduces to K-linear representations of partially ordered sets (abbr. **posets**), where K is the residue field of D (see [41], [29, Chapter 13], [34]). A similar reduction applies in the study of some categories of Cohen-Macaulay modules and categories of abelian groups (see [1], [3], [39], [40]).

Our principal idea in proving the main theorems of the paper is a reduction of the problem for lattices over three-partite subamalgams of tiled D-orders to a corresponding problem for K-linear socle projective representations of two-peak posets (that is, having exactly two maximal elements) with zero-relations and studied in [28], where $K = D/\mathfrak{p}$. Our reduction extends the reduction given in [32, Section 2] and will involve the reduction functors defined in [12] and [26], and the covering technique for bipartite stratified posets developed by the author in [28].

The reader is referred to [36] and to the author's survey article [34] for an elementary introduction and an explanation of the role of peak posets I, peak I-spaces and their relevance to amalgamation of D-orders and their representation type.

Throughout we shall denote by $(I; \preceq)$ a finite **poset**, that is, a finite partially ordered set $(I; \preceq)$ with the partial order $\preceq$. We shall write $i \prec j$ if $i \preceq j$ and $i \neq j$. For the sake of simplicity we write I instead of $(I, \preceq)$. We denote by $\max I$ the set of all maximal elements of I and I will be called an r-**peak poset** if $|\max I| = r$.

Given a poset I we denote by KI the incidence algebra of I [29], that is, the subalgebra of the full matrix algebra $\mathbb{M}_I(K)$ consisting of all $I \times I$ square matrices $\lambda = [\lambda_{pq}]_{p,q \in I}$ such that $\lambda_{pq} = 0$ if $p \not\preceq q$ in $(I; \preceq)$.

For $i \preceq j$ we denote by $e_{ij} \in KI$ the matrix having 1 at i-j-th position and zero elsewhere. Given j in I we denote by $e_j = e_{jj}$ the standard primitive idempotent of KI corresponding to j.

The algebra KI is basic and the standard matrix idempotents e_i, $i \in I$, form a complete set of primitive orthogonal idempotents of KI. Moreover, KI is of finite global dimension and the right socle of KI is isomorphic to a direct sum of copies of the right ideals e_pKI, $p \in \max I$, called the **right peak ideals** of KI (see [30]).

We shall denote by $\mathrm{mod}_{\mathrm{sp}}(KI)$ the category of **socle projective right KI-modules**, that is, the full subcategory of $\mathrm{mod}(KI)$ consisting of modules X such that the socle $\mathrm{soc}(X)$ of X is projective and isomorphic to a direct sum of copies of the right ideals e_pKI, $p \in \max I$.

We recall from [36] that a **zero-relation** in a poset $(I, \preceq)$ is a pair (i_0, j_0) of elements of I such that $i_0 \prec j_0$. A **set of zero-relations** in I is a set $\mathfrak{Z}$ satisfying the following two conditions:

(Z1) $\mathfrak{Z}$ consists of zero-relations (i_0, j_0) of I;

(Z2) If $(i_0, j_0) \in \mathfrak{Z}$ and $i_1 \preceq i_0 \preceq j_0 \preceq j_1$ then $(i_1, j_1) \in \mathfrak{Z}$.

A **right multipeak (or precisely r-peak) poset with zero-relations** is a pair $(I, \mathfrak{Z})$, where I is a poset, $r = |\max I|$, and $\mathfrak{Z}$ is a set of zero-relations satisfying the following condition:

(Z3) For every $i \in I \backslash \max I$ there exists $p \in \max I$ such that $(i, p) \notin \mathfrak{Z}$.

In case the set $\mathfrak{Z}$ is empty we shall write I instead of $(I, \mathfrak{Z})$.

A right multipeak poset $(I', \mathfrak{Z}')$ with zero-relations is said to be a peak subposet of $(I, \mathfrak{Z})$ if I' is a subposet of I, $\mathfrak{Z}'$ is the restriction of $\mathfrak{Z}$ to I' and $\max I' = I' \cap (\max I)$.

Let K be a field and let $(I, \mathfrak{Z})$ be a right r-peak poset with zero-relations. The **incidence K-algebra** of $(I, \mathfrak{Z})$ is the K-algebra

$$K(I, \mathfrak{Z}) = \{\lambda = [\lambda_{ij}]_{i,j \in I} \in KI;\ \lambda_{ij} = 0, \text{ for } \ (i,j) \in \mathfrak{Z}\} \subseteq KI \tag{2.1}$$

consisting of all $I \times I$ square matrices $\lambda = [\lambda_{ij}]_{i,j \in I} \in \mathbb{M}_I(K)$ such that $\lambda_{ij} = 0$ if $i \not\preceq j$ in $(I; \preceq)$, or if $(i, j) \in \mathfrak{Z}$. The addition in $K(I, \mathfrak{Z})$ is the usual matrix addition, whereas the multiplication of two matrices $\lambda = [\lambda_{ij}]_{i,j \in I}$ and $\lambda' = [\lambda'_{ij}]_{i,j \in I}$ in $K(I, \mathfrak{Z})$ is the matrix $\lambda'' = [\lambda''_{ij}]_{i,j \in I}$, where

$$\lambda''_{ij} = \begin{cases} \sum\limits_{i \preceq s \preceq j} \lambda_{is}\lambda'_{sj} & \text{if } i \preceq j \text{and } \ (i,j) \notin \mathfrak{Z}, \\ 0 & \text{if } i \not\preceq j \text{ or } \ (i,j) \in \mathfrak{Z}. \end{cases}$$

Note that in case the set $\mathfrak{Z}$ is empty we get $KI = K(I, \mathfrak{Z})$.

It is easy to see that the incidence algebra $K(I, \mathfrak{Z})$ is basic and the standard matrix idempotents e_i, $i \in I$, form a complete set of primitive orthogonal idempotents of $K(I, \mathfrak{Z})$.

We shall denote by $\operatorname{mod}_{\mathrm{sp}} K(I, \mathfrak{Z})$ the category of **socle projective right $K(I, \mathfrak{Z})$-modules**, that is, the full subcategory of $\operatorname{mod} K(I, \mathfrak{Z})$ consisting of modules X such that the socle $\operatorname{soc}(X)$ of X is projective and isomorphic to a direct sum of copies of the right ideals $e_pK(I, \mathfrak{Z})$, $p \in \max I$.

It is easy to see that the algebra $K(I, \mathfrak{Z})$ is a factor K-algebra of KI modulo the ideal generated by all matrices $e_{ij} \in KI$ such that $(i, j) \in \mathfrak{Z}$. By [36, Lemma 2.4], the global dimension of $K(I, \mathfrak{Z})$ is finite and the right socle of the algebra $K(I, \mathfrak{Z})$ is isomorphic to a direct sum of copies of the right ideals $e_pK(I,\mathfrak{Z})$, $p \in \max I$, called the **right peak ideals** of $K(I,\mathfrak{Z})$.

It is often useful to deal with filtered forms of socle projective K-linear representations of $(I,\mathfrak{Z})$. For this purpose we have introduced in [36] the following definition extending that one given in [30] for posets without zero-relations.

DEFINITION 2.2. *Let K be a field and let $(I,\mathfrak{Z})$ be a right multipeak poset with zero-relations. A* **peak $(I,\mathfrak{Z})$-space** *(or* **a filtered socle projective representation** *of $(I,\mathfrak{Z})$) over the field K is the system $\mathbf{M} = (M_j)_{j\in I}$ of finite dimensional K-vector spaces M_j satisfying the following four conditions.*

(a) *For any $j \in I$ the K-space M_j is a K-subspace of the space*

$$M^\bullet = \bigoplus_{p\in\max I} M_p.$$

(b) *The inclusion $M_p \subseteq M^\bullet$ is the usual p-coordinate embedding for any $p \in \max I$.*

(c) *$\pi_j(M_i) \subseteq M_j$ for all $i \prec j$ in I, where $\pi_j : M^\bullet \to M^\bullet$ is the composed K-linear endomorphism*

$$M^\bullet \xrightarrow{\pi'_j} \bigoplus_{\substack{j\preceq p\in\max I\\ (i,p)\notin\mathfrak{Z}}} M_p \hookrightarrow M^\bullet$$

of $M^\bullet$ and π'_j is the direct summand projection.

(d) *If $p \in \max I$ and either $i \npreceq p$ or $i \prec p$ and $(i,p) \in \mathfrak{Z}$ then $\pi_p(M_i) = 0$.*

A morphism $f : \mathbf{M} \to \mathbf{M}'$ from $\mathbf{M}$ to $\mathbf{M}'$ is a system $f = (f_p)_{p\in\max I}$ of K-linear maps $f_p : M_p \to M'_p$, $p \in \max I$, such that $(\bigoplus_{p\in\max I} f_p)(M_j) \subseteq M'_j$ for all $j \in I$. □

It follows from the conditions (c) and (d) of Definition 2.2 that $\pi_j(M_i) = 0$ if the relation $i \prec j$ holds in I and $(i,j) \in \mathfrak{Z}$.

We denote by $(I,\mathfrak{Z})$-spr the **category of peak I-spaces** (or filtered socle projective representations of $(I,\mathfrak{Z})$) over the field K. The direct sum and the indecomposablity in the category $(I,\mathfrak{Z})$-spr are defined in an obvious way.

In case the set $\mathfrak{Z}$ is empty the category $(I,\mathfrak{Z})$-spr is the category I-spr of peak I-spaces (or socle projective K-linear representations of I) introduced in [30].

It is easy to see that $(I, \mathfrak{Z})$-spr is an additive category with the finite unique decomposition property [29, Section 1.1], and the K-linear functor

$$\boldsymbol{\rho} : (I, \mathfrak{Z})\text{-spr} \xrightarrow{\simeq} \operatorname{mod}_{\mathrm{sp}} K(I, \mathfrak{Z}) \tag{2.3}$$

$\mathbf{M} \mapsto \widehat{\mathbf{M}} = (M_j;\ {}_j\pi_i)_{i \prec j}$, is an equivalence of categories, where ${}_j\pi_i : M_i \to M_j$ is a unique K-linear map making the diagram

$$\begin{array}{ccc} M_i & \subseteq & M^\bullet \\ \downarrow {}_j\pi_i & & \downarrow \pi_j \\ M_j & \subseteq & M^\bullet \end{array}$$

commutative. By [24] and [36, Corollary 2.9], the category $(I, \mathfrak{Z})$-spr is an additive K-category, $(I, \mathfrak{Z})$-spr has Auslander-Reiten sequences, source maps, sink maps, enough relative projective objects and enough relative injective objects (see [24]).

Following [24] and [30, (3.1)] we associate to any r-peak poset $(I, \mathfrak{Z})$ with zero-relations the integral **Tits quadratic form** $q_{(I,\mathfrak{Z})} : \mathbb{Z}^I \to \mathbb{Z}$ defined by the formula

$$q_{(I,\mathfrak{Z})}(z) = \sum_{j \in I} z_j^2 + \sum_{\substack{i \prec j \notin \max I \\ (i,j) \notin \mathfrak{Z}}} z_i z_j - \sum_{p \in \max I} \sum_{\substack{i \prec p \\ (i,p) \notin \mathfrak{Z}}} z_i z_p. \tag{2.4}$$

We recall from [36, Section 3] that a module X in $\operatorname{mod} K(I, \mathfrak{Z})$ is said to be **prinjective** if and only if there exists a short exact sequence

$$0 \longrightarrow P_1 \longrightarrow P_0 \longrightarrow X \longrightarrow 0$$

in $\operatorname{mod} K(I, \mathfrak{Z}))$, where P_0, P_1 are projective $K(I, \mathfrak{Z})$-modules and P_1 is semisimple of the form $P_1 = \bigoplus_{p \in \max I} (e_p K(I, \mathfrak{Z}))^{t_p}$, $t_p \geq 0$.

It is easy to see that the category $\operatorname{prin} K(I, \mathfrak{Z})$ of prinjective right $K(I, \mathfrak{Z})$-modules coincides with that one introduced in [24] (see also [29, Proposition 11.32]) and the study of socle projective $K(I, \mathfrak{Z})$-modules reduces to the study of the category $\operatorname{prin} K(I, \mathfrak{Z})$ via the functor (3.2) defined below.

Given a module X in $\operatorname{mod} K(I, \mathfrak{Z})$ we define the **coordinate vector** $\mathbf{cdn}(X) \in \mathbb{N}^I$ by the following formula (see [27], [24], [30])

$$(\mathbf{cdn}(X))(j) = \begin{cases} \dim_K(X_j) & \text{for } j \in \max I, \\ \dim_K(\operatorname{top} X) e_j & \text{for } j \in I \setminus \max I. \end{cases} \tag{2.5}$$

We view $\mathbf{cdn}(X)$ as a map $\mathbf{cdn}(X) : I \to \mathbb{N}$. Note that the projective cover $P(X)$ of X has the form

$$P(X) = \bigoplus_{j \in I \setminus \max I} (e_j K(I, \mathfrak{Z}))^{(\mathbf{cdn} X)(j)}$$

if X is an indecomposable object in the category $\operatorname{prin} K(I, \mathfrak{Z})$ or in the category $(I, \mathfrak{Z})$-spr $\cong \operatorname{mod}_{\mathrm{sp}}(K(I, \mathfrak{Z}))$, and X is not isomorphic to any of the simple projective modules $e_{p_1} K(I, \mathfrak{Z}), \ldots, e_{p_r} K(I, \mathfrak{Z})$.

By [24] and [36, 3.2] there exists an **adjustment functor**

$$\text{(2.6)} \qquad \boldsymbol{\Theta}_I : \operatorname{prin} K(I, \mathfrak{Z}) \longrightarrow \operatorname{mod}_{\mathrm{sp}} K(I, \mathfrak{Z})$$

with the properties listed below.

PROPOSITION 2.7. (a) *The adjustment functor* $\boldsymbol{\Theta}_I$ (2.6) *is full dense and* $\operatorname{Ker} \boldsymbol{\Theta}_I = [e_j K(I \setminus \max I, \mathfrak{Z}^-); j \in I \setminus \max I]$, *that is,* $\operatorname{Ker} \boldsymbol{\Theta}_I$ *consists of all homomorphisms in* $\operatorname{prin} K(I, \mathfrak{Z})$ *having a factorisation through a direct sum of copies of the projective* $K(I \setminus \max I, \mathfrak{Z}^-)$*-modules* $e_j K(I \setminus \max I, \mathfrak{Z}^-)$, $j \in I \setminus \max I$.

(b) *If* X *is an indecomposable module in* $\operatorname{prin} K(I, \mathfrak{Z})$ *and* $\boldsymbol{\Theta}_I(X) \neq 0$ *then* $\mathbf{cdn}\, X = \mathbf{cdn}\, \boldsymbol{\Theta}_I(X)$.

(c) *The adjustment functor* $\boldsymbol{\Theta}_I$ *preserves and reflects finite representation type, tame representation type, the polynomial growth, and induces an equivalence of categories*

$$(I, \mathfrak{Z})\text{-spr} \cong \operatorname{prin} K(I, \mathfrak{Z}) / [e_j K(I \setminus \max I, \mathfrak{Z}^-);\ j \in I \setminus \max I]$$

Proof. The statements (a), (b) and the final part of (c) follow from [24] and [36, Corollary 3.3]. The first part of (c) is a consequence of the fact that the adjustment functor $\boldsymbol{\Theta}_I$ lowers and lifts almost parameterizing families in the sense of [35, Definition 6.9]. To see this apply [35, Theorem 6.10] to the algebra $R = K(I, \mathfrak{Z})$ equipped with the bipartition (3.7) defined in Section 3 below. Note that $\operatorname{prin}(R)$ is the category $\operatorname{mod}_{pr}^{pr}(R)$ of propartite R-modules and $(I, \mathfrak{Z})$-spr $\cong \operatorname{mod}_{\mathrm{sp}} K(I, \mathfrak{Z})$ is the category $\operatorname{mod}_{pr}(R)$ in the notation of [35]. □

Assume that $(I, \mathfrak{Z})$ is a multipeak poset with zero-relations. Following [35, Definition 6.2] the category $\operatorname{prin} K(I, \mathfrak{Z})$ is said to be of **tame representation type** (resp. $(I, \mathfrak{Z})$-spr is of tame representation type) if for every vector $v \in \mathbb{N}^I$ there exist functors

$$\widehat{N}^{(1)}, \ldots, \widehat{N}^{(r)} : \ \mathrm{ind}_1(\mathrm{mod}(K[t])) \longrightarrow \mathrm{mod}\, K(I, \mathfrak{Z})$$

forming an **almost parameterizing family** (see also [29, Definition 14.13]) for the full subcategory $\mathrm{ind}_v(\mathrm{prin}\, K(I, \mathfrak{Z})$ (resp. for $\mathrm{ind}_v((I, \mathfrak{Z})\text{-spr}))$ of $\mathrm{mod}\, K(I, \mathfrak{Z})$ consisting of representatives of the isomorphism classes of indecomposable objects X in the category $\mathrm{prin}\, K(I, \mathfrak{Z})$ (resp. in the category $(I, \mathfrak{Z})\text{-spr} \cong \mathrm{mod}_{\mathrm{sp}}\, K(I, \mathfrak{Z}))$ such that $\mathbf{cdn}\, X = v$.

This means that $\widehat{N}^{(j)} = (-) \otimes_{K[t]} N^{(j)}$ and $N^{(j)}$ is a $K[t]$-$K(I, \mathfrak{Z})$-bimodule which is finitely generated as a $K[t]$-module for $j = 1, \ldots, r$ and $\mathrm{Im}\, \widehat{N}^{(1)} \cup \ldots \cup \mathrm{Im}\, \widehat{N}^{(r)}$ contains up to isomorphism all but finitely many objects in $\mathrm{ind}_v(\mathrm{prin}\, K(I, \mathfrak{Z}))$ (resp. in $\mathrm{ind}_v((I, \mathfrak{Z}))\text{-spr}) \cong \mathrm{ind}_v(\mathrm{mod}_{\mathrm{sp}}\, K(I, \mathfrak{Z})$.

The reader is referred to [35, Section 6] and [29, Definition 14.13] for various definitions and discussions of tame representation type and wild representation type for the categories $\mathrm{prin}\, K(I, \mathfrak{Z})$ and $(I, \mathfrak{Z})\text{-spr}) \cong \mathrm{mod}_{\mathrm{sp}}\, K(I, \mathfrak{Z})$. In particular, it follows from [35, Lemma 6.4] that our definition of tameness is equivalent to that one given in [29, Definition 14.13] (see also the proof of [30, Proposition 2.4]).

Note that [35, Theorem 6.5], applied to the algebra $R = K(I, \mathfrak{Z})$ equipped with the bipartition (3.7) in Section 3, yields the tame-wild dichotomy for the categories $\mathrm{prin}\, K(I, \mathfrak{Z})$ and $(I, \mathfrak{Z}))\text{-spr}$.

Let us finish this section with a discussion of the relations between the representation type of a poset with zero-relations $(I, \mathfrak{Z})$ and of its peak subposet with zero-relations $(L, \mathfrak{Z}')$. Following [29, Section 17.5] let us define the commutative diagram of categories and K-linear functors

$$\begin{array}{ccc}
\mathrm{prin}\, K(L, \mathfrak{Z}') & \underset{\mathrm{res}_L}{\overset{\widehat{T}_L}{\rightleftarrows}} & \mathrm{prin}\, K(I, \mathfrak{Z}) \\
\Big\downarrow \boldsymbol{\Theta}'_L & & \Big\downarrow \boldsymbol{\Theta}'_I \\
(L, \mathfrak{Z}')\text{-spr} & \xrightarrow{T_L} & I\text{-spr}
\end{array} \tag{2.8}$$

by the formulas $\widehat{T}_L(X) = X \otimes_{K(L,\mathfrak{Z}')} (e_L K(I, \mathfrak{Z}) e_{L \cup I^-})$ and $\mathrm{res}_L(Y) = Y e_L$, where $I^- = I \setminus \max I$, and for any subset J of I we set $e_J = \sum_{i \in J} e_i$. Note that there is an algebra isomorphism $K(L, \mathfrak{Z}') \cong e_L K(I, \mathfrak{Z}) e_L$. The functors $\boldsymbol{\Theta}'_L$ and $\boldsymbol{\Theta}'_I$ are the compositions of the adjustment functor (2.6) with the inverse to the equivalence (2.3) applied to L and I, respectively.

The functor T_L is defined by attaching to $\mathbf{M} = (M_t)_{t \in L}$ in $(L, \mathfrak{Z}')$-spr the object $T_L(\mathbf{M}) = (\widehat{\mathbf{M}}_{\mathbf{i}})_{\mathbf{i} \in \mathbf{I}}$ in $(I, \mathfrak{Z})$-spr, where $\widehat{M}_p = M_p$ for $p \in \max L$,

and

$$\widehat{M}_j = \sum_{\substack{j \succeq t \in L \\ (t,j) \notin \mathfrak{Z}}} \pi_j(M_t) \subseteq \bigoplus_{\substack{j \preceq p \in \max I \\ (j,p) \notin \mathfrak{Z}}} M_p \subseteq M^{\bullet}$$

for $j \in L$.

The following result extends and corrects [20, Lemma 3].

PROPOSITION 2.9. *Assume that $(I, \mathfrak{Z})$ is a multipeak poset with zero-relations, $(L, \mathfrak{Z}')$ is a peak subposet with zero-relations of $(I, \mathfrak{Z})$ and $\mathrm{res}_L : \mathrm{mod}(KI) \to \mathrm{mod}(KL)$, $X \mapsto Xe_L$, is the restriction functor.*

(a) *The diagram (2.8) is commutative up to a natural equivalence of K-linear functors, i.e. $T_L \circ \boldsymbol{\Theta}'_L \cong \boldsymbol{\Theta}'_I \circ \widehat{T}_L$. The K-linear functor $\widehat{T}_L$ is full, faithful and right exact, res_L is exact, $\widehat{T}_L$ is left adjoint to res_L and $\mathrm{res}_L \circ \widehat{T}_L \cong \mathrm{id}$. The functor T_L is full, faithful and $\mathrm{res}_L \circ T_L \cong \mathrm{id}$.*

(b) *Given a module X in $\mathrm{prin}\, K(L, \mathfrak{Z}')$ or an object $\mathbf{M}$ in $(L, \mathfrak{Z}')$-spr the number $\mathbf{cdn}(\widehat{T}_L(X))(i)$ equals $\mathbf{cdn}(X)(i)$, the number $\mathbf{cdn}(T_L(\mathbf{M}))(i)$ equals $\mathbf{cdn}(\mathbf{M})(i)$ if $i \in L$ and the numbers $\mathbf{cdn}(\widehat{T}_L(X))(i)$, $\mathbf{cdn}(T_L(\mathbf{M}))(i)$ are zero for $i \notin L$. The image of $\widehat{T}_L$ is the full subcategory of $\mathrm{prin}\, K(I, \mathfrak{Z})$ formed by objects Y such that $\mathbf{cdn}(Y)(j) = 0$ for $j \notin L$. The image of T_L is the full subcategory of $(I, \mathfrak{Z})$-spr formed by objects $\mathbf{M}$ such that $\mathbf{cdn}(\mathbf{M})(j) = 0$ for $j \notin L$.*

(c) *If any of the categories $\mathrm{prin}\, K(I, \mathfrak{Z})$, $(I, \mathfrak{Z})$-spr is of tame representation type (resp. of finite, linear, polynomial growth) (see [29, Section 14.4]) then the categories $\mathrm{prin}\, K(L, \mathfrak{Z}')$ and $(L, \mathfrak{Z}')$-spr are of tame representation type (resp. of finite, linear, polynomial growth).*

Proof. The commutativity of the diagram (2.8) follows by a straightforward analysis. For the idea of the proof of (a) and (b) we refer to [29, Theorem 17.46], [29, Proposition 5.14] and [19, Proposition 2.4].

In the prove of (c) we shall use the terminology and notation from [29, Section 14.4]). Assume that $v \in \mathbb{N}^L$ and let $\overline{v} \in \mathbb{N}^I$ be defined by $\overline{v}(i) = v(i)$ if $i \in L$ and $\overline{v}(i) = 0$ otherwise. Assume that the functors

$$\widehat{N}^{(l)} = (-) \otimes_{K[t]} N^{(l)} \ : \mathrm{ind}_1(\mathrm{mod}(K[t])) \longrightarrow \mathrm{mod}\, K(I, \mathfrak{Z})$$

$l = 1, \ldots, r$, form an almost parameterizing family for the category $\mathrm{ind}_{\overline{v}}((I, \mathfrak{Z})\text{-spr})$ of indecomposable peak $(I, \mathfrak{Z})$-spaces $\mathbf{M}$ such that $\mathbf{cdn}\, \mathbf{M} = \overline{v}$. It is clear, that the family of functors $\mathrm{res}_L \circ \widehat{N}(l) = (-) \otimes_{K[t]} N^{(l)} e_L$, $l = 1, \ldots, r$, is an almost parameterizing family for $\mathrm{ind}_v((L, \mathfrak{Z}')\text{-spr})$. Thus the statement follows for the category of peak spaces (see also [20, Lemma 2]). The proof for the category of prinjective modules is analogous. □

3. Proof of Theorem 1.7

We start by recalling from [36] the definition of a two-peak poset with zero-relations $(I^{*+}_{\Lambda^\bullet}, \mathfrak{z}_{\Lambda^\bullet})$ associated with the D-order $\Lambda^\bullet$ (1.4).

Suppose that Λ, Λ_1, Λ_2 and Λ_3 are tiled D-orders in (1.2). In order to define $(I^{*+}_{\Lambda^\bullet}, \mathfrak{z}_{\Lambda^\bullet})$ we consider the poset $(I_\Lambda; \preceq)$ (see [41]) , where

$$I_\Lambda = \{1, \dots, n\} \quad \text{and} \quad i \prec j \Leftrightarrow {}_iD_j = D. \tag{3.1}$$

First we associate with the D-order $\Lambda^\bullet$ (1.4) the following combinatorial object

$$I_{\Lambda^\bullet,\sigma} = (I_\Lambda, \preceq, I', C, I'', \sigma : I' \to I'') \tag{3.2}$$

where $(I_\Lambda, \preceq)$ is the poset (3.1), $I' = I_{\Lambda_1} = \{1, 2, \dots, n_1\}$, $I'' = I_{\Lambda_2} = \{n_1 + n_3 + 1, \dots, n-1, n\}$ and $C = I_{\Lambda_3} = \{n_1 + 1 \prec \dots \prec n_1 + n_3 - 1 \prec n_1 + n_3\}$ are viewed as subposets of I_Λ such that $I_\Lambda = I' \cup C \cup I''$ is a splitting decomposition of I_Λ in the sense of [29, Section 8.1], and $\sigma : I' \to I''$ is the poset isomorphism defined by the formula $\sigma(j) = n_1 + n_3 + j$. It is clear that $I_{\Lambda^\bullet,\sigma}$ is a bipartite stratified poset in the sense of [28] and [29, Section 17.8], or a completed poset in the sense of [23].

Let $C' = \{c'; c \in C\}$ be a chain isomorphic with C. We construct two one-peak enlargements

$$(C \cup I'')^* = C \cup I'' \cup \{*\} \qquad \text{and} \qquad (I' \cup C)^+ = I' \cup C' \cup \{+\}$$

of the posets $C \cup I''$ and $I' \cup C \equiv I' \cup C'$ by the unique maximal points $*$ and $+$, and by the new relations $i \prec *$ and $s \prec +$ for all $i \in C \cup I''$ and all $s \in I' \cup C'$.

Next we associate with the D-order $\Lambda^\bullet$ (1.4) the two-peak poset

$$I^{*+}_{\Lambda^\bullet} = (C \cup I'')^* \bigcup_{I'' \equiv I'} (I' \cup C)^+ \tag{3.3}$$

obtained from the disjoint union $(C \cup I'')^* \cup (I' \cup C)^+$ of $(C \cup I'')^*$ and $(I' \cup C)^+$ by making the identification $j \equiv \sigma(j)$ for any element $j \in I' \subseteq (I' \cup C)^+$.

It is easy to see that $I^{*+}_{\Lambda^\bullet}$ is a poset and $\max I^{*+}_{\Lambda^\bullet} = \{*, +\}$. The reader is referred to [32, 2.5] for illustrating examples.

We define the set $\mathfrak{z}_{\Lambda^\bullet}$ of zero-relations in $I^{*+}_{\Lambda^\bullet}$ to be the set of all pairs (c, c_1') such that $c \in C \subseteq (C \cup I'')^*$, $c_1' \in C' \subseteq (I' \cup C)^+$ and the relations $c \prec s$, $\sigma(s) \prec c_1$ hold in I_Λ for some $s \in I'$. Here we use the convention $+' = +$.

We call the pair

$$(3.4) \qquad (I^{*+}_{\Lambda^\bullet}, \mathfrak{z}_{\Lambda^\bullet})$$

a two-peak **poset with zero-relations associated with the** D**-order** $\Lambda^\bullet$.

A key role in the proof of Theorem 1.7 is playing by the following reduction result proved in [36, Theorem 4.14] and [37, Theorem 3.4 (iii)].

THEOREM 3.5. *Assume that D is a complete discrete valuation domain, $\mathfrak{p}$ is the unique maximal ideal of D and $K = D/\mathfrak{p}$. Let Λ be the D-order* (1.1) *with the three-partition* (1.2) *and $\Lambda_1 = \Lambda_2 \subseteq \mathbb{M}_{n_1}(D)$, $\Lambda_3 \subseteq \mathbb{M}_{n_3}(D)$ and n_1, n_3 as in Section 1. Let $\Lambda^\bullet$ be the subamalgam D-order* (1.4) *and let $(I^{*+}_{\Lambda^\bullet}, \mathfrak{z}_{\Lambda^\bullet})$ be the two-peak poset* (3.3) *with zero-relations associated with $\Lambda^\bullet$. Then the following statements hold.*

(a) *The Tits quadratic forms $q_{\Lambda^\bullet}$* (1.5) *and $q_{(I^{*+}_{\Lambda^\bullet}, \mathfrak{z}_{\Lambda^\bullet})}$* (see (2.4)) *coincide.*

(b) *There exists a K-linear reduction functor*

$$(3.6) \qquad \mathbb{H} : \mathrm{latt}(\Lambda^\bullet) \longrightarrow (I^{*+}_{\Lambda^\bullet}, \mathfrak{z}_{\Lambda^\bullet})\text{-spr} \cong \mathrm{mod}_{\mathrm{sp}} K(I^{*+}_{\Lambda^\bullet}, \mathfrak{z}_{\Lambda^\bullet})$$

with the following properties:

(i) *$\mathbb{H}$ is full, reflects isomorphisms, preserves the indecomposability and preserves and reflects the finite representation type and tame representation type.*

(ii) *$\mathrm{Im}\,\mathbb{H}$ consists up to isomorphism of all objects of $(I^{*+}_{\Lambda^\bullet}, \mathfrak{z})$-spr having no direct summand of one of the following two types: the simple projective representation $P_* = e_* K(I^{*+}_{\Lambda^\bullet}, \mathfrak{z})$ corresponding to the peak idempotent e_* and any of the hereditary sp-injective representations $\mathbf{H}^-_{n_3} \hookrightarrow \mathbf{H}^-_{n_3-1} \hookrightarrow \cdots \hookrightarrow \mathbf{H}^-_0$ defined in* [36, 4.12].

(iii) *The set $\mathrm{ind}_s(\mathrm{latt}(\Lambda^\bullet))$ of the isomorphism classes of indecomposable $\Lambda^\bullet$-lattices Y of D-rank s is finite for any $s \in \mathbb{N}$ if and only if the set $\mathrm{ind}_v((I^{*+}_{\Lambda^\bullet}, \mathfrak{z})\text{-spr})$ of the isomorphism classes of indecomposable objects X in $(I^{*+}_{\Lambda^\bullet}, \mathfrak{z})$-spr with $\mathbf{cdn}(X) = v$ is finite for any vector $v \in \mathbb{N}^{n_1+2n_3+2}$.*

Proof of Theorem 1.7. The implication (a)$\Rightarrow$(a′) is obvious.

(a$'$)$\Rightarrow$(b) By Theorem 3.5 and Corollary 2.4, the condition (a$'$) implies each of the following equivalent condition

(a1) For any vector $v \in \mathbb{N}^{n_1+2n_3+2}$ the set $\mathrm{ind}_v((I^{*+}_{\Lambda\bullet}, \mathfrak{z}_{\Lambda\bullet})\text{-spr})$ of the isomorphism classes of indecomposable objects X in $(I^{*+}_{\Lambda\bullet}, \mathfrak{z}_{\Lambda\bullet})$-spr with $v = \mathbf{cdn}(X)$ is finite.

(a2) For any vector $v \in \mathbb{N}^{n_1+2n_3+2}$ the set $\mathrm{ind}_v(\mathrm{prin}\, K(I^{*+}_{\Lambda\bullet}, \mathfrak{z}_{\Lambda\bullet}))$ of the isomorphism classes of indecomposable modules X in $\mathrm{prin}\, K(I^{*+}_{\Lambda\bullet}, \mathfrak{z}_{\Lambda\bullet})$ with $v = \mathbf{cdn}(X)$ is finite.

Throughout we shall view the algebra $R = K(I^{*+}_{\Lambda\bullet}, \mathfrak{z}_{\Lambda\bullet})$ as a bipartite K-algebra as follows

$$R = K(I^{*+}_{\Lambda\bullet}, \mathfrak{z}_{\Lambda\bullet}) = \begin{pmatrix} A & M \\ 0 & B \end{pmatrix} \tag{3.7}$$

where

$A = K(I^{*+}_{\Lambda\bullet} \setminus \max I^{*+}_{\Lambda\bullet}, \mathfrak{z}_{\Lambda\bullet}) \cong e_- R e_- \cong R/\mathrm{soc}(R_R)$,

$B = K(\max I^{*+}_{\Lambda\bullet}) \cong e_+ R e_+ \cong K \times K \times \cdots \times K \quad (|\max I^{*+}_{\Lambda\bullet}|\text{-times})$,

$e_- = \sum_{j \in I^{*+}_{\Lambda\bullet} \setminus \max I^{*+}_{\Lambda\bullet}} e_j$, $e_+ = \sum_{p \in \max I^{*+}_{\Lambda\bullet}} e_p$, the vector space

$$M = \bigoplus_{p \in \max I^{*+}_{\Lambda\bullet}} \ \bigoplus_{\substack{j \prec p \\ (j,p) \notin \mathfrak{z}_{\Lambda\bullet}}} e_{ip} K \ \cong \ e_- R e_+$$

is viewed as an A-B-bimodule in an obvious way and multiplication is given by the usual matrix multiplication formula $\begin{pmatrix} a & m \\ 0 & b \end{pmatrix} \begin{pmatrix} a' & m' \\ 0 & b' \end{pmatrix} = \begin{pmatrix} aa' & am' + mb' \\ 0 & bb' \end{pmatrix}$, for $a, a' \in A$, $b, b' \in B$ and $m, m' \in M$.

Throughout we shall identify any right R-module X with the system

$$X = (X', X'', \varphi) \tag{3.8}$$

where X' is the right A-module Xe_-, X'' is the right B-module Xe_+ and $\varphi : X' \otimes_A M \to X''$ is a B-module homomorphism defined by the formula $\varphi(xe_- \otimes m) = xe_- m$. The right R-module X is prinjective if and only if $X' = Xe_-$ is a projective A-module. It follows that in our case the category $\mathrm{prin}(R)$ is the category $\mathrm{mod}^{pr}_{pr}(R)$ of propartite right R-modules in the sense of [35].

Fix a vector $v \in \mathbb{N}^{n_1+2n_3+2}$. Following [35, Section 5] and [30], we associate with v and $(I^{*+}_{\Lambda\bullet}, \mathfrak{z}_{\Lambda\bullet})$ the affine K-space

$$\mathbf{prin}_v^{K(I^{*+}_{\Lambda\bullet}, \mathfrak{z}_{\Lambda\bullet})} = \mathrm{Hom}_B(P(v) \otimes_B M,\ Q(v)) \tag{3.9}$$

where

$$P(v) = \bigoplus_{j \notin \max I^{*+}_{\Lambda^\bullet}} (e_j A)^{v(j)}, \quad Q(v) = \bigoplus_{p \in \max I^{*+}_{\Lambda^\bullet}} (e_p B)^{v(p)}.$$

It is clear that $\mathbf{prin}_v^{K(I^{*+}_{\Lambda^\bullet}, \mathfrak{Z}_{\Lambda^\bullet})}$ is an irreducible algebraic K-variety (in Zariski topology). Consider the natural algebraic group action

$$\star : \mathbf{G}(v) \times \mathbf{prin}_v^{K(I^{*+}_{\Lambda^\bullet}, \mathfrak{Z}_{\Lambda^\bullet})} \longrightarrow \mathbf{prin}_v^{K(I^{*+}_{\Lambda^\bullet}, \mathfrak{Z}_{\Lambda^\bullet})} \tag{3.10}$$

where $\mathbf{G}(v) = \operatorname{Aut}_A(P(v)) \times \operatorname{Aut}_B(Q(v))$. It is not difficult to see that there is a bijection between the isomorphism classes of modules in the category $\operatorname{prin} K(I^{*+}_{\Lambda^\bullet}, \mathfrak{Z}_{\Lambda^\bullet})$ with $\mathbf{cdn}(X) = v$ and the $\mathbf{G}(v)$-orbits in the variety $\mathbf{prin}_v^{K(I^{*+}_{\Lambda^\bullet}, \mathfrak{Z}_{\Lambda^\bullet})}$ given by attaching to the prinjective module $X = (X', X'', \varphi)$ the composed B-homomorphism (compare with [35, Section 5] and [36, 3.3 and 3.6])

$$\widetilde{f}_X = [P(v) \otimes_A M \cong X' \otimes_A M \xrightarrow{\varphi} X'' \cong Q(v)] \tag{3.11}$$

It is easy to check that

$$\operatorname{Hom}_B(e_j A \otimes_A M, e_p B) \cong \begin{cases} K & \text{if } j \preceq p \text{ and } (j,p) \notin \mathfrak{Z}_{\Lambda^\bullet}, \\ 0 & \text{if } j \not\preceq p, \text{ or } (j,p) \in \mathfrak{Z}_{\Lambda^\bullet}. \end{cases}$$

Hence we conclude that the variety $\mathbf{prin}_v^{K(I^{*+}_{\Lambda^\bullet}, \mathfrak{Z}_{\Lambda^\bullet})}$ can be identified with the variety $\mathcal{M}_{\Lambda^\bullet}(v)$ (see (1.6)).

A direct calculation shows that $\operatorname{End}(e_j A) \cong K$ for all $j \notin \max I^{*+}_{\Lambda^\bullet}$, $\operatorname{End}(e_p B) \cong K$ for all $p \in \max I^{*+}_{\Lambda^\bullet}$ and

$$\operatorname{Hom}_A(e_j A, e_t A) \cong \begin{cases} K & \text{if } j \preceq t \text{ and } (j,t) \notin \mathfrak{Z}_{\Lambda^\bullet}, \\ 0 & \text{if } j \not\preceq t, \text{ or } (j,t) \in \mathfrak{Z}_{\Lambda^\bullet}. \end{cases}$$

for all $j, t \in I^{*+}_{\Lambda^\bullet} \setminus \max I^{*+}_{\Lambda^\bullet}$. Hence we easily conclude that there exits an algebraic group isomorphism $\mathbf{G}(v) \xrightarrow{\sim} G_{\Lambda^\bullet}(v)$ such that the following diagram is commutative (see [35, Section 5] and [36, Proposition 3.12])

$$\begin{array}{ccc} \mathbf{G}(v) \times \mathbf{prin}_v^{K(I^{*+}_{\Lambda^\bullet}, \mathfrak{Z}_{\Lambda^\bullet})} & \xrightarrow{*} & \mathbf{prin}_v^{K(I^{*+}_{\Lambda^\bullet}, \mathfrak{Z}_{\Lambda^\bullet})} \\ \cong\downarrow & & \cong\downarrow \\ G_{\Lambda^\bullet}(v) \times \mathcal{M}_{\Lambda^\bullet}(v) & \xrightarrow{\bullet} & \mathcal{M}_{\Lambda^\bullet}(v) \end{array}$$

for any coordinate vector $v \in \mathbb{N}^{n_1+2n_3+2}$, where $\bullet$ is the algebraic group action (1.6).

Consequently the map $X \mapsto \widetilde{f}_X$ (3.11) together with the isomorphisms above defines a bijection between the isomorphism classes of modules X in $\operatorname{prin} K(I^{*+}_{\Lambda^\bullet}, \mathfrak{z}_{\Lambda^\bullet})$ with $\mathbf{cdn}(X) = v$ and the $G_{\Lambda^\bullet}(v)$-orbits in $\mathcal{M}_{\Lambda^\bullet}(v)$. Hence in view of the unique decomposition property in the category $\operatorname{prin} K(I^{*+}_{\Lambda^\bullet}, \mathfrak{z}_{\Lambda^\bullet})$ the statement (b) easily follows (use the arguments applied in the proof of Theorem 10.1 in [29, p. 128] or [30, Theorem 3.1]).

The implication (b)$\Rightarrow$(c) can be proved by applying the well-known Tits arguments as follows (compare with [10], the proof of Theorem 10.1 in [29, p. 128-129], [30, Theorem 3.1]).

Consider the algebraic group action (1.6). Let $\mathcal{P}G_{\Lambda^\bullet}(v) = G_{\Lambda^\bullet}(v)/K^*$, where K^* is the multiplicative group of K embedded into $G_{\Lambda^\bullet}(v)$ diagonally. Since K^* acts trivially on $\mathcal{M}_{\Lambda^\bullet}(v)$ then $*$ induces an algebraic group action

$$* : \mathcal{P}G_{\Lambda^\bullet}(v) \times \mathcal{M}_{\Lambda^\bullet}(v) \longrightarrow \mathcal{M}_{\Lambda^\bullet}(v).$$

By (b), the affine K-space $\mathcal{M}_{\Lambda^\bullet}(v)$ has only finitely many $\mathcal{P}G_{\Lambda^\bullet}(v)$-orbits $\mathcal{O}_1, \ldots, \mathcal{O}_t$. Hence looking at the closure of $\mathcal{M}_{\Lambda^\bullet}(v) = \mathcal{O}_1 \cup \ldots \cup \mathcal{O}_t$ we get

$$\mathcal{M}_{\Lambda^\bullet}(v) = \overline{\mathcal{M}_{\Lambda^\bullet}(v)} = \overline{\mathcal{O}}_1 \cup \ldots \cup \overline{\mathcal{O}}_t$$

and since the variety $\mathcal{M}_{\Lambda^\bullet}(v)$ is irreducible then there exists j such that $\mathcal{M}_{\Lambda^\bullet}(v) = \overline{\mathcal{O}}_j$. Since $\mathcal{O}_j$ is open in $\overline{\mathcal{O}}_j$ (see [6], [15]) we get $\dim \mathcal{M}_{\Lambda^\bullet}(v) = \dim \mathcal{O}_j$ (the variety dimension). If $\mathcal{O}_j = \mathcal{P}G_{\Lambda^\bullet}(v) * u_0$ then we have a commutative diagram of rational maps

$$\begin{array}{ccccc} \mathcal{P}G_{\Lambda^\bullet}(v) & \hookrightarrow & \mathcal{P}G_{\Lambda^\bullet}(v) \times \mathcal{M}_{\Lambda^\bullet}(v), & g \mapsto & (g, u_0) \\ \downarrow & & \downarrow{*} & & \\ \mathcal{O}_j & \hookrightarrow & \mathcal{M}_{\Lambda^\bullet}(v) & & \end{array}$$

and therefore $\dim \mathcal{P}G_{\Lambda^\bullet}(v) \geq \dim \mathcal{O}_j = \dim \overline{\mathcal{O}}_j = \dim \mathcal{M}_{\Lambda^\bullet}(v)$ (see [6], [15]). The obvious inequality $\dim \mathcal{P}G_{\Lambda^\bullet}(v) < \dim G_{\Lambda^\bullet}(v)$ yields

$$\dim G_{\Lambda^\bullet}(v) - \dim \mathcal{M}_{\Lambda^\bullet}(v) > \dim \mathcal{P}G_{\Lambda^\bullet}(v) - \dim \mathcal{M}_{\Lambda^\bullet}(v) \geq 0$$

for any vector $v \in \mathbb{N}^{n_1+2n_3+2}$, and the implication (b)$\Rightarrow$(c) follows.

(c)$\Rightarrow$(d) We recall from Theorem 3.5 that the Tits quadratic form $q_{\Lambda^\bullet}$ (1.5) coincides with the Tits quadratic form $q_{(I^{*+}_{\Lambda^\bullet}, \mathfrak{z}_{\Lambda^\bullet})}$ (2.4) of the two-peak poset with zero-relations $(I^{*+}_{\Lambda^\bullet}, \mathfrak{z}_{\Lambda^\bullet})$.

It follows from the definition of the group $G_{\Lambda^\bullet}(v)$ and the variety $\mathcal{M}_{\Lambda^\bullet}(v)$ (see Introduction) that

$\dim G_{\Lambda^\bullet}(v) = \sum_{j\in I^{*+}_{\Lambda^\bullet}} v_j^2 + \sum_{\substack{i\prec j\notin \max I\\ (i,j)\notin 3_{\Lambda^\bullet}}} v_i v_j,$ and

$\dim \mathcal{M}_{\Lambda^\bullet}(v) = \sum_{p\in \max I^{*+}_{\Lambda^\bullet}} \sum_{\substack{i\prec p\\ (i,p)\notin 3_{\Lambda^\bullet}}} v_i v_p.$

This together with (c) yields

$$q_{\Lambda^\bullet}(v) = \dim G_{\Lambda^\bullet}(v) - \dim \mathcal{M}_{\Lambda^\bullet}(v) > 0$$

for any vector $v \in \mathbb{N}^{n_1+2n_3+2}$, and (d) follows.

The equivalences (a)$\Leftrightarrow$(d)$\Leftrightarrow$(g) were proved in [36, Theorem 1.6].

(d)$\Rightarrow$(e) It follows from [33, Theorem 3.12] and [11, Proposition 3.1] that the statement (e) does not hold if and only if there exists a full faithful and exact K-linear functor

$$T : \operatorname{mod}(\Gamma_2(K)) \longrightarrow \operatorname{prin}(R)$$

where $\Gamma_2(K) = \begin{pmatrix} K & K^2 \\ 0 & K \end{pmatrix}$ is the Kronecker K-algebra. Since the Kronecker K-algebra is of infinite representation type and (a) is equivalent with (a2) then (d)$\Rightarrow$(e) follows.

(e)$\Rightarrow$(d) Assume to the contrary that the Tits quadratic form $q_{\Lambda^\bullet} : \mathbb{Z}^{n_1+2n_3+2} \to \mathbb{Z}$ (1.4) is not weakly positive. By [36, Theorem 1.6], the poset $(I^{*+}_{\Lambda^\bullet}, 3_{\Lambda^\bullet})$ with zero-relations contains as a two-peak subposet $\mathcal{F}$ with zero-relations isomorphic with any of the forms $\mathcal{F}_1, \dots, \mathcal{F}_8$ presented in Theorem 1.7 (c). It follows from Proposition 2.9 that there exists a full faithful and exact K-linear functor

$$T : \operatorname{prin}(K\mathcal{F}) \longrightarrow \operatorname{prin}(R).$$

We claim that there exists a pair of modules U, V in $\operatorname{prin}(K\mathcal{F})$ satisfying the following conditions:

(i′) $\operatorname{End}(U) \cong \operatorname{End}(V) \cong K$,

(ii′) $\operatorname{Hom}_{K\mathcal{F}}(U, V) = 0$ and $\operatorname{Hom}_{K\mathcal{F}}(V, U) = 0$,

(iii′) $\dim_K \operatorname{Ext}^1_{K\mathcal{F}}(U, V) = 2$.

If $\mathcal{F} = \mathcal{F}_j$ and $j \neq 4$ the claim follows from [11, Theorem 4.4 (c)]. Assume now that $\mathcal{F} = \mathcal{F}_4$ and has the form

J :

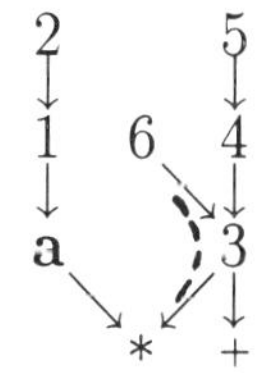

The set of zero-relations of J consists of one pair $(6, *)$. Denote by J'_a the two-peak subposet with zero-relations of J obtained from J by removing the point $\mathbf{a}$. Note that the Tits quadratic form $q_{J'_a} : \mathbb{Z}^8 \to \mathbb{Z}$ is defined by the formula

$$\begin{aligned} q_{J'_a}(z) &= \sum_{j=1}^{6} z_j^2 + z_*^2 + z_+^2 + z_1 z_2 + z_3(z_4 + z_5 + z_6) + z_4 z_5 \\ &\quad -z_*(z_1 + z_2 + z_3 + z_4 + z_5) \\ &\quad -z_+(z_3 + z_4 + z_5 + z_6) \end{aligned}$$

It is easy to see that every point of J'_a is separating (see [5], [24], [28, Section 4]) and according to [24, Remark 4.17] the category $\operatorname{prin}(KJ'_a)$ has a preprojective component (see [4], [14], [29]). By applying a slight modification of the Algorithm 4.4 in [21] one can construct a finite preprojective component of $\operatorname{prin}(KJ'_a)$. By a well-known criterion of Auslander for the finite representation type (see [4], [29, Section 11.8]), obviously extended to the case of prinjective modules, the category $\operatorname{prin}(KJ'_a)$ is of finite representation type and coincides with its preprojective component. By [24, Proposition 4.13] the map $Y \mapsto \mathbf{cdn}\, Y$ defines a bijection between the isomorphism classes of indecomposable modules in $\operatorname{prin}(KJ'_a)$ and the positive roots of the quadratic form $q_{J'_a}$. Note that the vector

$$v = (1, 1, 1, 2, 2, 1, 4, 3)$$

is a positive root of $q_{J'_a}$, that is, v satisfies the equation $q_{J'_a}(v) = 1$. It follows that there exists an indecomposable module V' in $\operatorname{prin}(KJ'_a)$ such that $\mathbf{cdn}\, V' = v$. Since the module V' is preprojective there is an algebra isomorphism $\operatorname{End}(V') \cong K$.

By Proposition 2.9 there exists a full faithful and exact K-linear functor

$$T' : \operatorname{prin}(KJ'_a) \longrightarrow \operatorname{prin}(K\mathcal{F})$$

Denote by $S(a)$ the simple prinjective KJ-module corresponding to the point $a \in J$. Let V be the image of V' under the functor T'. It is easy to check (like in the proof of [11, Lemma 4.2]) that $\mathrm{Hom}_{KJ}(S(a), V) = 0$, $\mathrm{Hom}_{KJ}(V, S(a)) = 0$ and

$$\dim_K \mathrm{Ext}^1_{KJ}(S(a), V) = \sum_{a \prec p \in \max J} v(p) - \sum_{i \prec a} v(i) = v_* - v_1 - v_2 = 4 - 1 - 1 = 2$$

It follows that the conditions (i′), (ii′) and (iii′) are satisfied for the KJ-modules $U = S(a)$ and $V = T'(V')$. Then the R-modules $U_1 = T(U)$ and $U_2 = T(V)$ satisfy the conditions (i), (ii) and (iii) of the statement (e). Moreover, the fully faithful exact functor T induces a K-linear monomorphism $\mathrm{Ext}^1_{KJ}(U, V) \to \mathrm{Ext}^1_R(U_1, U_2)$. It follows that

$$\dim_K \mathrm{Ext}^1_R(U_1, U_2) \geq \dim_K \mathrm{Ext}^1_{KJ}(U, V) = 2$$

contrary to our assumption in the statement (e). This finishes the proof of the implication (e)⇒(d).

(f)⇔(a) It follows from Theorem 3.5 that (a) holds if and only if the category $\mathrm{prin}\, K(I^{*+}_{\Lambda^\bullet}, \mathfrak{z}_{\Lambda^\bullet})$ is of finite representation type. Then the implication (f)⇐(a) is a consequence of [24, Proposition 4.13], because (a) implies (g) (see [36, Theorem 1.6]), the poset $C = I_{\Lambda_3}$ (3.2) is linearly ordered and therefore a simple analysis shows that every point of the poset with zero-relations $(I^{*+}_{\Lambda^\bullet}, \mathfrak{z}_{\Lambda^\bullet})$ is separating (see [5], [28, Section 4]). It follows that the algebra $K(I^{*+}_{\Lambda^\bullet}, \mathfrak{z}_{\Lambda^\bullet}) \cong KI^{*+}_{\Lambda^\bullet}/(\mathfrak{z}_{\Lambda^\bullet})$ has the separation property for radicals of indecomposable projective modules, and therefore there exists a preprojective component $\mathcal{P}_{(I^{*+}_{\Lambda^\bullet}, \mathfrak{z}_{\Lambda^\bullet})}$ in $\mathrm{prin}\, K(I^{*+}_{\Lambda^\bullet}, \mathfrak{z}_{\Lambda^\bullet})$. This follows from [28, Proposition 4.9] applied to the bipartite stratified poset $I_{\Lambda^\bullet, \sigma}$ (3.2), because the algebra $K(I^{*+}_{\Lambda^\bullet}, \mathfrak{z}_{\Lambda^\bullet})$ is obtained from $I_{\Lambda^\bullet, \sigma}$ by a construction required in [28, Proposition 4.9].

In order to prove the converse implication (f)⇒(a) we recall that the map $X \mapsto \mathbf{cdn}\, X$ defines an injection from the preprojective component $\mathcal{P}_{(I^{*+}_{\Lambda^\bullet}, \mathfrak{z}_{\Lambda^\bullet})}$ of $\mathrm{prin}\, K(I^{*+}_{\Lambda^\bullet}, \mathfrak{z}_{\Lambda^\bullet})$ to the set

$$\mathcal{R}^+_{\Lambda^\bullet} = \{v \in \mathbb{N}^{n_1 + 2n_3 + 2};\ q_{\Lambda^\bullet}(v) = 1\}$$

of positive roots of the quadratic form $q_{\Lambda^\bullet} : \mathbb{Z}^{n_1 + 2n_3 + 2} \to \mathbb{Z}$ (1.5) (the proof of [29, Corollary 11.96] extends to our case, see also [22]). Since the set $\mathcal{R}^+_{\Lambda^\bullet}$ is finite by (f) then the component $\mathcal{P}_{(I^{*+}_{\Lambda^\bullet}, \mathfrak{z}_{\Lambda^\bullet})}$ is finite and therefore the category $\mathrm{prin}\, K(I^{*+}_{\Lambda^\bullet}, \mathfrak{z}_{\Lambda^\bullet})$ is of finite representation type by a well-known criterion of Auslander for the finite representation type (see [4], [29, Section 11.8]). This implies the statement (a) and finishes the proof of the theorem. □

4. Subamalgam orders of tame lattice type

We finish the paper by presenting a characterisation of three-partite subamalgam D-orders (1.4) of tame lattice type by applying the results in [37]. In a special case the results were proved in [32].

Throughout this section K is an algebraically closed field and D is a complete discrete valuation domain which is a K-algebra such that $D/\mathfrak{p} \cong K$, where $\mathfrak{p}$ is the unique maximal ideal of D.

Assume that $(I, \mathfrak{Z})$ is a multipeak poset with zero-relations in the sense of Section 2.

Let us recall from [9], [29, Section 15.12] and [31, Section 3] the definition of D-orders of tame lattice type. Let Ω be an arbitrary D-order in a semisimple D_0-algebra C, where D is a complete discrete valuation domain which is an algebra over an algebraically closed field K and $D/\mathfrak{p} \cong K$. Then Ω is said to be of **tame lattice type** (or the category $\operatorname{latt}(\Omega)$ is said to be of tame representation type) if for any number $r \in \mathbb{N}$ there exists a non-zero polynomial $h \in K[y]$ and a family of additive functors

$$(-) \otimes_A M^{(1)}, \ldots, (-) \otimes_A M^{(s)} : \operatorname{ind}_1(A) \longrightarrow \operatorname{latt}(\Omega)$$

where $A = K[y, h^{-1}]$, $M^{(1)}, \ldots M^{(s)}$ are A-Ω-bimodules satisfying the following conditions:

(P0) The left A-modules ${}_AM^{(1)}, \ldots, {}_AM^{(s)}$ are flat.

(P1) All but finitely many indecomposable Ω-lattices of D-rank r are isomorphic to lattices in $\operatorname{Im}(-) \otimes_A M^{(1)} \cup \cdots \cup \operatorname{Im}(-) \otimes_A M^{(s)}$.

(P2) $M^{(1)}_\Omega, \ldots, M^{(s)}_\Omega$ viewed as D-modules are torsion-free.

(P3) ${}_AM^{(1)}_\Omega, \ldots, {}_AM^{(s)}_\Omega$ are finitely generated as A-Ω-bimodules.

Here we denote by $\operatorname{ind}_1(A)$ the full subcategory of $\operatorname{mod}(A)$ consisting of one dimensional A-modules.

Given an integer $r \geq 1$ we define $\boldsymbol{\mu}^1_{\operatorname{latt}(\Omega)}(r)$ to be the minimal number s of functors satisfying the conditions above.

The D-order Ω of tame lattice type is defined to be of **polynomial growth** [31, Section 3] if there exists an integer $g \geq 1$ such that the inequality $\boldsymbol{\mu}^1_{\mathrm{latt}(\Omega)}(r) \leq r^g$ holds for all integers $r \geq 2$ (compare with [29, p.291]).

The reader is referred to [9], [31, Section 3] and [31, p. 58] for various definitions and discussions of tame representation type and wild representation type for the category latt(Ω) of lattices over D-orders Ω.

It follows from [9] and [35] that the tame-wild dichotomy holds for the categories latt$(\Lambda^\bullet)$, prin $K(I, \mathfrak{Z})$ and $(I, \mathfrak{Z}))$-spr.

We recall from [36] that given a sequence $1 \leq i_1 < \cdots < i_s \leq n_1$, we say that the D-order Δ is a $(i_1, \ldots, i_s)$-**minor** D-**suborder** of Λ_1 in (1.2) if Δ is obtained from Λ_1 by omitting the i_jth row and the i_jth column for $j = 1, \ldots, s$.

A three-partite D-order Ω is said to be a **three-partite minor** D-**suborder** of $\Lambda^\bullet$ if Ω is a minor D-suborder of $\Lambda^\bullet$ obtained by omitting rows and columns simultaneously in parts Λ_1 and Λ_2, that is, we omit any i-th row and any i-th column of $\Lambda^\bullet$, where $1 \leq i \leq n_1$, and simultaneously we omit the $(n_1 + n_3 + i)$-th row and the $(n_1 + n_3 + i)$-th column of $\Lambda^\bullet$.

A three-partite subamalgam D-order $\Lambda^\bullet$ (1.4) is said to be **dominated** by a three-partite subamalgam D-order $\overline{\Lambda}^\bullet$ if $\Lambda^\bullet$ is a three-partite D-suborder of $\overline{\Lambda}^\bullet$ of the same size (1.3) and $\Lambda_1 = \overline{\Lambda}_1$, $\Lambda_2 = \overline{\Lambda}_2$, $\Lambda_3 = \overline{\Lambda}_3$, $\mathcal{X} \subseteq \overline{\mathcal{X}}$, $\mathcal{Y} \subseteq \overline{\mathcal{Y}}$ (see [36], [40, p.69]).

Following [36], given a matrix $\lambda \in \mathbb{M}_n(D)$ we define the **reflection transpose** $rt(\lambda) \in \mathbb{M}_n(D)$ of λ to be the transpose matrix of λ with respect to the non main diagonal. Given any D-order Λ (1.1) we define the **reflection transpose** of Λ (resp. of $\Lambda^\bullet$ (1.4)) to be the D-order $rt(\Lambda) = \{rt(\lambda);\ \lambda \in \Lambda\}$, (resp. $rt(\Lambda^\bullet) = \{rt(\lambda);\ \lambda \in \Lambda^\bullet\}$).

THEOREM 4.1. *Assume that K is an algebraically closed field and D is a complete discrete valuation domain which is a K-algebra such that $D/\mathfrak{p} \cong K$, where $\mathfrak{p}$ is the unique maximal ideal of D. Let Λ, $\Lambda^\bullet$, $q_{\Lambda^\bullet}$ and $* : G_{\Lambda^\bullet}(v) \times \mathcal{M}_{\Lambda^\bullet}(v) \to \mathcal{M}_{\Lambda^\bullet}(v)$, $v \in \mathbb{N}^{I^{*+}_{\Lambda^\bullet}} = \mathbb{N}^{n_1+2n_3+2}$ be as in Section 1. If the part $\mathcal{X}$ or the part $\mathcal{Y}$ of the D-order Λ in (1.2) consists of matrices with coefficients in $\mathfrak{p}$ then the following conditions are equivalent.*

(a) *The D-order $\Lambda^\bullet$ is of tame lattice type.*

(b) $\dim G_{\Lambda^\bullet}(v) \geq \dim \mathcal{M}_{\Lambda^\bullet}(v)$ *for every $v \in \mathbb{N}^{n_1+2n_3+2}$, where* dim *is the variety dimension.*

(c) *For every vector $v \in \mathbb{N}^{n_1+2n_3+2}$ the constructible subset* ind $\mathcal{M}_{\Lambda^\bullet}(v)$ *of $\mathcal{M}_{\Lambda^\bullet}(v)$ consisting of indecomposable $G_{\Lambda^\bullet}(v)$-orbits (see* [29, Sections 1.1 and 2.1]) *contains a constructible subset $C(v)$ (see* [15]) *such that*

$\operatorname{ind}\mathcal{M}_{\Lambda^\bullet}(v) = G_{\Lambda^\bullet}(v) \star C(v)$ *and* $\dim C(v) \leq 1$.

(d) *The Tits reduced quadratic form* $q_{\Lambda^\bullet} : \mathbb{Z}^{n_1+2n_3+2} \to \mathbb{Z}$ *is weakly non-negative, that is,* $q_{\Lambda^\bullet}(v) \geq 0$ *for all vectors* $v \in \mathbb{N}^{n_1+2n_3+2}$.

(e) *If* $R = KI^{*+}_{\Lambda^\bullet}/\mathfrak{Z}_{\Lambda^\bullet}$ (2.1) *is the factor of the incidence* K*-algebra* $KI^{*+}_{\Lambda^\bullet}$ *of the poset* $I^{*+}_{\Lambda^\bullet}$ (3.3) *modulo the ideal* $\mathfrak{Z}_{\Lambda^\bullet}$ *of zero-relations then*

$$\dim_K \operatorname{Ext}^1_R(U_1, U_2) \leq 2$$

for any pair of modules U_1, U_2 *in* $\operatorname{mod}(R)$ *satisfying the following conditions:*

(i) $\operatorname{End}(U_1) \cong \operatorname{End}(U_2) \cong K$,

(ii) $\operatorname{Hom}_R(U_1, U_2) = 0$ *and* $\operatorname{Hom}_R(U_2, U_1) = 0$,

(iii) *the kernel of the projective cover* $P(U_j) \to U_j$ *of* U_j *in* $\operatorname{mod}(R)$ *is a semisimple projective* R*-module for* $j = 1, 2$.

(f) *Either* $n_3 = 0$ *and the* D*-order* Λ_1 *in* (1.2) *does not contain minor* D*-suborders of one of the forms*

$$\Delta_0 = \begin{pmatrix} D & \mathfrak{p} & \mathfrak{p} \\ \mathfrak{p} & D & \mathfrak{p} \\ \mathfrak{p} & \mathfrak{p} & D \end{pmatrix}, \quad \Delta_1 = \begin{pmatrix} D & \mathfrak{p} & D \\ \mathfrak{p} & D & \mathfrak{p} \\ \mathfrak{p} & \mathfrak{p} & D \end{pmatrix},$$

$$\Delta_2 = \begin{pmatrix} D & D & \mathfrak{p} \\ \mathfrak{p} & D & \mathfrak{p} \\ \mathfrak{p} & \mathfrak{p} & D \end{pmatrix}, \quad \Delta_3 = \begin{pmatrix} D & \mathfrak{p} & \mathfrak{p} \\ \mathfrak{p} & D & D \\ \mathfrak{p} & \mathfrak{p} & D \end{pmatrix}$$

or else $n_3 \geq 1$, Λ_1 *is hereditary of the form* (1.3) *and three-partite subamalgam* D*-orders* $\Lambda^\bullet$ *and* $rt(\Lambda)^\bullet$ *do not contain three-partite minor* D*-suborders being dominated by any of the* 17 *three-partite subamalgam* D*-orders* $\Omega^\bullet_1, \ldots, \Omega^\bullet_{17}$ *of size at most* 15×15 *listed in* [37, Table 1].

In this case the structure of the Auslander-Reiten quiver of the category $\operatorname{latt}(\Lambda^\bullet)$ *can be described like in* [36, Theorem 6.1] *and the shape of it is presented in Section* 6 *of* [36, Figure 1].

Proof. By Theorem 3.5 (i) and Proposition 2.7 (c) the condition (a) is equivalent to each of the following two equivalent conditions:

(a′) The category $(I^{*+}_{\Lambda^\bullet}, \mathfrak{Z}_{\Lambda^\bullet})$-spr is of tame representation type.

(b′) The category $\operatorname{prin} K(I^{*+}_{\Lambda^\bullet}, \mathfrak{Z}_{\Lambda^\bullet})$ is of tame representation type.

(a)⇒(d) Since the Tits quadratic forms $q_{\Lambda^\bullet}$ (1.5) and $q_{(I^{*+}_{\Lambda^\bullet}, \mathfrak{Z}_{\Lambda^\bullet})}$ (see (2.4)) coincide (Theorem 3.5) and by the above (a)⇒(b′) then the implication (a)⇒(d) follows from [24, Proposition 4.2].

The equivalence (d)⇔(b) is a consequence of the equality

$$q_{\Lambda^\bullet}(v) = \dim G_{\Lambda^\bullet}(v) - \dim \mathcal{M}_{\Lambda^\bullet}(v)$$

established in the proof of Theorem 1.7 for any vector $v \in \mathbb{N}^{n_1+2n_3+2}$.

Since the implications (f)$\Leftrightarrow$(d)$\Rightarrow$(a) follow from [36, Theorem 1.5], the conditions (a), (b), (d) and (f) are equivalent.

(a)$\Rightarrow$(c) The constructibility of the subset $\operatorname{ind}\mathcal{M}_{\Lambda^\bullet}(v)$ of $\mathcal{M}_{\Lambda^\bullet}(v)$ consisting of the indecomposable $G_{\Lambda^\bullet}(v)$-orbits can be proved by the arguments applied in the proof of [18, Theorem 3.17 (a)]. In view of the remark above it remains to prove the implication (b′)$\Rightarrow$(c).

(b′)$\Rightarrow$(c) We view the algebra $R = K(I^{*+}_{\Lambda^\bullet}, \mathfrak{z}_{\Lambda^\bullet})$ as a bipartite K-algebra

$$R = K(I^{*+}_{\Lambda^\bullet}, \mathfrak{z}_{\Lambda^\bullet}) = \begin{pmatrix} A & M \\ 0 & B \end{pmatrix}$$

as in (3.7). Any right R-module X is viewed as a triple $X = (X', X'', \varphi)$, where X' is a right A-module, X'' is a right B-module and $\varphi : X' \otimes_A M \to X''$ is a B-module homomorphism (see (3.8)).

Let $v \in \mathbb{N}^{n_1+2n_3+2}$ be a fixed coordinate vector. Assume that a functor

$$\widehat{X} = (-) \otimes_{K[t]} X_R : \ \operatorname{mod}_1(K[t]) \longrightarrow \operatorname{mod}(R)$$

is a one-parameter family (in the sense of [29, Definition 14.13]) such that for all but a finite number of modules $S_\lambda = K[t]/(t-\lambda)$ in $\operatorname{mod}_1(K[t])$ the module $\widehat{X}(S_\lambda)$ is prinjective and $\mathbf{cdn}\,\widehat{X}(S_\lambda) = v$. We denote here by $\operatorname{mod}_1(K[t])$ the category of one-dimensional $K[t]$-modules. It follows from [16, Lemma 3.1] that under a suitable localization of X with respect to a polynomial in $K[t]$ the $K[t]$-R-bimodule X is isomorphic to a bimodule of the form (see (3.7))

$$X^\wedge = (K[t] \otimes_K x'_A,\ K[t] \otimes_K x''_B;\ \phi_X)$$

where x'_A and x''_B are finite dimensional modules over A and B respectively, x'_A is A-projective and $\phi_X : K[t] \otimes_K x'_A \otimes_A M_B \to K[t] \otimes_K x''_B$ is a B-module homomorphism.

It follows from the discussion in the proof of Theorem 1.7 in Section 3 that for any $\lambda \in K$ the B-module homomorphism $\mathrm{id}_{S_\lambda} \otimes \phi_X : x'_A \otimes_A M \to x''_B$ is an element of the algebraic K-variety $\mathbf{prin}_v^{K(I^{*+}_{\Lambda^\bullet}, \mathfrak{z}_{\Lambda^\bullet})} \cong \mathcal{M}_{\Lambda^\bullet}(v)$ and defines a unique element $\widehat{\phi}_X(\lambda)$ of the variety $\mathcal{M}_{\Lambda^\bullet}(v)$. It is easy to check that the map $K \to \mathcal{M}_{\Lambda^\bullet}(v)$, $\lambda \mapsto \widehat{\phi}_X(\lambda)$, is regular.

Assume now that

$$\widehat{X}_i = (-) \otimes_{K[t]} X_i : \ \operatorname{mod}_1(K[t]) \longrightarrow \operatorname{mod}(R)$$

i=1,...,r, is an almost parameterizing family for the indecomposable prinjective R-modules X with $\mathbf{cdn}\, X = v$ and let $\widehat{\phi}_1, \ldots, \widehat{\phi}_s$ be the elements of $\mathcal{M}_{\Lambda^\bullet}(v)$ corresponding to all indecomposable modules not lying in $\bigcup_{i=1}^r \operatorname{Im} \widehat{X}_i$ up to isomorphism. It is easy to check that (compare with [21, p. 101])

$$C(v) = \{\widehat{\phi}_1, \ldots, \widehat{\phi}_s\} \cup \{\widehat{\phi}_{X_i}(\lambda);\ \lambda \in K, i = 1, \ldots, r\} \subseteq \mathcal{M}_{\Lambda^\bullet}(v)$$

is a subset of $\operatorname{ind} \mathcal{M}_{\Lambda^\bullet}(v)$ and satisfies the conditions required in (c).

(c)⇒(a) For any vector $v \in \mathbb{N}^{n_1+2n_3+2}$ consider the algebraic group action $\star : \mathbf{G}(v) \times \mathbf{prin}_v^{K(I^{*+}_{\Lambda^\bullet}, 3_{\Lambda^\bullet})} \longrightarrow \mathbf{prin}_v^{K(I^{*+}_{\Lambda^\bullet}, 3_{\Lambda^\bullet})}$ (3.10). We recall from [35, Section 5], [36, Proposition 3.12] and from Section 3 that there exist a variety isomorphism $\mathbf{prin}^{K(I^{*+}_{\Lambda^\bullet}, 3_{\Lambda^\bullet})} \cong \mathcal{M}_{\Lambda^\bullet}(v)$ (see (1.6)) and an algebraic group isomorphism $\mathbf{G}(v) \xrightarrow{\sim} G_{\Lambda^\bullet}(v)$ such that the following diagram is commutative

$$\begin{array}{ccc}
\mathbf{G}(v) \times \mathbf{prin}_v^{K(I^{*+}_{\Lambda^\bullet}, 3_{\Lambda^\bullet})} & \xrightarrow{*} & \mathbf{prin}_v^{K(I^{*+}_{\Lambda^\bullet}, 3_{\Lambda^\bullet})} \\
\cong\downarrow & & \cong\downarrow \\
G_{\Lambda^\bullet}(v) \times \mathcal{M}_{\Lambda^\bullet}(v) & \xrightarrow{\bullet} & \mathcal{M}_{\Lambda^\bullet}(v)
\end{array}$$

where $\bullet$ is the algebraic group action (1.6). The map $X \mapsto \widetilde{f}_X$ (3.11) together with the isomorphisms above defines a bijection between the isomorphism classes of modules X in $\operatorname{prin} K(I^{*+}_{\Lambda^\bullet}, 3_{\Lambda^\bullet})$ with $\mathbf{cdn}(X) = v$ and the $G_{\Lambda^\bullet}(v)$ orbits in $\mathcal{M}_{\Lambda^\bullet}(v)$.

Assume to the contrary that (a) does not hold. Then the category $\operatorname{prin} K(I^{*+}_{\Lambda^\bullet}, 3_{\Lambda^\bullet})$ is not of tame representation type and by the tame-wild dichotomy it is of wild representation type. By the arguments used in the proof of [18, Theorem 3.14] there exist a vector $v \in \mathbb{N}^{n_1+2n_3+2}$ and a constructible subset U_2' of $\operatorname{ind} \mathbf{prin}_v^{K(I^{*+}_{\Lambda^\bullet}, 3_{\Lambda^\bullet})}$ such that $\dim U_2' = 2$. Let U_2 be the image of U_2' under the variety isomorphism $\mathbf{prin}_v^{K(I^{*+}_{\Lambda^\bullet}, 3_{\Lambda^\bullet})} \cong \mathcal{M}_{\Lambda^\bullet}(v)$. By [29, Lemma 15.15] applied to $V = \mathcal{M}_{\Lambda^\bullet}(v)$, $G = G_{\Lambda^\bullet}(v)$, $H = \{E\}$, and the constructible sets U_2 and $U_2 = C(v)$ (see the condition (e)) we get $2 = \dim U_2 \leq \dim C(v) \leq 1$; a contradiction (see also [18, Lemma 3.16]). This finishes the proof of the implication (c)⇒(a).

(a)⇒(e) Assume to the contrary that (e) does not hold. It follows from [25, Lemmas 1.5 and 8.6] and [33, Theorem 3.12] that the condition (e) does not hold if and only if there exists a full faithful and exact K-linear functor

$$F : \operatorname{mod}(\Gamma_3(K)) \longrightarrow \operatorname{prin} K(I^{*+}_{\Lambda^\bullet}, \mathfrak{z}_{\Lambda^\bullet})$$

where $\Gamma_3(K) = \begin{pmatrix} K & K^3 \\ 0 & K \end{pmatrix}$. Since the K-algebra $\Gamma_3(K)$ is of wild representation type (see [29, p. 286]) and (a) is equivalent with (b′) then (a)⇒(e) follows, because the tame-wild dichotomy [9] holds for the category $\operatorname{prin} K(I^{*+}_{\Lambda^\bullet}, \mathfrak{z}_{\Lambda^\bullet})$ by applying [35, Theorem 6.5] to the bipartite algebra $R = K(I^{*+}_{\Lambda^\bullet}, \mathfrak{z}_{\Lambda^\bullet})$ (see (3.7)).

(e)⇒(a). Assume to the contrary that (a) does not hold. Since (a) is equivalent with (b′) then the category $\operatorname{prin} K(I^{*+}_{\Lambda^\bullet}, \mathfrak{z}_{\Lambda^\bullet})$ is not of tame representation type and according to [37, Theorem 1.5] the two-peak poset $(I^{*+}_{\Lambda^\bullet}, \mathfrak{z}_{\Lambda^\bullet})$ with zero-relations associated with $\Lambda^\bullet$ in (3.4) contains as a two-peak subposet with zero-relations any of the following thirteen forms

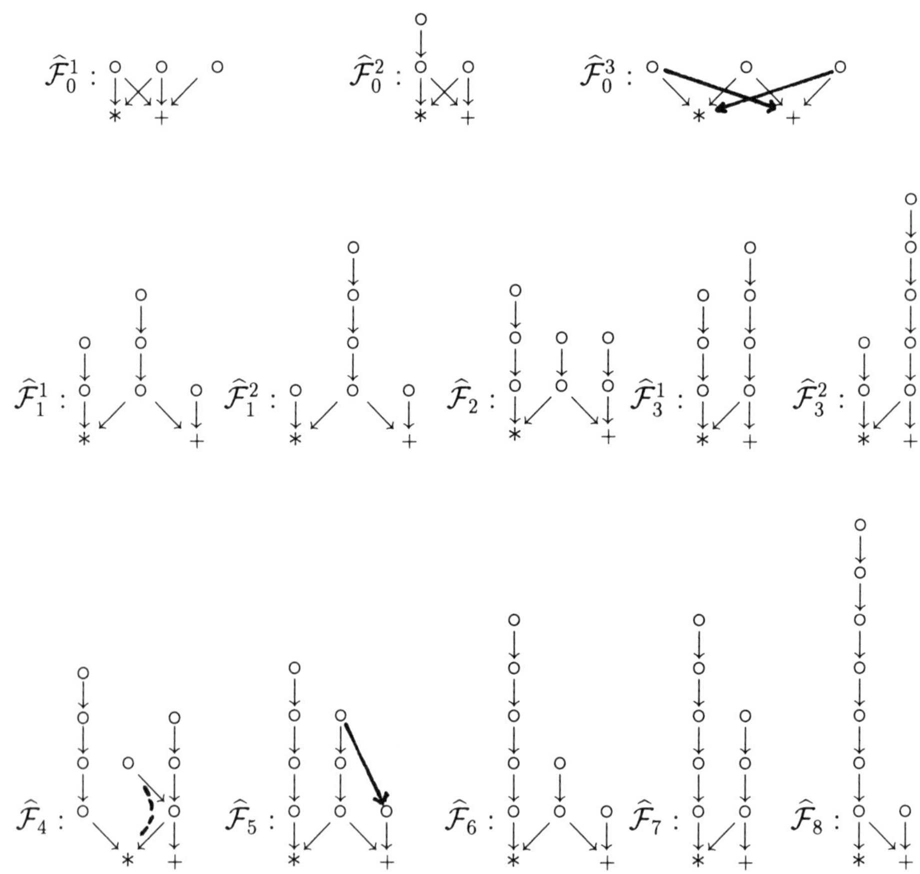

where the dotted line in $\widehat{\mathcal{F}}_4$ means a zero-relation.

If $R = K(I^{*+}_{\Lambda^\bullet}, 3_{\Lambda^\bullet})$ and $\widehat{\mathcal{F}} \subseteq (I^{*+}_{\Lambda^\bullet}, 3_{\Lambda^\bullet})$ is any of the above forms then according to Proposition 2.9 there exists a full faithful and exact K-linear functor

$$T : \mathrm{prin}(K\widehat{\mathcal{F}}) \longrightarrow \mathrm{prin}(R).$$

We claim that there exists a pair of modules U, V in $\mathrm{prin}(K\widehat{\mathcal{F}})$ satisfying the following conditions:

(i′) $\mathrm{End}(U) \cong \mathrm{End}(V) \cong K$,

(ii′) $\mathrm{Hom}_{K\widehat{\mathcal{F}}}(U, V) = 0$ and $\mathrm{Hom}_{K\widehat{\mathcal{F}}}(V, U) = 0$,

(iii′) $\dim_K \mathrm{Ext}^1_{K\widehat{\mathcal{F}}}(U, V) \geq 3$.

If $\widehat{\mathcal{F}} = \widehat{\mathcal{F}}_j$ and $j \neq 4$ the claim follows from [19, Lemmas 3.6 and 3.7] and their proof. Assume now that $\widehat{\mathcal{F}} = \widehat{\mathcal{F}}_4$. and $\widehat{\mathcal{F}}_4$ has the form

$J :$ 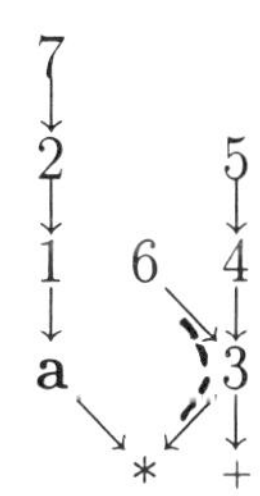

The set of zero-relations of J consists of one pair $(6, *)$. Denote by J'_a the two-peak subposet with zero-relations of J obtained from J by removing the point **a**.

It is easy to see that every point of J'_a is separating (see [5], [24], [28, Section 4]) and according to [24, Remark 4.17] the category $\mathrm{prin}(KJ'_a)$ has a preprojective component (see [4], [14], [29]). By applying a slight modification of the Algorithm 4.4 in [21] one can construct a unique preprojective component $\mathcal{P}_{J'_a}$ of $\mathrm{prin}(KJ'_a)$.

Following the proof of [19, Lemma 3.8] we shall find an indecomposable preprojective module V' in $\mathrm{prin}(KJ'_a)$ such that $\ell(\mathbf{cdn}\, V') = 3$, where $\ell : \mathbb{Z}^9 \to \mathbb{Z}$ is an additive function defined by the formula $\ell(v) = v_* - v_1 - v_2 - v_7$.

For this purpose we note that the beginning of the preprojective component $\mathcal{P}_{J'_a}$ of $\mathrm{prin}(KJ'_a)$ looks as follows

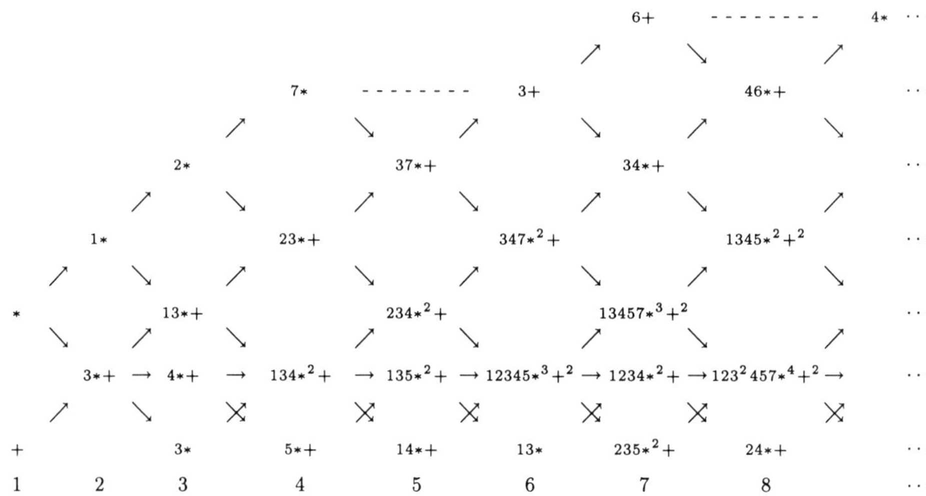

Here we write the coordinate vectors $w = \mathbf{cdn}\, W$ instead of indecomposable preinjective modules W, and we use the exponential notation of coordinate vectors introduced in [29, 11.88], that is, the vector $w = \mathbf{cdn}\, W = (w_1, \dots, w_9) \in \mathbb{N}^9$ is written in the form

$$w = \mathbf{cdn}\, W = 1^{w_1} 2^{w_2} \dots 9^{w_9}$$

where we omit j^{w_j} if $w_j = 0$, and we set $j^{w_j} = j$ if $w_j = 1$.

The value diagram of the additive function $\ell(v) = v_* - v_1 - v_2 - v_7$ over the component $\mathcal{P}_{J'_a}$ is the following (see also [33]

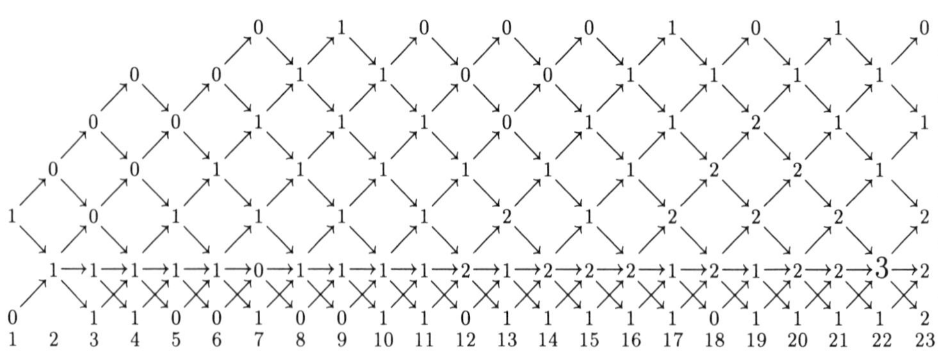

Since the above diagram contains the number 3 in the 22nd column then there exists an indecomposable preprojective module V' in $\mathrm{prin}\, K(J'_a)$ such that $\ell(\mathbf{cdn}\, V') = 3$. One can show by drawing the preprojective component above that

$$\mathbf{cdn}\, V' = 1^2 2^2 3^3 4^4 5^4 6^2 7^2 *^9 +^7 \equiv (2,2,3,4,4,2,2,9,7)$$

Since the module V' is preprojective there exists an algebra isomorphism $\mathrm{End}(V') \cong K$. By Proposition 2.9 there exists a full faithful and exact K-linear functor

$$T' : \mathrm{prin}(KJ'_a) \longrightarrow \mathrm{prin}(K\mathcal{F})$$

Denote by $S(a)$ the simple prinjective KJ-module corresponding to the point $a \in J$. Let $v = \mathbf{cdn}\, V' = (2,2,3,4,4,2,2,9,7)$ and denote by V the image of V' under the functor T'. It is easy to check like in the proof of [19, Lemma 3.7] and [11, Lemma 4.2] that $\mathrm{Hom}_{KJ}(S(a), V) = 0$, $\mathrm{Hom}_{KJ}(V, S(a)) = 0$ and

$$\dim_K \mathrm{Ext}^1_{KJ}(S(a), V) = \sum_{a \prec p \in \max J} v(p) - \sum_{i \prec a} v(i) = \ell(v) = 9 - 2 - 2 - 2 = 3$$

It follows that the conditions (i′), (ii′) and (iii′) are satisfied for the KJ-modules $U = S(a)$ and $V = T'(V')$. Then the R-modules $U_1 = T(U)$ and $U_2 = T(V)$ satisfy the conditions (i), (ii) and (iii) of the statement (e). Moreover, the fully faithful exact functor T induces a K-linear monomorphism $\mathrm{Ext}^1_{KJ}(U, V) \to \mathrm{Ext}^1_R(U_1, U_2)$. It follows that

$$\dim_K \mathrm{Ext}^1_R(U_1, U_2) \geq \dim_K \mathrm{Ext}^1_{KJ}(U, V) \geq 3$$

contrary to our assumption in the statement (e).

One can get an alternative proof of the existence of such a preprojective module V' in $\mathrm{prin}(KJ'_a)$ by applying [17, Corollary 3.17] to the additive function $\ell : \mathbb{Z}^9 \to \mathbb{Z}$, $\ell(v) = v_* - v_1 - v_2 - v_7$, as follows.

First we note that the Tits quadratic form $q_{J'_a} : \mathbb{Z}^9 \to \mathbb{Z}$ is the defined by the formula

$$\begin{aligned}
q_{J'_a}(z) &= \sum_{j=1}^{7} z_j^2 + z_*^2 + z_+^2 + z_1 z_2 + z_1 z_7 + z_2 z_7 + z_3(z_4 + z_5 + z_6) + z_4 z_5 \\
&\quad - z_*(z_1 + z_2 + z_3 + z_4 + z_5 + z_7) - z_+(z_3 + z_4 + z_5 + z_6) \\
&= (z_6 + \tfrac{1}{2} z_3 - \tfrac{1}{2} z_9)^2 + (z_1 + \tfrac{1}{2}(z_2 + z_7 - z_8))^2 \\
&\quad + \tfrac{3}{4}(z_2 + \tfrac{1}{3} z_7 - \tfrac{1}{3} z_8)^2 + (\tfrac{\sqrt{5}}{2\sqrt{2}} z_8 - \tfrac{2\sqrt{5}}{5\sqrt{2}}(z_3 + z_4 + z_5))^2 \\
&\quad + (\tfrac{\sqrt{3}}{2} z_9 - \tfrac{\sqrt{3}}{3}(\tfrac{1}{2} z_3 + z_4 + z_5))^2 + \tfrac{2}{3}(z_7 - \tfrac{1}{4} z_8)^2 \\
&\quad + \tfrac{4}{15}(z_3 - \tfrac{1}{4} z_4 - \tfrac{1}{4} z_5)^2 + \tfrac{1}{4}(z_4 + z_5)^2 \;\geq 0
\end{aligned}$$

and the bipartite algebra KJ'_a is prin-critical in the sense of [17], because $\mathrm{prin}(KJ'_a)$ has a preprojective component and it is easy to see that q_L is weakly positive for any subposet with zero-relations L of J'_a, and by [30, Proposition 4.13] the algebra KL is of finite prinjective type. Next we observe that $q_{J'_a}(\mu) = 0$, where $\mu = (1,1,1,2,2,1,1,4,3)$. Moreover the above formula for $q_{J'_a}(z)$ yields

$$\mathrm{Ker}\, q_{J'_a} = \{z \in \mathbb{Z}^9;\ q_{J'_a}(z) = 0\} = \mathbb{Z}\mu$$

Since $\ell(\mu) = 1 > 0$ then according to [17, Corollary 3.17] there exists an indecomposable preprojective module V' in $\mathrm{prin}(KJ'_a)$ such that $\ell(\mathbf{cdn}\, V') \geq 3$, and consequently $\dim_K \mathrm{Ext}^1_{KJ}(S(a), V) = \ell(\mathbf{cdn}\, V') \geq 3$, where $V = T'(V')$. This finishes the proof of the implication (e)⇒(a) and completes the proof of the theorem. □

Let us finish this section by presenting without proof a characterisation given in [38] of three-partite subamalgam D-orders (1.4) of tame lattice type and of polynomial growth.

THEOREM 4.2. *Let K be an algebraically closed field and let D, $\mathfrak{p}$, Λ, $\Lambda^\bullet$ and $\mathcal{X}$, $\mathcal{Y}$ be as in Theorem* 4.1. *The following conditions are equivalent.*

(a) *The D-order $\Lambda^\bullet$* (1.4) *is of tame lattice type and of polynomial growth.*

(b) *Either $n_3 \geq 1$, the D-order Λ_1 in* (1.2) *is hereditary of the form* (1.3) *and the three-partite D-orders $\Lambda^\bullet$ and $rt(\Lambda)^\bullet$ do not contain three-partite minor D-suborders being dominated by any of the* 17 *three-partite subamalgam D-orders $\Omega_1^\bullet, \ldots, \Omega_{17}^\bullet$ listed in* [37, Table 1], *or else $n_3 = 0$ and there exists at most one pair (i, j) such that $1 \leq i < j \leq n_1$ and ${}_iD_j = \mathfrak{p}$ in Λ* (1.1). □

References

[1] D. M. Arnold, Representations of partially ordered sets and abelian groups, *Contemporary Math.* 87(1989), 91-109.

[2] D. M. Arnold and M. Dugas, Block rigid almost completely decomposable groups and lattices over multiple pull-back rings, *J. Pure Appl. Algebra* 87(1993), 105-121.

[3] D. M. Arnold, F. Richman and C. Vinsonhaler, Representations of finite posets and valuated groups, *J. Algebra* 155(1993), 110-126.

[4] M. Auslander, I. Reiten and S. Smalø, *"Representation Theory of Artin Algebras"*, Cambridge Studies in Advanced Mathematics 36, Cambridge University Press, 1995.

[5] K. Bongartz, A criterion for finite representation type, *Math. Ann.* 269(1984), 1-12.

[6] A. Borel, *"Linear Algebraic Groups"*, W.A. Benjamin, New York - Amsterdam, 1969.

[7] Ch. W. Curtis and I. Reiner, *"Methods of Representation Theory"*, Vol. I, Wiley Classics Library Edition, New York, 1990.

[8] P. Dowbor and A. Skowroński, Galois coverings of representation-infinite algebras, *Comment. Math. Helv.* 62(1987), 311-337.

[9] J. A. Drozd and M. G. Greuel, Tame-wild dichotomy for Cohen-Macaulay modules, *Math. Ann.* 294(1992), 387-394.

[10] P. Gabriel, Représentations indécomposables, Séminaire Bourbaki, *in* Lecture Notes in Math. No. 431, Springer-Verlag, Berlin-Heidelberg-New York-Tokyo, 1975, pp. 143-169.

[11] R. Göbel and D. Simson, Embeddings of Kronecker modules into the category of prinjective modules and the endomorphism ring problem, *Colloquium Math.* 75 (1998), 213-244.

[12] E. L. Green and I. Reiner, Integral representations and diagrams, *Michigan Math. J.* 25 (1978), 53-84.

[13] J. Haeffner and L. Klingler, Special quasi-triads and integral group rings of finite representation type, I and II, *J. Algebra* 158(1993), 279-374.

[14] H.-J. von Höhne and D. Simson, Bipartite posets of finite prinjective type, *J. Algebra* 201(1998), 86-114.

[15] J. E. Humphreys, *"Linear Algebraic Groups"*, Graduate Texts in Mathematics 21, Springer-Verlag, 1975.

[16] S. Kasjan, Adjustment functors and tame representation type, *Comm. Algebra* 22(1994), 5587-5597.

[17] S. Kasjan, Minimal bipartite algebras of infinite prinjective type with prin-preprojective component, *Colloquium Math.* 76(1998), 295-317.

[18] S. Kasjan and D. Simson, Varieties of poset representations and minimal posets of wild prinjective type, *in* Proceedings of the Sixth International Conference on Representations of Algebras, *Canadian Mathematical Society Conference Proceedings,* AMS, Vol. 14, 1993, pp. 245-284.

[19] S. Kasjan and D. Simson, Fully wild prinjective type of posets and their quadratic forms, *J. Algebra* 172(1995), 506-529.

[20] S. Kasjan and D. Simson, Tame prinjective type and Tits form of two-peak posets I, *J. Pure Appl. Algebra* 106(1996), 307-330.

[21] S. Kasjan and D. Simson, A subbimodule reduction, a peak reduction functor and tame prinjective type, *Bull. Pol. Acad. Sci. Math.* 45(1997), 89-107.

[22] J. Kosakowska and D. Simson, On Tits form and prinjective representations of posets of finite prinjective type, *Comm. Algebra* 26(1998), 1613-1623.

[23] L. A. Nazarova and V. A. Roiter, Representations of completed posets, *Comment. Math. Helv.* 63(1988), 498-526.

[24] J. A. de la Peña and D. Simson, Prinjective modules, reflection functors, quadratic forms and Auslander-Reiten sequences, *Trans. Amer. Math. Soc.* 329(1992), 733-753.

[25] C. M. Ringel, Representations of K-species and bimodules, *J. Algebra* 41(1976), 269-302.

[26] C. M. Ringel and K. W. Roggenkamp, Diagrammatic methods in the representation theory of orders, *J. Algebra* 60(1979), 11-42.

[27] D. Simson, Vector space categories, right peak rings and their socle projective modules, *J. Algebra* 92(1985), 532-571.

[28] D. Simson, Right peak algebras of two-separate stratified posets, their Galois coverings and socle projective modules, *Comm. Algebra* 20(1992), 3541-3591.

[29] D. Simson, *"Lincar Representations of Partially Ordered Sets and Vector Space Categories"*, Algebra, Logic and Applications, Vol. 4, Gordon & Breach Science Publishers, New York, 1992.

[30] D. Simson, Posets of finite prinjective type and a class of orders, *J. Pure Appl. Algebra* 90(1993), 77-103.

[31] D. Simson, On representation types of module subcategories and orders, *Bull. Pol. Acad. Sci., Math.* 41(1993), 77-93.

[32] D. Simson, A reduction functor, tameness and Tits form for a class of orders, *J. Algebra* 174(1995), 430-452.

[33] D. Simson, Representation embedding problems, categories of extensions and prinjective modules, *in* Proceedings of the Seventh International Conference on Representations of Algebras, *Canadian Mathematical Society Conference Proceedings*, AMS, Vol. 18, 1996, 601-639.

[34] D. Simson, Socle projective representations of partially ordered sets and Tits quadratic forms with application to lattices over orders, *in* Proceedings of the Conference on Abelian Groups and Modules, Colorado Springs, August 1995, *Lecture Notes in Pure and Appl. Math.*, Vol. 182, 1996, pp. 73-111.

[35] D. Simson, Prinjective modules, propartite modules, representations of bocses and lattices over orders, *J. Math. Soc. Japan* 49(1997), 31-68.

[36] D. Simson, Three partite subamalgams of tiled orders of finite lattice type, *J. Pure Appl. Algebra* 133(1998), in press.

[37] D. Simson, A reduced Tits quadratic form and tameness of three-partite subamalgams of tiled orders, preprint, Toruń, 1997.

[38] D. Simson, Tame three-partite subamalgams of tiled orders of polynomial growth, preprint, Toruń, 1997.

[39] Y. Yoshino, Brauer-Thrall type theorem for maximal Cohen-Macaulay modules, *J. Math. Soc. Japan* 39(1987), 719-739.

[40] Y. Yoshino, *"Cohen-Macaulay Modules over Cohen-Macaulay Rings"*, London Math. Soc. Lecture Notes Series, Vol 146, Cambridge University Press, 1990.

[41] A. G. Zavadskij and V. V. Kirichenko, Torsion-free modules over prime rings,*in* Zap. Naychn. Sem. Leningrad. Otdel. Mat. Inst. Steklov. (LOMI), 57(1976), 100-116 (in Russian).

Contemporary Mathematics
Volume **229**, 1998

Degenerations in Module Varieties with Finitely Many Orbits

ANDRZEJ SKOWROŃSKI AND GRZEGORZ ZWARA

ABSTRACT. Let A be a finite dimensional K-algebra over an algebraically closed field K, $\mathbf{d}$ a vector in the Grothendieck group $K_0(A)$ of A with nonnegative coordinates, and $\mathrm{mod}_A(\mathbf{d})$ the affine variety of A-modules with the dimension-vector $\mathbf{d}$. We prove that if $\mathrm{Ext}^1_A(X,X) = 0$ for all indecomposable A-modules of dimension-vector $\leq \mathbf{d}$ then the degeneration order $\leq_{\mathrm{deg}}$ in $\mathrm{mod}_A(\mathbf{d})$ coincides with the extension order $\leq_{\mathrm{ext}}$. We describe also the structure of minimal degenerations in an arbitrary variety $\mathrm{mod}_A(\mathbf{d})$ provided the orders $\leq_{\mathrm{ext}}$ and $\leq_{\mathrm{deg}}$ coincide for all A-modules of dimension-vector $\leq \mathbf{d}$. As a consequence we get a set-theoretic structure of degenerations in the varieties of quiver representations with finitely many orbits.

0. Introduction

Throughout the paper K will denote a fixed algebraically closed field. By an algebra is meant an associative finite dimensional K-algebra with an identity, which we shall assume (without loss of generality) to be a basic and connected. For such an algebra A, there exists an isomorphism $A \simeq KQ/I$, where KQ is the path algebra of the Gabriel quiver $Q = Q_A$ of A and I is an admissible ideal of KQ. Denote by Q_0 the set of vertices of Q, by Q_1 the set of arrows of Q, and by $s, e : Q_1 \to Q_0$ the maps which assign to each arrow $\alpha \in Q_1$ its source $s(\alpha)$ and its end $e(\alpha)$, respectively. The ideal I is generated by a (finite) system of forms $\sum_{1\leq j\leq t} \lambda_j \alpha_{m_j,j} \dots \alpha_{1,j}$ (called K-linear relations in Q), where λ_j, $1 \leq j \leq t$, are elements of K and $\alpha_{m_j,j} \dots \alpha_{1,j}$, $1 \leq j \leq t$, are paths of length ≥ 2 in Q having a common source and a common end. Then the category mod_A of all finite dimensional (over K) left A-modules is equivalent to the category $\mathrm{rep}_K(Q,I)$ of all finite dimensional K-linear representations $V = (V_i, \varphi_\alpha)_{i\in Q_0,\alpha\in Q_1}$ of Q, where V_i, $i \in Q_0$, are finite dimensional K-vector spaces and $\varphi_\alpha : V_{s(\alpha)} \to V_{e(\alpha)}$, $\alpha \in Q_1$, are K-linear maps, satisfying the equations $\sum_{1\leq j\leq t} \lambda_j \varphi_{\alpha_{m_j,j}} \cdots \varphi_{\alpha_{1,j}} = 0$ for all K-linear relations $\sum_{1\leq j\leq t} \lambda_j \alpha_{m_j,j} \dots \alpha_{1,j} \in I$ (see [**7**, Section 4] for details). We shall identify mod_A with $\mathrm{rep}_K(Q,I)$ and call finite dimensional A-modules shortly A-modules. The Grothendieck group $K_0(A)$ of mod_A is then

1991 *Mathematics Subject Classification.* 14L30, 16G20, 16G60.
Research supported by the Polish Scientific Grant KBN No. 2 PO3A 012 14.
This paper is in final form and no version of it will be submitted for publication elsewhere.

identified with the group $\mathbb{Z}^{Q_0}$, and we may assign to each module $V = (V_i, \varphi_\alpha) \in \text{mod}_A = \text{rep}_K(Q, I)$ the dimension-vector $\mathbf{dim}\, V = (\dim_K V_i)_{i \in Q_0} \in \mathbb{Z}^{Q_0}$. Fix now a vector $\mathbf{d} = (d_i) \in K_0(A) = \mathbb{Z}^{Q_0}$ with nonnegative coordinates. Denote by $\text{mod}_A(\mathbf{d})$ the set of all representations V in $\text{rep}_K(Q, I)$ with $V_i = K^{d_i}$ for all $i \in Q_0$. A representation V in $\text{mod}_A(\mathbf{d})$ is given by $d_{e(\alpha)} \times d_{s(\alpha)}$-matrices $V(\alpha)$ determining the maps $\varphi_\alpha : K^{s(\alpha)} \to K^{e(\alpha)}$, $\alpha \in Q_1$, in the canonical bases of K^{d_i}, $i \in Q_0$. Moreover, the matrices $V(\alpha)$, $\alpha \in Q_1$, satisfy the relations

$$\sum_{1 \le j \le t} \lambda_j V(\alpha_{m_j,j}) \dots V(\alpha_{1,j}) = 0$$

for all K-linear relations $\sum_{1 \le j \le t} \lambda_j \alpha_{m_j,j} \dots \alpha_{1,j} \in I$. Therefore, $\text{mod}_A(\mathbf{d})$ is a closed subset of $\prod_{\alpha \in Q_1} K^{d_{e(\alpha)} d_{s(\alpha)}}$ in the Zariski topology, and so $\text{mod}_A(\mathbf{d})$ is an affine variety. We note that $\text{mod}_A(\mathbf{d})$ is not necessarily irreducible. Clearly, it is the case when $I = 0$. The affine (reductive) algebraic group $G(\mathbf{d}) = \prod_{i \in Q_0} \text{Gl}_{d_i}(K)$ acts on the variety $\text{mod}_A(\mathbf{d})$ by conjugation:

$$(g \cdot V)(\alpha) = g_{e(\alpha)} V(\alpha) g_{s(\alpha)}^{-1}$$

for $g = (g_i) \in G(\mathbf{d})$, $V \in \text{mod}_A(\mathbf{d})$, $\alpha \in Q_1$. We shall identify an A-module V of dimension-vector $\mathbf{d}$ with the corresponding point of the variety $\text{mod}_A(\mathbf{d})$. Then it is easy to see that two A-modules M and N in $\text{mod}_A(\mathbf{d})$ are isomorphic if and only if M and N belong to the same $G(\mathbf{d})$-orbit.

One of the important problems of the geometric classification of finite dimensional modules is to study the degenerations of modules of a fixed dimension-vector. For $M, N \in \text{mod}_A(\mathbf{d})$, we say that N is a degeneration of M if N belongs to the closure $\overline{G(\mathbf{d})M}$ of the orbit $G(\mathbf{d})M$, and we denote this fact by $M \le_{\text{deg}} N$. Then $\le_{\text{deg}}$ is a partial order on the set of isomorphism classes of A-modules ($G(\mathbf{d})$-orbits in $\text{mod}_A(\mathbf{d})$) of a given dimension-vector $\mathbf{d}$. There are two other partial orders $\le_{\text{ext}}$ and $\le$ on the isomorphism classes of A-modules defined in terms of extensions and homomorphism as follows [**1**], [**6**], [**10**]:

- $M \le_{\text{ext}} N \Leftrightarrow$ there are modules M_i, U_i, V_i and short exact sequences $0 \to U_i \to M_i \to V_i \to 0$ in mod_A such that $M = M_1$, $M_{i+1} = U_i \oplus V_i$ and $N = M_{s+1}$ for some natural number s.
- $M \le N \Leftrightarrow [M, X] \le [N, X]$ holds for all modules X from mod_A.

Here and later on we abbreviate $\dim_K \text{Hom}_A(U, V)$ by $[U, V]$. We note that $\le$ is a partial order by a result of Auslander [**2**]. Furthermore, by a result due to Auslander and Reiten [**3**], $M \le N$ is equivalent to the inequalities $[X, M] \le [X, N]$ for all modules X from mod_A. Then for modules M and N in a variety $\text{mod}_A(\mathbf{d})$ the following implications hold:

$$M \le_{\text{ext}} N \Rightarrow M \le_{\text{deg}} N \Rightarrow M \le N.$$

Unfortunately, the reverse implications are not true in general and it is an interesting open problem to find out when they are. However, it has been proved by Bongartz [**5**], [**6**] and the second named author [**15**] that for the path algebras $A = KQ$ of Dynkin and extended Dynkin quivers Q, the partial orders $\leq_{\text{ext}}$, $\leq_{\text{deg}}$, $\leq$ coincide for all A-modules of any fixed dimension-vector. Moreover, by the main result of [**14**], if an algebra A is representation-finite, then $\leq_{\text{deg}}$ and $\leq$ also coincide for all A-modules of any fixed dimension-vector. On the other hand, there are very simple examples [**10**] of representation-finite algebras A for which the partial orders $\leq_{\text{ext}}$ and $\leq_{\text{deg}}$ do not coincide.

Here, we are concerned with the problem of when for a given algebra A (possibly of wild representation type) and a fixed dimension-vector $\mathbf{d} \in K_0(A)$, the partial orders $\leq_{\text{ext}}$ and $\leq_{\text{deg}}$ coincide for all modules in the variety $\text{mod}_A(\mathbf{d})$. Note that the modules in $\text{mod}_A(\mathbf{d})$ are direct sums of indecomposable A-modules of dimension-vector $\leq \mathbf{d}$. In particular, the variety $\text{mod}_A(\mathbf{d})$ has a finite number of $G(\mathbf{d})$-orbits if and only if there are only finitely many $G(\mathbf{d}')$-orbits in $\text{mod}_A(\mathbf{d}')$ for all $\mathbf{d}' \in K_0(A)$ with $\mathbf{d}' \leq \mathbf{d}$. It is known that $\text{mod}_A(\mathbf{d})$ has only finitely many $G(\mathbf{d})$-orbits if $\text{Ext}_A^1(X,X) = 0$ for any indecomposable A-module X with $\mathbf{dim}\, X \leq \mathbf{d}$. Indeed, if $\text{Ext}_A^1(X,X) = 0$ and $\mathbf{e} = \mathbf{dim}\, X$, then the closure of $G(\mathbf{e})X$ in $\text{mod}_A(\mathbf{e})$ is an irreducible component (see [**9**, (2.7)]), and our claim easily follows. We may now formulate our first result.

THEOREM 1. *Let A be an algebra and $\mathbf{d} \in K_0(A)$ a nonnegative vector such that $\text{Ext}_A^1(X,X) = 0$ for any indecomposable A module X with $\mathbf{dim}\, X \leq \mathbf{d}$. Then the partial orders $\leq_{\text{ext}}$ and $\leq_{\text{deg}}$ coincide for all modules from $\text{mod}_A(\mathbf{d})$.*

The above result is an extension of [**14**, Theorem 2] which asserts that the partial orders $\leq_{\text{ext}}$ and $\leq$ coincide for all A-modules of any fixed dimension-vector provided $\text{Ext}_A^1(X,X) = 0$ for all indecomposable finite dimensional A-modules. The proof of this fact presented in [**14**] essentially uses that under the above stronger assumption the algebra A is representation-finite, and hence cannot be applied in the situation considered in Theorem 1.

As an application of Theorem 1, an argument due to Tits, and a deep result by Kac [8] we obtain our second result which gives a positive answer for the question raised by A. Zelevinsky during the visit of the first named author in Boston (October 1997).

THEOREM 2. *Let A be the path algebra KQ of a finite quiver Q and $\mathbf{d} \in K_0(A)$ be a nonnegative vector such that $\text{mod}_A(\mathbf{d})$ has only finitely many $G(\mathbf{d})$-orbits. Then the partial orders $\leq_{\text{ext}}$ and $\leq_{\text{deg}}$ coincide for all modules from $\text{mod}_A(\mathbf{d})$.*

Our third result describes the set-theoretic structure of minimal degenerations provided the partial orders $\leq_{\text{ext}}$ and $\leq_{\text{deg}}$ coincide for all modules of dimension-vector smaller or equal than a fixed one.

THEOREM 3. *Let A be an algebra and $\mathbf{d} \in K_0(A)$ be a nonnegative vector. Assume that the partial orders $\leq_{\text{ext}}$ and $\leq_{\text{deg}}$ coincide for all A-modules of dimension-vector $\leq \mathbf{d}$. Consider two modules M and N in $\text{mod}_A(\mathbf{d})$. Then N is a minimal degeneration of M if and only if there is an exact sequence $0 \to U \to M' \to V \to 0$ of A-modules with the following properties:*

(a) *U and V are indecomposable A-modules with $M \simeq M' \oplus U^{p-1} \oplus V^{q-1} \oplus X$, $N \simeq U^p \oplus V^p \oplus X$, and $U \oplus V$ and $M' \oplus X$ have no common nonzero direct summands.*

(b) *$U \oplus V$ is a minimal degeneration of M'.*

(c) *Any common indecomposable direct summand $W \not\simeq V$ of M and N satisfies $[W, M] = [W, N]$.*

(d) *Any common indecomposable direct summand $W \not\simeq U$ of M and N satisfies $[M, W] = [N, W]$.*

Here, U, V, M', p and q are uniquely determined by M and N.

Under the stronger assumption that the three partial orders $\leq_{\text{ext}}$, $\leq_{\text{deg}}$ and $\leq$ coincide for all A-modules of dimension-vector $\leq \mathbf{d}$, the above theorem follows from a result by Bongartz [**4**, Theorem 4]. But without assumptions that $\leq_{\text{ext}}$ and $\leq$ coincide, in our proof of Theorem 3 some new arguments are necessary. We note also that there are important classes of (representation-infinite) algebras such that $\leq_{\text{ext}}$ and $\leq_{\text{deg}}$ coincide for all modules of any fixed dimension-vector but $\leq_{\text{deg}}$ and $\leq$ do not coincide. For example, one can prove that it is the case for all tubular algebras (in the sense of Ringel [**11**]).

1. Preliminary results

In this section we collect some results concerning the degenerations of modules which we will use in the proofs of Theorems 1 and 3. Throughout this section A denotes a fixed algebra. We start with the following sufficient condition for degenerations.

LEMMA 4. *Let $E : 0 \to M' \to M \to M'' \to 0$ be an exact sequence of A-modules and let φ be an endomorphism of M'. Then M degenerates to the middle term N of the pushout sequence φE. In particular, M degenerates to $M' \oplus M''$.*

PROOF. See [**6**, Lemma 1.1].

Next result is the following important fact proved by Riedtmann.

PROPOSITION 5. *Let* $0 \to N \to M \oplus Z \to Z \to 0$ *(respectively,* $0 \to Z' \to Z' \oplus M \to N \to 0$*) be an exact sequence of* A*-modules. Then* M *degenerates to* N.

PROOF. See [**10**, Proposition 3.4].

Motivated by the above proposition, the second named author introduced in [**16**] a new order on the isomorphism classes of A-modules having the same dimension-vector. Namely, for $M, N \in \text{mod}_A(\mathbf{d})$, $0 \leq \mathbf{d} \in K_0(A)$, one defines

- $M \leq_R N \Leftrightarrow$ there exists an exact sequence $0 \to N \to M \oplus Z \to Z \to 0$ for some A-module Z.

Then we have the following facts.

THEOREM 6. *For any* $0 \leq \mathbf{d} \in K_0(A)$, $\leq_R$ *is a partial order on the isomorphism classes (orbits) of modules in* $\text{mod}_A(\mathbf{d})$. *Moreover, for* $M, N \in \text{mod}_A(\mathbf{d})$, *we have* $M \leq_R N$ *if and only if there exists an exact sequence* $0 \to Z' \to Z' \oplus M \to N \to 0$ *for some* A*-module* Z'.

PROOF. See [**16**, Theorems 2.2 and 2.3].

Therefore, for two modules M and N in a variety $\text{mod}_A(\mathbf{d})$ the following implications hold

$$M \leq_{\text{ext}} N \Rightarrow M \leq_R N \Rightarrow M \leq_{\text{deg}} N \Rightarrow M \leq N.$$

The following result proved recently in [**17**, Theorems 1 and 4] will be crucial in our proof of Theorem 1.

THEOREM 7. *Let* M *and* N *be two modules in a module variety* $\text{mod}_A(\mathbf{d})$. *Then the following hold.*

(i) *If* $M \leq_{\text{deg}} N$ *then* $M \leq_R N$.
(ii) *If* N *is a minimal degeneration of* M *but* $M <_{\text{ext}} N$ *does not hold then there exist* A*-modules* X, Y, Z *such that* $M \simeq X \oplus Y$, $N \simeq X \oplus Z$, $Y <_R Z$, *and* Z *is indecomposable.*

For the proof of Theorem 3 we need some notation and additional results. Recall that for A-modules U and V, $[U, V]$ is the abbreviation of $\dim_K \text{Hom}_A(U, V)$. Then following [**10**] for any A-modules M and N of the same dimenion-vector, we define two function $\delta_{M,N}$ and $\delta'_{M,N}$ on A-modules X as follows

$$\delta_{M,N}(X) = [N, X] - [M, X], \qquad \delta'_{M,N}(X) = [X, N] - [X, M].$$

Further, for an A-module M and an indecomposable A-module Z, we denote by $\mu(M, Z)$ the multiplicity of Z as a direct summand of M. For a nonprojective indecomposable A-module U we denote by $E(U)$ the middle term of an Auslander-Reiten sequence

$$0 \to D\text{Tr}\, U \to E(U) \to U \to 0.$$

In the case U is an indecomposable projective A-module, we define $E(U)$ to be the radical of U. For A-modules U and V we denote by $\mathrm{rad}(U,V)$ the Jacobson radical in $\mathrm{Hom}_A(U,V)$. Finally, for an A-module X and a positive integer p, X^p will denote the direct sum of p copies of X. The following results are known.

LEMMA 8. *Let M and N be two A-modules with $\mathbf{dim}\, M = \mathbf{dim}\, N$ and U be an indecomposable A-module. Then*

$$\mu(N,U) - \mu(M,U) = \delta_{M,N}(U) + \delta'_{M,N}(U) - \delta_{M,N}(E(U)).$$

PROOF. See [**13**, Lemma 1.5].

LEMMA 9. *Let M and N be two A-modules with $\mathbf{dim}\, M = \mathbf{dim}\, N$. If $M <_{\deg} N$, then $\delta_{M,N}(N) > 0$ and $\delta'_{M,N}(N) > 0$.*

PROOF. See [**14**, Lemma 2.11].

We end this section with an important result for degenerations proved by Bongartz.

THEOREM 10. *Let Y and Z be two A-modules with $\mathbf{dim}\, Y = \mathbf{dim}\, Z$ and X an A-module. Then the following hold.*

(i) *If $[X,Y] = [X,Z]$ or $[Y,X] = [Z,X]$, then $X \oplus Y \leq_{\deg} X \oplus Z$ if and only if $Y \leq_{\deg} Z$.*
(ii) *If $[X,Y] = [X,Z]$ and $[Y,X] = [Z,X]$, then the map $L \mapsto X \oplus L$ induces an isomorphism between the partially ordered sets*

$$\langle Y, Z\rangle = \{L \mid Y \leq_{\deg} L \leq_{\deg} Z\}$$

and

$$\langle X\oplus Y, X\oplus Z\rangle = \{R \mid X\oplus Y \leq_{\deg} R \leq_{\deg} X\oplus Z\}.$$

PROOF. See [**4**, Theorem 2] and [**6**, Corollary 2.5].

2. Proof of Theorem 1

Let $\mathbf{d}$ be a nonnegative vector in $K_0(A)$ such that $\mathrm{Ext}^1_A(X,X) = 0$ for any indecomposable A-module X with $\mathbf{dim}\, X \leq \mathbf{d}$. Let M and N be modules from $\mathrm{mod}_A(\mathbf{d})$. We know that always $M \leq_{\mathrm{ext}} N$ implies $M \leq_{\deg} N$. In order to prove the reverse implication, we may assume that N is a minimal degeneration of M but $M <_{\mathrm{ext}} N$ does not hold. Then it follows from Theorem 7 that there exist A-modules X, Y, Z such that $M \simeq X \oplus Y$, $N \simeq X \oplus Z$, $Y <_R Z$, and Z is indecomposable. Clearly, Y and Z are modules in $\mathrm{mod}_A(\mathbf{e})$ for some $\mathbf{e} \leq \mathbf{d}$.

In particular, invoking our assumption, we get $\mathrm{Ext}^1_A(Z,Z)=0$. Since $Y <_R Z$ there are (see Theorem 6) nonsplittable exact sequences

$$\Sigma : 0 \longrightarrow Z \longrightarrow Y \oplus U \xrightarrow{f} U \longrightarrow 0$$
$$\Delta : 0 \longrightarrow V \longrightarrow V \oplus Y \xrightarrow{g} Z \longrightarrow 0$$

for some A-modules U and V. Observe that in the exact sequence of K-vector spaces

$$0 \longrightarrow \mathrm{Hom}_A(U,Z) \longrightarrow \mathrm{Hom}_A(Y \oplus U, Z) \xrightarrow{f_*} \mathrm{Hom}_A(Z,Z)$$

induced by Σ, f_* is not an epimorphism, and consequently

$$[Z,Z]-[Y,Z]=[U,Z]+[Z,Z]-[Y\oplus U,Z]>0.$$

This implies that

$$[Z,Z]+[V,Z]-[V\oplus Y,Z]=[Z,Z]-[Y,Z]>0.$$

Therefore, in the exact sequence

$$0 \longrightarrow \mathrm{Hom}_A(Z,Z) \longrightarrow \mathrm{Hom}_A(V\oplus Y,Z) \longrightarrow \mathrm{Hom}_A(V,Z) \longrightarrow \mathrm{Ext}^1_A(Z,Z)$$

induced by Δ, we have $\mathrm{Ext}^1_A(Z,Z)\neq 0$, a contradiction. Hence, $M \leq_{\deg} N$ implies $M \leq_{\mathrm{ext}} N$, and this finishes our proof.

3. Proof of Theorem 2

It is well-known that if a finite quiver $Q=(Q_0,Q_1)$ admits only a finite number of isomorphism classes of K-linear representations of a dimension-vector $\mathbf{d}\in\mathbb{N}^{Q_0}$, then the full subquiver of Q given by all nonzero coordinates of $\mathbf{d}$ is a tree. Hence we may assume that $A=KQ$ for a finite connected tree $Q=(Q_0,Q_1)$. Consider the Tits form $q_A:\mathbb{Z}^{Q_0}\to\mathbb{Z}$ defined for $\mathbf{x}\in\mathbb{Z}^{Q_0}$ by

$$q_A(\mathbf{x})=\sum_{i\in Q_0}x_i^2-\sum_{\alpha\in Q_1}x_{s(\alpha)}x_{e(\alpha)}.$$

Thus for any $\mathbf{d}\in\mathbb{N}^{Q_0}$ we have $q_A(\mathbf{d})=\dim G(\mathbf{d})-\dim\mathrm{mod}_A(\mathbf{d})$. On the other hand, it follows from [**11**, 2.4] that for any module M in $\mathrm{mod}_A(\mathbf{d})$ the equality

$$q_A(\mathbf{d})=\dim_K\mathrm{End}_A(M)-\dim_K\mathrm{Ext}^1_A(M,M)$$

holds. Moreover, since the stabilizer of any module $M\in\mathrm{mod}_A(\mathbf{d})$ coincides with with the automorphism space $\mathrm{Aut}_A(M)$, we have also the following equality $\dim G(\mathbf{d})M=\dim G(\mathbf{d})-\dim_K\mathrm{End}_A(M)$.

Assume now that for a vector $\mathbf{d} \in \mathbb{N}^{Q_0}$, the number of $G(\mathbf{d})$-orbits in $\operatorname{mod}_A(\mathbf{d})$ is finite. Let X be an indecomposable A-module with $\mathbf{dim}\, X = \mathbf{e} \leq \mathbf{d}$. We shall show that $\operatorname{Ext}_A^1(X, X) = 0$. First we apply an argument due to Tits. It follows from our assumption that the number of $G(\mathbf{e})$-orbits in $\operatorname{mod}_A(\mathbf{e})$ is also finite. Since $\operatorname{mod}_A(\mathbf{e})$ is irreducible there exists a module $Y \in \operatorname{mod}_A(\mathbf{e})$ such that $\operatorname{mod}_A(\mathbf{e}) = \overline{G(\mathbf{e})Y}$. Then

$$\dim G(\mathbf{e}) > \dim \overline{G(\mathbf{e})Y} = \dim \operatorname{mod}_A(\mathbf{e}),$$

and so $q_A(\mathbf{e}) = \dim G(\mathbf{e}) - \dim \operatorname{mod}_A(\mathbf{e}) \geq 1$. As a consequence of a deep result by Kac [**8**, Theorem 2] we know that, for any indecomposable A-module Z, $q_A(\mathbf{dim}\, Z) \leq 1$ holds, and if $q_A(\mathbf{x}) = 1$ for some $\mathbf{x} \in \mathbb{N}^{Q_0}$, then there exists exactly one (up to isomorphism) indecomposable A-module N with $\mathbf{dim}\, N = \mathbf{x}$. Therefore, in our situation, we get $q_A(\mathbf{e}) = 1$ and X is a unique indecomposable A-module of dimension-vector $\mathbf{e}$. Further, since $1 = \dim G(\mathbf{e}) - \overline{\dim G(\mathbf{e})Y}$, we conclude that $\operatorname{End}_A(Y) \simeq K$, and hence $\operatorname{Ext}_A^1(Y, Y) = 0$, by the above homological formula for $q_A(\mathbf{dim}\, Y)$. In particular, Y is also indecomposable. Hence $X \simeq Y$ and $\operatorname{Ext}_A^1(X, X) = 0$. Consequently, Theorem 2 follows now from Theorem 1.

We note also that the arguments applied above prove the following fact. For a finite quiver $Q = (Q_0, Q_1)$ and $\mathbf{d} \in \mathbb{N}^{Q_0}$, the affine variety $\operatorname{mod}_{KQ}(\mathbf{d})$ has only finitely many $G(\mathbf{d})$-orbits if and only if the inequality $q_{KQ}(\mathbf{e}) > 0$ holds for any nonzero vector $\mathbf{e} \in \mathbb{N}^{Q_0}$ with $\mathbf{e} \leq \mathbf{d}$. Finally, consider the following wild quiver Δ of type $\widetilde{\widetilde{\mathbb{E}}}_8$

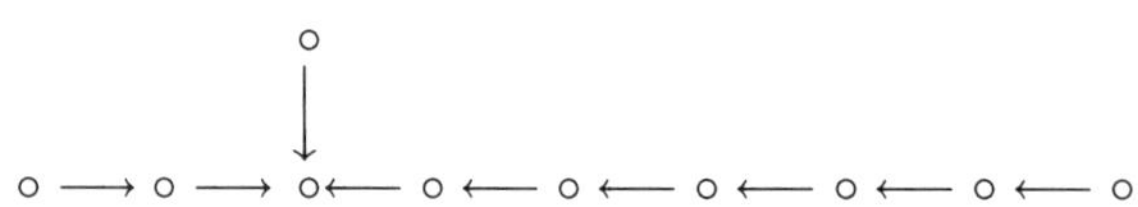

and the dimension-vector $\mathbf{d} = 24\overset{3}{6}443211 \in \mathbb{N}^{\Delta_0}$. Then it is easy to check that $q_{K\Delta}(\mathbf{e}) > 0$ for any $0 \neq \mathbf{e} \in \mathbb{N}^{\Delta_0}$ with $\mathbf{e} \leq \mathbf{d}$. Hence $\operatorname{mod}_{K\Delta}(\mathbf{d})$ has only finitely many $G(\mathbf{d})$-orbits, and consequently the partial orders $\leq_{\text{ext}}$ and $\leq_{\text{deg}}$ coincide for all modules in $\operatorname{mod}_A(\mathbf{d})$. On the other hand, since $K\Delta$ is of wild representation type, we know by [**12**, Corollary 2] that there are degenerations of indecomposable $K\Delta$-modules to indecomposable $K\Delta$-modules. Therefore, for some dimension-vectors $\mathbf{c} \in \mathbb{N}^{\Delta_0}$, the partial orders $\leq_{\text{ext}}$ and $\leq_{\text{deg}}$ do not coincide in $\operatorname{mod}_{K\Delta}(\mathbf{c})$.

4. Proof of Theorem 3

First we shall prove the following lemma.

LEMMA 11. *Let A be an algebra, M and A-module, U and V indecomposable A-modules, and assume that there exists a nonsplittable exact sequence*

$$\Omega: 0 \longrightarrow U \xrightarrow{f} M \xrightarrow{g} V \longrightarrow 0.$$

If the induced degeneration $M <_{\deg} U \oplus V$ is minimal, then

$$\delta_{M,U\oplus V}(U) = 1 = \delta'_{M,U\oplus V}(V).$$

PROOF. By symmetry of our claim, we prove only that $\delta_{M,U\oplus V}(U) = 1$. Consider the following exact sequence induced by Ω

$$0 \longrightarrow \operatorname{Hom}_A(V,U) \longrightarrow \operatorname{Hom}_A(M,U) \xrightarrow{f^*} \operatorname{Hom}_A(U,U) \xrightarrow{\partial} \operatorname{Ext}^1_A(V,U).$$

Since Ω is not splittable, f^* is not an epimorphism and hence

$$\delta_{M,U\oplus V}(U) = [V,U] - [M,U] + [U,U] > 0.$$

Suppose that $\delta_{M,U\oplus V}(U) > 1$. Observe that $[U,U] - \dim_K \operatorname{rad}(U,U) = 1$, because U is indecomposable and A is a finite dimensional K-algebra over an algebraically closed field K. Then there exists $e \in \operatorname{rad}(U,U)$ such that $\partial(e) \neq 0$, and we get the following commutative diagram with exact nonsplittable rows

$$\begin{array}{ccccccccc}
\Omega: 0 & \longrightarrow & U & \xrightarrow{f} & M & \longrightarrow & V & \longrightarrow & 0 \\
 & & \downarrow{\scriptstyle e} & & \downarrow & & \| & & \\
\Sigma: 0 & \longrightarrow & U & \longrightarrow & N & \longrightarrow & V & \longrightarrow & 0.
\end{array}$$

Applying Lemma 4 we get $M \leq_{\deg} N$. Further, since the sequence Ω is not splittable and U is indecomposable, we have $f \in \operatorname{rad}(U,M)$, and so $\begin{bmatrix} e \\ f \end{bmatrix} \in \operatorname{rad}(U, U \oplus M)$. Then the induced (pushout) exact sequence

$$0 \longrightarrow U \xrightarrow{\begin{bmatrix} e \\ f \end{bmatrix}} U \oplus M \longrightarrow N \longrightarrow 0$$

is nonsplittable, and consequently $M \not\simeq N$. Thus we get a chain of degenerations $M <_{\deg} N <_{\deg} U \oplus V$, because Σ is not splittable. But this contradicts the minimality of the degeneration $M <_{\deg} U \oplus V$. Therefore, $\delta_{M,U\oplus V}(U) = 1$, and this finishes our proof.

Let now A be an algebra and $\mathbf{d} \geq 0$ a vector from $K_0(A)$ such that the partial orders $\leq_{\text{ext}}$ and $\leq_{\deg}$ coincide for all A-modules of dimension-vector $\leq \mathbf{d}$. Let

M and N be modules in $\operatorname{mod}_A(\mathbf{d})$. Assume that N is a minimal degeneration of M. It follows then from our assumption that $M <_{\text{ext}} N$, and hence there exists an exact sequence

$$0 \longrightarrow N_1 \longrightarrow M \longrightarrow N_2 \longrightarrow 0$$

with $N \simeq N_1 \oplus N_2$, for some A-modules N_1 and N_2. Splitting off exact sequences of the form

$$\begin{gathered} 0 \longrightarrow D \xrightarrow{\sim} D \longrightarrow 0 \longrightarrow 0, \\ 0 \longrightarrow 0 \longrightarrow E \xrightarrow{\sim} E \longrightarrow 0, \end{gathered}$$

we get an exact sequence

$$\Omega : 0 \longrightarrow U \xrightarrow{f} M' \xrightarrow{g} V \longrightarrow 0$$

such that $M \simeq M' \oplus U^{p-1} \oplus V^{q-1} \oplus X$, $N \simeq U^p \oplus V^q \oplus X$, $f \in \operatorname{rad}(U, M')$, $g \in \operatorname{rad}(M', V)$, and neither U nor V is a direct summand of X, for some A-modules U, V, M', X and integers $p, q \geq 1$. Clearly, Ω gives a minimal degeneration $M' <_{\deg} U \oplus V$. We shall prove now that U and V are indecomposable. By symmetry it is enough to show that U is indecomposable. Suppose that $U = B \oplus C$ for some nonzero A-modules B and C. Then we obtain the following commutative diagram with exact rows and columns

$$\begin{array}{ccccccccc}
 & & 0 & & 0 & & & & \\
 & & \downarrow & & \downarrow & & & & \\
 & & B & = & B & & & & \\
 & & \downarrow{\scriptstyle i} & & \downarrow{\scriptstyle \alpha} & & & & \\
0 & \longrightarrow & U & \xrightarrow{f} & M' & \xrightarrow{g} & V & \longrightarrow & 0 \\
 & & \downarrow{\scriptstyle r} & & \downarrow{\scriptstyle \gamma} & & \| & & \\
0 & \longrightarrow & C & \xrightarrow{\beta} & M'' & \xrightarrow{\omega} & V & \longrightarrow & 0. \\
 & & \downarrow & & \downarrow & & & & \\
 & & 0 & & 0 & & & &
\end{array}$$

Since $f \in \operatorname{rad}(U, M')$, we conclude that α and β are not sections. Hence, the exact sequences

$$\begin{gathered} 0 \longrightarrow B \xrightarrow{\alpha} M' \xrightarrow{\gamma} M'' \longrightarrow 0 \\ 0 \longrightarrow C \xrightarrow{\beta} M'' \xrightarrow{\omega} V \longrightarrow 0 \end{gathered}$$

are not splittable. Therefore, we get $M' <_{\text{deg}} B \oplus M'' <_{\text{deg}} B \oplus C \oplus V$ with $B \oplus C \oplus V \simeq U \oplus V$, a contradiction with the minimality of $M' <_{\text{deg}} U \oplus V$. This shows that U is indecomposable. Suppose now that U is a direct summand of M'. Then $M' \simeq U \oplus Y$ for some A-module Y, and we obtain a nonsplittable exact sequence of the form

$$0 \longrightarrow U \longrightarrow U \oplus Y \longrightarrow V \longrightarrow 0.$$

Thus $Y <_{\text{deg}} V$ by Proposition 5. Moreover, $\mathbf{dim}\, Y = \mathbf{dim}\, V \leq \mathbf{dim}\, N \leq \mathbf{d}$, and invoking our assumption we get $Y <_{\text{ext}} V$. But this leads to a contradiction because V is indecomposable. Similarly one proves that V is not a direct summand of M'. Therefore, the required property (a) holds. In order to prove the necessity part of Theorem 3, it remains only to show (by symmetry of (c) and (d)) that (c) holds. This is done by an argument due to Markolf (see [**4**, p. 594]). For completeness of our considerations we will repeat it here. Suppose that $[W, N] > [W, M]$ for some common indecomposable direct summand W of M and N nonisomorphic to V. Then the last map in the following exact sequence

$$\begin{aligned} 0 \to \mathrm{Hom}_A(W, U) \to \mathrm{Hom}_A(W, M') &\to \mathrm{Hom}_A(W, V) \\ &\to \mathrm{Ext}^1_A(W, U) \to \mathrm{Ext}^1_A(W, M') \end{aligned}$$

induced by Ω is not injective. Therefore, we find a nonsplittable exact sequence in $\mathrm{Ext}^1_A(W, U)$ whose pushout under $f : U \to M'$ splits. We then get the following commutative diagram with exact rows and columns

$$\begin{array}{ccccccccc} & & 0 & & 0 & & & & \\ & & \downarrow & & \downarrow & & & & \\ 0 & \longrightarrow & U & \longrightarrow & Y & \longrightarrow & W & \longrightarrow & 0 \\ & & \downarrow{\scriptstyle f} & & \downarrow & & \| & & \\ 0 & \longrightarrow & M' & \longrightarrow & M' \oplus W & \longrightarrow & W & \longrightarrow & 0. \\ & & \downarrow{\scriptstyle g} & & \downarrow & & & & \\ & & V & = & V & & & & \\ & & \downarrow & & \downarrow & & & & \\ & & 0 & & 0 & & & & \end{array}$$

Further, since V is not a direct summand of $M' \oplus W$, we then get

$$M' \oplus W <_{\text{ext}} V \oplus Y <_{\text{ext}} V \oplus U \oplus W.$$

Moreover, since neither U nor V is a direct summand of M', we conclude that there is an A-module Z such that $M \simeq M' \oplus W \oplus Z$ and $N \simeq U \oplus V \oplus W \oplus Z$. This implies

$$M <_{\text{ext}} V \oplus Y \oplus Z <_{\text{ext}} N,$$

which again contradicts to the minimality of $M <_{\text{deg}} N$. This finishes the proof of the necessity part.

Assume now that there exists a short exact sequence

$$\Omega : 0 \longrightarrow U \longrightarrow M' \longrightarrow V \longrightarrow 0$$

and the conditions (a)–(d) hold. From Theorem 10(ii) and our assumptions (c) and (d), we may assume that $X = 0$. Clearly, by (a) and (b) we have $M \leq_{\text{deg}} N$. Suppose that this degeneration is not minimal. Then there is an A-module L such that $\mathbf{dim}\, L = \mathbf{d}$,

$$M = M' \oplus U^{p-1} \oplus V^{q-1} <_{\text{deg}} L <_{\text{deg}} U^p \oplus V^q = N$$

and the right degeneration is minimal. Since by (b) the degeneration $M' <_{\text{deg}} U \oplus V$, induced by the exact sequence Ω, is minimal, applying Lemma 11 we get the equalities

$$\delta_{M,N}(U) = \delta_{M',U\oplus V}(U) = 1 = \delta'_{M',U\oplus V}(V) = \delta'_{M,N}(V).$$

We shall prove now that $U \simeq V$. Suppose it is not the case. Assume $p > 1$. Then the property (c) implies $\delta'_{M,N}(U) = 0$. Applying now Lemma 8 we get

$$\begin{aligned} 1 &= \mu(N,U) - \mu(M,U) = \delta_{M,N}(U) + \delta'_{M,N}(U) - \delta_{M,N}(E(U)) \\ &= 1 + 0 - \delta_{M,N}(E(U)), \end{aligned}$$

and so $\delta_{M,N}(E(U)) = 0$. Moreover, $M <_{\text{deg}} L <_{\text{deg}} N$ implies $M < L < N$, and hence $0 \leq \delta_{M,L} \leq \delta_{M,N}$. Thus $\delta_{M,L}(E(U)) = 0$. Invoking again Lemma 8 we obtain

$$\begin{aligned} \mu(L,U) - \mu(M,U) &= \delta_{M,L}(U) + \delta'_{M,L}(U) - \delta_{M,L}(E(U)) \\ &= \delta_{M,L}(U) + \delta'_{M,L}(U) \geq 0. \end{aligned}$$

Hence, $L = U^{p-1} \oplus L'$ for some A-module L'. Further, since $[U^{p-1}, M] \leq [U^{p-1}, L] \leq [U^{p-1}, N]$ and $[U^{p-1}, M] = [U^{p-1}, N]$, by (c), we obtain $[U^{p-1}, M] = [U^{p-1}, L] = [U^{p-1}, N]$. Applying now Theorem 10(i) we get the following degenerations

$$M' \oplus V^{q-1} <_{\text{deg}} L' <_{\text{deg}} U \oplus V^q.$$

Therefore, we may assume $p = 1$, and by duality also $q = 1$. But then $M' <_{\deg} L' <_{\deg} U \oplus V$, which contradicts (b). Therefore $U \simeq V$.

Applying again Lemma 8 we get the equalities

$$\begin{aligned} 2 - \mu(N,U) - \mu(M,U) &= \delta_{M,N}(U) + \delta'_{M,N}(U) - \delta_{M,N}(E(U)) \\ &= \delta_{M,N}(U) + \delta'_{M,N}(V) - \delta_{M,N}(E(U)) = 1 + 1 - \delta_{M,N}(E(U)), \end{aligned}$$

and hence $\delta_{M,N}(E(U)) = 0$. Since $\delta_{M,N} = \delta_{M,L} + \delta_{L,N}$ and $\delta_{M,L} \geq 0$, $\delta_{L,N} \geq 0$, we then get $\delta_{M,L}(E(U)) = 0 = \delta_{L,N}(E(U))$. Applying now Lemma 9 to the degeneration $L <_{\deg} N = U^{p+q}$, we get $\delta_{L,N}(U^{p+q}) > 0$ and $\delta'_{L,N}(U^{p+q}) > 0$. Thus $\delta_{L,N}(U) > 0$ and $\delta'_{L,N}(U) > 0$. But $\delta_{L,N}(U) \leq \delta_{M,N}(U) = 1$, and consequently $\delta_{L,N}(U) = 1$. Similarly, we get $\delta'_{L,N}(U) = 1$. Invoking Lemma 8 again we obtain

$$\mu(N,U) - \mu(L,U) = \delta_{L,N}(U) + \delta'_{L,N}(U) - \delta_{L,N}(E(U)) = 1 + 1 - 0 = 2.$$

This gives $\mu(L,U) = p+q-2$, and hence $L = U^{p+q-2} \oplus L'$ for some A-module L'. Obviously we have

$$\delta_{M,L}(U) = \delta_{M,N}(U) - \delta_{L,N}(U) = 1 - 1 = 0,$$

and then, by Theorem 10(i), $M = M' \oplus U^{p+q-2} <_{\deg} L = L' \oplus U^{p+q-2}$ implies $M' <_{\deg} L'$. On the other hand, applying the (already proved) necessity part to the minimal degeneration $L = U^{p+q-2} \oplus L' <_{\deg} N = U^{p+q}$ we get a nonsplittable exact sequence of the form

$$0 \longrightarrow U \longrightarrow L' \longrightarrow U \longrightarrow 0.$$

But then we get the following chain of degenerations

$$M' <_{\deg} L' < U^2 = U \oplus V$$

which contradicts our assumption (b) on the minimality of $M' <_{\deg} U \oplus V$. Therefore, the degeneration $M <_{\deg} N$ is minimal. This finishes the proof.

References

[1] S. Abeasis and A. del Fra, *Degenerations for representations of quiver of type* $\mathbb{A}_m$, J. Algebra **93** (1985), 376–412.

[2] M. Auslander, *Representation theory of finite dimensional algebras*, Contemp. Math. **13** (1982), 27–39.

[3] M. Auslander and I. Reiten, *Modules determined by their composition factors*, Illinois J. Math. **29** (1985), 280–301.

[4] K. Bongartz, *Minimal singularities for representations of Dynkin quivers*, Comment. Math. Helvetici **69** (1994), 575–611.

[5] K. Bongartz, *Degenerations for representations of tame quivers*, Ann. Sci. École Normale Sup. **28** (1995), 647–668.
[6] K. Bongartz, *On degenerations and extensions of finite dimensional modules*, Advances Math. **121** (1996), 245–287.
[7] P. Gabriel, *Auslander-Reiten sequences and representation-finite algebras*, Lecture Notes in Math. **831** (Springer-Verlag, 1980), 1–71.
[8] V. G. Kac, *Infinite root systems, representations of graphs and invariant theory*, Invent. Math. **56** (1980), 57–92.
[9] H. Kraft, *Geometrische Methoden in der Invariantheorie* (Vieweg Verlag, 1984).
[10] C. Riedtmann, *Degenerations for representations of quivers with relations*, Ann. Sci. École Normale Sup. **4** (1986), 275–301.
[11] C. M. Ringel, *Tame algebras and integral quadratic forms*, Lecture Notes in Math. **1099** (Springer-Verlag, 1994).
[12] A. Skowroński and G. Zwara, *Degenerations for indecomposable modules and tame algebras*, Ann. Sci. École Normale Sup. **31** (1998), 153–180.
[13] G. Zwara, *Degenerations for modules over representation-finite biserial algebras*, J. Algebra **198** (1997), 563–581.
[14] G. Zwara, *Degenerations for modules over representation-finite algebras*, Proc. Amer. Math. Soc., in press.
[15] G. Zwara, *Degenerations for representations of extended Dynkin quivers*, Comment. Math. Helvetici **73** (1998), 71–88.
[16] G. Zwara, *A degeneration-like order for modules*, Archiv Math., in press.
[17] G. Zwara, *Degenerations of finite dimensional modules are given by extensions*, Preprint (Toruń, 1998).

FACULTY OF MATHEMATICS AND INFORMATICS, NICHOLAS COPERNICUS UNIVERSITY
CHOPINA 12/18, 87-100 TORUŃ, POLAND
E-mail address: skowron@mat.uni.torun.pl, gzwara@mat.uni.torun.pl

Selected Titles in This Series

(Continued from the front of this publication)

204 **Yair Censor and Simeon Reich, Editors,** Recent developments in optimization theory and nonlinear analysis, 1997

203 **Hanna Nencka and Jean-Pierre Bourguignon, Editors,** Geometry and nature: In memory of W. K. Clifford, 1997

202 **Jean-Louis Loday, James D. Stasheff, and Alexander A. Voronov, Editors,** Operads: Proceedings of Renaissance Conferences, 1997

201 **J. R. Quine and Peter Sarnak, Editors,** Extremal Riemann surfaces, 1997

200 **F. Dias, J.-M. Ghidaglia, and J.-C. Saut, Editors,** Mathematical problems in the theory of water waves, 1996

199 **G. Banaszak, W. Gajda, and P. Krasoń, Editors,** Algebraic K-theory, 1996

198 **Donald G. Saari and Zhihong Xia, Editors,** Hamiltonian dynamics and celestial mechanics, 1996

197 **J. E. Bonin, J. G. Oxley, and B. Servatius, Editors,** Matroid theory, 1996

196 **David Bao, Shiing-shen Chern, and Zhongmin Shen, Editors,** Finsler geometry, 1996

195 **Warren Dicks and Enric Ventura,** The group fixed by a family of injective endomorphisms of a free group, 1996

194 **Seok-Jin Kang, Myung-Hwan Kim, and Insok Lee, Editors,** Lie algebras and their representations, 1996

193 **Chongying Dong and Geoffrey Mason, Editors,** Moonshine, the Monster, and related topics, 1996

192 **Tomek Bartoszyński and Marion Scheepers, Editors,** Set theory, 1995

191 **Tuong Ton-That, Kenneth I. Gross, Donald St. P. Richards, and Paul J. Sally, Jr., Editors,** Representation theory and harmonic analysis, 1995

190 **Mourad E. H. Ismail, M. Zuhair Nashed, Ahmed I. Zayed, and Ahmed F. Ghaleb, Editors,** Mathematical analysis, wavelets, and signal processing, 1995

189 **S. A. M. Marcantognini, G. A. Mendoza, M. D. Morán, A. Octavio, and W. O. Urbina, Editors,** Harmonic analysis and operator theory, 1995

188 **Alejandro Adem, R. James Milgram, and Douglas C. Ravenel, Editors,** Homotopy theory and its applications, 1995

187 **G. W. Brumfiel and H. M. Hilden,** $SL(2)$ representations of finitely presented groups, 1995

186 **Shreeram S. Abhyankar, Walter Feit, Michael D. Fried, Yasutaka Ihara, and Helmut Voelklein, Editors,** Recent developments in the inverse Galois problem, 1995

185 **Raúl E. Curto, Ronald G. Douglas, Joel D. Pincus, and Norberto Salinas, Editors,** Multivariable operator theory, 1995

184 **L. A. Bokut′, A. I. Kostrikin, and S. S. Kutateladze, Editors,** Second International Conference on Algebra, 1995

183 **William C. Connett, Marc-Olivier Gebuhrer, and Alan L. Schwartz, Editors,** Applications of hypergroups and related measure algebras, 1995

182 **Selman Akbulut, Editor,** Real algebraic geometry and topology, 1995

181 **Mila Cenkl and Haynes Miller, Editors,** The Čech Centennial, 1995

180 **David E. Keyes and Jinchao Xu, Editors,** Domain decomposition methods in scientific and engineering computing, 1994

179 **Yoshiaki Maeda, Hideki Omoro, and Alan Weinstein, Editors,** Symplectic geometry and quantization, 1994

178 **Hélène Barcelo and Gil Kalai, Editors,** Jerusalem Combinatorics '93, 1994

177 **Simon Gindikin, Roe Goodman, Frederick P. Greenleaf, and Paul J. Sally, Jr., Editors,** Representation theory and analysis on homogeneous spaces, 1994

(See the AMS catalog for earlier titles)